Modern Power System Analysis

Series Foreword

The McGraw-Hill Companies has been a leader in providing trusted information and analysis for well over a century. From the Industrial Revolution to the Internet Revolution, The McGraw-Hill Companies has filled a critical need for information and insight by helping individuals and businesses in the field of Engineering.

As early as 1910, the McGraw-Hill book company was making a difference on college campuses with the publication of its first series, Electrical Engineering Texts, outlined and edited by Professor Harry E. Clifford of Harvard University. McGraw-Hill's Electrical Engineering textbooks have shaped engineering curricula worldwide. I am thrilled that I have been invited to be the Global Series Editor in Electrical Engineering, helping to shape how the next generation of electrical engineering students around the globe will learn.

As advances in networking and communications bring the global academic community even closer together, it is essential that textbooks recognize and respond to this shift. It is in this spirit that we will publish textbooks in the McGraw-Hill Core Concepts in Electrical Engineering Series. The series will offer textbooks for the global electrical engineering curriculum that are reasonably priced, innovative, dynamic, and will cover fundamental subject areas studied by Electrical and Computer Engineering students. Written with a global perspective and presenting the latest in technological advances, these books will give students of all backgrounds a solid foundation in key engineering subjects.

Modern Power Systems Analysis by D.P. Kothari of the Indian Institute, Delhi, and I.J. Nagrath of the Birla Institute of Technology, is a textbook for undergraduate engineering students. This book provides a fundamental description of the operation, control and analysis of electric power systems. Included are key topics such as security, stability and operation of power systems. Matlab is introduced and eighteen examples of the use of Matlab for solving power system problems are provided. About 120 solved problems are provided throughout the book.

This book has been reviewed and assessed for use in engineering classrooms at all levels. Like *Modern Power Systems Analysis,* each book in the Core Concepts series presents a comprehensive, straightforward, and accurate treatment of an important subject in Electrical & Computer Engineering. With their clear approach, contemporary technology, and international perspective, Core Concepts books are an unmistakable choice for professors wanting understandable, concise engineering textbooks that adhere to the standards of The McGraw-Hill Companies.

—**Richard C. Dorf,**
University of California, Davis
Series Editor, The Core Concepts Series in Electrical and Computer Engineering

Modern Power System Analysis

D P Kothari

Professor, Centre of Energy Studies
Deputy Director (Admin.)
Indian Institute of Technology
Delhi

I J Nagrath

Adjunct Professor, and Former Deputy Director,
Birla Institute of Technology and Science
Pilani

Boston Burr Ridge, IL Dubuque, IA New York
San Francisco St. Louis Bangkok Bogotá Caracas Kuala Lumpur
Lisbon London Madrid Mexico City Milan Montreal New Delhi
Santiago Seoul Singapore Sydney Taipei Toronto

MODERN POWER SYSTEM ANALYSIS

Published by McGraw-Hill, a business unit of The McGraw-Hill Companies, Inc., 1221 Avenue of the Americas, New York, NY 10020. Copyright © 2008 by The McGraw-Hill Companies, Inc. All rights reserved. No part of this publication may be reproduced or distributed in any form or by any means, or stored in a database or retrieval system, without the prior written consent of The McGraw-Hill Companies, Inc., including, but not limited to, in any network or other electronic storage or transmission, or broadcast for distance learning.

Some ancillaries, including electronic and print components, may not be available to customers outside the United States.

This book is printed on acid-free paper.

1 2 3 4 5 6 7 8 9 0 BKM/BKM 0 9 8 7 6

ISBN 978–0–07–340455–4
MHID 0–07–340455–1

Publisher: *Suzanne Jeans*
Senior Sponsoring Editor: *Michael S. Hackett*
Developmental Editor: *Rebecca Olson*
Executive Marketing Manager: *Michael Weitz*
Project Coordinator: *Melissa M. Leick*
Senior Production Supervisor: *Kara Kudronowicz*
Associate Media Producer: *Christina Nelson*
Senior Designer: *David W. Hash*
Cover Designer: *Rokusek Design*
(USE) Cover Image: *©PhotoDisc*
Compositor: *Techbooks*
Printer: *Book-mart Press*

This book was previously published by Tata McGraw-Hill Publishing Company Limited, New Delhi, India, copyright © 2003, 1989, 1980.

Library of Congress Cataloging-in-Publication Data

Kothari, D. P.
    Modern power system analysis / D. P. Kothari, I. J. Nagrath. — 1st ed.
        p.   cm.
    Includes index.
    ISBN 978–0–07–340455–4 — ISBN 0–07–340455–1 (alk. paper)
    1. Electric power systems. I. Nagrath, I. J. II. Title.

TK1005.K685   2008
621.319'1—dc22   2006041933
                                                      CIP

Information contained in this work has been obtained by McGraw-Hill from sources believed to be reliable. However, neither McGraw-Hill nor its authors guarantee the accuracy or completeness of any information published herein and neither McGraw-Hill nor its authors shall be responsible for any errors, omissions, or damages arising out of use of this information. The work is published with the understanding that McGraw-Hill and its authors are supplying information but are not attempting to render engineering or other professional services. If such services are required, the assistance of an appropriate professional should be sought.

Preface

The overall energy situation has changed considerably in recent years, and this has generated great interest in non-conventional and renewable energy sources, energy conservation and management, power reforms and restructuring and distributed and dispersed generation. Modern power systems have grown larger and spread over larger geographical area with many interconnections between neighbouring systems. Optimal planning, operation and control of such large-scale systems require advanced computer-based techniques many of which are explained in a student-oriented and reader-friendly manner by means of numerical examples throughout this book. Electric utility engineers will also be benefitted by the book as it will prepare them more adequately to face the new challenges. The style of writing is amenable to self-study. The wide range of topics facilitates versatile selection of chapters and sections for completion in the semester time frame.

The book is designed for a two-semester course at the undergraduate level. With a judicious choice of advanced topics, some institutions may also find it useful for a first course for postgraduates.

The reader is expected to have a prior grounding in circuit theory and electrical machines. He should also have been exposed to Laplace transform, linear differential equations, optimisation techniques and a first course in control theory. Matrix analysis is applied throughout the book. However, a knowledge of simple matrix operations would suffice and these are summarised in an appendix for quick reference.

The digital computer being an indispensable tool for power system analysis, computational algorithms for various system studies such as load flow, fault level analysis, stability, etc. have been included at appropriate places in the book. It is suggested that where computer facilities exist, students should be encouraged to build computer programs for these studies using the algorithms provided. Further, the students can be asked to pool the various programs for more advanced and sophisticated studies, e.g. optimal scheduling. An important novel feature of the book is the inclusion of the latest and practically useful topics like unit commitment, generation reliability, optimal thermal scheduling, optimal hydro-thermal scheduling and decoupled load flow in a text which is primarily meant for undergraduates.

The introductory chapter contains a discussion on various methods of electrical energy generation and their techno-economic comparison. A glimpse is given into the future of electrical energy. The reader is also exposed to the Indian power scenario with facts and figures.

Chapters 2 and 3 give the transmission line parameters and these are included for the sake of completeness of the text. Chapter 4 on the representation of power system components gives the steady state models of the synchronous machine and the circuit models of composite power systems along with the per unit method.

Chapter 5 deals with the performance of transmission lines. The load flow problem is introduced right at this stage through the simple two-bus system and basic concepts of watt and var control are illustrated. A brief treatment of circle diagrams is included as this forms an excellent teaching air for putting across the concept of load flow and line compensation. *ABCD* constants are generally well covered in the circuit theory course and are, therefore, relegated to an appendix.

Chapter 6 gives power network modelling and load flow analysis, while Chapter 7 gives optimal system operation with both approximate and rigorous treatment.

Chapter 8 deals with load frequency control wherein both conventional and modern control approaches have been adopted for analysis and design. Voltage control is briefly discussed.

Chapters 9–11 discuss fault studies (abnormal system operation). The synchronous machine model for transient studies is heuristically introduced to the reader.

Chapter 12 emphasizes the concepts of various types of stability in a power system. In particular the concepts of transient stability are well illustrated through the equal area criterion. The classical numerical solution technique of the swing equation as well as the algorithm for large system stability are advanced.

Chapter 13 deals with power system security. Contingency analysis and sensitivity factors are described. An analytical framework is developed to control bulk power systems in such a way that security is enhanced. Everything seems to have a propensity to fail. Power systems are no exception. Power system security practices try to control and operate power systems in a defensive posture so that the effects of these inevitable failures are minimized.

Chapter 14 is an introduction to the use of state estimation in electric power systems. We have selected Least Squares Estimation to give basic solution. External system equivalencing and treatment of bad data are also discussed.

The economics of power transmission has always lured the planners to transmit as much power as possible through existing transmission lines. Difficulty of acquiring the right of way for new lines (the corridor crisis) has always motivated the power engineers to develop compensatory systems. Therefore, Chapter 15 addresses compensation in power systems. Both series and shunt compensation of lines have been thoroughly discussed. Concepts of SVS, STATCOM and FACTS have been briefly introduced.

Chapter 16 covers the important topic of load forecasting technique. Knowing load is absolutely essential for solving any power system problem.

Chapter 17 deals with the important problem of voltage stability. Mathematical formulation, analysis, and state-of-art future trends and challenges are discussed.

MATLAB and SIMULINK, ideal programs for power system analysis are included in this book as an appendix along with 18 solved examples illustrating their use in solving representative power system problems. The help rendered by Shri Sunil Bhat of VNIT, Nagpur in writing this appendix is thankfully acknowledged.

Every concept and technique presented is well supported through examples employing mainly a two-bus structure, through sometimes three- and four-bus illustrations have also been used. A large number of problems have answers included at the end of each chapter. These have been selected so that apart from providing a drill they help the reader develop a deeper insight and illustrate some points beyond what is directly covered by the text.

The internal organisation of various chapters is flexible and permits the teacher to adapt them to the particular needs of the class and curriculum. If desired, some of the advanced level topics could be bypassed without loss of continuity. The style of writing is specially adapted to self-study. Exploiting this fact, a teacher will have enough time at his/her disposal to extend the coverage of this book to suit his/her particular syllabus and to include tutorial work on the numerous examples suggested in the text.

The authors are indebted to their colleagues at the Birla Institute of Technology and Science, Pilani and the Indian Institute of Technology, Delhi for the encouragement and various useful suggestions they received from them while writing this book. They are grateful to the authorities of the Birla Institute of Technology and Science, Pilani and the Indian Institute of Technology, Delhi for providing facilities necessary for writing the book. The authors welcome any constructive criticism of the book and will be grateful for any appraisal by the readers.

I J NAGRATH
D P KOTHARI

About the Authors

D P Kothari is a Professor, at the Centre for Energy Studies, and Deputy Director (Administration) Indian Institute of Technology, Delhi. He has also been the Head of the Centre for Energy Studies (1995–97) and Principal (1997–98), Visvesvaraya Regional Engineering College, Nagpur. Earlier (1982–83 and 1989), he was a visiting fellow at RMIT, Melbourne, Australia. He obtained his B.E., M.E., and Ph.D. degrees from BITS, Pilani. A fellow of the Institution Engineers (India), Prof. Kothari has published/presented 450 papers in national and international journals/conferences. He has authored/co-authored more than 15 books, including *Power System Engineering, Electric Machines, 2/e, Power System Transients, Theory and Problems of Electric Machines, 2/e.,* and *Basic Electrical Engineering.* His research interests include power system control, optimisation, reliability and energy conservation.

I J Nagrath is an Adjunct Professor, at BITS Pilani and retired as Professor of Electrical Engineering and Deputy Director of Birla Institute of Technology and Science, Pilani. He obtained his B.E. in Electrical Engineering from the University of Rajasthan in 1951 and M.S. from the University of Wisconsin in 1956. He has co-authored several successful books which include *Electric Machines 2/e, Power System Engineering, Signals and Systems* and *Systems: Modelling and Analysis.* He has also published several research papers in prestigious national and international journals.

Contents

1

Introduction

1.1 A PERSPECTIVE

Electric energy is an essential ingredient for the industrial and all-round development of any country. It is a coveted form of energy, because it can be generated centrally in bulk and transmitted economically over long distances. Further, it can be adapted easily and efficiently to domestic and industrial applications, particularly for lighting purposes and mechanical work*, e.g. drives. The per capita consumption of electrical energy is a reliable indicator of a country's state of development—figures for 2001 are 425 kWh for India and 5600 kWh for UK and 13000 kWh for USA.

Conventionally, electric energy is obtained by conversion from fossil fuels (coal, oil, natural gas), and nuclear and hydro sources. Heat energy released by burning fossil fuels or by fission of nuclear material is converted to electricity by first converting heat energy to the mechanical form through a thermocycle and then converting mechanical energy through generators to the electrical form. Thermocycle is basically a low efficiency process—highest efficiencies for modern large size plants range up to 40%, while smaller plants may have considerably lower efficiencies. The earth has fixed non-replenishable resources of fossil fuels and nuclear materials, with certain countries over-endowed by nature and others deficient. Hydro energy, though replenishable, is also limited in terms of power. The world's increasing power requirements can only be partially met by hydro sources. Furthermore, ecological and biological factors place a stringent limit on the use of hydro sources for power production. (The USA has already developed around 50% of its hydro potential and hardly any further expansion is planned because of ecological considerations.)

---

* Electricity is a very inefficient agent for heating purposes, because it is generated by the low efficiency thermocycle from heat energy. Electricity is used for heating purposes for only very special applications, say an electric furnace.

With the ever increasing per capita energy consumption and exponentially rising population, technologists already see the end of the earth's non-replenishable fuel resources*. The oil crisis of the 1970s has dramatically drawn attention to this fact. In fact, we can no longer afford to use oil as a fuel for generation of electricity. In terms of bulk electric energy generation, a distinct shift is taking place across the world in favour of coal and in particular nuclear sources for generation of electricity. Also, the problems of air and thermal pollution caused by power generation have to be efficiently tackled to avoid ecological disasters. A coordinated worldwide action plan is, therefore, necessary to ensure that energy supply to humanity at large is assured for a long time and at low economic cost. Some of the factors to be considered and actions to be taken are:

Curtailment of energy consumption

The energy consumption of most developed countries has already reached a level, which this planet cannot afford. There is, in fact, a need to find ways and means of reducing this level. The developing countries, on the other hand, have to intensify their efforts to raise their level of energy production to provide basic amenities to their teeming millions. Of course, in doing so they need to constantly draw upon the experiences of the developed countries and guard against obsolete technology.

Intensification of efforts to develop alternative sources of energy including unconventional sources like solar, tidal energy, etc.

Distant hopes are pitched on fusion energy but the scientific and technological advances have a long way to go in this regard. Fusion when harnessed could provide an inexhaustible source of energy. A break-through in the conversion from solar to electric energy could provide another answer to the world's steeply rising energy needs.

Recycling of nuclear wastes

Fast breeder reactor technology is expected to provide the answer for extending nuclear energy resources to last much longer.

Development and application of antipollution technologies

In this regard, the developing countries already have the example of the developed countries whereby they can avoid going through the phases of

---

*Varying estimates have been put forth for reserves of oil, gas and coal and fissionable materials. At the projected consumption rates, oil and gases are not expected to last much beyond 50 years; several countries will face serious shortages of coal after 2200 A.D. while fissionable materials may carry us well beyond the middle of the next century. These estimates, however, cannot be regarded as highly dependable.

intense pollution in their programmes of energy development. Bulk power generating stations are more easily amenable to control of pollution since centralized one-point measures can be adopted.

Electric energy today constitutes about 30% of the total annual energy consumption on a worldwide basis. This figure is expected to rise as oil supply for industrial uses becomes more stringent. Transportation can be expected to go electric in a big way in the long run, when non-conventional energy resources are well developed or a breakthrough in fusion is achieved.

To understand some of the problems that the power industry faces let us briefly review some of the characteristic features of generation and transmission. Electricity, unlike water and gas, cannot be stored economically (except in very small quantities—in batteries), and the electric utility can exercise little control over the load (power demand) at any time. The power system must, therefore, be capable of matching the output from generators to the demand at any time at a specified voltage and frequency. The difficulty encountered in this task can be imagined from the fact that load variations over a day comprises three components—a steady component known as **base load**; a varying component whose daily pattern depends upon the time of day; weather, season, a popular festival, etc.; and a purely randomly varying component of relatively small amplitude. Figure 1.1 shows a typical daily load curve. The characteristics of a daily load curve on a gross basis are indicated by **peak load** and the time of its occurrence and load factor defined as

$$\frac{\text{average load}}{\text{maximum (peak) load}} = \text{less than unity}$$

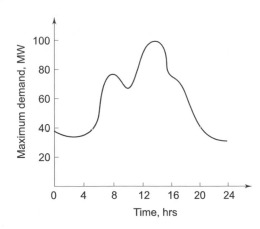

Fig. 1.1 Typical daily load curve

The average load determines the energy consumption over the day, while the peak load along with considerations of standby capacity determines plant capacity for meeting the load.

A high load factor helps in drawing more energy from a given installation. As individual load centres have their own characteristics, their peaks in general have a time diversity, which when utilized through transmission interconnection, greatly aids in jacking up load factors at an individual plant— excess power of a plant during light load periods is evacuated through long distance high voltage transmission lines, while a heavily loaded plant receives power.

Diversity Factor

This is defined as the sum of individual maximum demands on the consumers, divided by the maximum load on the system. This factor gives the time diversification of the load and is used to decide the installation of sufficient generating and transmission plant. If all the demands came at the same time, i.e. unity diversity factor, the total installed capacity required would be much more. Luckily, the factor is much higher than unity, especially for domestic loads.

A high diversity factor could be obtained by:

1. Giving incentives to farmers and/or some industries to use electricity in the night or light load periods.
2. Using day-light saving as in many other countries.
3. Staggering the office timings.
4. Having different time zones in the country like USA, Australia, etc.
5. Having two-part tariff in which consumer has to pay an amount dependent on the maximum demand he makes, plus a charge for each unit of energy consumed. Sometimes consumer is charged on the basis of kVA demand instead of kW to penalize loads of low power factor.
   Two other factors used frequently are:

Plant capacity factor

$$= \frac{\text{Actual energy produced}}{\begin{array}{c} \text{maximum possible energy that could have been produced} \\ \text{(based on installed plant capacity)} \end{array}}$$

$$= \frac{\text{Average demand}}{\text{Installed capacity}}$$

Plant use factor

$$= \frac{\text{Actual energy produced (kWh)}}{\text{plant capacity (kW)} \times \text{Time (in hours) the plant has been in operation}}$$

Tariffs

The cost of electric power is normally given by the expression ($a + b \times$ kW $+ c \times$ kWh) per annum, where a is a fixed charge for the utility, independent of the power output; b depends on the maximum demand on the system and hence on the interest and depreciation on the installed power station; and c depends on the units produced and therefore on the fuel charges and the wages of the station staff.

Tariff structures may be such as to influence the load curve and to improve the load factor.

Tariff should consider the pf (power factor) of the load of the consumer. If it is low, it takes more current for the same kWs and hence T and D (transmission and distribution) losses are correspondingly increased. The power station has to install either pf correcting (improvement) devices such as synchronous capacitors, SVC (Static Var Compensator) or voltage regulating equipment to maintain the voltages within allowed limits and thus total cost increases. One of the following alternatives may be used to avoid low pf:

(i) to charge the consumers based on kVA rather than kW.

(ii) a pf penalty clause may be imposed on the consumer.

(iii) the consumer may be asked to use shunt capacitors for improving the power factor of his installations.

Example 1.1

A factory to be set up is to have a fixed load of 760 kW at 0.8 pf. The electricity board offers to supply energy at the following alternate rates:

(a) LV supply at Rs 32/kVA max demand/annum + 10 paise/kWh

(b) HV supply at Rs 30/kVA max demand/annum + 10 paise/kWh.

The HV switchgear costs Rs 60/kVA and switchgear losses at full load amount to 5%. Interest, depreciation charges for the switchgear are 12% of the capital cost. If the factory is to work for 48 hours/week, determine the more economical tariff.

Solution $\qquad$ Maximum demand $= \dfrac{760}{0.8} = 950$ kVA

Loss in switchgear $= 5\%$

$\therefore \qquad\qquad$ Input demand $= \dfrac{950}{0.95} = 1000$ kVA

Cost of switchgear $= 60 \times 1000 = $ Rs 60,000

Annual charges on depreciation $= 0.12 \times 60,000 = $ Rs 7,200

Annual fixed charges due to maximum demand corresponding to tariff (b)

$$= 30 \times 1,000 = \text{Rs } 30,000$$

Annual running charges due to kWh consumed

$$= 1000 \times 0.8 \times 48 \times 52 \times 0.10$$

$$= \text{Rs } 1,99,680$$

Total charges/annum = Rs 2,36,880

Max. demand corresponding to tariff(a) = 950 kVA

Annual fixed charges = 32 × 950 = Rs 30,400

Annual running charges for kWh consumed

$$= 950 \times 0.8 \times 48 \times 52 \times 0.10$$
$$= \text{Rs } 1,89,696$$
$$\text{Total} = \text{Rs } 2,20,096$$

Therefore, tariff (a) is economical.

Example 1.2

A region has a maximum demand of 500 MW at a load factor of 50%. The load duration curve can be assumed to be a triangle. The utility has to meet this load by setting up a generating system, which is partly hydro and partly thermal. The costs are as under:

Hydro plant: Rs 600 per kW per annum and operating expenses at 3p per kWh.

Thermal plant: Rs 300 per kW per annum and operating expenses at 13p per kWh.

Determine the capacity of hydro plant, the energy generated annually by each, and overall generation cost per kWh.

Solution

Total energy generated per year = 500 × 1000 × 0.5 × 8760

$$= 219 \times 10^7 \text{ kWh}$$

Figure 1.2 shows the load duration curve. Since operating cost of hydro plant is low, the base load would be supplied from the hydro plant and peak load from the thermal plant.

Let the hydro capacity be P kW and the energy generated by hydro plant E kWh/year.

Thermal capacity = (5,00,000 – P) kW
Thermal energy = $(219 \times 10^7 - E)$ kWh

Annual cost of hydro plant

$$= 600 P + 0.03E$$

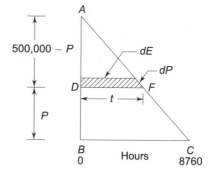

Fig. 1.2   Load duration curve

Annual cost of thermal plant = $300(5,00,000 - P) + 0.13(219 \times 10^7 - E)$

Total cost $C = 600P + 0.03E + 300(5,00,000 - P)$

$$+ 0.13(219 \times 10^7 - E)$$

For minimum cost, $\dfrac{dC}{dP} = 0$

$$\therefore 600 + 0.03\,\frac{dE}{dP} - 300 - 0.13\frac{dE}{dP} = 0$$

or $$dE = 3000\ dP$$

But $$dE = dP \times t$$

$$\therefore \qquad t = 3000\ \text{hours}$$

From triangles ADF and ABC,

$$\frac{5,00,000 - P}{5,00,000} = \frac{3000}{8760}$$

$$\therefore \qquad P = 328,\ \text{say 330 MW}$$

$$\text{Capacity of thermal plant} = 170\ \text{MW}$$

$$\text{Energy generated by thermal plant} = \frac{170 \times 3000 \times 1000}{2}$$

$$= 255 \times 10^6\ \text{kWh}$$

$$\text{Energy generated by hydro plant} = 1935 \times 10^6\ \text{kWh}$$

$$\text{Total annual cost} = \text{Rs } 340.20 \times 10^6/\text{year}$$

$$\text{Overall generation cost} = \frac{340.20 \times 10^6}{219 \times 10^7} \times 100$$

$$= 15.53\ \text{paise/kWh}$$

Example 1.3

A generating station has a maximum demand of 25 MW, a load factor of 60%, a plant capacity factor of 50%, and a plant use factor of 72%. Find (a) the daily energy produced, (b) the reserve capacity of the plant, and (c) the maximum energy that could be produced daily if the plant, while running as per schedule, were fully loaded.

Solution

$$\text{Load factor} = \frac{\text{average demand}}{\text{maximum demand}}$$

$$0.60 = \frac{\text{average demand}}{25}$$

$$\therefore \qquad \text{Average demand} = 15\ \text{MW}$$

$$\text{Plant capacity factor} = \frac{\text{average demand}}{\text{installed capacity}}$$

$$0.50 = \frac{15}{\text{installed capacity}}$$

$$\therefore \qquad \text{Installed capacity} = \frac{15}{0.5} = 30\ \text{MW}$$

$\therefore$ Reserve capacity of the plant = installed capacity – maximum demand
$$= 30 - 25 = 5 \text{ MW}$$
Daily energy produced = average demand × 24 = 15 × 24
$$= 360 \text{ MWh}$$
Energy corresponding to installed capacity per day
$$= 24 \times 30 = 720 \text{ MWh}$$
Maximum energy that could be produced

$$= \frac{\text{actual energy produced in a day}}{\text{plant use factor}}$$

$$= \frac{360}{0.72} = 500 \text{ MWh/day}$$

Example 1.4

From a load duration curve, the following data are obtained:

Maximum demand on the system is 20 MW. The load supplied by the two units is 14 MW and 10 MW. Unit No. 1 (base unit) works for 100% of the time, and Unit No. 2 (peak load unit) only for 45% of the time. The energy generated by Unit 1 is 1×10^8 units, and that by Unit 2 is 7.5×10^6 units. Find the load factor, plant capacity factor and plant use factor of each unit, and the load factor of the total plant.

Solution

$$\text{Annual load factor for Unit 1} = \frac{1 \times 10^8 \times 100}{14,000 \times 8760} = 81.54\%$$

The maximum demand on Unit 2 is 6 MW.

$$\text{Annual load factor for Unit 2} = \frac{7.5 \times 10^6 \times 100}{6000 \times 8760} = 14.27\%$$

Load factor of Unit 2 for the time it takes the load

$$= \frac{7.5 \times 10^6 \times 100}{6000 \times 0.45 \times 8760}$$

$$= 31.71\%$$

Since no reserve is available at Unit No. 1, its capacity factor is the same as the load factor, i.e. 81.54%. Also since Unit 1 has been running throughout the year, the plant use factor equals the plant capacity factor i.e. 81.54%.

$$\text{Annual plant capacity factor of Unit 2} = \frac{7.5 \times 10^6 \times 100}{10 \times 8760 \times 100} = 8.56\%$$

$$\text{Plant use factor of Unit 2} = \frac{7.5 \times 10^6 \times 100}{10 \times 0.45 \times 8760 \times 100} = 19.02\%$$

The annual load factor of the total plant $= \dfrac{1.075 \times 10^8 \times 100}{20,000 \times 8760} = 61.35\%$

Comments The various plant factors, the capacity of base and peak load units can thus be found out from the load duration curve. The load factor of the peak load unit is much less than that of the base load unit, and thus the cost of power generation from the peak load unit is much higher than that from the base load unit.

Example 1.5

There are three consumers of electricity having different load requirements at different times. Consumer 1 has a maximum demand of 5 kW at 6 p.m. and a demand of 3 kW at 7 p.m. and a daily load factor of 20%. Consumer 2 has a maximum demand of 5 kW at 11 a.m., a load of 2 kW at 7 p.m. and an average load of 400 W. Consumer 3 has an average load of 1 kW and his maximum demand is 3 kW at 7 p.m. Determine: (a) the diversity factor, (b) the load factor and average load of each consumer, and (c) the average load and load factor of the combined load.

Solution

(a) Consumer 1   MD 5 kW   3 kW   LF
                       at 6 pm   at 7 pm   20%
       Consumer 2   MD 5 kW   2 kW   Average load
                       at 11 am   at 7 pm   1.2 kW
       Consumer 3   MD 3 kW   Average load
                       at 7 pm   1 kW

Maximum demand of the system is 8 kW at 7 p.m.

Sum of the individual maximum demands = 5 + 5 + 3 = 13 kW

∴   Diversity factor = 13/8 = 1.625

(b) Consumer 1 Average load 0.2 × 5 = 1 kW,   LF= 20%

Consumer 2 Average load 1.2 kW, $LF = \dfrac{1.2}{5} \times 1000 = 24\%$

Consumer 3 Average load 1 kW, $LF = \dfrac{1}{3} \times 100 = 33.3\%$

(c) Combined average load = 1 + 1.2 + 1 = 3.2 kW

∴   Combined load factor $= \dfrac{3.2}{8} \times 100 = 40\%$

Load Forecasting

As power plant planning and construction require a gestation period of four to eight years or even longer for the present day super power stations, energy and load demand forecasting plays a crucial role in power system studies.

This necessitates long range forecasting. While sophisticated probabilistic methods exist in literature [5, 16, 28], the simple extrapolation technique is quite adequate for long range forecasting. Since weather has a much more influence on residential than the industrial component, it may be better to prepare forecast in constituent parts to obtain total. Both power and energy forecasts are made. Multi factors involved render forecasting an involved process requiring experience and high analytical ability.

Yearly forecasts are based on previous year's loading for the period under consideration updated by factors such as general load increases, major loads and weather trends.

In short-term load forecasting, hour-by-hour predictions are made for the particular day under consideration. A minor forecast error on low side might necessitate the use of inefficient, oil-fired turbine generators or "peaking units" which are quite costly. On the other hand, a high side forecast error would keep excessive generation in hot reserve. Accuracy of the order of 1% is desirable.A temperature difference of 2°C can vary the total load by 1%. This indicates the importance of reliable weather forecast to a good load forecast. The short term forecast problem is not a simple one as often random factors such as unexpected storms, strikes, the sudden telecast of a good TV programme can upset the predictions. Regression analysis is often used for obtaining a short term load forecast which is very important and is required before solving unit commitment and economic load despatch problems discussed in Chapter 7. Owing to the great importance of load forecasting (an important input-before solving almost all power system problems), a full chapter is added in this book describing various methods of load forecasting (Chapter 16).

In India, energy demand and installed generating capacity are both increasing exponentially (so is population growth—a truly formidable combination). Power demand* has been roughly doubling every ten years as in many other countries. In 2001, the total installed generation capacity in India is 100,000 MW. As per the present indications, by the time we enter the 2nd decade of the 21st century it would be nearing 2,00,000 MW—a stupendous task indeed. This, in turn, would require a corresponding development in coal resources.

1.2 STRUCTURE OF POWER SYSTEMS

Generating stations, transmission lines and the distribution systems are the main components of an electric power system. Generating stations and a distribution system are connected through transmission lines, which also

---

* 38% of the total power required in India is for industrial consumption. Generation of electricity in India was around 530 billion kWh in 2000–2001 A.D. compared to less than 200 billion kWh in 1986–87.

connect one power system (grid, area) to another. A distribution system connects all the loads in a particular area to the transmission lines.

For economical and technological reasons (which will be discussed in detail in later chapters), individual power systems are organized in the form of electrically connected areas or regional grids (also called power pools). Each area or regional grid operates technically and economically independently, but these are eventually interconnected* to form a national grid (which may even form an international grid) so that each area is contractually tied to other areas in respect to certain generation and scheduling features. India is now heading for a national grid.

The siting of hydro stations is determined by the natural water power sources. The choice of site for coal fired thermal stations is more flexible. The following two alternatives are possible.

1. Power stations may be built close to coal mines (called pit head stations) and electric energy is evacuated over transmission lines to the load centres.

2. Power stations may be built close to the load centres and coal is transported to them from the mines by rail road.

In practice, however, power station siting will depend upon many factors— technical, economical and environmental. As it is considerably cheaper to transport bulk electric energy over extra high voltage (EHV) transmission lines than to transport equivalent quantities of coal over rail road, the recent trends in India (as well as abroad) is to build super (large) thermal power stations near coal mines. Bulk power can be transmitted to fairly long distances over transmission lines of 400 kV and above. However, the country's coal resources are located mainly in the eastern belt and some coal fired stations will continue to be sited in distant western and southern regions.

As nuclear stations are not constrained by the problems of fuel transport and air pollution, a greater flexibility exists in their siting, so that these stations are located close to load centres while avoiding high density pollution areas to reduce the risks, however remote, of radioactivity leakage.

---

*Interconnection has the economic advantage of reducing the reserve generation capacity in each area. Under conditions of sudden increase in load or loss of generation in one area, it is immediately possible to borrow power from adjoining interconnected areas. Interconnection causes larger currents to flow on transmission lines under faulty condition with a consequent increase in capacity of circuit breakers. Also, the synchronous machines of all interconnected areas must operate stably and in a synchronized manner. The disturbance caused by a short circuit in one area must be rapidly disconnected by circuit breaker openings before it can seriously affect adjoining areas. It permits the construction of larger and more economical generating units and the transmission of large chunk of power from the generating plants to major load centres. It provides capacity savings by seasonal exchange of power between areas having opposing winter and summer requirements. It permits capacity savings from time zones and random diversity. It facilitates transmission of off-peak power. It also gives the flexibility to meet unexpected emergency loads.

In India, as of now, about 75% of electric power used is generated in thermal plants (including nuclear). 23% from mostly hydro stations and 2%. come from renewables and others. Coal is the fuel for most of the steam plants, the rest depends upon oil/natural gas and nuclear fuels.

Electric power is generated at a voltage of 11 to 25 kV which then is stepped up to the transmission levels in the range of 66 to 400 kV (or higher). As the transmission capability of a line is proportional to the square of its voltage, research is continuously being carried out to raise transmission voltages. Some of the countries are already employing 765 kV. The voltages are expected to rise to 800 kV in the near future. In India, several 400 kV lines are already in operation. One 800 kV line has just been built.

For very long distances (over 600 km), it is economical to transmit bulk power by DC transmission. It also obviates some of the technical problems associated with very long distance AC transmission. The DC voltages used are 400 kV and above, and the line is connected to the AC systems at the two ends through a transformer and converting/inverting equipment (silicon controlled rectifiers are employed for this purpose). Several DC transmission lines have been constructed in Europe and the USA. In India two HVDC transmission line (bipolar) have already been commissioned and several others are being planned. Three back to back HVDC systems are in operation.

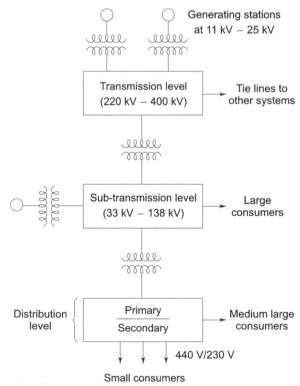

Fig. 1.3 Schematic diagram depicting power system structure

The first stepdown of voltage from transmission level is at the bulk power substation, where the reduction is to a range of 33 to 132 kV, depending on the transmission line voltage. Some industries may require power at these voltage levels. This stepdown is from the transmission and grid level to subtransmission level.

The next stepdown in voltage is at the distribution substation. Normally, two distribution voltage levels are employed:

1. The primary or feeder voltage (11 kV)
2. The secondary or consumer voltage (415 V three phase/230 V single phase).

The distribution system, fed from the distribution transformer stations, supplies power to the domestic or industrial and commercial consumers.

Thus, the power system operates at various voltage levels separated by transformer. Figure 1.3 depicts schematically the structure of a power system.

Though the distribution system design, planning and operation are subjects of great importance, we are compelled, for reasons of space, to exclude them from the scope of this book.

1.3   CONVENTIONAL SOURCES OF ELECTRIC ENERGY

Thermal (coal, oil, nuclear) and hydro generations are the main conventional sources of electric energy. The necessity to conserve fossil fuels has forced scientists and technologists across the world to search for unconventional sources of electric energy. Some of the sources being explored are solar, wind and tidal sources. The conventional and some of the unconventional sources and techniques of energy generation are briefly surveyed here with a stress on future trends, particularly with reference to the Indian electric energy scenario.

Thermal Power Stations—Steam/Gas-based

The heat released during the combustion of coal, oil or gas is used in a boiler to raise steam. In India heat generation is mostly coal based except in small sizes, because of limited indigenous production of oil. Therefore, we shall discuss only coal-fired boilers for raising steam to be used in a turbine for electric generation.

The chemical energy stored in coal is transformed into electric energy in thermal power plants. The heat released by the combustion of coal produces steam in a boiler at high pressure and temperature, which when passed through a steam turbine gives off some of its internal energy as mechanical energy. The axial-flow type of turbine is normally used with several cylinders on the same shaft. The steam turbine acts as a prime mover and drives the electric generator (alternator). A simple schematic diagram of a coal fired thermal plant is shown in Fig. 1.4.

The efficiency of the overall conversion process is poor and its maximum value is about 40% because of the high heat losses in the combustion gases and the large quantity of heat rejected to the condenser which has to be given off in cooling towers or into a stream/lake in the case of direct condenser cooling. The steam power station operates on the Rankine cycle, modified to include superheating, feed-water heating, and steam reheating. The thermal efficiency (conversion of heat to mechanical energy) can be increased by using steam at the highest possible pressure and temperature. With steam turbines of this size, additional increase in efficiency is obtained by reheating the steam after it has been partially expanded by an external heater. The reheated steam is then returned to the turbine where it is expanded through the final states of bleeding.

To take advantage of the principle of economy of scale (which applies to units of all sizes), the present trend is to go in for larger sizes of units. Larger units can be installed at much lower cost per kilowatt. They are also cheaper to operate because of higher efficiency. They require lower labour and maintenance expenditure. According to Chaman Kashkari [3] there may be a saving of as high as 15% in capital cost per kilowatt by going up from a 100 to 250 MW unit size and an additional saving in fuel cost of about 8% per kWh. Since larger units consume less fuel per kWh, they produce less air, thermal and waste pollution, and this is a significant advantage in our concern for environment. The only trouble in the case of a large unit is the tremendous shock to the system when outage of such a large capacity unit occurs. This shock can be tolerated so long as this unit size does not exceed 10% of the on-line capacity of a large grid.

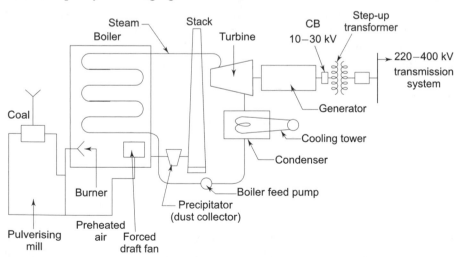

Fig. 1.4 Schematic diagram of a coal fired steam plant

In India, in 1970s the first 500 MW superthermal unit had been commissioned at Trombay. Bharat Heavy Electricals Limited (BHEL) has produced several turbogenerator sets of 500 MW capacity. Today's maximum generator unit size is (nearly 1200 MW) limited by the permissible current densities used in rotor and stator windings. Efforts are on to develop *super conducting* machines where the winding temperature will be nearing absolute zero. Extreme high current and flux densities obtained in such machines could perhaps increase unit sizes to several GWs which would result in better generating economy.

Air and thermal pollution is always present in a coal fired steam plant. The air polluting agents (consisting of particulates and gases such as NOX, CO, CO_2, oxides of sulphur, etc.) are emitted via the exhaust gases and thermal pollution is due to the rejected heat transferred from the condenser to cooling water. Cooling towers are used in situations where the stream/lake cannot withstand the thermal burden without excessive temperature rise. The problem of air pollution can be minimized through scrubbers and electro-static precipitators and by resorting to minimum emission dispatch [32] and Clean Air Act has already been passed in Indian Parliament.

Fluidized-bed Boiler

The main problem with coal in India is its high ash content (up to 40% max). To solve this, *fluidized bed combustion technology* is being developed and perfected. The fluidized-bed boiler is undergoing extensive development and is being preferred due to its lower pollutant level and better efficiency. Direct ignition of pulverized coal is being introduced but initial oil firing support is needed.

Cogeneration

Considering the tremendous amount of waste heat generated in thermal power generation, it is advisable to save fuel by the simultaneous generation of electricity and steam (or hot water) for industrial use or space heating. Now called cogeneration, such systems have long been common, here and abroad. Currently, there is renewed interest in these because of the overall increase in energy efficiencies which are claimed to be as high as 65%.

Cogeneration of steam and power is highly energy efficient and is particularly suitable for chemicals, paper, textiles, food, fertilizer and petroleum refining industries. Thus these industries can solve energy shortage problem in a big way. Further, they will not have to depend on the grid power which is not so reliable. Of course they can sell the extra power to the government for use in deficient areas. They may also sell power to the neighbouring industries, a concept called *wheeling power*.

As on 31.12.2000, total co-generation potential in India is 19,500 MW and actual achievement is 273 MW as per MNES (Ministry of Non-Conventional Energy Sources, Government of India) Annual Report 2000–01.

There are two possible ways of cogeneration of heat and electricity: (i) Topping cycle, (ii) Bottoming cycle. In the topping cycle, fuel is burnt to produce electrical or mechanical power and the waste heat from the power generation provides the process heat. In the bottoming cycle, fuel first produces process heat and the waste heat from the processes is then used to produce power.

Coal-fired plants share environmental problems with some other types of fossil-fuel plants; these include "acid rain" and the "greenhouse" effect.

Gas Turbines

With increasing availability of natural gas (methane) (recent finds in Bangladesh) primemovers based on gas turbines have been developed on the lines similar to those used in aircraft. Gas combustion generates high temperatures and pressures, so that the efficiency of the gas turbine is comparable to that of steam turbine. Additional advantage is that exhaust gas from the turbine still has sufficient heat content, which is used to raise steam to run a conventional steam turbine coupled to a generator. This is called combined-cycle gas-turbine (CCGT) plant. The schematic diagram of such a plant is drawn in Fig. 1.5.

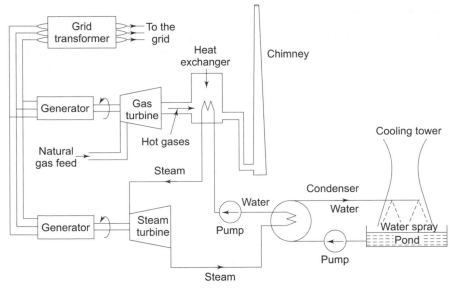

Fig. 1.5 CCGT power station

The CCGT plant has a fast start of 2–3 min for the gas turbine and about 20 minutes for the steam turbine. Local storage tanks of gas can be used in case of gas supply interruption. The unit can take up to 10% overload for short periods of time to take care of any emergency.

CCGT unit produces 55% of CO_2 produced by a coal/oil-fired plant. Units are now available for a fully automated operation for 24h or to meet the peak demands.

In Delhi (India) a CCGT unit of 34MW is installed at Indraprastha Power Station.

There are currently many installations using gas turbines in the world with 100 MW generators. A 6×30 MW gas turbine station has already been put up in Delhi. A gas turbine unit can also be used as synchronous compensator to help maintain flat voltage profile in the system.

Hydroelectric Power Generation

The oldest and cheapest method of power generation is that of utilizing the potential energy of water. The energy is obtained almost free of running cost and is completely pollution free. Of course, it involves high capital cost because of the heavy civil engineering construction works involved. Also it requires a long gestation period of about five to eight years as compared to four to six years for steam plants. Hydroelectric stations are designed, mostly, as multipurpose projects such as river flood control, storage of irrigation and drinking water, and navigation. A simple block diagram of a hydro plant is given in Fig. 1.6. The vertical difference between the upper reservoir and tail race is called the *head*.

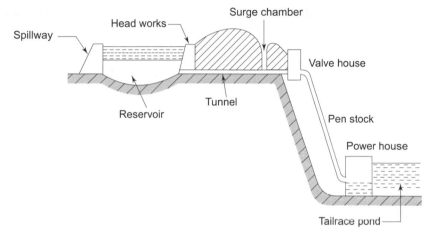

Fig. 1.6   A typical layout for a storage type hydro plant

Hydro plants are of different types such as run-of-river (use of water as it comes), pondage (medium head) type, and reservoir (high head) type. The reservoir type plants are the ones which are employed for bulk power generation. Often, cascaded plants are also constructed, i.e., on the same water stream where the discharge of one plant becomes the inflow of a downstream plant.

The utilization of energy in tidal flows in channels has long been the subject of research. The technical and economic difficulties still prevail. Some of the major sites under investigation are: Bhavnagar, Navalakhi (Kutch), Diamond Harbour and Ganga Sagar. The basin in Kandala (Gujrat) has been estimated to have a capacity of 600 MW. There are of course intense siting problems of the basin. Total potential is around 9000 MW out of which 900 MW is being planned.

A tidal power station has been constructed on the La Rance estuary in northern France where the tidal height range is 9.2 m and the tidal flow is estimated to be 18,000 m³/sec.

Different types of turbines such as Pelton, Francis and Kaplan are used for storage, pondage and run-of-river plants, respectively. Hydroelectric plants are capable of starting quickly—almost in five minutes. The rate of taking up load on the machines is of the order of 20 MW/min [18]. Further, no losses are incurred at standstill. Thus, hydro plants are ideal for meeting peak loads. The time from start up to the actual connection to the grid can be as short as 2 min.

The power available from a hydro plant is

$$P = 981 \rho WH \text{ W}$$

where

$W =$ discharge m^3/s through turbine

$\rho =$ density 1000 kg/m^2

$H =$ head (m)

Problems peculiar to hydro plant which inhibit expansion are:

1. Silting—reportedly Bhakra dead storage has silted fully in 30 years
2. Seepage
3. Ecological damage to region
4. Displacement of human habitation from areas behind the dam which will fill up and become a lake.
5. These cannot provide base load, must be used for peak shaving and energy saving in coordination with thermal plants.

India also has a tremendous potential (5000 MW) of having large number of *micro* (< 1 MW), *mini* (< 1–5 MW), and *small* (< 15 MW) *hydel plants* in Himalayan region, Himachal, UP, Uttaranchal and JK which must be fully exploited to generate cheap and clean power for villages situated far away from the grid power*. At present 500 MW capacity is under construction.

In areas where sufficient hydro generation is not available, peak load may be handled by means of pumped storage. This consists of an upper and lower reservoirs and reversible turbine-generator sets, which can also be used as motor-pump sets. The upper reservoir has enough storage for about six hours of full load generation. Such a plant acts as a conventional hydro plant during the peak load period, when production costs are the highest. The turbines are driven by water from the upper reservoir in the usual manner. During the light load period, water in the lower reservoir is pumped back into the upper one so as to be ready for use in the next cycle of the peak load period. The generators in this period change to synchronous motor action and drive the turbines which now work as pumps. The electric power is supplied to the sets from the general power network or adjoining thermal plant. The overall efficiency of the sets is normally as high as 60–70%. The pumped storage scheme, in fact, is analogous to the charging and discharging of a battery. It has the added advantage that the synchronous machines can be used as synchronous condensers for VAR compensation of the power network, if required. In a way, from the point of view of the thermal sector of the system,

* Existing capacity (small hydro) is 1341 MW as on June 2001. Total estimated potential is 15000 MW.

the pumped storage scheme "shaves the peaks" and "fills the troughs" of the daily load demand curve.

Some of the existing pumped storage plants are 300 MW Srisailem in AP and 80 MW at Bhira in Maharashtra.

Nuclear Power Stations

With the end of coal reserves in sight in the not too distant future, the immediate practical alternative source of large scale electric energy generation is nuclear energy. In fact, the developed countries have already switched over in a big way to the use of nuclear energy for power generation. In India, at present, this source accounts for only 3% of the total power generation with nuclear stations at Tarapur (Maharashtra), Kota (Rajasthan), Kalpakkam (Tamil Nadu), Narora (UP) and Kakrapar (Gujarat). Several other nuclear power plants will be commissioned by 2012. In future, it is likely that more and more power will be generated using this important resource (it is planned to raise nuclear power generation to 10,000 MW by the year 2010).

When Uranium-235 is bombarded with neutrons, fission reaction takes place releasing neutrons and heat energy. These neutrons then participate in the chain reaction of fissioning more atoms of ^{235}U. In order that the freshly released neutrons be able to fission the uranium atoms, their speeds must be reduced to a critical value. Therefore, for the reaction to be sustained, nuclear fuel rods must be embedded in neutron speed reducing agents (like graphite, heavy water, etc.) called moderators. For reaction control, rods made of neutron-absorbing material (boron-steel) are used which, when inserted into the reactor vessel, control the amount of neutron flux thereby controlling the rate of reaction. However, this rate can be controlled only within a narrow range. The schematic diagram of a nuclear power plant is shown in Fig. 1.7. The heat released by the nuclear reaction is transported to a heat exchanger via primary coolant (CO_2, water, etc.). Steam is then generated in the heat exchanger, which is used in a conventional manner to generate electric energy by means of a steam turbine. Various types of reactors are being used in practice for power plant purposes, viz., advanced gas reactor (AGR), boiling water reactor (BWR), and heavy water moderated reactor, etc.

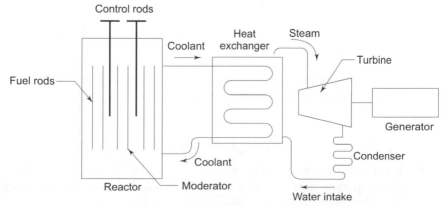

Fig. 1.7 Schematic view of a nuclear power plant

CANDU reactor—Natural uranium (in oxide form), pressurized heavy water moderated—is adopted in India. Its schematic diagram is shown in Fig. 1.8.

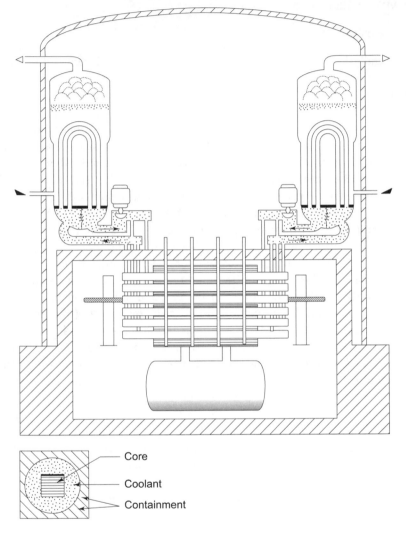

Fig. 1.8 CANDU reactor-pressurized heavy water moderated—adopted in India

The associated merits and problems of nuclear power plants as compared to conventional thermal plants are mentioned below.

Merits

1. A nuclear power plant is totally free of air pollution.
2. It requires little fuel in terms of volume and weight, and therefore poses no transportation problems and may be sited, independently of nuclear

fuel supplies, close to load centres. However, safety considerations require that they be normally located away from populated areas.

Demerits

1. Nuclear reactors produce radioactive fuel waste, the disposal of which poses serious environmental hazards.
2. The rate of nuclear reaction can be lowered only by a small margin, so that the load on a nuclear power plant can only be permitted to be marginally reduced below its full load value. Nuclear power stations must, therefore, be realiably connected to a power network, as tripping of the lines connecting the station can be quite serious and may required shutting down of the reactor with all its consequences.
3. Because of relatively high capital cost as against running cost, the nuclear plant should operate continuously as the base load station. Wherever possible, it is preferable to support such a station with a pumped storage scheme mentioned earlier.
4. The greatest danger in a fission reactor is in the case of loss of coolant in an accident. Even with the control rods fully lowered quickly called *scram operation*, the fission does continue and its after-heat may cause vaporizing and dispersal of radioactive material.

The world uranium resources are quite limited, and at the present rate may not last much beyond 50 years. However, there is a redeeming feature. During the fission of ^{235}U, some of the neutrons are absorbed by the more abundant uranium isotope ^{238}U (enriched uranium contains only about 3% of ^{235}U while most of its is ^{238}U) converting it to plutonium (^{239}U), which in itself is a fissionable material and can be extracted from the reactor fuel waste by a fuel reprocessing plant. Plutonium would then be used in the next generation reactors (fast breeder reactors—FBRs), thereby considerably extending the life of nuclear fuels. The FBR technology is being intensely developed as it will extend the availability of nuclear fuels at predicted rates of energy consumption to several centuries.

Figure 1.9 shows the schematic diagram of an FBR. It is essential that for breeding operation, conversion ratio (fissile material generated/fissile material consumed) has to be more than unity. This is achieved by fast moving neutrons so that no moderator is needed. The neutrons do slow down a little through collisions with structural and fuel elements. The energy density/kg of fuel is very high and so the core is small. It is therefore necessary that the coolant should possess good thermal properties and hence liquid sodium is used. The fuel for an FBR consists of 20% plutonium plus 8% uranium oxide. The coolant, liquid sodium, leaves the reactor at 650°C at atmospheric pressure. The heat so transported is led to a secondary sodium circuit which transfers it to a heat exchanger to generate steam at 540°C.

With a breeder reactor the release of plutonium, an extremely toxic material, would make the environmental considerations most stringent.

An experimental fast breeder test reactor (FBTR) (40 MW) has been built at Kalpakkam alongside a nuclear power plant. FBR technology is expected to reduce the cost of electric energy so as to compare favourably with that of conventional thermal plants.

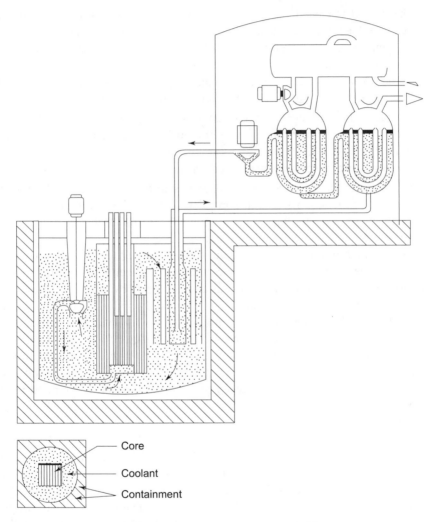

Core

Coolant

Containment

Fig. 1.9   Fast breeder reactor (FBR)

An important advantage of FBR technology is that it can also use thorium (as fertile material) which gets converted to ^{233}U which is fissionable. This holds great promise for India as we have one of the world's largest deposits of thorium—about 450000 tons in form of sand dunes in Kerala and along the Gopalpur Chatrapur coast of Orissa. We have merely 1 per cent of the world's coal resources and some traces of oil and gas. Thus, nuclear energy is best suited for India, with poor quality coal, inadequate hydro potential, plentiful reserves of uranium (70,000 tons) and thorium, and many years of nuclear engineering experience. The present cost of nuclear generation, comparable with coal-fired power plant, can be further reduced by standardising plant design and shifting from heavy water reactor to light water reactor technology.

Typical power densities (MW/m^3) in fission reactor cores are: gas cooled 0.53, high temperature gas cooled 7.75, heavy water 18.0, boiling water 29.0, pressurized water 54.75, fast breeder reactor 760.0.

Fusion

Energy is produced in this process by the combination of two light nuclei to form a single heavier one under sustained conditions of extremely high temperatures (in millions of degree centigrade). Fusion is futuristic. Generation of electricity via fusion would solve the long-term energy needs of the world with minimum environmental problems. A commercial reactor is expected by 2010 AD. Considering radioactive wastes, the impact of fusion reactors would be much less than the fission reactors.

In case of success in fusion technology sometime in the distant future or a breakthrough in the pollution-free solar energy, FBRs would become obsolete. However, there is an intense need today to develop FBR technology as an insurance against failure to develop these two technologies.

In the past few years, serious doubts have been raised about the safety claims of nuclear power plants. There have been as many as 150 near disaster nuclear accidents from the Three-mile accident in USA to the recent Chernobyl accident in the former USSR. There is a fear that all this may put the nuclear energy development in reverse gear. If this happens there could be serious energy crisis in the third world countries which have pitched their hopes on nuclear energy to meet their burgeoning energy needs. France (with 78% of its power requirement from nuclear sources) and Canada are possibly the two countries with a fairly clean record of nuclear generation. India needs to watch carefully their design, construction and operating strategies as it is committed to go in a big way for nuclear generation and hopes to achieve a capacity of 10,000 MW by 2010 AD. As per Indian nuclear scientists, our heavy water-based plants are most safe. But we must adopt more conservative strategies in design, construction and operation of nuclear plants.

World scientists have to adopt of different reaction safety strategy—may be to discover additives to automatically inhibit reaction beyond critical rather than by mechanically inserted control rods which have possibilities of several primary failure events.

Magnetohydrodynamic (MHD) Generation

In thermal generation of electric energy, the heat released by the fuel is converted to rotational mechanical energy by means of a thermocycle. The mechanical energy is then used to rotate the electric generator. Thus two stages of energy conversion are involved in which the heat to mechanical energy conversion has inherently low efficiency. Also, the rotating machine has its associated losses and maintenance problems. In MHD technology, electric energy is directly generated by the hot gases produced by the combustion of fuel without the need for mechanical moving parts.

In a MHD generator, electrically conducting gas at a very high temperature is passed in a strong magnetic field, thereby generating electricity. High temperature is needed to ionize the gas, so that it has good electrical conductivity. The conducting gas is obtained by burning a fuel and injecting a seeding materials such as potassium carbonate in the products of combustion. The principle of MHD power generation is illustrated in Fig. 1.10. About 50% efficiency can be achieved if the MHD generator is operated in tandem with a conventional steam plant.

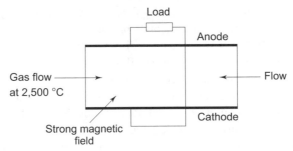

Fig. 1.10   The principle of MHD power generation

Though the technological feasibility of MHD generation has been established, its economic feasibility is yet to be demonstrated. India had started a research and development project in collaboration with the former USSR to install a pilot MHD plant based on coal and generating 2 MW power. In Russia, a 25 MW MHD plant which uses natural gas as fuel had been in operation for some years. In fact with the development of CCGT (combined cycle gas turbine) plant, MHD development has been put on the shelf.

Geothermal Power Plants

In a geothermal power plant, heat deep inside the earth act as a source of power. There has been some use of geothermal energy in the form of steam coming from underground in the USA, Italy, New Zealand, Mexico, Japan, Philippines and some other countries. In India, feasibility studies of 1 MW station at Puggy valley in Ladakh is being carried out. Another geothermal field has been located at Chumantang. There are a number of hot springs in India, but the total exploitable energy potential seems to be very little.

The present installed geothermal plant capacity in the world is about 500 MW and the total estimated capacity is immense provided heat generated in the volcanic regions can be utilized. Since the pressure and temperatures are low, the efficiency is even less than the conventional fossil fuelled plants, but the capital costs are less and the fuel is available free of cost.

1.4 RENEWABLE ENERGY SOURCES

To protect environment and for sustainable development, the importance of renewable energy sources cannot be overemphasized. It is an established and accepted fact that renewable and non-conventional forms of energy will play an increasingly important role in the future as they are cleaner and easier to use and environmentally benign and are bound to become economically more viable with increased use.

Because of the limited availability of coal, there is considerable international effort into the development of alternative/new/non-conventional/renewable/clean sources of energy. Most of the new sources (some of them in fact have been known and used for centuries now!) are nothing but the manifestation of solar energy, e.g., wind, sea waves, ocean thermal energy conversion (OTEC) etc. In this section, we shall discuss the possibilities and potentialities of various methods of using solar energy.

Wind Power

Winds are essentially created by the solar heating of the atmosphere. Several attempts have been made since 1940 to use wind to generate electric energy and development is still going on. However, technoeconomic feasibility has yet to be satisfactorily established.

Wind as a power source is attractive because it is plentiful, inexhaustible and non-polluting. Further, it does not impose extra heat burden on the environment. Unfortunately, it is non-steady and undependable. Control equipment has been devised to start the wind power plant whenever the wind speed reaches 30 km/h. Methods have also been found to generate constant frequency power with varying wind speeds and consequently varying speeds of wind mill propellers. Wind power may prove practical for small power needs in isolated sites. But for maximum flexibility, it should be used in conjunction with other methods of power generation to ensure continuity.

For wind power generation, there are three types of operations:

1. Small, 0.5–10 kW for isolated single premises
2. Medium, 10–100 kW for communities
3. Large, 1.5 MW for connection to the grid.

The theoretical power in a wind stream is given by

$$P = 0.5 \ \rho A V^3 \ \text{W}$$

where ρ = density of air (1201 g/m^3 at NTP)

   V = mean air velocity (m/s) and

   A = swept area (m^2).

For a rotor of 17 m diameter and a velocity of 48 km/h the theoretical power is 265 kW and the practical would be roughly half of this value.

There are some distinctive energy end-use features of wind power systems:

1. Most wind power sites are in remote rural, island or marine areas.
2. Rural grid systems are likely to be 'weak' in these areas, since they carry relatively low voltage supplies (e.g. 33 kV).
3. There are always periods without wind.

In India, wind power plants have been installed in Gujarat, Orissa, Maharashtra and Tamil Nadu, where wind blows at speeds of 30 km/h during summer. On the whole, the wind power potential of India has been estimated to be substantial and is around 45000 MW. The installed capacity as on Dec. 2000 is 1267 MW, the bulk of which is in Tamil Nadu (60%). The corresponding world figure is 14000 MW, the bulk of which is in Europe (70%).

Solar Energy

The average incident solar energy received on earth's surface is about 600 W/m^2 but the actual value varies considerably. It has the advantage of being free of cost, non-exhaustible and completely pollution-free. On the other hand, it has several drawbacks—energy density per unit area is very low, it is available for only a part of the day, and cloudy and hazy atmospheric conditions greatly reduce the energy received. Therefore, harnessing solar energy for electricity generation, challenging technological problems exist, the most important being that of the collection and concentration of solar energy and its conversion to the electrical form through efficient and comparatively economical means.

At present, two technologies are being developed for conversion of solar energy to the electrical form. In one technology, collectors with concentrators are employed to achieve temperatures high enough (700°C) to operate a heat engine at reasonable efficiency to generate electricity. However, there are considerable engineering difficulties in building a single tracking bowl with a diameter exceeding 30 m to generate perhaps 200 kW. The scheme involves large and intricate structures involving huge capital outlay and as of today is far from being competitive with conventional electricity generation.

The solar power tower [15] generates steam for electricity production. There is a 10 MW installation of such a tower by the Southern California Edison Co. in USA using 1818 plane mirrors, each 7 m × 7 m reflecting direct radiation to the raised boiler.

Electricity may be generated from a Solar Pond by using a special 'low temperature' heat engine coupled to an electric generator. A solar pond at Ein Borek in Israel produces a steady 150 kW from 0.74 hectare at a busbar cost of about $ 0.10/kWh.

Solar power potential is unlimited, however, total capacity of about 2000 MW is being planned.

Total solar energy potential in India is 5×10^{15} kWh/yr. Up to 31.12.2000, 462000 solar cookers, $55 \times 10^4 \text{m}^2$ solar thermal system collector area, 47 MW of SPV power, 270 community lights, 278000 solar lanterns (PV domestic lighting units), 640 TV (solar), 39000 PV street lights and 3370 water pumps were installed. Village power plants (stand-alone) of 1.1 MW capacity and 1.1 MW of grid connected solar power plants were in operation. As per one estimate [36], solar power will overtake wind in 2040 and would become the world's overall largest source of electricity by 2050.

Direct Conversion to Electricity (Photovoltaic Generation)

This technology converts solar energy to the electrical form by means of silicon wafer photoelectric cells known as "Solar Cells". Their theoretical efficiency is about 25% but the practical value is only about 15%. But that does not matter as solar energy is basically free of cost. The chief problem is the cost and maintenance of solar cells. With the likelihood of a breakthrough in the large scale production of cheap solar cells with amorphous silicon, this technology may compete with conventional methods of electricity generation, particularly as conventional fuels become scarce.

Solar energy could, at the most, supplement up to 5–10% of the total energy demand. It has been estimated that to produce 10^{12} kWh per year, the necessary cells would occupy about 0.1% of US land area as against highways which occupy 1.5% (in 1975) assuming 10% efficiency and a daily insolation of 4 kWh/m^2.

In all solar thermal schemes, storage is necessary because of the fluctuating nature of sun's energy. This is equally true with many other unconventional sources as well as sources like wind. Fluctuating sources with fluctuating loads complicate still further the electricity supply.

Wave Energy

The energy content of sea waves is very high. In India, with several hundreds of kilometers of coast line, a vast source of energy is available. The power in the wave is proportional to the square of the amplitude and to the period of the motion. Therefore, the long period (~ 10 s), large amplitude (~ 2m) waves are of considerable interest for power generation, with energy fluxes commonly averaging between 50 and 70 kW/m width of oncoming wave. Though the engineering problems associated with wave-power are formidable, the amount of energy that can be harnessed is large and development work is in progress (also see the section on Hydroelectric Power Generation, page 17). Sea wave power estimated potential is 20000 MW.

Ocean Thermal Energy Conversion (OTEC)

The ocean is the world's largest solar collector. Temperature difference of 20°C between warm, solar absorbing surface water and cooler 'bottom' water can occur. This can provide a continually replenished store of thermal energy which is in principle available for conversion to other energy forms. OTEC refers to the conversion of some of this thermal energy into work and thence into electricity. Estimated potential of ocean thermal power in India is 50,000 MW.

A proposed plant using sea temperature difference would be situated 25 km east of Miami (USA), where the temperature difference is 17.5°C.

Biofuels

The material of plants and animals is called *biomass*, which may be transformed by chemical and biological processes to produce intermediate *biofuels* such as methane gas, ethanol liquid or charcoal solid. Biomass is burnt to provide heat for cooking, comfort heat (space heat), crop drying, factory processes and raising steam for electricity production and transport. In India potential for bio-Energy is 17000 MW and that for agricultural waste is about 6000 MW. There are about 2000 community biogas plants and family size biogas plants are 3.1×10^6. Total biomass power harnessed so far is 222 MW.

Renewable energy programmes are specially designed to meet the growing energy needs in the rural areas for promoting decentralized and hybrid development so as to stem growing migration of rural population to urban areas in search of better living conditions. It would be through this integration of energy conservation efforts with renewable energy programmes that India would be able to achieve a smooth transition from fossil fuel economy to sustainable renewable energy based economy and bring "Energy for all" for equitable and environmental friendly sustainable development.

1.5 ENERGY STORAGE

There is a lot of problem in storing electricity in large quantities. Energy which can be converted into electricity can be stored in a number of ways. Storage of any nature is however very costly and its economics must be worked out properly. Various options available are: pumped storage, compressed air, heat, hydrogen gas, secondary batteries, flywheels and superconducting coils.

As already mentioned, gas turbines are normally used for meeting peak loads but are very expensive. A significant amount of storage capable of instantaneous use would be better way of meeting such peak loads, and so far the most important way is to have a pumped storage plant as discussed earlier. Other methods are discussed below very briefly.

Secondary Batteries

Large scale battery use is almost ruled out and they will be used for battery powered vehicles and local fluctuating energy sources such as wind mills or solar. The most widely used storage battery is the lead acid battery, invented by Plante in 1860. Sodium-sulphur battery (200 Wh/kg) and other combinations of materials are also being developed to get more output and storage per unit weight.

Fuel Cells

A fuel cell converts chemical energy of a fuel into electricity directly, with no intermediate combustion cycle. In the fuel cell, hydrogen is supplied to the negative electrode and oxygen (or air) to the positive. Hydrogen and oxygen are combined to give water and electricity. The porous electrodes allow hydrogen ions to pass. The main reason why fuel cells are not in wide use is their cost (> \$ 2000/kW). Global electricity generating capacity from full cells will grow from just 75 MW in 2001 to 15000 MW by 2010. US, Germany and Japan may take lead for this. '

Hydrogen Energy Systems

Hydrogen can be used as a medium for energy transmission and storage. Electrolysis is a well-established commercial process yielding pure hydrogen. H_2 can be converted very efficiently back to electricity by means of fuel cells. Also the use of hydrogen as fuel for aircraft and automobiles could encourage its large scale production, storage and distribution.

1.6   GROWTH OF POWER SYSTEMS IN INDIA

India is fairly rich in natural resources like coal and lignite; while some oil reserves have been discovered so far, intense exploration is being undertaken in various regions of the country. India has immense water power resources also of which only around 25% have so far been utilised, i.e., only 25000 MW has so far been commissioned up to the end of 9th plan. As per a recent report of the CEA (Central Electricity Authority), the total potential of hydro power is 84,040 MW at 60% load factor. As regards nuclear power, India is deficient in uranium, but has rich deposits of thorium which can be utilised at a future date in fast breeder reactors. Since independence, the country has made tremendous progress in the development of electric energy and today it has the largest system among the developing countries.

When India attained independence, the installed capacity was as low as 1400 MW in the early stages of the growth of power system, the major portion of generation was through thermal stations, but due to economical reasons, hydro development received attention in areas like Kerala, Tamil Nadu, Uttar Pradesh and Punjab.

In the beginning of the First Five Year Plan (1951–56), the total installed capacity was around 2300 MW (560 MW hydro, 1004 MW thermal, 149 MW through oil stations and 587 MW through non-utilities). For transporting this power to the load centres, transmission lines of up to 110 kV voltage level were constructed.

Fig. 1.11   Map of India showing five regional projected energy requirement in MkWh and park load in MW for year 2011–12.

The emphasis during the Second Plan (1956–61) was on the development of basic and heavy industries and thus there was a need to step up power generation. The total installed capacity which was around 3420 MW at the end of the First Five Year Plan became 5700 MW at the end of the Second Five Year Plan. The introduction of 230 kV transmission voltage came up in Tamil Nadu and Punjab. During this Plan, totally about 1009 circuit kilometres were energized. In 1965-66, the total installed capacity was increased to 10,170 MW. During the Third Five Year Plan (1961–66) transmission growth took place very rapidly, with a nine-fold expansion in voltage level below 66 kV. Emphasis was on rural electrification. A significant development in this phase was the emergence of an inter-state grid system. The country was divided into five regions, each with a regional electricity board, to promote integrated operation of the constituent power systems. Figure 1.11 shows these five regions of the country with projected energy requirement and peak load in the year 2011-12 [19].

During the Fourth Five Plan, India started generating nuclear power. At the Tarapur Nuclear Plant 2 × 210 MW units were commissioned in April-May 1969. This station uses two boiling water reactors of American design. By August 1972, the first unit of 220 MW of the Rajasthan Atomic Power Project, Kota (Rajasthan), was added to the nuclear generating capability. The total generating capacity at Kota is 430 MW with nuclear reactors of Canadian design which use natural uranium as fuel and heavy water as a moderator and coolant. The third nuclear power station of 2 × 235 MW has been commissioned at Kalpakkam (Tamil Nadu). This is the first nuclear station to be completely designed, engineered and constructed by Indian scientists and engineers. A reactor research centre has been set up near the Madras Atomic Power Station to carry out study in fast breeder reactor technology. The fourth nuclear power plant has been set up at Narora in Uttar Pradesh. It has two units of 235 MW each. The fifth is in Kaiga in Karnataka and sixth in Gujarat near Surat, Kakrapar (440 MW). Several other nuclear power plants will be commissioned by 2012.

The growth of generating capacity so far and future projection for 2011–2012 A.D. are given in Table 1.1.

Table 1.1   Growth of Installed Capacity in India (In MW)

Year	Hydro	Nuclear	Thermal	Diesel	Total
1970–71	6383	420	7503	398	14704
1978–79	11378	890	16372	—	28640
1984–85	14271	1095	27074	—	42240
2000–01	25141	2720	71060	≈2700 MW renewable	101630

Pattern of utlization of electrical energy in 1997–98 was: Domestic 20.69%, commercial 6.91%, irrigation 30.54%, industry 35.22% and others is 6.65%. It is expected to remain more or less same in 2004–05.

To be self sufficient in power, BHEL has plants spread out all over the country and these turn out an entire range of power equipment, viz. turbo sets, hydro sets, turbines for nuclear plants, high pressure boilers, power transformers, switch gears, etc. Each plant specializes in a range of equipment. BHEL's first 500 MW turbo-generator was commissioned at Singrauli. Today BHEL is considered one of the major power plant equipment manufacturers in the world.

1.7   ENERGY CONSERVATION

Energy conservation is the cheapest new source of energy. We should resort to various conservation measures such as cogeneration (discussed earlier), and use energy efficient motors to avoid wasteful electric uses. We can achieve considerable electrical power savings by reducing unnecessary high lighting levels, oversized motors, etc. A 9 W compact fluorescent lamp (CFL) may be used instead of 40 W fluorescent tube or 60 W lamp, all having the same lumens output. The pay-back period for any additional initial cost is less than a year. Everyone should be made aware through print or electronic media how consumption levels can be reduced without any essential lowering of comfort. Rate restructuring can have incentives in this regard. There is no consciousness on energy accountability yet and no sense of urgency as in developed countries.

Transmission and distribution losses should not exceed 20%. This can be achieved by employing series/shunt compensation, power factor improvement methods, static var compensators, HVDC option and FACTS (flexible ac technology) devices/controllers.

Gas turbine combined with steam turbine is employed for peak load shaving. This is more efficient than normal steam turbine and has a quick automated start and shut down. It improves the load factor of the steam station.

Energy storage can play an important role where there is time or rate mismatch between supply and demand of energy. This has been discussed in Section 1.5. Pumped storage (hydro) scheme has been considered in Section 1.3.

Industry

In India corporate sector is required by law, to include in their annual report, the measures taken for energy conservation. One of the key steps for energy conservation is a compulsory *energy audit*. This will put a finger on the places and items where there is wasteful use. Energy efficient drives and regenerative braking [1] should be employed.

In India where most areas have large number of sunny days hot water for bath and kitchen by solar water heaters is becoming common for commercial buildings, hotels even hospitals.

In India where vast regions are deficient in electric supply and are subjected to long hours of power shedding mostly random, the use of small diesel/petrol generators and inverters are very common in commercial and domestic use. These are highly wasteful energy devices. By proper *planned maintenance* the downtime of existing large stations can be cut down. Plant utilization factors of existing plants must be improved. Maintenance must be on schedule rather than an emergency. Maintenance manpower training should be placed on war footing. These actions will also improve the load factor of most power stations, which would indirectly contribute to energy conservation.

Load Management

As mentioned earlier by various 'load management' schemes. It is possible to shift demand away from peak hours (Section 1.1.). A more direct method would be the control of the load either through modified tariff structure that encourage the individual customers to readjust their own electric use schedules or direct electrical control of appliance in the form of remote timer controlled on/off switches with the least inconvenience to the customer. Various systems for load management are described in Ref. [27]. Ripple control has been tried in Europe. Remote kWh meter reading by carrier systems is being tried. Most of the potential for load control lies in the domestic sector. Power companies are now planning the introduction of system-wide load management schemes.

1.8 DEREGULATION

For over one hundred years, the electric power industry worldwide operated as a *regulated* industry. In any area there was only one company or government agency (mostly state-owned) that produced, transmitted, distributed and sold electric power and services. Deregulation as a concept came in early 1990s. It brought in changes designed to encourage competition.

Restructuring involves disassembly of the power industry and reassembly into another form or functional organisation. *Privatisation* started sale by a government of its state-owned electric utility assets, and operating economy, to private companies. In some cases, deregulation was driven by privatization needs. The state wants to sell its electric utility investment and change the rules (deregulation) to make the electric industry more palatable for potential investors, thus raising the price it could expect from the sale. *Open access* is nothing but a common way for a government to encourage competition in the electric industry and tackle monopoly. The consumer is assured of good quality power supply at competitive price.

The structure for deregulation is evolved in terms of Genco (Generation Company), Transco (Transmission Company) and ISO (Independent System Operator). It is expected that the optimal bidding will help Genco to maximize its payoffs. The consumers are given choice to buy energy from different retail energy suppliers who in turn buy the energy from Genco in a power market. (independent power producer, IPP).

The restructuring of the electricity supply industry that normally accompanies the introduction of competition provides a fertile ground for the growth of embedded generation, i.e. generation that is connected to the distribution system rather than to the transmission system.

The earliest reforms in power industries were initiated in Chile. They were followed by England, the USA, etc. Now India is also implementing the restructuring. Lot of research is needed to clearly understand the power system operation under deregulation. The focus of research is now shifting towards finding the optimal bidding methods which take into account local optimal dispatch, revenue adequacy and market uncertainties.

India has now enacted the Electricity Regulatory Commission's Act, 1998 and the Electricity (Laws) Amendment Act, 1998. These laws enable setting up of Central Electricity Regulatory Commission (CERC) at central level and State Electricity Regulatory Commissions (SERC) at state level. The main purpose of CERC is to promote efficiency, economy and competition in bulk electricity supply. Orissa, Haryana, Andhra Pradesh, etc. have started the process of restructuring the power sector in their respective states.

1.9 DISTRIBUTED AND DISPERSED GENERATION

Distributed Generation (DG) entails using many small generators of 2–50 MW output, installed at various strategic points throughout the area, so that each provides power to a small number of consumers nearby. These may be solar, mini/micro hydel or wind turbine units, highly efficient gas turbines, small combined cycle plants, since these are the most economical choices.

Dispersed generation referes to use of still smaller generating units, of less than 500 kW output and often sized to serve individual homes or businesses. Micro gas turbines, fuel cells, diesel, and small wind and solar PV generators make up this category.

Dispersed generation has been used for decades as an emergency backup power source. Most of these units are used only for reliability reinforcement. Now-a-days inverters are being increasingly used in domestic sector as an emergency supply during black outs.

The distributed/dispersed generators can be stand alone/autonomous or grid connected depending upon the requirement.

At the time of writing this (2001) there still is and will probably always be some economy of scale favouring large generators. But the margin of economy decreased considerably in last 10 years [23]. Even if the power itself costs a bit more than central station power, there is no need of transmission lines, and perhaps a reduced need for distribution equipment as well. Another major advantage of dispersed generation is its modularity, portability and relocatability. Dispersed generators also include two new types of fossil fuel units—fuel cells and microgas turbines.

The main challenge today is to upgrade the existing technologies and to promote development, demonstration, scaling up and commercialization of new and emerging technologies for widespread adaptation. In the rural sector main thrust areas are biomass briquetting, biomass-based cogeneration, etc. In solar PV (Photovoltaic), large size solar cells/modules based on crystalline silicon thin films need to be developed. Solar cells efficiency is to be improved to 15% to be of use at commercial level. Other areas are development of high efficiency inverters. Urban and industrial wastes are used for various energy applications including power generation which was around 17 MW in 2002.

There are already 32 million improved chulhas. If growing energy needs in the rural areas are met by decentralised and hybrid energy systems (distributed/dispersed generation), this can stem growing migration of rural population to urban areas in search of better living conditions. Thus, India will be able to achieve a smooth transition from fossil fuel economy to sustainable renewable-energy based economy and bring "Energy for all" for equitable, environment-friendly, and sustainable development.

1.10 ENVIRONMENTAL ASPECTS OF ELECTRIC ENERGY GENERATION

As far as environmental and health risks involved in nuclear plants of various kinds are concerned, these have already been discussed in Section 1.3. The problems related to large hydro plants have also been dwelled upon in Section 1.3. Therefore, we shall now focus our attention on fossil fuel plant including gas-based plants.

Conversion of one form of energy or another to electrical form has unwanted side effects and the pollutants generated in the process have to be disposed off. Pollutants know no geographical boundary, as result the pollution issue has become a nightmarish problem and strong national and international pressure groups have sprung up and they are having a definite impact on the development of energy resources. Governmental awareness has created numerous legislation at national and international levels, which power engineers have to be fully conversant with in practice of their profession and survey and planning of large power projects. Lengthy, time consuming procedures at government level, PIL (public interest litigation) and demonstrative protests have delayed several projects in several countries. This has led to favouring of small-size projects and redevelopment of existing sites. But with the increasing gap in electric demand and production, our country has to move forward for several large thermal, hydro and nuclear power projects.

Emphasis is being laid on conservation issues, curtailment of transmission losses, theft, subsidized power supplies and above all on *sustainable development* with *appropriate technology* wherever feasible. It has to be particularly assured that no irreversible damage is caused to environment which would affect the living conditions of the future generations. Irreversible damages like ozone layer holes and global warming caused by increase in CO_2 in the atmosphere are already showing up.

Atmospheric Pollution

We shall treat here only pollution as caused by thermal plants using coal as feedstock. Certain issues concerning this have already been highlighted in Section 1.3. The fossil fuel based generating plants form the backbone of power generation in our country and also globally as other options (like nuclear and even hydro) have even stronger hazards associated with them.

Also it should be understood that pollution in large cities like Delhi is caused more by vehicular traffic and their emission. In Delhi of course Inderprastha and Badarpur power stations contribute their share in certain areas.

Problematic pollutants in emission of coal-based generating plants are.

SO_2

NO_x, nitrogen oxides

CO

CO_2

Certain hydrocarbons

Particulates

Though the account that follows will be general, it needs to be mentioned here that Indian coal has comparatively low sulphur content but a very high ash content which in some coals may be as high as 53%.

A brief account of various pollutants, their likely impact and methods of abatements are presented as follows.

Oxides of Sulphur (SO_2)

Most of the sulphur present in the fossil fuel is oxidized to SO_2 in the combustion chamber before being emitted by the chimney. In atmosphere it gets further oxidized to H_2SO_4 and metallic sulphates which are the major source of concern as these can cause acid rain, impaired visibility, damage to buildings and vegetation. Sulphate concentrations of $9-10$ $\mu g/m^3$ of air aggravate asthma, lung and heart disease. It may also be noted that although sulphur does not accumulate in air, it does so in soil.

Sulphur emission can be controlled by:

- Use of fuel with less than 1% sulphur; generally not a feasible solution.
- Use of chemical reaction to remove sulphur in the form of sulphuric acid, from combustion products by limestone scrubbers or fluidized bed combustion.
- Removing sulphur from the coal by gasification or floatation processes.

It has been noticed that the byproduct sulphur could off-set the cost of sulphur recovery plant.

Oxides of Nitrogen (NO_x)

Of these NO_2, nitrogen oxides, is a major concern as a pollutant. It is soluble in water and so has adverse affect on human health as it enters the lungs on inhaling and combining with moisture converts to nitrous and nitric acids, which damage the lungs. At levels of 25–100 parts per million NO_x can cause acute bronchitis and pneumonia.

Emission of NO_x can be controlled by fitting advanced technology burners which can assure more complete combustion, thereby reducing these oxides from being emitted. These can also be removed from the combustion products by absorption process by certain solvents going on to the stock.

Oxides of Carbon (CO, CO₂)

CO is a very toxic pollutant but it gets converted to CO_2 in the open atmosphere (if available) surrounding the plant. On the other hand CO_2 has been identified as a major cause of global warming. It is not yet a serious problem in developing countries.

Hydrocarbons

During the oxidation process in combustion chamber certain light weight hydrocarbon may be formed. The compounds are a major source of photochemical reaction that adds to depletion of ozone layer.

Particulates (fly ash)

Dust content is particularly high in the Indian coal. Particulates come out of the stack in the form of fly ash. It comprises fine particles of carbon, ash and other inert materials. In high concentrations, these cause poor visibility and respiratory diseases.

Concentration of pollutants can be reduced by dispersal over a wider area by use of high stacks. *Precipitators* can be used to remove particles as the flue gases rise up the stack. If in the stack a vertical wire is strung in the middle and charged to a high negative potential, it emits electrons. These electrons are captured by the gas molecules thereby becoming negative ions. These ions accelerate towards the walls, get neutralized on hitting the walls and the particles drop down the walls. Precipitators have high efficiency up to 99% for large particles, but they have poor performance for particles of size less than 0.1 μm in diameter. The efficiency of precipitators is high with reasonable sulphur content in flue gases but drops for low sulphur content coals; 99% for 3% sulphur and 83% for 0.5% sulphur.

Fabric filters in form of *bag houses* have also been employed and are located before the flue gases enter the stack.

Thermal Pollution

Steam from low-pressure turbine has to be liquefied in a *condenser* and reduced to lowest possible temperature to maximize the thermodynamic efficiency. The best efficiency of steam-cycle practically achievable is about 40%. It means that 60% of the heat in steam at the cycle end must be removed. This is achieved by following two methods.

1. *Once through* circulation through condenser cooling tubes of sea or river water where available. This raises the temperature of water in these two sources and threatens sea and river life around in sea and downstream in river. These are serious environmental objections and many times cannot be overruled and also there may be legislation against it.

2. *Cooling towers* Cool water is circulated round the condenser tube to remove heat from the exhaust steam in order to condense it. The

circulating water gets hot in the process. It is pumped to cooling tower and is sprayed through nozzles into a rising volume of air. Some of the water evaporates providing cooling. The latent heat of water is 2×10^6 J/kg and cooling can occur fast. But this has the disadvantage of raising the humidity to high (undesirable) levels in the surrounding areas. Of course the water evaporated must be made up in the system by adding fresh water from the source.

Closed cooling towers where condensate flows through tubes and air is blown in these tubes avoids the humidity problem but at a very high cost. In India only *wet towers* are being used.

Electromagnetic Radiation from Overhead Lines

Biological effects of electromagnetic radiation from power lines and even cables in close proximity of buildings have recently attracted attention and have also caused some concern. Power frequency (50 or 60 Hz) and even their harmonics are not considered harmful. Investigations carried out in certain advanced countries have so far proved inconclusive. The electrical and electronics engineers, while being aware of this controversy, must know that many other environmental agents are moving around that can cause far greater harm to human health than does electromagnetic radiation.

As a piece of information it may be quoted that directly under an overhead line of 400 kV, the electric field strength is 11000 V/m and magnetic flux density (depending on current) may be as much as 40 μT. Electric field strength in the range of 10000–15000 V/m is considered safe.

Visual and Audible Impacts

These environmental problems are caused by the following factors.

1. Right of way acquires land underneath. Not a serious problem in India at present. Could be a problem in future.

2. Lines converging at a large substation mar the beauty of the landscape around. Underground cables as alternative are too expensive a proposition except in congested city areas.

3. Radio interference (RI) has to be taken into account and countered by various means.

4. Phenomenon of *corona* (a sort of electric discharge around the high tension line) produces a hissing noise which is audible when habitation is in close proximity. At the towers great attention must be paid to tightness of joints, avoidance of sharp edges and use of earth screen shielding to limit audible noise to acceptable levels.

5. Workers inside a power plant are subjected to various kinds of noise (particularly near the turbines) and vibration of floor. To reduce this noise to tolerable level foundations and vibration filters have to be designed properly and simulation studies carried out. The worker must be given regular medical examinations and sound medical advice.

1.11 POWER SYSTEM ENGINEERS AND POWER SYSTEM STUDIES

The power system engineer of the first decade of the twenty-first century has to face a variety of challenging tasks, which he can meet only by keeping abreast of the recent scientific advances and the latest techniques. On the planning side, he or she has to make decisions on how much electricity to generate—where, when, and by using what fuel. He has to be involved in construction tasks of great magnitude both in generation and transmission. He has to solve the problems of planning and coordinated operation of a vast and complex power network, so as to achieve a high degree of economy and reliability. In a country like India, he has to additionally face the perennial problem of power shortages and to evolve strategies for energy conservation and load management.

For planning the operation, improvement and expansion of a power system, a power system engineer needs load flow studies, short circuit studies, and stability studies. He has to know the principles of economic load despatch and load frequency control. All these problems are dealt with in the next few chapters after some basic concepts in the theory of transmission lines are discussed. The solutions to these problems and the enormous contribution made by digital computers to solve the planning and operational problems of power systems is also investigated.

1.12 USE OF COMPUTERS AND MICROPROCESSORS

The first methos for solving various power system problems were AC and DC network analysers developed in early 1930s. AC analysers were used for load flow and stability studies whereas DC were preferred for short-circuit studies.

Analogue computers were developed in 1940s and were used in conjunction with AC network analyser to solve various problems for off-line studies. In 1950s many analogue devices were developed to control the on-line functions such as generation control, frequency and tie-line control.

The 1950s also saw the advent of digital computers which were first used to solve a load flow problem in 1956. Power system studies by computers gave greater flexibility, accuracy, speed and economy. Till 1970s, there was a widespread use of computers in system analysis. With the entry of micro-processors in the arena, now, besides main frame computers, mini, micro and personal computers are all increasingly being used to carry out various power system studies and solve power system problems for off-line and on-line applications.

Off-line applications include research, routine evaluation of system performance and data assimilation and retrieval. It is mainly used for planning and analysing some new aspects of the system. On-line and real time applications include data-logging and the monitoring of the system state.

A large central computer is used in central load despatch centres for economic and secure control of large integrated systems. Microprocessors and computers installed in generating stations control various local processes such as starting up of a generator from the cold state, etc. Table 1.2 depicts the time scale of various hierarchical control problems to be solved by computers/ microprocessors. Some of these problems are tackled in this book.

Table 1.2

Time scale	Control Problems
Milliseconds	Relaying and system voltage control and excitation control
2 s–5 minutes	AGC (Automatic generation control)
10 min–few hours	ED (Economic despatch)
– do–	Security analysis
few hours–1 week	UC (Unit commitment)
1 month-6 months	Maintenance scheduling
1 yr–10 years	System planning (modification/extension)

1.13 PROBLEMS FACING INDIAN POWER INDUSTRY AND ITS CHOICES

The electricity requirements of India have grown tremendously and the demand has been running ahead of supply. Electricity generation and transmission processes in India are very inefficient in comparison with those of some developed countries. As per one estimate, in India generating capacity is utilized on an average for 3600 hours out of 8760 hours in a year, while in Japan it is used for 5100 hours. If the utilization factor could be increased, it should be possible to avoid power cuts. The transmission loss in 1997–98 on a national basis was 23.68% consisting of both technical losses in transmission lines and transformers, and also non-technical losses caused by energy thefts and meters not being read correctly. It should be possible to achieve considerable saving by reducing this loss to 15% by the end of the Tenth Five Year Plan by using well known ways and means and by adopting sound commercial practices. Further, every attempt should be made to improve system load factors by flattening the load curve by giving proper tariff incentives and taking other administrative measures. As per the Central Electricity Authority's (CEA) sixteenth annual power survey of India report, the all India load factor up to 1998–99 was of the order of 78%. In future it is likely to be 71%. By 2001, 5.07 lakh of villages (86%) have been electrified and 117 lakh of pumpsets have been energized.

Assuming a very modest average annual energy growth of 5%, India's electrical energy requirement in the year 2010 will be enormously high. A difficult and challenging task of planning, engineering and constructing new power stations is imminent to meet this situation. The government has built several super thermal stations such as at Singrauli (Uttar Pradesh), Farakka (West Bengal), Korba (Madhya Pradesh), Ramagundam (Andhra Pradesh) and Neyveli (Tamil Nadu), Chandrapur (Maharashtra) all in coal mining areas, each with a capacity in the range of 2000 MW*. Many more super thermal plants would be built in future. Intensive work must be conducted on boiler furnaces to burn coal with high ash content. National Thermal Power Corporation (NTPC) is in charge of these large scale generation projects.

Hydro power will continue to remain cheaper than the other types for the next decade. As mentioned earlier, India has so far developed only around 18% of its estimated total hydro potential of 89000 MW. The utilization of this perennial source of energy would involve massive investments in dams, channels and generation-transmission system. The Central Electricity Authority, the Planning Commission and the Ministry of Power are coordinating to work out a perspective plan to develop all hydroelectric sources by the end of this century to be executed by the National Hydro Power Corporation (NHPC). NTPC has also started recently development of hydro plants.

Nuclear energy assumes special significance in energy planning in India. Because of limited coal reserves and its poor quality, India has no choice but to keep going on with its nuclear energy plans. According to the Atomic Energy Commission, India's nuclear power generation will increase to 10000 MW by year 2010. Everything seems to be set for a take off in nuclear power production using the country's thorium reserves in breeder reactors.

In India, concerted efforts to develop solar energy and other non-conventional sources of energy need to be emphasized, so that the growing demand can be met and depleting fossil fuel resources may be conserved. To meet the energy requirement, it is expected that the coal production will have to be increased to more than 450 million tons in 2004–2005 as compared to 180 million tonnes in 1988.

A number of 400 kV lines are operating successfully since 1980s as mentioned already. This was the first step in working towards a national grid. There is a need in future to go in for even higher voltages (800 kV). It is expected that by the year 2011–12, 5400 ckt km of 800 kV lines and 48000 ckt km of 400 kV lines would be in operation. Also lines may be series and shunt compensated to carry huge blocks of power with greater stability. There is a need for constructing HVDC (High Voltage DC) links in the country since DC lines can carry considerably more power at the same voltage and require fewer conductors. A 400 kV Singrauli—Vindhyachal of 500 MW capacity first HVDC back-to-back scheme has been commissioned by NPTC (National Power Transmission Corporation) followed by first point-to-point bulk EHVDC transmission of 1500 MW at + 500 kV over a distance of 915 km from Rihand to Delhi. Power Grid recently commissioned on 14 Feb. 2003 a

---

* NTPC has also built seven gas-based combined cycle power stations such as Anta and Auraiya.

2000 MW Talcher-Kolar ± 500 kV HVDC bipole transmission system thus enabling excess power from East to flow to South. 7000 ckt km of + 500 kV HVDC line is expected by 2011–12.

At the time of writing, the whole energy scenario is so clouded with uncertainty that it would be unwise to try any quantitative predictions for the future. However, certain trends that will decide the future developments of electric power industry are clear.

Generally, unit size will go further up from 500 MW. A higher voltage (765/1200 kV) will come eventually at the transmission level. There is little chance for six-phase transmission becoming popular though there are few such lines in USA. More of HVDC lines will come in operation. As population has already touched the 1000 million mark in India, we may see a trend to go toward underground transmission in urban areas.

Public sector investment in power has increased from Rs 2600 million in the First Plan to Rs 242330 million in the Seventh Plan (1985–90). Shortfall in the Sixth Plan has been around 26%. There have been serious power shortages and generation and availability of power in turn have lagged too much from the industrial, agricultural and domestic requirements. Huge amounts of funds (of the order of Rs. 1893200 million) will be required if we have to achieve power surplus position by the time we reach the terminal year to the XI Plan (2011-2012). Otherwise achieving a target of 975 billion units of electric power will remain an utopian dream.

Power grid is planning creation of transmission highways to conserve Right-of-way. Strong national grid is being developed in phased manner. In 2001 the interregional capacity was 5000 MW. It is expected that by 2011-12, it will be 30000 MW. Huge investment is planned to the tune of US $ 20 billion in the coming decade. Present figures for HVDC is 3136 ckt km, 800 kV is 950 ckt km, 400 kV is 45500 ckt km and 220/132 kV is 215000 ckt km. State-of-the art technologies which are, being used in India currently are HVDC bipole, HVDC back-to-back, SVC (Static Var Compensator), FACTs (Flexible AC Transmissions) devices etc. Improved O and M (Operation and Maintenance) technologies which are being used today are hotline maintenance, emergency restoration system, thermovision scanning, etc.

Because of power shortages, many of the industries, particularly power-intensive ones, have installed their own captive power plants.* Currently 20% of electricity generated in India comes from the captive power plants and this is bound to go up in the future. Consortium of industrial consumers should be encouraged to put up coal-based captive plants. Import should be liberalized to support this activity.

---

* Captive diesel plants (and small diesel sets for commercial and domestic uses) are very uneconomical from a national point of view. Apart from being lower efficiency plants they use diesel which should be conserved for transportation sector.

With the ever increasing complexity and growth of power networks and their economic and integrated operation, several central/regional automatic load despatch centres with real time computer control have been established. In future it is envisaged that using SCADA (Supervisory Control and Data Acquisition) etc. will be possible to achieve nationwide on-line monitoring and real time control of power system. It may also be pointed out that this book will also help in training and preparing the large number of professionals trained in computer aided power system operation and control that would be required to handle vast expansion planned in power system in the coming decades.

REFERENCES

Books

1. Nagrath, I.J. and D.P. Kothari, *Electric Machines*, Tata McGraw-Hill, New Delhi, 2nd edn, 1997.
2. Eilgerd, O.I., *Basic Electric Power Engineering*, Reading, Mass., 1977.
3. Kashkari, C., *Energy Resources, Demand and Conservation with Special Reference to India*, Tata McGraw-Hill, New Delhi, 1975.
4. Parikh, Kirit, *Second India Studies—Energy,* Macmillian, New Delhi, 1976.
5. Sullivan, R.L., *Power System Planning*, McGraw-Hill, New York, 1977.
6. S. Krotzki, B.G.A. and W.A. Vopat, *Power Station Engineering and Economy*, McGraw-Hill, New York, 1960.
7. Car, T.H., *Electric Power Stations*, vols I and II, Chapman and Hall, London, 1944.
8. Central Electricity Generating Board, *Modern Power Station Practice*, 2nd edn, Pergamon, 1976.
9. Golding, E.W., *The Generation of Electricity by Wind Power*, Chapman and Hall, London, 1976.
10. McMillan, J.T., et. al., *Energy Resources and Supply*, Wiley, London, 1976.
11. Bennet, D.J., *The Elements of Nuclear Power*, Longman, 1972.
12. Berkowitz, D.A., *Power Generation and Environmental Change*, M.I.T. Press, Cambridge, Mass., 1972.
13. Steinberg, M.J. and T.H. Smith, *Economy-loading of Power Plants and Electric Systems*, Wiley, New York, 1943.
14. Power System Planning and Operations: Future Problems and Research Needs, *EPRI EL-377-SR*, February 1977.
15. Twidell, J.W. and A.D. Weir, *Renewable Energy Resources*, E. and F. N, Spon, London, 1986.
16. Mahalanabis, A.K., D.P. Kothari and S.I. Ahson, *Computer Aided Power System Analysis and Control*, Tata McGraw-Hill, New Delhi, 1988.
17. Robert Noyes (Ed.), *Cogeneration of Steam and Electric Power*, Noyes Dali Corp., USA, 1978.
18. Weedy, B.M. and B.J. Cory, *Electric Power Systems*, 4th edn, Wiley, New York, 1998.
19. CEA 12 *Annual Survey of Power Report*, Aug. 1985; 14th Report, March 1991; 16th Electric Power Survey of India, Sept 2000.

20. Kothari, D.P. and D.K. Sharma (Eds), *Energy Engineering. Theory and Practice*, S. Chand, 2000.
21. Kothari, D.P. and I.J. Nagrath, *Basic Electrical Engineering*, 2nd edn, Tata McGraw-Hill, New Delhi, 2002. (Ch. 15).
22. Wehenkel, L.A. *Automatic Learning Techniques in Power Systems*, Norwell MA: Kluwer, 1997.
23. Philipson, L and H. Lee Willis, *Understanding Electric Utilities and Deregulation*, Marcel Dekker Inc, NY, 1999.

Papers

24. Kusko, A., 'A Prediction of Power System Development, 1968–2030', *IEEE Spectrum*, Apl. 1968, 75.
25. Fink, L. and K. Carlsen, *'Operating under Stress and Strain'*, *IEEE Spectrum*, Mar. 1978.
26. Talukdar, S.N., et. al., 'Methods for Assessing energy Management Options', *IEEE Trans.*, Jan. 1981, PAS-100, no. 1, 273.
27. Morgen, M.G. and S.N. Talukdar, 'Electric Power Load Management: Some Technological, Economic, Regularity and Social Issues', *Proc. IEEE*, Feb. 1979, vol. 67, no. 2, 241.
28. Sachdev, M.S.,'Load Forecasting—Bibliography, *IEEE Trans.*, PAS-96, 1977, 697.
29. Sporn, P., 'Our Environment—Options on the Way into the Future', *ibid,* May 1977, 49.
30. Kothari D.P., Energy Problems Facing the Third World, *Seminar to the Bio-Physics Workshop*, 8 Oct., 1986 Trieste Italy
31. Kothari, D.P, 'Energy System Planning and Energy Conservation', presented at *XXIV National Convention a of IIIE*, New Delhi, Feb. 1982.
32. Kothari, D.P. et. al., 'Minimization of Air Pollution due to Thermal Plants', *JIE* (India), Feb. 1977, 57, 65.
33. Kothari, D.P. and J. Nanda, 'Power Supply Scenario in India' 'Retrospects and Prospects', *Proc. NPC Cong.*, on Captive Power Generation, New Delhi, Mar. 1986.
34. National Solar Energy Convention, Organised by SESI, 1-3 Dec. 1988, Hyderabad.
35. Kothari, D.P., "Mini and Micro Hydropower Systems in India", invited chapter in the book, Energy Resources and Technology, Scientific Publishers, 1992, pp 147–158.
36. *Power Line*, vol. 5. no. 9, June 2001.
37. United Nations*: Electricity Costs and Tariffs: A General Study*; 1972.
38. Shikha, T.S. Bhatti and D.P. Kothari, "Wind as an Eco-friendly Energy Source to meet the Electricity Needs of SAARC Region", Proc. Int. Conf. (ICME 2001), BUET, Dhaka, Bangladesh. Dec. 2001, pp 11–16.
39. Bansal, R. C., D.P. Kothari & T. S. Bhatti, "On Some of the Design Aspects of Wind Energy Conversion Systems", *Int. J. of Energy Conversion and Managment*, Vol. 43, 16, Nov. 2002, pp. 2175–2187.
40. D. P. Kothari and Amit Arora, *"Fuel Cells in Transporation-Beyond Batteries"*, Proc. Nut. Conf. on Transportation Systems, IIT Delhi, April 2002, pp. 173–176.
41. Saxena, Anshu, D. P. Kothari et al, *"Analysis of Multimedia and Hypermedia for Computer Simulation and Growth"*, EJEISA, UK, Vol 3, 1 Sep 2001, 14–28.

2

Inductance and Resistance of Transmission Lines

2.1 INTRODUCTION

The four parameters which affect the performance of a transmission line as an element of a power system are inductance, capacitance, resistance and conductance. Shunt conductance, which is normally due to leakage over line insulators, is almost always neglected in overhead transmission lines. This chapter deals with the series line parameters, i.e. inductance and resistance. These parameters are uniformly distributed along the line and they together form the series impedance of the line.

Inductance is by far the most dominant line parameter from a power system engineer's viewpoint. As we shall see in later chapters, it is the inductive reactance which limits the transmission capacity of a line.

2.2 DEFINITION OF INDUCTANCE

Voltage induced in a circuit is given by

$$e = \frac{d\psi}{dt} \text{ V} \tag{2.1}$$

where ψ represents the flux linkages of the circuit in weber-turns (Wb-T). This can be written in the form

$$e = \frac{d\psi}{di} \cdot \frac{di}{dt} = L\frac{di}{dt} \text{ V} \tag{2.2}$$

where $L = \dfrac{d\psi}{di}$ is defined as the inductance of the circuit in henrys, which in general may be a function of i. In a linear magnetic circuit, i.e., a circuit with constant permeability, flux linkages vary linearly with current such that the inductance is constant given by

$$L = \frac{\psi}{i} \ H$$

or

$$\psi = Li \ \text{Wb-T} \tag{2.3}$$

If the current is alternating, the above equation can be written as

$$\lambda = LI \tag{2.4}$$

where λ and I are the rms values of flux linkages and current respectively. These are of course in phase.

Replacing $\frac{d}{dt}$ in Eq. (2.1) by $j\omega$, we get the steady state AC voltage drop due to alternating flux linkages as

$$V = j\omega LI = j\omega\lambda \ \text{V} \tag{2.5}$$

On similar lines, the mutual inductance between two circuits is defined as the flux linkages of one circuit due to current in another, i.e.,

$$M_{12} = \frac{\lambda_{12}}{I_2} \ \text{H} \tag{2.6}$$

The voltage drop in circuit 1 due to current in circuit 2 is

$$V_1 = j\omega M_{12}I_2 = j\omega\lambda_{12} \ \text{V} \tag{2.7}$$

The concept of mutual inductance is required while considering the coupling between parallel lines and the influence of power lines on telephone lines.

2.3   FLUX LINKAGES OF AN ISOLATED CURRENT-CARRYING CONDUCTOR

Transmission lines are composed of parallel conductors which, for all practical purposes, can be considered as infinitely long. Let us first develop expressions for flux linkages of a long isolated current-carrying cylindrical conductor with return path lying at infinity. This system forms a single-turn circuit, flux linking which is in the form of circular lines concentric to the conductor. The total flux can be divided into two parts, that which is internal to the conductor and the flux external to the conductor. Such a division is helpful as the internal flux progressively links a smaller amount of current as we proceed inwards towards the centre of the conductor, while the external flux always links the total current inside the conductor.

Flux Linkages due to Internal Flux

Figure 2.1 shows the cross-sectional view of a long cylindrical conductor carrying current I.

The mmf round a concentric closed circular path of radius y internal to the conductor as shown in the figure is

$$\oint H_y.ds = I_y \ \text{(Ampere's law)} \tag{2.8}$$

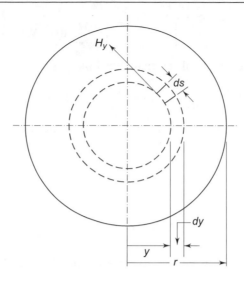

Fig. 2.1 Flux linkages due to internal flux (cross-sectional view)

where

H_y = magnetic field intensity (AT/m)

I_y = current enclosed (A)

By symmetry, H_y is constant and is in direction of ds all along the circular path. Therefore, from Eq. (2.8) we have

$$2\pi y H_y = I_y \tag{2.9}$$

Assuming uniform current density*

$$I_y = \left(\frac{\pi y^2}{\pi r^2}\right)I = \left(\frac{y^2}{r^2}\right)I \tag{2.10}$$

From Eqs. (2.9) and (2.10), we obtain

$$H_y = \frac{yI}{2\pi r^2} \text{ AT/m} \tag{2.11}$$

The flux density B_y, y metres from the centre of the conductors is

$$B_y = \mu H_y = \frac{\mu y I}{2\pi r^2} \text{ Wb/m}^2 \tag{2.12}$$

where μ is the permeability of the conductor.

Consider now an infinitesimal tubular element of thickness dy and length one metre. The flux in the tubular element d$\phi = B_y$ dy webers links the fractional turn ($I_y/I = y^2/r^2$) resulting in flux linkages of

---

*For power frequency of 50 Hz, it is quite reasonable to assume uniform current density. The effect of non-uniform current density is considered later in this chapter while treating resistance.

$$d\lambda = \left(\frac{y^2}{r^2}\right)d\phi = \left(\frac{y^2}{r^2}\right)\frac{\mu y I}{2\pi r^2}dy \quad \text{Wb-T/m} \tag{2.13}$$

Integrating, we get the total internal flux linkages as

$$\lambda_{\text{int}} = \int_0^r \frac{\mu I}{2\pi r^4} y^3 dy = \frac{\mu I}{8\pi} \quad \text{Wb-T/m} \tag{2.14}$$

For a relative permeability $\mu_r = 1$ (non-magnetic conductor), $\mu = 4\pi \times 10^{-7}$ H/m, therefore

$$\lambda_{\text{int}} = \frac{I}{2}\times 10^{-7} \quad \text{Wb-T/m} \tag{2.15}$$

and

$$L_{\text{int}} = \frac{1}{2}\times 10^{-7} \quad \text{H/m} \tag{2.16}$$

Flux Linkage due to Flux Between Two Points External to Conductor

Figure 2.2 shows two points P_1 and P_2 at distances D_1 and D_2 from a conductor which carries a current of I amperes. As the conductor is far removed from the return current path, the magnetic field external to the conductor is concentric circles around the conductor and therefore all the flux between P_1 and P_2 lines within the concentric cylindrical surfaces passing through P_1 and P_2.

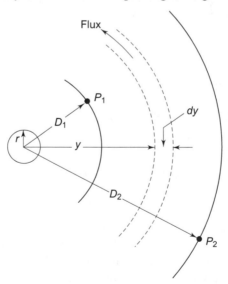

Fig. 2.2 Flux linkages due to flux between external points P_1, P_2

Magnetic field intensity at distance y from the conductor is

$$H_y = \frac{I}{2\pi y} \quad \text{AT/m}$$

The flux $d\phi$ contained in the tubular element of thickness dy is

$$d\phi = \frac{\mu I}{2\pi y} dy \text{ Wb/m length of conductor}$$

The flux $d\phi$ being external to the conductor links all the current in the conductor which together with the return conductor at infinity forms a single return, such that its flux linkages are given by

$$d\lambda = 1 \times d\phi = \frac{\mu I}{2\pi y} dy$$

Therefore, the total flux linkages of the conductor due to flux between points P_1 and P_2 is

$$\lambda_{12} = \int_{D_1}^{D_2} \frac{\mu I}{2\pi y} dy = \frac{\mu}{2\pi} I \ln \frac{D_2}{D_1} \text{ Wb-T/m}$$

where ln stands for natural logarithm*.

Since $\mu_r = 1$, $\mu = 4\pi \times 10^{-7}$

$$\therefore \qquad \lambda_{12} = 2 \times 10^{-7} I \ln \frac{D_2}{D_1} \text{ Wb/m} \qquad (2.17)$$

The inductance of the conductor contributed by the flux included between points P_1 and P_2 is then

$$L_{12} = 2 \times 10^{-7} \ln \frac{D_2}{D_1} \text{ H/m} \qquad (2.18)$$

or

$$L_{12} = 0.461 \log \frac{D_2}{D_1} \text{ mH/km} \qquad (2.19)$$

Flux Linkages due to Flux up to an External Point

Let the external point be at distance D from the centre of the conductor. Flux linkages of the conductor due to external flux (from the surface of the conductor up to the external point) is obtained from Eq. (2.17) by substituting $D_1 = r$ and $D_2 = D$, i.e.,

$$\lambda_{ext} = 2 \times 10^{-7} I \ln \frac{D}{r} \qquad (2.20)$$

Total flux linkages of the conductor due to internal and external flux are

$$\lambda = \lambda_{int} + \lambda_{ext}$$

$$= \frac{I}{2} \times 10^{-7} + 2 \times 10^{-7} I \ln \frac{D}{r}$$

---

*Throughout the book ln denotes natural logarithm (base e), while log denotes logarithm to base 10.

$$= 2 \times 10^{-7} I \left(\frac{1}{4} + \ln \frac{D}{r} \right)$$

$$= 2 \times 10^{-7} I \ln \frac{D}{re^{-1/4}}$$

Let $\qquad r' = re^{-1/4} = 0.7788r$

∴ $\qquad \lambda = 2 \times 10^{-7} I \ln \frac{D}{r'}$ Wb-T/m $\qquad\qquad$ (2.21a)

Inductance of the conductor due to flux up to an external point is therefore

$$L = 2 \times 10^{-7} \ln \frac{D}{r'} \text{ H/m} \qquad\qquad (2.21b)$$

Here r' can be regarded as the radius of a fictitious conductor with no internal inductance but the same total inductance as the actual conductor.

2.4 INDUCTANCE OF A SINGLE-PHASE TWO-WIRE LINE

Consider a simple two-wire line composed of solid round conductors carrying currents I_1 and I_2 as shown in Fig. 2.3. In a single-phase line,

$\qquad I_1 + I_2 = 0$

or

$\qquad I_2 = - I_1$

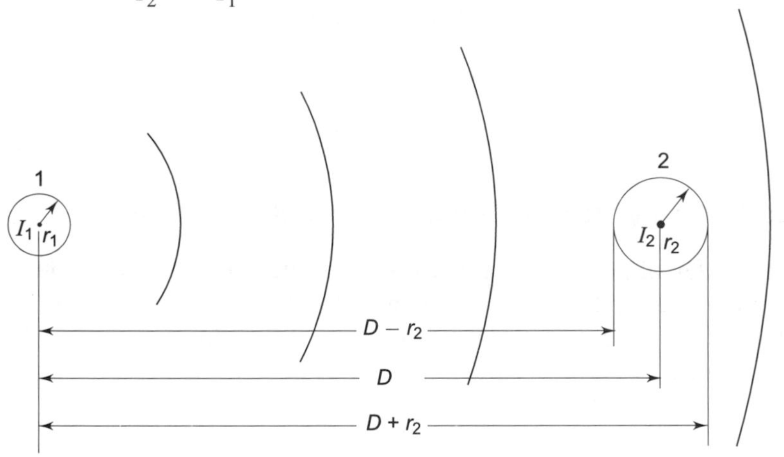

Fig. 2.3 Single-phase two-wire line and the magnetic field due to current in conductor 1 only

It is important to note that the effect of earth's presence on magnetic field geometry* is insignificant. This is so because the relative permeability of earth is about the same as that of air and its electrical conductivity is relatively small.

---

*The electric field geometry will, however, be very much affected as we shall see later while dealing with capacitance.

To start with, let us consider the flux linkages of the circuit caused by current in conductor 1 only. We make three observations in regard to these flux linkages:

1. External flux from r_1 to $(D - r_2)$ links all the current I_1 in conductor 1.

2. External flux from $(D - r_2)$ to $(D + r_2)$ links a current whose magnitude progressively reduces from I_1 to zero along this distance, because of the effect of negative current flowing in conductor 2.

3. Flux beyond $(D + r_2)$ links a net current of zero.

For calculating the total inductance due to current in conductor 1, a simplifying assumption will now be made. If D is much greater than r_1 and r_2 (which is normally the case for overhead lines), it can be assumed that the flux from $(D - r_2)$ to the centre of conductor 2 links all the current I_1 and the flux from the centre of conductor 2 to $(D + r_2)$ links zero current*.

Based on the above assumption, the flux linkages of the circuit caused by current in conductor 1 as per Eq. (2.21a) are

$$\lambda_1 = 2 \times 10^{-7} I_1 \ln \frac{D}{r'_1} \tag{2.22a}$$

The inductance of the conductor due to current in conductor 1 only is then

$$L_1 = 2 \times 10^{-7} \ln \frac{D}{r'_1} \tag{2.22b}$$

Similarly, the inductance of the circuit due to current in conductor 2 is

$$L_2 = 2 \times 10^{-7} \ln \frac{D}{r'_2} \tag{2.23}$$

Using the superposition theorem, the flux linkages and likewise the inductances of the circuit caused by current in each conductor considered separately may be added to obtain the total circuit inductance. Therefore, for the complete circuit

$$L = L_1 + L_2 = 4 \times 10^{-7} \ln \frac{D}{\sqrt{r'_1 r'_2}} \text{ H/m} \tag{2.24}$$

If $r'_1 = r'_2 = r'$; then

$$L = 4 \times 10^{-7} \ln D/r' \text{ H/m} \tag{2.25a}$$

$$L = 0.921 \log D/r' \text{ mH/km} \tag{2.25b}$$

Transmission lines are infinitely long compared to D in practical situations and therefore the end effects in the above derivation have been neglected.

2.5 CONDUCTOR TYPES

So far we have considered transmission lines consisting of single solid cylindrical conductors for forward and return paths. To provide the necessary flexibility for stringing, conductors used in practice are always stranded except

---

*Kimbark [19] has shown that the results based on this assumption are fairly accurate even when D is not much larger than r_1 and r_2.

for very small cross-sectional areas. Stranded conductors are composed of strands of wire, electrically in parallel, with alternate layers spiralled in opposite direction to prevent unwinding. The total number of strands (N) in concentrically stranded cables with total annular space filled with strands of uniform diameter (d) is given by

$$N = 3x^2 - 3x + 1 \qquad (2.26a)$$

where x is the number of layers wherein the single central strand is counted as the first layer. The overall diameter (D) of a stranded conductor is

$$D = (2x - 1)d \qquad (2.26b)$$

Aluminium is now the most commonly employed conductor material. It has the advantages of being cheaper and lighter than copper though with less conductivity and tensile strength. Low density and low conductivity result in larger overall conductor diameter, which offers another incidental advantage in high voltage lines. Increased diameter results in reduced electrical stress at conductor surface for a given voltage so that the line is *corona free*. The low tensile strength of aluminium conductors is made up by providing central strands of high tensile strength steel. Such a conductor is known as aluminium conductor steel reinforced (ACSR) and is most commonly used in overhead transmission lines. Figure 2.4 shows the cross-sectional view of an ACSR conductor with 24 strands of aluminium and 7 strands of steel.

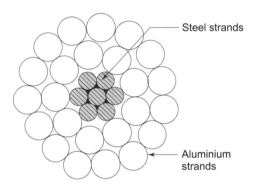

Fig. 2.4   Cross-sectional view of ACSR-7 steel strands, 24 aluminium strands

In extra high voltage (EHV) transmission line, *expanded* ACSR conductors are used. These are provided with paper or hessian between various layers of strands so as to increase the overall conductor diameter in an attempt to reduce electrical stress at conductor surface and prevent corona. The most effective way of constructing corona-free EHV lines is to provide several conductors per phase in suitable geometrical configuration. These are known as *bundled conductors* and are a common practice now for EHV lines.

2.6 FLUX LINKAGES OF ONE CONDUCTOR IN A GROUP

As shown in Fig. 2.5, consider a group of n parallel round conductors carrying phasor currents I_1, I_2,..., I_n whose sum equals zero. Distances of these conductors from a remote point P are indicated as D_1, D_2,..., D_n. Let us obtain an expression for the total flux linkages of the ith conductor of the group considering flux up to the point P only.

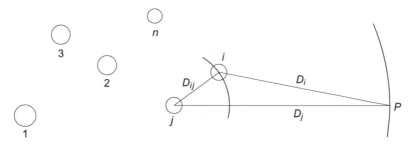

Fig. 2.5 Arbitrary group of n parallel round conductors carrying currents

The flux linkages of ith conductor due to its own current I_i (self linkages) are given by [see Eq. (2.21)]

$$\lambda_{ii} = 2 \times 10^{-7} I_i \ln \frac{D_i}{r'_i} \text{ Wb-T/m} \tag{2.27}$$

The flux linkages of conductor i due to current in conductor j [refer to Eq. (2.17)] is

$$\lambda_{ij} = 2 \times 10^{-7} I_j \ln \frac{D_j}{D_{ij}} \text{ Wb-T/m} \tag{2.28}$$

where D_{ij} is the distance of ith conductor from jth conductor carrying current I_j. From Eq. (2.27) and by repeated use of Eq. (2.28), the total flux linkages of conductor i due to flux up to point P are

$$\lambda_i = \lambda_{i1} + \lambda_{i2} + \dots + \lambda_{ii} + \dots + \lambda_{in}$$

$$= 2 \times 10^7 \left(I_1 \ln \frac{D_1}{D_{i1}} + I_2 \ln \frac{D_2}{D_{i2}} + \dots + I_i \ln \frac{D_i}{r'_i} \right.$$

$$\left. + \dots + I_n \ln \frac{D_n}{D_{in}} \right)$$

The above equation can be reorganized as

$$\lambda_i = 2 \times 10^{-7} \left[\left(I_1 \ln \frac{1}{D_{i1}} + I_2 \ln \frac{1}{D_{i2}} + \dots + I_i \ln \frac{1}{r'_i} + \dots + I_n \ln \frac{1}{D_{in}} \right) \right.$$

$$+ (I_1 \ln D_1 + I_2 \ln D_2 + \dots + I_i \ln D_i + \dots + I_n \ln D_n) \tag{2.29}$$

But, $I_n = - (I_1 + I_2 + \dots + I_{n-1})$.

Substituting for I_n in the second term of Eq. (2.29) and simplifying, we have

$$\lambda_i = 2 \times 10^{-7} \left[\left(I_1 \ln \frac{1}{D_{i1}} + I_2 \ln \frac{1}{D_{i2}} + ... + I_i \ln \frac{1}{r_i'} \right. \right.$$

$$\left. + ... + I_n \ln \frac{1}{D_{in}} \right) + \left(I_1 \ln \frac{D_1}{D_n} + I_2 \ln \frac{D_2}{D_n} + ... + I_i \ln \frac{D_i}{D_n} \right.$$

$$\left. \left. \cdots + I_{n-1} \ln \frac{D_{n-1}}{D_n} \right] \right)$$

In order to account for total flux linkages of conductor i, let the point P now recede to infinity. The terms such as $\ln D_1/D_n$, etc. approach $\ln 1 = 0$. Also for the sake of symmetry, denoting r_i' as D_{ii}, we have

$$\lambda_i = 2 \times 10^{-7} \left(I_1 \ln \frac{1}{D_{i1}} + I_2 \ln \frac{1}{D_{i2}} + I_2 \ln \frac{1}{D_{ii}} \right.$$

$$\left. + ... + I_n \ln \frac{1}{D_{in}} \right) \text{ Wb-T/m} \tag{2.30}$$

2.7 INDUCTANCE OF COMPOSITE CONDUCTOR LINES

We are now ready to study the inductance of transmission lines composed of composite conductors. Future 2.6 shows such a single-phase line comprising composite conductors A and B with A having n parallel filaments and B having m' parallel filaments. Though the inductance of each filament will be somewhat different (their resistances will be equal if conductor diameters are chosen to be uniform), it is sufficiently accurate to assume that the current is equally divided among the filaments of each composite conductor. Thus, each filament of A is taken to carry a current I/n, while each filament of conductor B carries the return current of $- I/m'$.

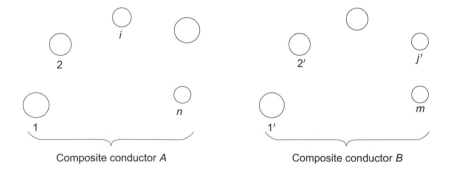

Fig. 2.6 Single-phase line consisting of two composite conductors

Applying Eq. (2.30) to filament i of conductor A, we obtain its flux linkages as

$$\lambda_i = 2\times10^{-7}\frac{I}{n}\left(\ln\frac{1}{D_{i1}}+\ln\frac{1}{D_{i2}}+\ldots+\ln\frac{1}{D_{ii}}+\ldots+\ln\frac{1}{D_{in}}\right)$$

$$-2\times10^{-7}\frac{I}{m'}\left(\ln\frac{1}{D_{i1'}}+\ln\frac{1}{D_{i2'}}+\ldots+\ln\frac{1}{D_{im'}}\right)$$

$$= 2\times10^{-7}I\ln\frac{(D_{i1'}D_{i2'}\ldots D_{im'})^{1/m'}}{(D_{i1}D_{i2}\ldots D_{ii}\ldots D_{in})^{1/n}} \quad \text{Wb-T/m}$$

The inductance of filament i is then

$$L_i =\frac{\lambda_i}{1/n} = 2n\times10^{-7}\ln\frac{(D_{i1'}\ldots D_{ij'}\ldots D_{im'})^{1/m'}}{(D_{i1}D_{i2}\ldots D_{ii}\ldots D_{in})^{1/n}} \quad \text{H/m} \qquad (2.31)$$

The average inductance of the filaments of composite conductor A is

$$L_{\text{avg}} = \frac{L_1+L_2+L_3+\ldots+L_n}{n}$$

Since conductor A is composed of n filaments electrically in parallel, its inductance is

$$L_A =\frac{L_{\text{avg}}}{n} = \frac{L_1+L_2+\ldots+L_n}{n^2} \qquad (2.32)$$

Using the expression for filament inductance from Eq. (2.31) in Eq. (2.32), we obtain

$$L_A = 2\times10^{-7}\ln\frac{[(D_{11'}\ldots D_{1j'}\ldots D_{1m'})\ldots(D_{i1'}\ldots D_{ij'}\ldots D_{im'})\ldots(D_{nl'}\ldots D_{nj'}\ldots D_{nm'})]^{1/m'n}}{[(D_{11}\ldots D_{1j}\ldots D_{1n})\ldots(D_{il}\ldots D_{ij}\ldots D_{in})\ldots(D_{n1}\ldots D_{ni}\ldots D_{nn})]^{1/n^2}} \quad \text{H/m} \quad (2.33)$$

The numerator of the argument of the logarithm in Eq. (2.33) is the $m'n$th root of the $m'n$ terms, which are the products of all possible mutual distances from the n filaments of conductor A to m' filaments of conductor B. It is called *mutual geometric mean distance* (mutual GMD) between conductors A and B and is abbreviated as D_m. Similarly, the denominator of the argument of the logarithm in Eq. (2.33) is the n^2th root of n^2 product terms (n sets of n product terms each). Each set of n product term pertains to a filament and consists of $r'(D_{ii})$ for that filament and $(n-1)$ distances from that filament to every other filament in conductor A. The denominator is defined as the *self geometric mean distance* (self GMD) of conductor A, and is abbreviated as D_{sA}. Sometimes, self GMD is also called *geometric mean radius* (GMR).

In terms of the above symbols, we can write Eq (2.33) as

$$L_A = 2 \times 10^{-7} \ln \frac{D_m}{D_{sA}} \text{ H/m} \tag{2.34a}$$

$$= 0.461 \log \frac{D_m}{D_{sA}} \text{ mH/km} \tag{2.34b}$$

Note the similarity of the above relation with Eq. (2.22b), which gives the inductance of one conductor of a single-phase line for the special case of two solid, round conductors. In Eq. (2.22b) r'_1 is the self GMD of a single conductor and D is the mutual GMD of two single conductors.

The inductance of the composite conductor B is determined in a similar manner, and the total inductance of the line is

$$L = L_A + L_B \tag{2.35}$$

Example 2.1

A conductor is composed of seven identical copper strands, each having a radius r, as shown in Fig. 2.7. Find the self GMD of the conductor.

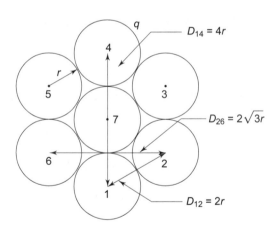

Fig. 2.7 Cross-section of a seven-strand conductor

Solution The self GMD of the seven strand conductor is the 49th root of the 49 distances. Thus

$$D_s = ((r')^7 (D_{12}^2 D_{26}^2 D_{14} D_{17})^6 (2r)^6)^{1/49}$$

Substituting the values of various distances,

$$D_s = ((0.7788r)^7 (2^2 r^2 \times 3 \times 2^2 r^2 \times 2^2 r \times 2r \times 2r)^6)^{1/49}$$

or

$$D_s = \frac{2r(3(0.7788))^{1/7}}{6^{1/49}} = 2.177r$$

Example 2.2

The outside diameter of the single layer of aluminium strands of an ACSR conductor shown in Fig. 2.8 is 5.04 cm. The diameter of each strand is 1.68 cm. Determine the 50 Hz reactance at 1 m spacing; neglect the effect of the central strand of steel and advance reasons for the same.

Solution   The conductivity of steel being much poorer than that of aluminium and the internal inductance of steel strands being μ-times that of aluminium strands, the current conducted by the central strands of steel can be assumed to be zero.

Diameter of steel strand = $5.04 - 2 \times 1.68 = 1.68$ cm.

Thus, all strands are of the same diameter, say d. For the arrangement of strands as given in Fig. 2.8a,

$$D_{12} = D_{16} = d$$

$$D_{13} = D_{15} = \sqrt{3}\,d$$

$$D_{14} = 2d$$

$$D_s = \left(\left[\left(\frac{d'}{2} \right) d^2 (\sqrt{3}\,d)^2 (2d) \right]^6 \right)^{1/36}$$

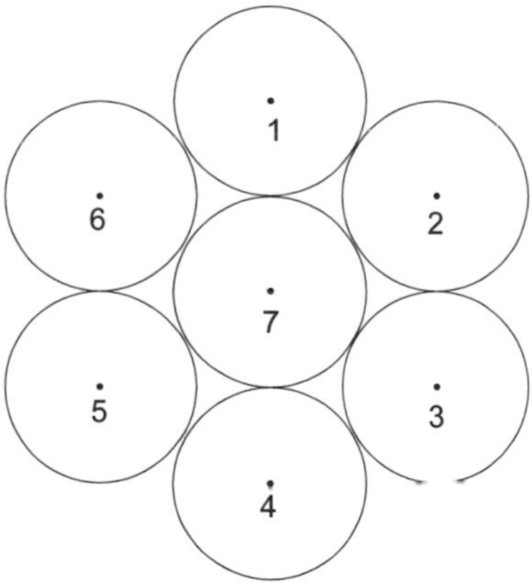

(a) Cross-section of ACSR conductor

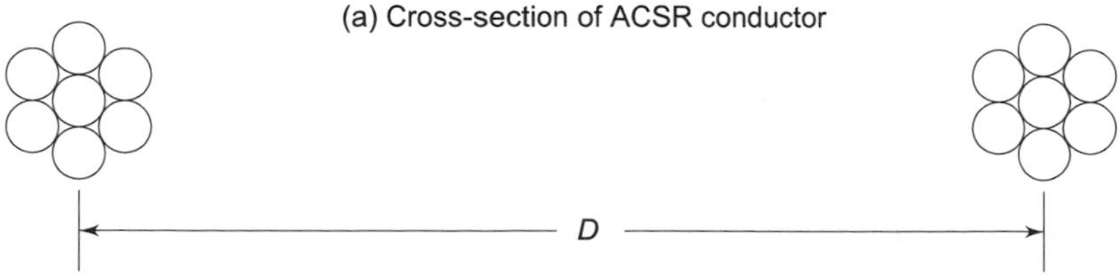

(b) Line composed of two ACSR conductors

Fig. 2.8

Substituting $d' = 0.7788d$ and simplifying

$$D_s = 1.155d = 1.155 \times 1.68 = 1.93 \text{ cm}$$

$$D_m \simeq D \text{ since } D \gg d$$

Now, the inductance of each conductor is

$$L = 0.461 \log \frac{100}{1.93} = 0.789 \text{ mH/km}$$

Loop inductance $= 2 \times 0.789 = 1.578$ mH/km

Loop reactance $= 1.578 \times 314 \times 10^{-3} = 0.495$ ohms/km

Example 2.3

The arrangement of conductors of a single-phase transmission line is shown in Fig. 2.9, wherein the forward circuit is composed of three solid wires 2.5 mm in radius and the return circuit of two-wires of radius 5 mm placed symmetrically with respect to the forward circuit. Find the inductance of each side of the line and that of the complete line.

Solution The mutual GMD between sides A and B is

$$D_m = ((D_{14}D_{15})\ (D_{24}D_{25})\ (D_{34}D_{35}))^{1/6}$$

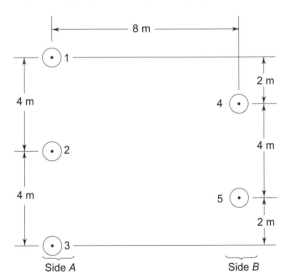

Fig. 2.9 Arrangement of conductors for Example 2.3

From the figure it is obvious that

$$D_{14} = D_{24} = D_{25} = D_{35} = \sqrt{68} \text{ m}$$
$$D_{15} = D_{34} = 10 \text{ m}$$
$$D_m = (68^2 \times 100)^{1/6} = 8.8 \text{ m}$$

The self GMD for side A is

$$D_{sA} = ((D_{11}D_{12}D_{13})(D_{21}D_{22}D_{23})(D_{31}D_{32}D_{33}))^{1/9}$$

Here,

$$D_{11} = D_{22} = D_{33} = 2.5 \times 10^{-3} \times 0.7788 \text{ m}$$

Substituting the values of various interdistances and self distances in D_{sA}, we get

$$D_{sA} = ((2.5 \times 10^{-3} \times 0.7788)^3 \times 4^4 \times 8^2)^{1/9}$$
$$= 0.367 \text{ m}$$

Similarly,

$$D_{sB} = ((5 \times 10^{-3} \times 0.7788)^2 \times 4^2)^{1/4}$$
$$= 0.125 \text{ m}$$

Substituting the values of D_m, D_{sA} and D_{sB} in Eq. (2.25b), we get the various inductances as

$$L_A = 0.461 \log\frac{8.8}{0.367} = 0.635 \text{ mH/km}$$

$$L_B = 0.461 \log\frac{8.8}{0.125} = 0.85 \text{ mH/km}$$

$$L = L_A + L_B = 1.485 \text{ mH/km}$$

If the conductors in this problem are each composed of seven identical strands as in Example 2.1, the problem can be solved by writing the conductor self distances as

$$D_{ii} = 2.177 r_i$$

where r_i is the strand radius.

2.8 INDUCTANCE OF THREE-PHASE LINES

So far we have considered only single-phase lines. The basic equations developed can, however, be easily adapted to the calculation of the inductance of three-phase lines. Figure 2.10 shows the conductors of a three-phase line with unsymmetrical spacing.

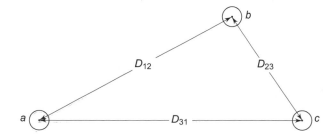

Fig. 2.10 Cross-sectional view of a three-phase line with unsymmetrical spacing

Assume that there is no neutral wire, so that

$$I_a + I_b + I_c = 0$$

Unsymmetrical spacing causes the flux linkages and therefore the inductance of each phase to be different resulting in unbalanced receiving-end voltages even when sending-end voltages and line currents are balanced. Also voltages will be induced in adjacent communication lines even when line currents are balanced. This problem is tackled by exchanging the positions of the conductors at regular intervals along the line such that each conductor occupies the original position of every other conductor over an equal distance. Such an exchange of conductor positions is called *transposition*. A complete transposition cycle is shown in Fig. 2.11. This arrangement causes each conductor to have the same average inductance over the transposition cycle. Over the length of one transposition cycle, the total flux linkages and hence net voltage induced in a nearby telephone line is zero.

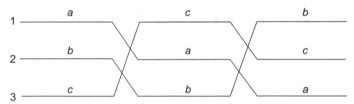

Fig. 2.11 A complete transposition cycle

To find the average inductance of each conductor of a transposed line, the flux linkages of the conductor are found for each position it occupies in the transposed cycle. Applying Eq. (2.30) to conductor a of Fig. 2.11, for section 1 of the transposition cycle wherein a is in position 1, b is in position 2 and c is in position 3, we get

$$\lambda_{a1} = 2 \times 10^{-7} \left(I_a \ln \frac{1}{r'_a} + I_b \ln \frac{1}{D_{12}} + I_c \ln \frac{1}{D_{31}} \right) \text{ Wb-T/m}$$

For the second section

$$\lambda_{a2} = 2 \times 10^{-7} \left(I_a \ln \frac{1}{r'_a} + I_b \ln \frac{1}{D_{23}} + I_c \ln \frac{1}{D_{12}} \right) \text{ Wb-T/m}$$

For the third section

$$\lambda_{a3} = 2 \times 10^{-7} \left(I_a \ln \frac{1}{r'_a} + I_b \ln \frac{1}{D_{13}} + I_c \ln \frac{1}{D_{23}} \right) \text{ Wb-T/m}$$

Average flux linkages of conductor a are

$$\lambda_a = \frac{\lambda_{a1} + \lambda_{a2} + \lambda_{a3}}{3}$$

$$= 2 \times 10^{-7} \left(I_a \ln \frac{1}{r'_a} I_b \ln \frac{1}{(D_{12}D_{23}D_{31})^{1/3}} + I_c \ln \frac{1}{(D_{12}D_{23}D_{31})^{1/3}} \right)$$

But, $I_b + I_c = -I_a$, hence

$$\lambda_a = 2 \times 10^{-7} I_a \ln \frac{(D_{12}D_{23}D_{31})^{1/3}}{r'_a}$$

Let

$$D_{eq} = (D_{12}D_{23}D_{31})^{1/3} = \text{equivalent equilateral spacing}$$

Then

$$L_a = 2 \times 10^{-7} \ln \frac{D_{eq}}{r'_a} = 2 \times 10^{-7} \ln \frac{D_{eq}}{D_s} \quad \text{H/m} \qquad (2.36)$$

This is the same relation as Eq. (2.34a) where $D_m = D_{eq}$, the mutual GMD between the three-phase conductors. If $r_a = r_b = r_c$, we have

$$L_a = L_b = L_c$$

It is not the present practice to transpose the power lines at regular intervals. However, an interchange in the position of the conductors is made at switching stations to balance the inductance of the phases. For all practical purposes the dissymmetry can be neglected and the inductance of an untransposed line can be taken equal to that of a transposed line.

If the spacing is equilateral, then

$$D_{eq} = D$$

and

$$L_a = 2 \times 10^{-7} \ln \frac{D}{r'_a} \quad \text{H/m} \qquad (2.37)$$

If $r_a = r_b = r_c$, it follows from Eq. (2.37) that

$$L_a = L_b = L_c$$

Example 2.4

Show that over the length of one transposition cycle of a power line, the total flux linkages of a nearby telephone line are zero, for balanced three-phase currents.

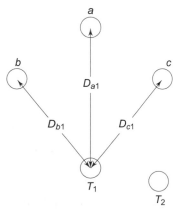

Fig. 2.12 Effect of transposition on Induced voltage of a telephone line

Solution Referring to Fig. 2.12, the flux linkages of the conductor t_1 of the telephone line are

$$\lambda_{t1} = 2 \times 10^{-7}\left(I_a \ln \frac{1}{D_{a1}} + 1_b \ln \frac{1}{D_{b1}} + I_c \ln \frac{1}{D_{c1}} \right) \text{ Wb-T/m} \quad (2.38)$$

Similarly,

$$\lambda_{t2} = 2 \times 10^{-7}\left(I_a \ln \frac{1}{D_{a2}} + I_b \ln \frac{1}{D_{b2}} + I_c \ln \frac{1}{D_{c2}} \right) \text{ Wb-T/m} \quad (2.39)$$

The net flux linkages of the telephone line are

$$\lambda_t = \lambda_{t1} - \lambda_{t2}$$

$$= 2 \times 10^{-7}\left(I_a \ln \frac{D_{a2}}{D_{a1}} + I_b \ln \frac{D_{b2}}{D_{b1}} + I_c \ln \frac{D_{c2}}{D_{c1}} \right) \text{ Wb-T/m} \quad (2.40)$$

The emf induced in the telephone line loop is

$$E_t = 2\pi f \lambda_t \text{ V/m}$$

Under balanced load conditions, λ_t is not very large because there is a cancellation to a great extent of the flux linkages due to I_a, I_b and I_c. Such cancellation does not take place with harmonic currents which are multiples of three and are therefore in phase. Consequently, these frequencies, if present, may be very troublesome.

If the power line is fully transposed with respect to the telephone line

$$\lambda_{t1} = \frac{\lambda_{t1}(\text{I}) + \lambda_{t1}(\text{II}) + \lambda_{t1}(\text{III})}{3}$$

where $\lambda_{t1}(\text{I})$, $\lambda_{t1}(\text{II})$ and $\lambda_{t1}(\text{III})$ are the flux linkages of the telephone line t_1 in the three transposition sections of the power line.

Writing for $\lambda_{t1}(\text{I})$, $\lambda_{t2}(\text{II})$ and $\lambda_{t3}(\text{III})$ by repeated use of Eq. (2.38), we have

$$\lambda_{t1} = 2 \times 10^{-7}(I_a + I_b + I_c) \ln \frac{1}{(D_{a1}D_{b1}D_{c1})^{1/3}}$$

Similarly,

$$\lambda_{t2} = \frac{\lambda_{t2}(\text{I}) + \lambda_{t2}(\text{II}) + \lambda_{t2}(\text{III})}{3}$$

$$= 2 \times 10^{-7}(I_a + I_b + I_c) \ln \frac{1}{(D_{a2}D_{b2}D_{c2})^{1/3}}$$

$$\therefore \qquad \lambda_t = 2 \times 10^{-7}(I_a + I_b + I_c) \ln \frac{(D_{a2}D_{b2}D_{c2})^{1/3}}{(D_{a1}D_{b2}D_{c3})^{1/3}} \qquad (2.41)$$

If $I_a + I_b + I_c = 0$, $\lambda_t = 0$, i.e. voltage induced in the telephone loop is zero over one transposition cycle of the power line.

It may be noted here that the condition $I_a + I_b + I_c = 0$ is not satisfied for

(i) power frequency L-G (line-to-ground fault) currents, where

$$I_a + I_b + I_c = 3I_0$$

(ii) third and multiple of third harmonic currents under healthy condition, where

$$I_a(3) + I_b(3) + I_c(3) = 3I(3)$$

∴ $$E_t(3) = 6\pi f \lambda_t(3)$$

The harmonic line currents are troublesome in two ways:

(i) Induced emf is proportional to the frequency.

(ii) Higher frequencies come within the audible range.

Thus there is need to avoid the presence of such harmonic currents on power line from considerations of the performance of nearby telephone lines.

It has been shown above that voltage induced in a telephone line running parallel to a power line is reduced to zero if the power line is transposed and provided it carries balanced currents. It was also shown that power line transposition is ineffective in reducing the induced telephone line voltage when power line currents are unbalanced or when they contain third harmonics. Power line transposition apart from being ineffective introduces mechanical and insulation problems. It is, therefore, easier to eliminate induced voltages by transposing the telephone line instead. In fact, the reader can easily verify that even when the power line currents are unbalanced or when they contain harmonics, the voltage induced over complete transposition cycle (called a *barrel*) of a telephone line is zero. Some induced voltage will always be present on a telephone line running parallel to a power line because in actual practice transposition is never completely symmetrical. Therefore, when the lines run parallel over a considerable length, it is a good practice to transpose both power and telephone lines. The two transposition cycles are staggered and the telephone line is transposed over shorter lengths compared to the power line.

Example 2.5

A three-phase, 50 Hz, 15 km long line has four No. 4/0 wires (1 cm dia) spaced horizontally 1.5 m apart in a plane. The wires in order are carrying currents I_a, I_b and I_c, and the fourth wire, which is a neutral, carries zero current. The currents are:

$$I_a = -30 + j50 \text{ A}$$
$$I_b = -25 + j55 \text{ A}$$
$$I_c = 55 - j105 \text{ A}$$

The line is untransposed.

(a) From the fundamental consideration, find the flux linkages of the neutral. Also find the voltage induced in the neutral wire.

(b) Find the voltage drop in each of the three-phase wires.

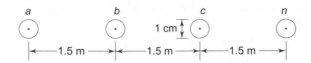

Fig. 2.13 Arrangement of conductors for Example 2.5

Solution (a) From Fig. 2.13,

$$D_{an} = 4.5 \text{ m}, \; D_{bn} = 3 \text{ m}, \; D_{cn} = 1.5 \text{ m}$$

Flux linkages of the neutral wire n are

$$\lambda_n = 2 \times 10^{-7} \left(I_a \ln \frac{1}{D_{an}} + I_b \ln \frac{1}{D_{bn}} + I_c \ln \frac{1}{D_{cn}} \right) \text{Wb-T/m}$$

Substituting the values of D_{an}, D_{bn} and D_{cn}, and simplifying, we get

$$\lambda_n = -2 \times 10^{-7} (1.51 \, I_a + 1.1 \, I_b + 0.405 \, I_c) \text{ Wb-T/m}$$

Since $I_c = -(I_a + I_b)$ (this is easily checked from the given values),

$$\lambda_n = -2 \times 10^{-7} (1.105 I_a + 0.695 I_b) \text{ Wb–T/m}$$

The voltage induced in the neutral wire is then

$$V_n = j\omega\lambda_n \times 15 \times 10^3 \text{ V}$$
$$= -j314 \times 15 \times 10^3 \times 2 \times 10^{-7} (1.105 \, I_a + 0.695 \, I_b) \text{ V}$$

or $\qquad V_n = -j0.942 \, (1.105 \, I_a + 0.695 \, I_b) \text{ V}$

Substituting the values of I_a and I_b, and simplifying

$$V_n = 0.942 \times 106 = 100 \text{ V}$$

(b) From Eq. (2.30), the flux linkages of the conductor a are

$$\lambda_a = 2 \times 10^{-7} \left(I_a \ln \frac{1}{r'_a} + I_b \ln \frac{1}{D} + I_c \ln \frac{1}{2D} \right) \text{ Wb-T/m}$$

The voltage drop/metre in phase a can be written as

$$\Delta V_a = 2 \times 10^{-7} j\omega \left(I_a \ln \frac{1}{r'_a} + I_b \ln \frac{1}{D} + I_c \ln \frac{1}{2D} \right) \text{ V/m}$$

Since $I_c = -(I_a + I_b)$, and further since $r_a = r_b = r_c = r$, the expression for ΔV_a can be written in simplified form

$$\Delta V_a = 2 \times 10^{-7} j\omega \left(I_a \ln \frac{2D}{r'} + I_b \ln 2 \right) \text{ V/m}$$

Similarly, voltage drop/metre of phases b and c can be written as

$$\Delta V_b = 2 \times 10^{-7} j\omega I_b \ln \frac{D}{r'}$$

$$\Delta V_c = 2 \times 10^{-7} j\omega \left(I_b \ln 2 + I_c \ln \frac{2D}{r'} \right)$$

Using matrix notation, we can present the result in compact form

$$\begin{bmatrix} \Delta V_a \\ \Delta V_b \\ \Delta V_c \end{bmatrix} = 2 \times 10^{-7} j\omega \begin{bmatrix} \ln 2D/r' & \ln 2 & 0 \\ 0 & \ln D/r' & 0 \\ 0 & \ln 2 & \ln 2D/r' \end{bmatrix} \begin{bmatrix} I_a \\ I_b \\ I_c \end{bmatrix}$$

The voltage drop of phase a is calculated below:

$$\Delta V_a = j2 \times 10^{-7} \times 314 \times 15 \times 10^3 \left(\ln \frac{300}{0.39} (-30 + j50) + 0.693(-25 + j55) \right)$$

$$= -(348.6 + j204) \quad V$$

Example 2.6

A single-phase 50 Hz power line is supported on a horizontal cross-arm. The spacing between the conductors is 3 m. A telephone line is supported symmetrically below the power line as shown in Fig. 2.14. Find the mutual inductance between the two circuits and the voltage induced per kilometre in the telephone line if the current in the power line is 100 A. Assume the telephone line current to be zero.

Solution   Flux linkages of conductor T_1

$$\lambda_{t1} = 2 \times 10^{-7} \left(I \ln \frac{1}{D_1} - I \ln \frac{1}{D_2} \right) = 2 \times 10^{-7} I \ln \frac{D_2}{D_1}$$

Flux linkages of conductor T_2

$$\lambda_{t2} = 2 \times 10^{-7} I \ln \frac{D_1}{D_2}$$

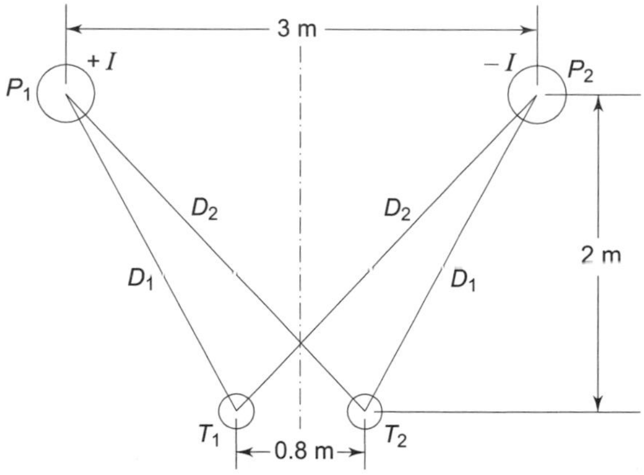

Fig. 2.14   Power and telephone lines for Example 2.6

Total flux linkage of the telephone circuit

$$\lambda_t = \lambda_{t1} - \lambda_{t2} = 4 \times 10^{-7} I \ln \frac{D_2}{D_1}$$

$$M_{pt} = 4 \times 10^{-7} \ln \frac{D_2}{D_1} \text{ H/m}$$

$$= 0.921 \log \frac{D_2}{D_1} \text{ mH/km}$$

$$D_1 = (1.1^2 + 2^2)^{1/2} = (5.21)^{1/2}$$

$$D_2 = (1.9^2 + 2^2)^{1/2} = (7.61)^{1/2}$$

$$M_{pt} = 0.921 \log \left(\frac{761}{521} \right)^{1/2} = 0.0758 \text{ mH/km}$$

Voltage induced in the telephone circuit $V_t = j\omega M_{pt} I$

$$|V_t| = 314 \times 0.0758 \times 10^{-3} \times 100 = 2.379 \quad \text{V/km}$$

2.9 DOUBLE-CIRCUIT THREE-PHASE LINES

It is common practice to build double-circuit three-phase lines so as to increase transmission reliability at somewhat enhanced cost. From the point of view of power transfer from one end of the line to the other (*see* Sec. 12.3), it is desirable to build the two lines with as low an inductance/phase as possible. In order to achieve this, self GMD (D_s) should be made high and mutual GMD (D_m) should be made low. Therefore, the individual conductors of a phase should be kept as far apart as possible (for high self GMD), while the distance between phases be kept as low as permissible (for low mutual GMD).

Figure 2.15 shows the three sections of the transposition cycle of two parallel circuit three-phase lines with vertical spacing (it is a very commonly used configuration).

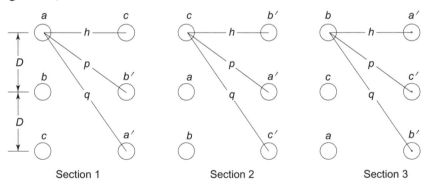

Fig. 2.15 Arrangement of conductors of a double-circuit three-phase line

It may be noted here that conductors a and a' in parallel compose phase a and similarly b and b' compose phase b and c and c' compose phase c. In order to achieve high D_s the conductors of two phases are placed diametrically opposite to each other and those of the third phase are horizontally opposite to each other. (The reader can try other configurations to verify that these will lead to low D_s.) Applying the method of GMD, the equivalent equilateral spacing is

$$D_{eq} = (D_{ab}D_{bc}D_{ca})^{1/3} \tag{2.42}$$

where $\quad D_{ab}$ = mutual GMD between phases a and b in section 1 of the transposition cycle

$$= (DpDp)^{1/4} = (Dp)^{1/2}$$

D_{bc} = mutual GMD between phases b and c in section 1 of the transposition cycle

$$= (Dp)^{1/2}$$

D_{ca} = mutual GMD between phases c and a in section 1 of the transposition cycle

$$= (2Dh)^{1/2}$$

Hence $\quad D_{eq} = 2^{1/6}D^{1/2}p^{1/3}h^{1/6} \tag{2.43}$

It may be noted here that D_{eq} remains the same in each section of the transposition cycle, as the conductors of each parallel circuit rotate cyclically, so do D_{ab}, D_{bc} and D_{ca}. The reader is advised to verify this for sections 2 and 3 of the transposition cycle in Fig. 2.15.

Self GMD in section 1 of phase a (i.e., conductors a and a') is

$$D_{sa} = (r'qr'q)^{1/4} = (r'q)^{1/2}$$

Self GMD of phases b and c in section 1 are respectively

$$D_{sb} = (r'hr'h)^{1/4} = (r'h)^{1/2}$$
$$D_{sc} = (r'qr'q)^{1/4} = (r'q)^{1/2}$$

$\therefore$ Equivalent self GMD $D_s = (D_{sa}D_{sb}D_{sc})^{1/3}$

$$= (r')^{1/2}q^{1/3}h^{1/6} \tag{2.44}$$

Because of the cyclic rotation of conductors of each parallel circuit over the transposition cycle, D_s also remains the same in each transposition section. The reader should verify this for sections 2 and 3 in Fig. 2.15.

The inductance per phase is

$$L = 2 \times 10^{-7} \ln \frac{D_{eq}}{D_S}$$

$$= 2 \times 10^{-7} \ln \frac{2^{1/6}D^{1/2}p^{1/3}h^{1/6}}{(r')^{1/2}q^{1/3}h^{1/6}}$$

$$= 2 \times 10^{-7} \ln \left[2^{1/6}\left(\frac{D}{r'}\right)^{1/2}\left(\frac{p}{q}\right)^{1/3} \right] \text{ H/phase/m} \tag{2.45}$$

The self inductance of each circuit is given by

$$L_s = 2 \times 10^{-7} \ln \frac{(2)^{1/3}D}{r'}$$

Equation (2.45) can now be written as

$$L = \frac{1}{2}\left[2\times10^{-7}\ln\frac{(2)^{1/3}D}{r'}+2\times10^{-7}\ln\left(\frac{p}{q}\right)^{2/3}\right] \quad (2.46)$$

$$= \frac{1}{2}(L_s + M)$$

where M is the mutual inductance between the two circuits, i.e.

$$M = 2\times10^{-7}\ln\left(\frac{p}{q}\right)^{2/3}$$

This is a well known result for the two coupled curcuits connected in parallel (at similar polarity ends).

If $h \gg D, \left(\dfrac{p}{q}\right) \to 1$ and $M \to 0$, i.e. the mutual impedance between the circuits

becomes zero. Under this condition

$$L = 1\times10^{-7}\ln\frac{3\sqrt{2}\,D}{r'} \quad (2.47)$$

The GMD method, though applied above to a particular configuration of a double circuit, is valid for any configuration as long as the circuits are electrically parallel.

While the GMD method is valid for fully transposed lines, it is commonly applied for untransposed lines and is quite accurate for practical purposes.

2.10 BUNDLED CONDUCTORS

It is economical to transmit large chunks of power over long distances by employing EHV lines. However, the line voltages that can be used are severely limited by the phenomenon of corona. Corona, in fact, is the result of ionization of the atmosphere when a certain field intensity (about 3,000 kV/m at NTP) is reached. Corona discharge causes communication interference and associated power loss which can be severe in bad weather conditions. Critical line voltage for formation of corona can be raised considerably by the use of bundled conductors—a group of two or more conductors per phase. This increase in critical corona voltage is dependent on number of conductors in the group, the clearance between them and the distance between the groups forming the separate phases[*]. Reichman [11] has shown that the spacing of conductors in a bundle affects voltage gradient and the optimum spacing is of the order of

---

[*] The bundle usually comprises two, three or four conductors arranged in configurations illustrated in Fig. 2.16. The current will not divide equally among the conductors of the bundle unless conductors within the bundle are fully transposed. The GMD method is still fairly accurate for all practical purposes.

8–10 times the conductor's diameter, irrespective of the number of conductors in the bundle.

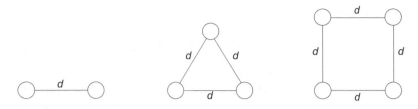

Fig. 2.16 Configuration of bundled conductors

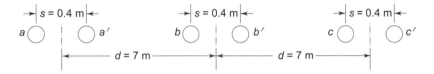

Fig. 2.17 Bundled conductor three-phase line

Further, because of increased self GMD[*] line inductance is reduced considerably with the incidental advantage of increased transmission capacity of the line.

Example 2.7

Find the inductive reactance in ohms per kilometer at 50 Hz of a three-phase bundled conductor line with two conductors per phase as shown in Fig. 2.17. All the conductors are ACSR with radii of 1.725 cm.

Even though the power lines are not normally transposed (except when they enter and leave a switching station), it is sufficiently accurate to assume complete transposition (of the bundles as well as of the conductors within the bundle) so that the method of GMD can be applied.

The mutual GMD between bundles of phases a and b

$$D_{ab} = (d\ (d + s)\ (d - s)\ d\ )^{1/4}$$

Mutual GMD between bundles of phases b and c

$$D_{bc} = D_{ab}\ (\text{by symmetry})$$

Mutual GMD between bundles of phases c and a

$$D_{ca} = (2d\ (2d + s\ )\ (2d - s)2d)^{1/4}$$

$$D_{ca} = (D_{ab}D_{bc}D_{ca}\ )^{1/3}$$

$$= (4d^6(d + s)^2(d - s)^2(2d + s)(2d - s))^{1/12}$$

$$= (4(7)^6(7.4)^2(6.6)^2(14.4)(13.6))^{1/12}$$

$$= 8.81\ \text{m}$$

---

[*] The more the number of conductors in a bundle, the more is the self GMD.

$$D_s = (r'sr's)^{1/4} = (r's)^{1/2} = (0.7788 \times 1.725 \times 10^{-2} \times 0.4)^{1/2}$$
$$= 0.073 \text{ m}$$

Inductive reactance per phase

$$X_L = 314 \times 0.461 \times 10^{-3} \log\frac{8.81}{0.073}$$
$$= 0.301 \text{ ohm/km}$$

In most cases, it is sufficiently accurate to use the centre to centre distances between bundles rather than mutual GMD between bundles for computing D_{eq}. With this approximation, we have for the example in hand

$$D_{eq} = (7 \times 7 \times 14)^{1/3} = 8.82 \text{ m}$$

$$X_L = 314 \times 0.461 \times 10^{-3} \log\frac{8.82}{0.073}$$

$$= 0.301 \text{ ohm/km}$$

Thus the approximate method yields almost the same reactance value as the exact method. It is instructive to compare the inductive reactance of a bundled conductor line with an equivalent (on heuristic basis) single conductor line. For the example in hand, the equivalent line will have $d = 7$ m and conductor diameter (for same total cross-sectional area) as $\sqrt{2} \times 1.725$ cm

$$X_L = 314 \times 0.461 \times 10^{-3} \log \frac{(7 \times 7 \times 14)^{1/3}}{0.7788 \times \sqrt{2} \times 1.725 \times 10^{-3}}$$

$$= 0.531 \text{ ohm/km}$$

This is 76.41% higher than the corresponding value for a bundled conductor line. As already pointed out, lower reactance of a bundled conductor line increases its transmission capacity.

2.11 RESISTANCE

Though the contribution of line resistance to series line impedance can be neglected in most cases, it is the main source of line power loss. Thus while considering transmission line economy, the presence of line resistance must be considered.

The effective AC resistance is given by

$$R = \frac{\text{average power loss in conductor in watts}}{I^2} \text{ ohms} \quad (2.48)$$

where I is the rms current in the conductor in amperes.
Ohmic or DC resistance is given by the formula

$$R_O = \frac{\rho l}{A} \text{ ohms} \quad (2.49)$$

where $\quad \rho$ = resistivity of the conductor, ohm-m
$\quad l$ = length, m

A = cross-sectional area, m^2

The effective resistance given by Eq. (2.48) is equal to the DC resistance of the conductor given by Eq. (2.49) only if the current distribution is uniform throughout the conductor.

For small changes in temperature, the resistance increases with temperature in accordance with the relationship

$$R_t = R \ (1 + \alpha_0 t) \tag{2.50}$$

where R = resistance at temperature 0°C

α_0 = temperature coefficient of the conductor at 0°C

Equation (2.50) can be used to find the resistance R_{t2} at a temperature t_2, if resistance R_{t1} at temperature t_1 is known

$$\frac{R_{t2}}{R_{t1}} = \frac{1/\alpha_0 + t_2}{1/\alpha_0 + t_1} \tag{2.51}$$

2.12 SKIN EFFECT AND PROXIMITY EFFECT

The distribution of current throughout the cross-section of a conductor is uniform only when DC is passing through it. On the contrary when AC is flowing through a conductor, the current is non-uniformly distributed over the cross-section in a manner that the current density is higher at the surface of the conductor compared to the current density at its centre. This effect becomes more pronounced as frequency is increased. This phenomenon is called *skin effect*. It causes larger power loss for a given rms AC than the loss when the same value of DC is flowing through the conductor. Consequently, the effective conductor resistance is more for AC then for DC. A qualitative explanation of the phenomenon is as follows.

Imagine a solid round conductor (a round shape is considered for convenience only) to be composed of annular filaments of equal cross-sectional area. The flux linking the filaments progressively decreases as we move towards the outer filaments for the simple reason that the flux inside a filament does not link it. The inductive reactance of the imaginary filaments therefore decreases outwards with the result that the outer filaments conduct more AC than the inner filaments (filaments being parallel). With the increase of frequency the non-uniformity of inductive reactance of the filaments becomes more pronounced, so also the non-uniformity of current distribution. For large solid conductors the skin effect is quite significant even at 50 Hz. The analytical study of skin effect requires the use of Bessel's functions and is beyond the scope of this book.

Apart from the skin effect, non-uniformity of current distribution is also caused by *proximity effect*. Consider a two-wire line as shown in Fig. 2.18. Each line conductor can be divided into sections of equal cross-sectional area (say three sections). Pairs aa', bb' and cc' can form three loops in parallel. The flux linking loop aa' (and therefore its inductance) is the least and it increases somewhat for loops bb' and cc'. Thus the density of AC flowing through the conductors is highest at the inner edges (aa') of the conductors and is the least at the outer edges (cc'). This type of non-uniform AC current distribution becomes more pronounced as the distance between conductors is reduced. Like skin effect, the non-uniformity of current distribution caused by proximity effect also increases the effective conductor resistance. For normal spacing of overhead lines, this effect is always of a negligible order. However, for underground cables where conductors are located close to each other, proximity effect causes an appreciable increase in effective conductor resistance.

Fig. 2.18

Both skin and proximity effects depend upon conductor size, frequency, distance between conductors and permeability of conductor material.

PROBLEMS

2.1 Derive the formula for the internal inductance in H/m of a hollow conductor having inside radius r_1 and outside radius r_2 and also determine the expression for the inductance in H/m of a single-phase line consisting of the hollow conductors described above with conductors spaced a distance D apart.

2.2 Calculate the 50 Hz inductive reactance at 1 m spacing in ohms/km of a cable consisting of 12 equal strands around a nonconducting core. The diameter of each strand is 0.25 cm and the outside diameter of the cable is 1.25 cm.

2.3 A concentric cable consists of two thin-walled tubes of mean radii r and R respectively. Derive an expression for the inductance of the cable per unit length.

2.4 A single-phase 50 Hz circuit comprises two single-core lead-sheathed cables laid side by side; if the centres of the cables are 0.5 m apart and each sheath has a mean diameter of 7.5 cm, estimate the longitudinal voltage induced per km of sheath when the circuit carries a current of 800 A.

2.5 Two long parallel conductors carry currents of $+I$ and $-I$. What is the magnetic field intensity at a point P, shown in Fig. P-2.5?

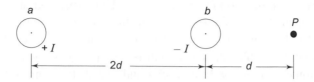

Fig. P-2.5

2.6 Two three-phase lines connected in parallel have self-reactances of X_1 and X_2. If the mutual reactance between them is X_{12}, what is the effective reactance between the two ends of the line?

2.7 A single-phase 50 Hz power line is supported on a horizontal cross-arm. The spacing between conductors is 2.5 m. A telephone line is also supported on a horizontal cross-arm in the same horizontal plane as the power line. The conductors of the telephone line are of solid copper spaced 0.6 m between centres. The distance between the nearest conductors of the two lines is 20 m. Find the mutual inductance between the circuits and the voltage per kilometre induced in the telephone line for 150 A current flowing over the power line.

2.8 A telephone line runs parallel to an untrasposed three-phase transmission line, as shown in Fig. P-2.8. The power line carries balanced current of 400 A per phase. Find the mutual inductance between the circuits and calculate the 50 Hz voltage induced in the telephone line per km.

Fig. P-2.8   Telephone line parallel to a power line

2.9 A 500 kV line has a bundling arrangement of two conductors per phase as shown in Fig. P-2.9.

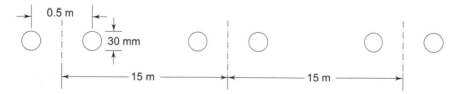

Fig. P-2.9   500 kV, three-phase bundled conductor line

Compute the reactance per phase of this line at 50 Hz. Each conductor carries 50% of the phase current. Assume full transposition.

2.10 An overhead line 50 kms in length is to be constructed of conductors 2.56 cm in diameter, for single-phase transmission. The line reactance must not exceed 31.4 ohms. Find the maximum permissible spacing.

2.11 In Fig. P-2.11 which depicts two three-phase circuits on a steel tower there is symmetry about both the horizontal and vertical centre lines. Let each three-phase circuit be transposed by replacing a by b and then by c, so that the reactances of the three-phases are equal and the GMD method of reactance calculations can be used. Each circuit remains on its own side of the tower. Let the self GMD of a single conductor be 1 cm. Conductors a and a' and other corresponding phase conductors are connected in parallel. Find the reactance per phase of the system.

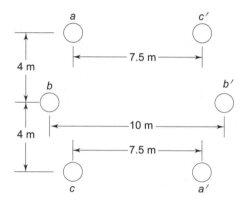

Fig. P-2.11

2.12 A double-circuit three-phase line is shown in Fig. P-2.12. The conductors a, a'; b, b' and c, c' belong to the same phase respectively. The radius of each conductor is 1.5 cm. Find the inductance of the double-circuit line in mH/km/phase.

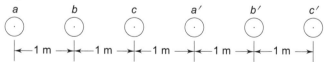

Fig. P-2.12   Arrangement of conductors for a double-circuit three-phase line

2.13 A three-phase line with equilateral spacing of 3 m is to be rebuilt with horizontal spacing ($D_{13} = 2D_{12} = 2D_{23}$). The conductors are to be fully transposed. Find the spacing between adjacent conductors such that the new line has the same inductance as the original line.

2.14 Find the self GMD of three arrangements of bundled conductors shown in Fig. 2.16 in terms of the total cross-sectional area A of conductors (same in each case) and the distance d between them.

REFERENCES

Books

1. *Electrical Transmission and Distribution Book*, Westinghouse Electric and Manufacturing Co., East Pittsburgh, Pennsylvania, 1964.
2. Waddicor, H., *Principles of Electric Power Transmission*, 5th edn, Chapman and Hall, London, 1964.
3. Nagrath, I.J. and D.P. Kothari, *Electric Machines*, 2nd edn, Tata McGraw-Hill, New Delhi, 1997.
4. Stevenson, W.D., *Elements of Power System Analysis*, 4th edn, McGraw-Hill, New York, 1982.
5. Edison Electric Institute, *EHV Transmission Line Reference* Book, 1968.
6. The Aluminium Association, *Aluminium Electrical Conductor Handbook*, New York, 1971.
7. Woodruff, L.F., *Principles of Electric Power Transmission*, John Wiley & Sons, New York, 1947.
8. Gross, C.A., *Power System Analysis*, Wiley, New York, 1979.
9. Weedy, B.M. and B.J. Cory *Electric Power Systems*, 4th edn, Wiley, New York, 1998.
10. Kimbark, E.W., *Electrical Transmission of Power and Signals*, John Wiley, New York, 1949.

Paper

11. Reichman, J., 'Bundled Conductor Voltage Gradient Calculations," *AIEE Trans.* 1959, Pt III, 78; 598.

3

Capacitance of
Transmission Lines

3.1 INTRODUCTION

The capacitance together with conductance forms the shunt admittance of a transmission line. As mentioned earlier the conductance is the result of leakage over the surface of insulators and is of negligible order. When an alternating voltage is applied to the line, the line capacitance draws a leading sinusoidal current called the *charging current* which is drawn even when the line is open circuited at the far end. The line capacitance being proportional to its length, the charging current is negligible for lines less than 100 km long. For longer lines the capacitance becomes increasingly important and has to be accounted for.

3.2 ELECTRIC FIELD OF A LONG STRAIGHT CONDUCTOR

Imagine an infinitely long straight conductor far removed from other conductors (including earth) carrying a uniform charge of q coulomb/metre length. By symmetry, the equipotential surfaces will be concentric cylinders, while the lines of electrostatic stress will be radial. The electric field intensity at a distance y from the axis of the conductor is

$$\varepsilon = \frac{q}{2\pi k y}\,\text{V/m}$$

where k is the permittivity* of the medium.

As shown in Fig. 3.1 consider two points P_1 and P_2 located at distances D_1 and D_2 respectively from the conductor axis. The potential difference V_{12} (between P_1 and P_2) is given by

---

* In SI units the permittivity of free space is $k_0 = 8.85 \times 10^{-12}$ F/m. Relative permittivity for air is $k_r = k/k_0 = 1$.

V_{12} (between P_1 and P_2) is given by

$$V_{12} = \oint \varepsilon \mathrm{d}y = \oint \frac{q}{2\,\pi ky}\,\mathrm{d}y \text{ V}$$

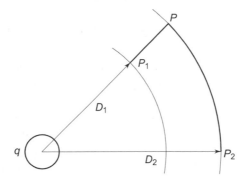

Fig. 3.1   Electric field of a long straight conductor

As the potential difference is independent of the path, we choose the path of integration as $P_1 P P_2$ shown in thick line. Since the path PP_2 lies along an equipotential, V_{12} is obtained simply by integrating along $P_1 P$, i.e.

$$V_{12} = \int_{D_1}^{D_2} \frac{q}{2\,\pi ky}\,\mathrm{d}y - \frac{q}{2\,\pi k} \ln \frac{D_2}{D_1} \text{ V} \qquad (3.1)$$

3.3   POTENTIAL DIFFERENCE BETWEEN TWO CONDUCTORS OF A GROUP OF PARALLEL CONDUCTORS

Figure 3.2 shows a group of parallel charged conductors. It is assumed that the conductors are far removed from the ground and are sufficiently removed from each other, i.e. the conductor radii are much smaller than the distances between them. The spacing commonly used in overhead power transmission lines always meets these assumptions. Further, these assumptions imply that the charge on each conductor remains uniformly distributed around its periphery and length.

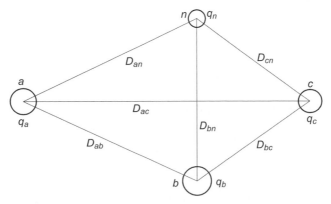

Fig. 3.2   A group of parallel charged conductors

The potential difference between any two conductors of the group can then be obtained by adding the contributions of the individual charged conductors by repeated application of Eq. (3.1). So, the potential difference between conductors a and b (voltage drop from a to b) is

$$V_{ab} = \frac{1}{2\pi k}\left(q_a \ln\frac{D_{ab}}{r_a} + q_b \ln\frac{r_b}{D_{ba}} + q_c \ln\frac{D_{cb}}{D_{ca}} + \ldots q_n \ln\frac{D_{nb}}{D_{na}}\right) \text{ V} \qquad (3.2)$$

Each term in Eq. (3.2) is the potential drop from a to b caused by charge on one of the conductors of the group. Expressions on similar lines could be written for voltage drop between any two conductors of the group.

If the charges vary sinusoidally, so do the voltages (this is the case for AC transmission line), the expression of Eq. (3.2) still applies with charges/metre length and voltages regarded as phasor quantities. Equation (3.2) is thus valid for instantaneous quantities and for sinusoidal quantities as well, wherein all charges and voltages are phasors.

3.4 CAPACITANCE OF A TWO-WIRE LINE

Consider a two-wire line shown in Fig. 3.3 excited from a single-phase source. The line develops equal and opposite sinusoidal charges on the two conductors which can be represented as phasors q_a and q_b so that $q_a = -q_b$.

Fig. 3.3 Cross-sectional view of a two-wire line

The potential difference V_{ab} can be written in terms of the contributions made by q_a and q_b by use of Eq. (3.2) with associated assumptions (i.e. D/r is large and ground is far away). Thus,

$$V_{ab} = \frac{1}{2\pi k}\left(q_a \ln\frac{D}{r_a} + q_b \ln\frac{r_b}{D}\right) \qquad (3.3)$$

Since $\qquad q_a = -q_b$, we have

$$V_{ab} = \frac{q_a}{2\pi k}\ln\frac{D^2}{r_a r_b}$$

The line capacitance C_{ab} is then

$$C_{ab} = \frac{q_a}{V_{ab}} = \frac{\pi k}{\ln\left(D/(r_a r_b)^{1/2}\right)} \text{ F/m length of line} \qquad (3.4a)$$

or

$$C_{ab} = \frac{0.0121}{\log (D / (r_a r_b)^{1/2})} \; \mu F/km \qquad (3.4b)$$

If $r_a = r_b = r$,

$$C_{ab} = \frac{0.0121}{\log (D/r)} \; \mu F/km \qquad (3.4c)$$

The associated line charging current is

$$I_c = j\omega C_{ab} V_{ab} \; A/km \qquad (3.5)$$

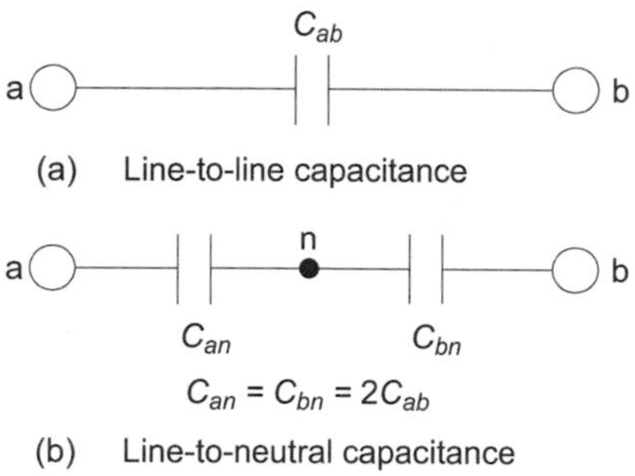

(a)   Line-to-line capacitance

(b)   Line-to-neutral capacitance

Fig. 3.4

As shown in Figs. 3.4 (a) and (b) the line-to-line capacitance can be equivalently considered as two equal capacitances is series. The voltage across the lines divides equally between the capacitances such that the neutral point n is at the ground potential. The capacitance of each line to neutral is then given by

$$C_n = C_{an} = C_{bn} = 2C_{ab} = \frac{0.0242}{\log (D/r)} \; \mu F/km \qquad (3.6)$$

The assumptions inherent in the above derivation are:

(i) The charge on the surface of each conductor is assumed to be uniformly distributed, but this is strictly not correct.

If non-uniformity of charge distribution is taken into account, then

$$C_n = \frac{0.0242}{\log \left(\dfrac{D}{2r} + \left(\dfrac{D^2}{4r^2} - 1 \right)^{1/2} \right)} \; \mu F/km \qquad (3.7)$$

If $D/2r \gg 1$, the above expression reduces to that of Eq. (3.6)and the error caused by the assumption of uniform charge distribution is negligible.

(ii) The cross-section of both the conductors is assumed to be circular, while in actual practice stranded conductors are used. The use of the radius of the circumscribing circle for a stranded conductor causes insignificant error.

3.5 CAPACITANCE OF A THREE-PHASE LINE WITH EQUILATERAL SPACING

Figure 3.5 shows a three-phase line composed of three identical conductors of radius r placed in equilateral configuration.

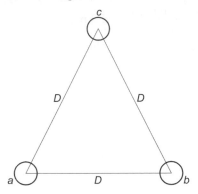

Fig. 3.5 Cross-section of a three-phase line with equilateral spacing

Using Eq. (3.2) we can write the expressions for V_{ab} and V_{ac} as

$$V_{ab} = \frac{1}{2\pi k}\left(q_a \ln \frac{D}{r} + q_b \ln \frac{r}{D} + q_c \ln \frac{D}{D}\right) \qquad (3.8)$$

$$V_{ac} = \frac{1}{2\pi k}\left(q_a \ln \frac{D}{r} + q_b \ln \frac{D}{D} + q_c \ln \frac{r}{D}\right) \qquad (3.9)$$

Adding Eqs. (3.8) and (3.9), we get

$$V_{ab} + V_{ac} = \frac{1}{2\pi k}\left[2q_a \ln \frac{D}{r} + (q_b + q_c) \ln \frac{r}{D}\right] \qquad (3.10)$$

Since there are no other charges in the vicinity, the sum of charges on the three conductors is zero. Thus $q_b + q_c = -q_a$, which when substituted in Eq. (3.10) yields

$$V_{ab} + V_{ac} = \frac{3q_a}{2\pi k} \ln \frac{D}{r} \qquad (3.11)$$

With balanced three-phase voltages applied to the line, it follows from the phasor diagram of Fig. 3.6 that

$$V_{ab} + V_{ac} = 3V_{an} \qquad (3.12)$$

Substituting for $(V_{ab} + V_{ac})$ from Eq. (3.12) in Eq. (3.11), we get

$$V_{an} = \frac{q_a}{2\pi k} \ln \frac{D}{r} \qquad (3.13)$$

The capacitance of line to neutral immediately follows as

$$C_n = \frac{q_a}{V_{an}} = \frac{2\pi k}{\ln (D/r)} \qquad (3.14a)$$

For air medium ($k_r = 1$),

$$C_n = \frac{0.0242}{\log (D/r)} \ \mu F/km \qquad (3.14b)$$

The line charging current of phase a is

$$I_a \ (\text{line charging}) = j\omega C_n V_{an} \qquad (3.15)$$

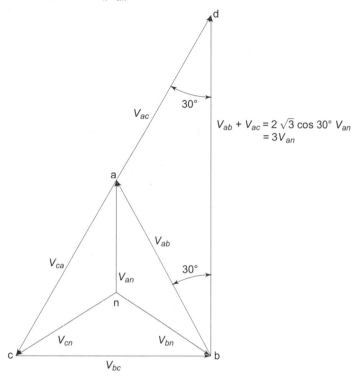

$$V_{ab} + V_{ac} = 2 \sqrt{3} \cos 30° \ V_{an}$$
$$= 3 V_{an}$$

Fig. 3.6 Phasor diagram of balanced three-phase voltages

3.6 CAPACITANCE OF A THREE-PHASE LINE WITH UNSYMMETRICAL SPACING

Figure 3.7 shows the three identical conductors of radius r of a three-phase line with unsymmetrical spacing. It is assumed that the line is fully transposed. As the conductors are rotated cyclically in the three sections of the transposition cycle, correspondingly three expressions can be written for V_{ab}. These expressions are:

For the first section of the transposition cycle

$$V_{ab} = \frac{1}{2\pi k} \left(q_{a1} \ln \frac{D_{12}}{r} + q_{b1} \ln \frac{r}{D_{12}} + q_{c1} \ln \frac{D_{23}}{D_{31}} \right) \qquad (3.16a)$$

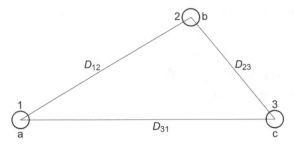

Fig. 3.7 Cross-section of a three-phase line with asymmetrical spacing (fully transposed)

For the second section of the transposition cycle

$$V_{ab} = \frac{1}{2\pi k}\left(q_{a2}\ln\frac{D_{23}}{r} + q_{b2}\ln\frac{r}{D_{23}} + q_{c2}\ln\frac{D_{31}}{D_{12}}\right) \quad (3.16b)$$

For the third section of the transposition cycle

$$V_{ab} = \frac{1}{2\pi k}\left(q_{a3}\ln\frac{D_{31}}{r} + q_{b3}\ln\frac{r}{D_{31}} + q_{c3}\ln\frac{D_{12}}{D_{23}}\right) \quad (3.16c)$$

If the voltage drop along the line is neglected, V_{ab} is the same in each transposition cycle. On similar lines three such equations can be written for $V_{bc} = V_{ab} \angle -120°$. Three more equations can be written equating to zero the summation of all line charges in each section of the transposition cycle. From these nine (independent) equations, it is possible to determine the nine unknown charges. The rigorous solution though possible is too involved.

With the usual spacing of conductors sufficient accuracy is obtained by assuming

$$q_{a1} = q_{a2} = q_{a3} = q_a \; ; \; q_{b1} = q_{b2} = q_{b3} = q_b;$$
$$q_{c1} = q_{c2} = q_{c3} = q_c \quad (3.17)$$

This assumption of equal charge/unit length of a line in the three sections of the transposition cycle requires, on the other hand, three different values of V_{ab} designated as V_{ab1}, V_{ab2} and V_{ab3} in the three sections. The solution can be considerably simplified by taking V_{ab} as the average of these three voltages, i.e.

$$V_{ab}(\text{avg}) = \frac{1}{3}(V_{ab1} + V_{ab2} + V_{ab3})$$

or
$$V_{ab} = \frac{1}{6\pi k}\left[q_a\ln\left(\frac{D_{12}D_{23}D_{31}}{r^3}\right) + q_b\ln\left(\frac{r^3}{D_{12}D_{23}D_{31}}\right) \right.$$
$$\left. + q_c\ln\left(\frac{D_{12}D_{23}D_{31}}{D_{12}D_{23}D_{31}}\right)\right]$$

$$= \frac{1}{2\pi k}\left(q_a \ln \frac{D_{eq}}{r} + q_b \ln \frac{r}{D_{eq}}\right) \tag{3.18}$$

where $D_{eq} = (D_{12}D_{23}D_{31})^{1/3}$

Similarly,

$$V_{ac} = \frac{1}{2\pi k}\left(q_a \ln \frac{D_{eq}}{r} + q_c \ln \frac{r}{D_{eq}}\right) \tag{3.19}$$

Adding Eqs. (3.18) and (3.19), we get

$$V_{ab} + V_{ac} = \frac{1}{2\pi k}\left(q_a \ln \frac{D_{eq}}{r} + (q_b + q_c) \ln \frac{r}{D_{eq}}\right) \tag{3.20}$$

As per Eq. (3.12) for balanced three-phase voltages

$$V_{ab} + V_{ac} = 3V_{an}$$

and also $(q_b + q_c) = -q_a$

Use of these relationships in Eq. (3.20) leads to

$$V_{an} = \frac{q_a}{2\pi k} \ln \frac{D_{eq}}{r} \tag{3.21}$$

The capacitance of line to neutral of the transposed line is then given by

$$C_n = \frac{q_a}{V_{an}} = \frac{2\pi k}{\ln (D_{eq}/r)} \quad \text{F/m to neutral} \tag{3.22a}$$

For air medium ($k_r = 1$)

$$C_n = \frac{0.0242}{\log (D_{eq}/r)} \quad \mu\text{F/km to neutral} \tag{3.22b}$$

It is obvious that for equilateral spacing $D_{eq} = D$, the above (approximate) formula gives the exact result presented earlier.

The line charging current for a three-phase line in phasor form is

$$I_a \text{ (line charging)} = j\omega C_n V_{an} \text{ A/km} \tag{3.23}$$

3.7 EFFECT OF EARTH ON TRANSMISSION LINE CAPACITANCE

In calculating the capacitance of transmission lines, the presence of earth was ignored, so far. The effect of earth on capacitance can be conveniently taken into account by the method of images.

Method of Images

The electric field of transmission line conductors must conform to the presence of the earth below. The earth for this purpose may be assumed to be a perfectly conducting horizontal sheet of infinite extent which therefore acts like an equipotential surface.

The electric field of two long, parallel conductors charged $+q$ and $-q$ per unit is such that it has a zero potential plane midway between the conductors as shown in Fig. 3.8. If a conducting sheet of infinite dimensions is placed at the zero potential plane, the electric field remains undisturbed. Further, if the conductor carrying charge $-q$ is now removed, the electric field above the conducting sheet stays intact, while that below it vanishes. Using these well known results in reverse, we may equivalently replace the presence of ground below a charged conductor by a fictitious conductor having equal and opposite charge and located as far below the surface of ground as the overhead conductor above it—such a fictitious conductor is the *mirror image* of the overhead conductor. This method of creating the same electric field as in the presence of earth is known as the *method of images* originally suggested by Lord Kelvin.

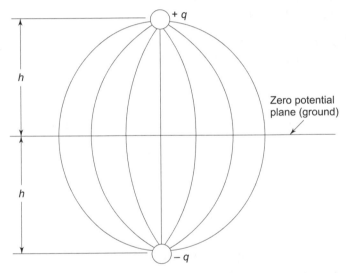

Fig. 3.8   Electric field of two long, parallel, oppositely charged conductors

Capacitance of a Single-Phase Line

Consider a single-phase line shown in Fig. 3.9. It is required to calculate its capacitance taking the presence of earth into account by the method of images described above. The equation for the voltage drop V_{ab} as determined by the two charged conductors a and b, and their images a' and b' can be written as follows:

$$V_{ab} = \frac{1}{2\pi k}\left[q_a \ln \frac{D}{r} + q_b \ln \frac{r}{D} + q_{a'} \ln \frac{(4h^2 + D^2)^{1/2}}{2h} \right.$$

$$\left. + q_{b'} \ln \frac{2h}{(4h^2 + D^2)^{1/2}} \right] \tag{3.24}$$

Substituting the values of different charges and simplifying, we get

$$V_{ab} = \frac{q}{\pi k} \ln \frac{2hD}{r(4h^2 + D^2)^{1/2}} \qquad (3.25)$$

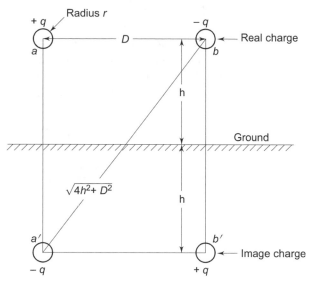

Fig. 3.9 Single-phase transmission line with images

It immediately follows that

$$C_{ab} = \frac{\pi k}{\ln \dfrac{D}{r(1+(D^2/4h^2))^{1/2}}} \quad \text{F/m line-to-line} \qquad (3.26a)$$

and

$$C_n = \frac{2\,\pi k}{\ln \dfrac{D}{r(1+(D^2/4h^2))^{1/2}}} \quad \text{F/m to neutral} \qquad (3.26b)$$

It is observed from the above equation that the presence of earth modifies the radius r to $r(1 + (D^2/4h^2))^{1/2}$. For h large compared to D (this is the case normally), the effect of earth on line capacitance is of negligible order.

Capacitance of a Three-Phase Line

The method of images can similarly be applied for the calculation of capacitance of a three-phase line, shown in Fig. 3.10. The line is considered to be fully transposed. The conductors a, b and c carry the charges q_a, q_b and q_c and occupy positions 1, 2, and 3, respectively, in the first section of the transposition cycle. The effect of earth is simulated by image conductors with charges $-q_a$, $-q_b$ and $-q_c$ respectively, as shown.

The equations for the three sections of the transposition cycle can be written for the voltage drop V_{ab} as determined by the three charged conductors and their images. With conductor a in position 1, b in position 2, and c in position 3,

$$V_{ab} = \frac{1}{2\pi k}\left[q_a\left(\ln\frac{D_{12}}{r} - \ln\frac{h_{12}}{h_1}\right) + q_b\left(\ln\frac{r}{D_{12}} - \ln\frac{h_2}{h_{12}}\right)\right.$$
$$\left. + q_c\left(\ln\frac{D_{23}}{D_{31}} - \ln\frac{h_{23}}{h_{31}}\right)\right] \qquad (3.27)$$

Similar equations for V_{ab} can be written for the second and third sections of the transposition cycle. If the fairly occurate assumption of constant charge per unit length of the conductor throughout the transmission cycle is made, the average value of V_{ab} for the three sections of the cycle is given by

$$V_{ab} = \frac{1}{2\pi k}\left[q_a\left(\ln\frac{D_{eq}}{r} - \ln\frac{(h_{12}h_{23}h_{31})^{1/3}}{(h_1 h_2 h_3)^{1/3}}\right)\right.$$
$$\left. + q_b\left(\ln\frac{r}{D_{eq}} - \ln\frac{(h_1 h_2 h_3)^{1/3}}{(h_{12}h_{23}h_{31})^{1/3}}\right)\right] \qquad (3.28)$$

where $D_{eq} = (D_{12}D_{23}D_{31})^{1/3}$

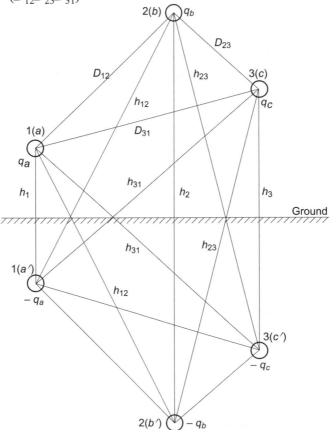

Fig. 3.10 Three-phase line with images

The equation for the average value of the phasor V_{ac} is found in a similar manner. Proceeding on the lines of Sec. 3.6 and using $V_{ab} + V_{ac} = 3V_{an}$ and $q_a + q_b + q_c = 0$, we ultimately obtained the following expression for the capacitance to neutral.

$$C_n = \frac{2\pi k}{\ln \dfrac{D_{eq}}{r} - \ln \left(\dfrac{(h_{12}h_{23}h_{31})^{1/3}}{(h_1 h_2 h_3)^{1/3}} \right)} \quad \text{F/m to neutral} \quad (3.29a)$$

or

$$C_n = \frac{0.0242}{\log \dfrac{D_{eq}}{r} - \log \dfrac{(h_{12}h_{23}h_{31})^{1/3}}{(h_1 h_2 h_3)^{1/3}}} \quad \mu\text{F/km to neutral} \quad (3.29b)$$

Comparing Eqs. (3.22a) and (3.29a), it is evident that the effect of earth is to increase the capacitance of a line. If the conductors are high above earth compared to the distances among them, the effect of earth on the capacitance of three-phase lines can be neglected.

Example 3.1

Calculate the capacitance to neutral/km of a single-phase line composed of No. 2 single strand conductors (radius = 0.328 cm) spaced 3 m apart and 7.5 m above the ground. Compare the results obtained by Eqs. (3.6), (3.7) and (3.26b).

Solution   (1) Neglecting the presence of earth [Eq. (3.6)]

$$C_n = \frac{0.0242}{\log \dfrac{D}{r}} \ \mu\text{F/km}$$

$$= \frac{0.0242}{\log \dfrac{300}{0.328}} = 0.00817 \ \mu\text{F/km}$$

By the rigorous relationship [(Eq. (3.7)]

$$C_n = \frac{0.0242}{\log \left[\dfrac{D}{2r} + \left(\dfrac{D^2}{4r^2} - 1 \right)^{1/2} \right]} \ \mu\text{F/km}$$

Since $\dfrac{D}{r} = 915$, the effect of non-uniformity of charge distribution is almost negligible.

$\therefore$ $\qquad\qquad C_n = 0.00817 \ \mu\text{F/km}$

(2) Considering the effect of earth and neglecting non-uniformity of charge distribution [Eq. (3.26b)]

$$C_n = \frac{0.0242}{\log \dfrac{D}{r(1+(D^2/4h^2))^{1/2}}}$$

$$\frac{D}{r\sqrt{1.04}} = \frac{300}{0.328\sqrt{1.04}} = 897$$

$$C_n = \frac{0.0242}{2.953} = 0.0082 \ \mu\text{F/km}$$

Note: The presence of earth increases the capacitance by approximately 3 parts in 800.

Example 3.2

A three-phase 50 Hz transmission line has flat horizontal spacing with 3.5 m between adjacent conductors. The conductors are No. 2/0 hard-drawn seven strand copper (outside conductor diameter = 1.05 cm). The voltage of the line is 110 kV. Find the capacitance to neutral and the charging current per kilometre of line.

Solution $D_{eq} = (3.5 \times 3.5 \times 7)^{1/3} = 4.4$ m

$$C_n = \frac{0.0242}{\log(D_{eq}/r)} = \frac{0.0242}{\log(440/0.525)}$$

$$= 0.00826 \ \mu\text{F/km}$$

$$X_n = \frac{1}{\omega C_n} = \frac{10^6}{314 \times 0.00826}$$

$$= 0.384 \times 10^6 \ \Omega/\text{km to neutral}$$

Charging current $= \dfrac{V_n}{X_n} = \dfrac{(110/\sqrt{3}) \times 1000}{0.384 \times 10^6}$

$$= 0.17 \ \text{A/km}$$

Example 3.3

The six conductors of a double-circuit three-phase line having an overall radius of 0.865×10^{-2} m are arranged as shown in Fig. 3.11. Find the capacitive reactance to neutral and charging current per kilometre per conductor at 110 kV, 50 Hz.

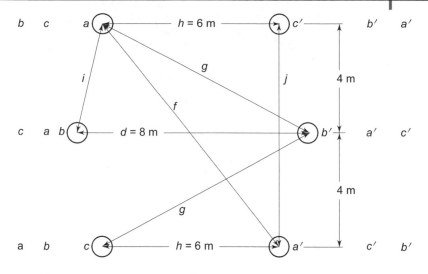

Fig. 3.11 Cross-section of a double-circuit three-phase line

Solution As in Sec. 3.6, assume that the charge per conductor on each phase is equal in all the three sections of the transposition cycle. For section I of the transposition cycle

$$V_{ab}(\text{I}) = \frac{1}{2\pi k}\left[q_a\left(\ln\frac{i}{r} + \ln\frac{g}{f}\right) + q_b\left(\ln\frac{r}{i} + \ln\frac{d}{g}\right)\right.$$

$$\left. + q_c\left(\ln\frac{i}{j} + \ln\frac{g}{h}\right)\right] \qquad (3.30)$$

For section II of the transposition cycle

$$V_{ab}(\text{II}) = \frac{1}{2\pi k}\left[q_a\left(\ln\frac{i}{r} + \ln\frac{g}{d}\right) + q_b\left(\ln\frac{r}{i} + \ln\frac{f}{g}\right)\right.$$

$$\left. + q_c\left(\ln\frac{j}{i} + \ln\frac{h}{g}\right)\right] \qquad (3.31)$$

For section III of the transposition cycle

$$V_{ab}(\text{III}) = \frac{1}{2\pi k}\left[q_a\left(\ln\frac{j}{r} + \ln\frac{h}{f}\right) + q_b\left(\ln\frac{r}{j} + \ln\frac{f}{h}\right)\right.$$

$$\left. + q_c\left(\ln\frac{i}{j} + \ln\frac{g}{g}\right)\right] \qquad (3.32)$$

Average value of V_{ab} over the transposition cycle is given by

$$V_{ab}(\text{avg}) = \frac{1}{6\pi k}\left[q_a\ln\left(\frac{ig\ ig\ jh}{rf\ rd\ rf}\right) + q_b\ln\left(\frac{rd\ rf\ rf}{ig\ ig\ jh}\right)\right]$$

$$= \frac{1}{2\pi k}(q_a - q_b) \ln\left(\frac{i^2 g^2 jh}{r^3 f^2 d}\right)^{1/3} \tag{3.33}$$

Similarly

$$V_{ac}\ (avg) = \frac{1}{2\pi k}(q_a - q_c) \ln\left(\frac{i^2 g^2 jh}{r^3 f^2 d}\right)^{1/3} \tag{3.34}$$

Now

$$V_{ab} + V_{ac} = 3V_{an} = \frac{1}{2\pi k}(2q_a - q_b - q_c) \ln\left(\frac{i^2 g^2 jh}{r^3 f^2 d}\right)^{1/3} \tag{3.35}$$

$$3V_{an} = \frac{3q_a}{2\pi k} \ln\left(\frac{i^2 g^2 jh}{r^3 f^2 d}\right)^{1/3}$$

Capacitance to neutral per conductor $= \dfrac{2\pi k}{\ln\left(\dfrac{i^2 g^2 jh}{r^3 f^2 d}\right)^{1/3}}$ $\tag{3.36}$

Total capacitance to neutral for two conductors in parallel

$$C_n = \frac{4\pi k}{\ln\left(\dfrac{i^2 g^2 jh}{r^3 f^2 d}\right)^{1/3}} \text{ F/m} \tag{3.37}$$

Now $h = 6$ m; $d = 8$ m; $j = 8$ m. Referring to Fig. 3.12, we can write

$$i = \left[\left(\frac{j}{2}\right)^2 + \left(\frac{d-h}{2}\right)^2\right]^{1/2} = \sqrt{17}\text{ m}$$

$$f = (j^2 + h^2)^{1/2} = 10\text{ m}$$

$$g = (7^2 + 4^2)^{1/2} = \sqrt{65}\text{ m}$$

Conductor radius (overall) $= 0.865 \times 10^{-2}$ m
Substituting the values for various distances, we have

$$C_n = \frac{4\pi \times 1 \times 8.85 \times 10^{-12} \times 10^6 \times 1000}{\ln\left[\dfrac{17 \times 65 \times 8 \times 6}{100 \times 8}\left(\dfrac{100}{0.865}\right)^3\right]^{1/3}} \ \mu\text{F / km}$$

$$= 0.0181\ \mu\text{F/km}$$

$$\omega C_n = 314 \times 0.0181 \times 10^{-6}$$

$$= 5.68 \times 10^{-6}\ \mho/\text{km}$$

Charging current/phase $= \dfrac{110 \times 1000}{\sqrt{3}} \times 5.68 \times 10^{-6} = 0.361$ A/km

$\therefore$ Charging current/conductor $= \dfrac{0.361}{2}$

$$= 0.1805 \text{ A/km}$$

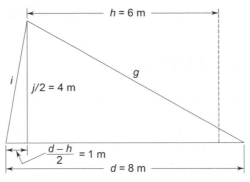

Fig. 3.12

3.8 METHOD OF GMD (MODIFIED)

A comparison of various expressions for inductance and capacitance of transmission lines [e.g. Eqs. (2.22b) and (3.6)] brings out the fact that the two are similar except that in inductance expressions we have to use the fictitious conductor radius $r' = 0.7788r$, while in the expressions for capacitance actual conductor radius r is used. This fact suggests that the method of GDM would be applicable in the calculations for capacitance as well provided it is modified by using the outer conductor radius for finding D_s, the self geometric mean distance.

Example 3.3 can be conveniently solved as under by using the modified GMD method.

For the first section of the transposition cycle mutual GMD is

$$D_{ab} = ((ig)\ (ig))^{1/4} = (ig)^{1/2}$$
$$D_{bc} = (ig)^{1/2}$$
$$D_{ca} = (jh)^{1/2}$$
$$\therefore \quad D_{eq} = (D_{ab}D_{bc}D_{ca})^{1/3} = [(i^2g^2jh)^{1/3}]^{1/2}$$

In the first section of the transposition cycle self GMD is

$$D_{sa} = (rf\ rf\ )^{1/4} = (rf)^{1/2}$$
$$D_{sb} = (rd)^{1/2}$$
$$D_{sc} = (rf)^{1/2}$$
$$D_s = (D_{sa}D_{sb}D_{sc})^{1/3} = [(r^3f^2d)^{1/3}]^{1/2}$$

Now
$$C_n = \frac{2\,\pi k}{\ln \dfrac{D_{eq}}{D_s}} = \frac{2\,\pi k}{\ln\left[\left(\dfrac{i^2g^2jh}{r^3f^2d}\right)^{1/3}\right]^{1/2}}$$

$$= \frac{4\,\pi k}{\ln\left(\dfrac{i^2g^2jh}{r^3f^2d}\right)^{1/3}} \quad \text{F/m}$$

This result obviously checks with the fundamentally derived expression in Example 3.3.

3.9 BUNDLED CONDUCTORS

A bundled conductor line is shown in Fig. 3.13. The conductors of any one bundle are in parallel, and it is assumed that the charge per bundle divides equally among the conductors of the bundle as $D_{12} \gg d$. Also $D_{12} - d \approx D_{12} + d \approx D_{12}$ for the same reason. The results obtained with these assumptions are fairly accurate for usual spacings. Thus if the charge on phase a is q_a, the conductors a and a' have a charge of $q_a/2$ each; similarly the charge is equally divided for phases b and c.

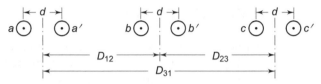

Fig. 3.13 Cross-section of a bundled conductor three-phase transmission line

Now, writing an equation for the voltage from conductor a to conductor b, we get

$$V_{ab} = \frac{1}{2\pi k}\left[0.5q_a\left(\ln\frac{D_{12}}{r} + \ln\frac{D_{12}}{d} \right) + 0.5q_b\left(\ln\frac{r}{D_{12}} + \ln\frac{d}{D_{12}} \right) \right.$$
$$\left. + 0.5q_c\left(\ln\frac{D_{23}}{D_{31}} + \ln\frac{D_{23}}{D_{31}} \right)\right] \tag{3.38}$$

or

$$V_{ab} = \frac{1}{2\pi k}\left(q_a \ln\frac{D_{12}}{\sqrt{rd}} + q_b \ln\frac{\sqrt{rd}}{D_{12}} + q_c \ln\frac{D_{23}}{D_{31}} \right) \tag{3.39}$$

Considering the line to be transposed and proceeding in the usual manner, the final result will be

$$C_n = \frac{0.0242}{\log\left(D_{eq}/\sqrt{rd} \right)} \; \mu\text{F/km to neutral} \tag{3.40}$$

where $D_{eq} = (D_{12}D_{23}D_{31})^{1/3}$

It is obvious from Eq. (2.42) that the method of modified GMD is equally valid in this case (as it should be).

PROBLEMS

3.1 Derive an expression for the charge (complex) value per metre length of conductor a of an untransposed three-phase line shown in Fig. P-3.1. The applied voltage is balanced three-phase, 50 Hz. Take the voltage of phase a as reference phasor. All conductors have the same radii. Also find the charging current of phase a. Neglect the effect of ground.

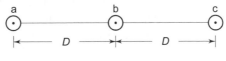

Fig. P-3.1

3.2 A three-phase double-circuit line is shown in Fig. P-3.2. The diameter of each conductor is 2.0 cm. The line is transposed and carries balanced load. Find the capacitance per phase to neutral of the line.

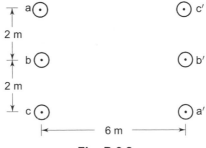

Fig. P-3.2

3.3 A three-phase, 50 Hz overhead line has regularly transposed conductors equilaterally spaced 4 m apart. The capacitance of such a line is 0.01 μF/km. Recalculate the capacitance per kilometre to neutral when the conductors are in the same horizontal plane with successive spacing of 4 m and are regularly transposed.

3.4 Consider the 500 kV, three-phase bundled conductor line as shown in Fig. P-2.9. Find the capacitive reactance to neutral in ohms/km at 50 Hz.

3.5 A three-phase transmission line has flat, horizontal spacing with 2 m between adjacent conductors. The radius of each conductor is 0.25 cm. At a certain instant the charges on the centre conductor and on one of the outside conductors are identical and voltage drop between these identically charged conductors is 775 V. Neglect the effect of ground, and find the value of the identical charge in coulomb/km at the instant specified.

3.6 Find the 50 Hz susceptance to neutral per kilometre of a double-circuit three phase line with transposition as shown in Fig. P-3.6. Given $D = 7$m and radius of each of the six conductors is 1.38 cm.

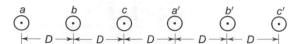

Fig. P-3.6 Double circuit three-phase line with flat spacing

3.7 A single conductor power cable has a conductor of No. 2 solid copper (radius = 0.328 cm). Paper insulation separating the conductor from the concentric lead sheath has a thickness of 2.5 mm and a relative permittivity of 3.8. The thickness of the lead sheath is 2 mm. Find the capacitive reactance per kilometre between the inner conductor and the lead sheath.

3.8 Find the capacitance of phase to neutral per kilometre of a three-phase line having conductors of 2 cm diameter placed at the corners of a triangle with sides 5 m, 6 m and 7 m respectively. Assume that the line is fully transposed and carries balanced load.

3.9 Derive an expression for the capacitance per metre length between two long parallel conductors, each of radius, r, with axes separated by a distance D, where $D \gg r$, the insulating medium being air. Calculate the maximum potential difference permissible between the conductors, if the electric field strength between them is not to exceed 25 kV/cm, r being 0.3 cm and D = 35 cm.

REFERENCES

Books

1. Stevenson, W.D., *Elements of Power System Analysis*, 4th edn, McGraw-Hill, New York, 1982.
2. Cotton, H., and H. Barber, *The Transmission and Distribution of Electrical Energy*, 3rd edn, Hodder and Stoughton, 1970.
3. Starr, A.T., *Generation, Transmission and Utilization of Electric Power*, Pitman, 1962.

Papers

4. Parton, J.E. and A. Wright, "Electric Stresses Associated with Bundle Conductors", *International Journal of Electrical Engineering Education* 1965, 3 : 357.
5. Stevens, R.A. and D.M. German, "The Capacitance and Inductance of Overhead Transmission Lines", *International Journal of Electrical Engineering Education*, 1965, 2 : 71.

4

Representation of Power System Components

4.1 INTRODUCTION

A complete diagram of a power system representing all the three phases becomes too complicated for a system of practical size, so much so that it may no longer convey the information it is intended to convey. It is much more practical to represent a power system by means of simple symbols for each component resulting in what is called a one-line diagram.

Per unit system leads to great simplification of three-phase networks involving transformers. An impedance diagram drawn on a per unit basis does not require ideal transformers to be included in it.

An important element of a power system is the synchronous machine, which greatly influences the system behaviour during both steady state and transient conditions. The synchronous machine model in steady state is presented in this chapter. The transient model of the machine will be presented in Chapter 9.

4.2 Single-Phase Solution of Balanced Three-Phase Networks

The solution of a three-phase network under balanced conditions is easily carried out by solving the single-phase network corresponding to the reference phase. Figure 4.1 shows a simple, balanced three-phase network. The generator and load neutrals are therefore at the same potential, so that $I_n = 0$. Thus the neutral impedance Z_n does not affect network behaviour. For the reference phase a

$$E_a = (Z_G + Z_L)I_a \tag{4.1}$$

The currents and voltages in the other phases have the same magnitude but are progressively shifted in phase by 120°. Equation (4.1) corresponds to the single-phase network of Fig. 4.2 whose solution completely determines the solution of the three-phase network.

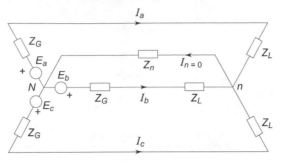

Fig. 4.1 Balanced three-phase network

Fig. 4.2 Single-phase equivalent of a balanced three-phase network of Fig. 4.1

Consider now the case where a three-phase transformer forms part of a three-phase system. If the transformer is Y/Y connected as shown in Fig. 4.3a, in the single-phase equivalent of the three-phase circuit it can be obviously represented by a single-phase transformer (as in Fig. 4.3b) with primary and secondary pertaining to phase a of the three-phase transformer.

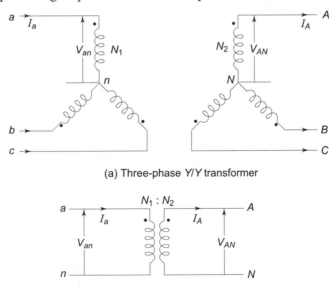

(a) Three-phase Y/Y transformer

(b) Single-phase equivalent of 3-phase Y/Y transformer

Fig. 4.3

If the transformer is Y/Δ connected as in Fig. 4.4a, the delta side has to be replaced by an equivalent star connection as shown dotted so as to obtain the single-phase equivalent of Fig. 4.4b. An important fact has, however, to be observed here. On the delta side the voltage to neutral V_{AN} and line current I_A have a certain phase angle shift[*] from the star side values V_{an} and I_a (90° for the phase labelling shown). In the single-phase equivalent (V_{AN}, I_A) are respectively in phase with (V_{an}, I_a). Since both phase voltage and line current shift through the same phase angle from star to delta side, the transformer per phase impedance and power flow are preserved in the single-phase equivalent. In most analytical studies, we are merely interested in the magnitude of voltages and currents so that the single-phase equivalent of Fig. 4.4b is an acceptable proposition. Wherever proper phase angles of currents and voltages are needed, correction can be easily applier after obtaining the solution through a single-phase transformer equivalent.

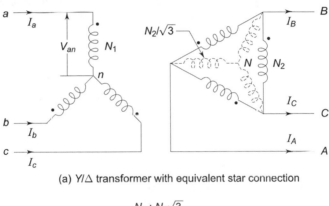

(a) Y/Δ transformer with equivalent star connection

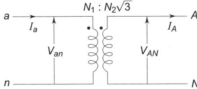

(b) Single-phase equivalent of Y/Δ transformer

Fig. 4.4

It may be noted here that irrespective of the type of connection, the transformation ratio of the single-phase equivalent of a three-phase transformer is the same as the line-to-line transformation ratio.

---

[*] See Section 10.3.

4.3 ONE-LINE DIAGRAM AND IMPEDANCE OR REACTANCE DIAGRAM

A one-line diagram of a power system shows the main connections and arrangements of components. Any particular component may or may not be shown depending on the information required in a system study, e.g. circuit breakers need not be shown in a load flow study but are a must for a protection study. Power system networks are represented by one-line diagrams using suitable symbols for generators, motors, transformers and loads. It is a convenient practical way of network representation rather than drawing the actual three-phase diagram which may indeed be quite cumbersome and confusing for a practical size power network. Generator and transformer connections—star, delta, and neutral grounding are indicated by symbols drawn by the side of the representation of these elements. Circuit breakers are represented as rectangular blocks. Figure 4.5 shows the one-line diagram of a simple power system. The reactance data of the elements are given below the diagram.

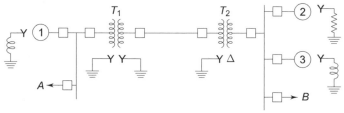

Fig. 4.5 One-line representation of a simple power system

Generator No. 1 : 30 MVA, 10.5 kV, $X'' = 1.6$ ohms
Generator No. 2 : 15 MVA, 6.6 kV, $X'' = 1.2$ ohms
Generator No. 3 : 25 MVA, 6.6 kV, $X'' = 0.56$ ohms
Transformer T_1 (3 phase) : 15 MVA, 33/11 kV, $X = 15.2$ ohms per phase on high tension side
Transformer T_2 (3 phase): 15 MVA, 33/6.2 kV, $X = 16$ ohms per phase on high tension side
Transmission line: 20.5 ohms/phase
Load A : 15 MW, 11 kV, 0.9 lagging power factor
Load B : 40 MW, 6.6 kV, 0.85 lagging power factor
    Note: Generators are specified in three-phase MVA, line-to-line voltage and per phase reactance (equivalent star). Transformers are specified in three-phase MVA, line-to-line transformation ratio, and per phase (equivalent star) impedance on one side. Loads are specified in three-phase MW, line-to-line voltage and power factor.

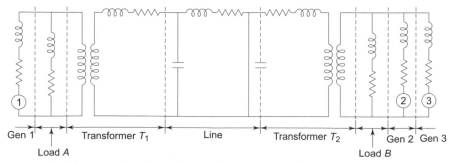

Fig. 4.6 Impedance diagram of the power system of Fig. 4.5

The impedance diagram on single-phase basis for use under balanced operating conditions can be easily drawn from the one-line diagram. For the system of Fig. 4.5 the impedance diagram is drawn in Fig. 4.6. Single-phase transformer equivalents are shown as ideal transformers with transformer impedances indicated on the appropriate side. Magnetizing reactances of the transformers have been neglected. This is a fairly good approximation for most power system studies. The generators are represented as voltage sources with series resistance and inductive reactance (synchronous machine model will be discussed in Sec. 4.6). The transmission line is represented by a π-model (to be discussed in Chapter 5). Loads are assumed to be passive (not involving rotating machines) and are represented by resistance and inductive reactance in series. Neutral grounding impedances do not appear in the diagram as balanced conditions are assumed.

Three voltage levels (6.6, 11 and 33 kV) are present in this system. The analysis would proceed by transforming all voltages and impedances to any selected voltage level, say that of the transmission line (33 kV). The voltages of generators are transformed in the ratio of transformation and all impedances by the square of ratio of transformation. This is a very cumbersome procedure for a large network with several voltage levels. The per unit method discussed below is the most convenient for power system analysis and will be used throughout this book.

4.4 PER UNIT (PU) SYSTEM

It is usual to express voltage, current, voltamperes and impedance of an electrical circuit in per unit (or percentage) of base or reference values of these quantities. The per unit[*] value of any quantity is defined as:

$$\frac{\text{the actual value in any units}}{\text{the base or reference value in the same units}}$$

The per unit method is particularly convenient in power systems as the various sections of a power system are connected through transformers and have different voltage levels.

Consider first a single-phase system. Let

Base voltamperes = $(VA)_B$ VA

Base voltage = V_B V

Then

Base current $I_B = \dfrac{(VA)_B}{V_B}$ A   (4.2a)

---

[*]Per cent value = per unit value × 100.
Per cent value is not convenient for use as the factor of 100 has to be carried in computations.

Base impedance $Z_B = \dfrac{V_B}{I_B} = \dfrac{V_B^2}{(VA)_B}$ ohms   (4.2b)

If the actual impedance is Z (ohms), its per unit value is given by

$$Z(\text{pu}) = \dfrac{Z}{Z_B} = \dfrac{Z(\text{ohms}) \times (VA)_B}{V_B^2}$$ (4.3)

For a power system, practical choice of base values are:

Base megavoltamperes = $(MVA)_B$

or

Base kilovoltamperes = $(kVA)_B$

Base kilovolts = $(kV)_B$

Base current $I_B = \dfrac{1{,}000 \times (MVA)_B}{(kV)_B} = \dfrac{(kVA)_B}{(kV)_B}$ A   (4.4)

Base impedance $Z_B = \dfrac{1{,}000 \times (kV)_B}{I_B}$

$$= \dfrac{(kV)_B^2}{(MVA)_B} = \dfrac{1{,}000 \times (kV)_B^2}{(kVA)_B} \text{ ohms}$$ (4.5)

Per unit impedance $Z\,(\text{pu}) = \dfrac{Z\,(\text{ohms}) \times (MVA)_B}{(kV)_B^2}$ (4.6)

$$= \dfrac{Z(\text{ohms}) \times (kVA)_B}{(kV)_B^2 \times 1{,}000}$$

In a three-phase system rather than obtaining the per unit values using per phase base quantities, the per unit values can be obtained directly by using three-phase base quantities. Let

     Three-phase base megavoltamperes = $(MVA)_B$
     Line-to-line base kilovolts = $(kV)_B$
     Assuming star connection (equivalent star can always be found),

Base current $I_B = \dfrac{1{,}000 \times (MVA)_B}{\sqrt{3}\,(kV)_B}$ A   (4.7)

Base impedance $Z_B = \dfrac{1{,}000 \times (kV)_B}{\sqrt{3}\,I_B}$

$$= \dfrac{(kV)_B^2}{(MVA)_B} = \dfrac{1{,}000 \times (kV)_B^2}{(kVA)_B} \text{ ohms}$$ (4.8)

Per unit impedance Z (pu) $= \dfrac{Z \,(\text{ohms}) \times (\text{MVA})_B}{(\text{kV})_B^2}$ (4.9)

$$= \dfrac{Z(\text{ohms}) \times (\text{kVA})_B}{(\text{kV})_B^2 \times 1{,}000}$$

When MVA base is changed from $(\text{MVA})_{B,\,\text{old}}$ to $(\text{MVA})_{B,\,\text{new}}$, and kV base is changed from $(\text{kV})_{B,\,\text{old}}$ to $(\text{kV})_{B,\,\text{new}}$, the new per unit impedance from Eq. (4.9) is given by

$$Z \,(\text{pu})_{\text{new}} = Z \,(\text{pu})_{\text{old}} \times \dfrac{(\text{MVA})_{B,\,\text{new}}}{(\text{MVA})_{B,\,\text{old}}} \times \dfrac{(\text{kV})_{B,\text{old}}^2}{(\text{kV})_{B,\,\text{new}}^2}$$ (4.10)

Per Unit Representation of a Transformer

It has been said in Section 4.2 that a three-phase transformer forming part of a three-phase system can be represented by a single-phase transformer in obtaining per phase solution of the system. The delta connected winding of the transformer is replaced by an equivalent star so that the transformation ratio of the equivalent single-phase transformer is always the line-to-line voltage ratio of the three-phase transformer.

Figure 4.7a represents a single-phase transformer in terms of primary and secondary leakage reactances Z_p and Z_s and an ideal transformer of ratio $1 : a$. The magnetizing impedance is neglected. Let us choose a voltampere base of $(\text{VA})_B$ and voltage bases on the two sides of the transformer in the ratio of transformation, i.e.

$$\dfrac{V_{1B}}{V_{2B}} = \dfrac{1}{a}$$ (4.11a)

(a) Representation of single-phase transformer (magnetizing impedance neglected)

(b) Per unit equivalent circuit of single-phase transformer

Fig. 4.7

Therefore, $\quad \dfrac{I_{1B}}{I_{2B}} = a$ (as $(VA)_B$ is common) $\hfill$ (4.11b)

$$Z_{1B} = \dfrac{V_{1B}}{I_{1B}}, \ Z_{2B} = \dfrac{V_{2B}}{I_{2B}} \hfill (4.11c)$$

From Fig. 4.7a we can write

$$V_2 = (V_1 - I_1 Z_p)\, a - I_2 Z_S \hfill (4.12)$$

We shall convert Eq. (4.12) into per unit form

$$V_2(\text{pu})V_{2B} = [V_1(\text{pu})V_{1B} - I_1(\text{pu})I_{1B}Z_p(\text{pu})Z_{1B}]a$$
$$-I_2(\text{pu})I_{2B}Z_s(\text{pu})Z_{2B}$$

Dividing by V_{2B} throughout and using base relations (4.11a, b, c), we get

$$V_2(\text{pu}) = V_1(\text{pu}) - I_1(\text{pu})Z_p(\text{pu}) - I_2(\text{pu})Z_s(\text{pu}) \hfill (4.13)$$

Now $\quad \dfrac{I_1}{I_2} = \dfrac{I_{1B}}{I_{2B}} = a$

or $\quad \dfrac{I_1}{I_{1B}} = \dfrac{I_2}{I_{2B}}$

$\therefore \quad I_1(\text{pu}) = I_2(\text{pu}) = I\ (\text{pu})$

Equation (4.13) can therefore be written as

$$V_2(\text{pu}) = V_1(\text{pu}) - I(\text{pu})Z(\text{pu}) \hfill (4.14)$$

where $\quad Z(\text{pu}) = Z_p(\text{pu}) + Z_s(\text{pu})$

Equation (4.14) can be represented by the simple equivalent circuit of Fig. 4.7b which does not require an ideal transformer. Considerable simplification has therefore been achieved by the per unit method with a common voltampere base and voltage bases on the two sides in the ratio of transformation.

$Z(\text{pu})$ can be determined directly from the equivalent impedance on primary or secondary side of a transformer by using the appropriate impedance base.

On primary side:

$$Z_1 = Z_p + Z_s/a^2$$

$$Z_1(\text{pu}) = \dfrac{Z_1}{Z_{1B}} = \dfrac{Z_p}{Z_{1B}} + \dfrac{Z_s}{Z_{1B}} \times \dfrac{1}{a^2}$$

But $\quad a^2 Z_{1B} = Z_{2B}$

$\therefore \quad Z_1(\text{pu}) = Z_p(\text{pu}) + Z_s(\text{pu}) = Z(\text{pu}) \hfill (4.15)$

On secondary side:

$$Z_2 = Z_s + a^2 Z_p$$

$$Z_2(\text{pu}) = \frac{Z_2}{Z_{2B}} = \frac{Z_s}{Z_{2B}} + a^2 \frac{Z_p}{Z_{2B}}$$

or $$Z_2(\text{pu}) = Z_s(\text{pu}) + Z_p(\text{pu}) = Z(\text{pu}) \tag{4.16}$$

Thus the per unit impedance of a transformer is the same whether computed from primary or secondary side so long as the voltage bases on the two sides are in the ratio of transformation (equivalent per phase ratio of a three-phase transformer which is the same as the ratio of line-to-line voltage rating).

The pu transformer impedance of a three-phase transformer is conveniently obtained by direct use of three-phase MVA base and line-to-line kV base in relation (4.9). Any other impedance on either side of a transformer is converted to pu value just like Z_p or Z_s.

Per Unit Impedance Diagram of a Power System

From a one-line diagram of a power system we can directly draw the impedance diagram by following the steps given below:

1. Choose an appropriate common MVA (or kVA) base for the system.
2. Consider the system to be divided into a number of sections by the transformers. Choose an appropriate kV base in one of the sections. Calculate kV bases of other sections in the ratio of transformation.
3. Calculate per unit values of voltages and impedances in each section and connect them up as per the topology of the one-line diagram. The result is the single-phase per unit impedance diagram.

The above steps are illustrated by the following examples.

Example 4.1

Obtain the per unit impedance (reactance) diagram of the power system of Fig. 4.5.

Solution   The per phase impedance diagram of the power system of Fig. 4.5 has been drawn in Fig. 4.6. We shall make some further simplifying assumptions.

1. Line capacitance and resistance are neglected so that it is represented as a series reactance only.
2. We shall assume that the impedance diagram is meant for short circuit studies. Current drawn by static loads under short circuit conditions can be neglected. Loads A and B are therefore ignored.

Let us convert all reactances to per unit form. Choose a common three-phase MVA base of 30 and a voltage base of 33 kV line-to-line on the transmission line. Then the voltage base in the circuit of generator 1 is 11 kV line-to-line and that in the circuits of generators 2 and 3 is 6.2 kV.

The per unit reactances of various components are calculated below:

Transmission line: $\dfrac{20.5 \times 30}{(33)^2} = 0.564$

Transformer T_1: $\dfrac{15.2 \times 30}{(33)^2} = 0.418$

Transformer T_2: $\dfrac{16 \times 30}{(33)^2} = 0.44$

Generator 1: $\dfrac{1.6 \times 30}{(11)^2} = 0.396$

Generator 2: $\dfrac{1.2 \times 30}{(6.2)^2} = 0.936$

Generator 3: $\dfrac{0.56 \times 30}{(6.2)^2} = 0.437$

The reactance diagram of the system is shown in Fig. 4.8.

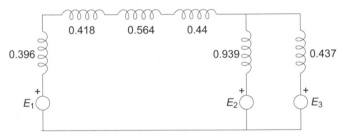

Fig. 4.8   Reactance diagram of the system of Fig. 4.5 (loads neglected)

E_1, E_2 and E_3 are per unit values of voltages to which the generators are excited. Quite often in a short circuit study, these will be taken as $1 \angle 0°$ pu (no load condition).

Example 4.2

The reactance data of generators and transformers are usually specified in pu (or per cent) values, based on equipment ratings rather than in actual ohmic values as given in Example 4.1; while the transmission line impedances may be given in actual values. Let us re-solve Example 4.1 by assuming the following pu values of reactances

Transformer T_1:   0.209
Transformer T_2:   0.220
Generator G_1:   0.435
Generator G_2:   0.413
Generator G_3:   0.3214

With a base MVA of 30, base voltage of 11 kV in the circuit of generator 1 and base voltage of 6.2 kV in the circuit of generators 2 and 3 as used in

Example 4.1, we now calculate the pu values of the reactances of transformers and generators as per relation (4.10):

Transformer T_1: $\qquad 0.209 \times \dfrac{30}{15} = 0.418$

Transformer T_2: $\qquad 0.22 \times \dfrac{30}{15} = 0.44$

Generator 1: $\qquad 0.435 \times \dfrac{(10.5)^2}{(11)^2} = 0.396$

Generator 2: $\quad 0.413 \times \dfrac{30}{15} \times \dfrac{(6.6)^2}{(6.2)^2} = 0.936$

Generator 3: $\quad 0.3214 \times \dfrac{30}{25} \times \dfrac{(6.6)^2}{(6.2)^2} = 0.437$

Obviously these values are the same as obtained already in Example 4.1.

4.5 Complex Power

Consider a single-phase load fed from a source as in Fig. 4.9. Let

$$V = |V| \angle \delta$$
$$I = |I| \angle (\delta - \theta)$$

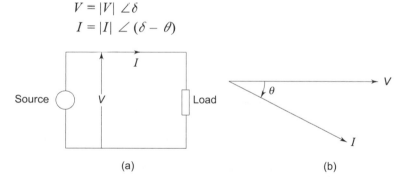

(a)   (b)

Fig. 4.9 Complex power flow in a single-phase load

When θ is positive, the current lags behind voltage. This is a convenient choice of sign of θ in power systems where loads have mostly lagging power factors.

Complex power flow in the direction of current indicated is given by

$$S = VI^*$$
$$= |V|\,|I| \angle \theta$$
$$= |V|\,|I| \cos\theta + j|V|\,|I| \sin\theta = P + jQ \qquad (4.17)$$

or

$$|S| = (P^2 + Q^2)^{1/2}$$

Here

$$S = \text{complex power (VA, kVA, MVA)}$$

$|S|$ = apparent power (VA, kVA, MVA); it signifies rating of equipments (generators, transformers)

$P = |V|\,|I|\,\cos\theta$ = real (active) power (watts, kW, MW)

$Q = |V|\,|I|\,\sin\theta$ = reactive power

= voltamperes reactive (VAR)

= kilovoltamperes reactive (kVAR)

= megavoltamperes reactive (MVAR)

It immediately follows from Eq. (4.17) that Q, the reactive power, is positive for lagging current (lagging power factor load) and negative for leading current (leading power factor load). With the direction of current indicated in Fig. 4.9, $S = P + jQ$ is supplied by the source and is absorbed by the load.

Equation (4.17) can be represented by the phasor diagram of Fig. 4.10 where

$$\theta = \tan^{-1}\frac{Q}{P} = \text{positive for lagging current} \qquad (4.18)$$

= negative for leading current

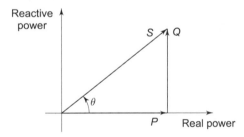

Fig. 4.10 Phasor representation of complex power (lagging pf load)

If two (or more) loads are in parallel as in Fig. 4.11

$$S = VI^* = V(I_1^* + I_2^*)$$
$$= VI_1^* + VI_2^*$$
$$= S_1 + S_2 = (P_1 + P_2) + j(Q_1 + Q_2) \qquad (4.19)$$

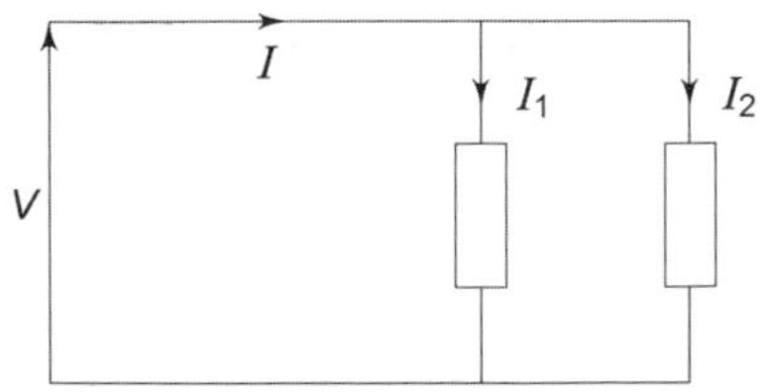

Fig. 4.11 Two loads in parallel

As per Eq. (4.19), Kirchhoff's current law applies to complex power (also applies separately to real and reactive powers).

In a series RL load carrying current I,

$$V = I \ (R + jX_L)$$
$$S = VI^* = I^2R + jI^2X_L$$
$\therefore$
$$P = I^2R = \text{active power absorbed by load}$$
$$Q = I^2X_L = \text{reactive power absorbed by load}$$

In case of a series RC load carrying current I,

$$P = I^2R$$
$$Q = -I^2X_C \text{ (reactive power absorbed is negative)}$$

Consider now a balanced three-phase load represented in the form of an equivalent star as shown in Fig. 4.12. The three-phase complex power fed into load is given by

$$S = 3V_P I_L^* = 3 \ |V_P| \ \angle \delta_P I_L^* = \sqrt{3} \ |V_L| \ \angle \delta_P I_L^* \qquad (4.20)$$

If

$$I_L = |I_L| \ \angle \ (\delta_P - \theta)$$

Then

$$S = \sqrt{3} \ |V_L| \ |I_L| \ \angle \ \theta$$
$$= \sqrt{3} \ |V_L| \ |I_L| \ \cos \theta + j\sqrt{3} \ |V_L| \ |I_L| \ \sin \theta - P + jQ \ (4.21)$$

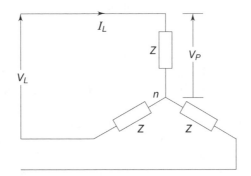

Fig. 4.12 Complex power fed to three-phase load

Here

$$|S| = \sqrt{3} \ |V_L| \ |I_L|$$
$$P = \sqrt{3} \ |V_L| \ |I_L| \ \cos \theta$$
$$Q = \sqrt{3} \ |V_L| \ |I_L| \ \sin \theta$$

where

$$\theta = \text{power factor angle}$$

If V_L, the line voltage, is expressed in kV; and I_L, the line current in amperes, S is in kVA; and if the line current is in kiloamperes, S is in MVA.

In terms of load impedance Z,

$$I_L = \frac{V_P}{Z} = \frac{|V_L|\angle\delta_P}{\sqrt{3}\,Z}$$

Substituting for I_L in Eq. (4.20)

$$S = \frac{|V_L|^2}{Z^*} \qquad\qquad (4.22\text{a})$$

If V_L is in kV, S is now given in MVA. Load impedance Z if required can be calculated from

$$Z = \frac{|V_L|^2}{S^*} = \frac{|V_L|^2}{P - jQ} \qquad\qquad (4.22\text{b})$$

4.6 SYNCHRONOUS MACHINE

The synchronous machine is the most important element of a power system. It converts mechanical power into electrical form and feeds it into the power network or, in the case of a motor, it draws electrical power from the network and converts it into the mechanical form. The machine excitation which is controllable determines the flow of VARs into or out of the machine. Books on electrical machines [1–5] may be consulted for a detailed account of the synchronous machine. We shall present here a simplified circuit model of the machine which with suitable modifications wherever necessary (under transient conditions) will be adopted throughout this book.

Figure 4.13 shows the schematic cross-sectional diagram of a three-phase synchronous generator (alternator) having a two pole structure. The stator has a balanced three-phase winding—aa', bb' and cc'. The winding shown is a concentrated one, while the winding in an actual machine is distributed across the stator periphery. The rotor shown is a cylindrical[*] one (round rotor or non-salient pole rotor) with rotor winding excited by the DC source. The rotor winding is so arranged on rotor periphery that the field excitation produces nearly sinusoidally distributed flux/pole (ϕ_f) in the air gap. As the rotor rotates, three-phase emfs are produced in stator winding. Since the machine is a balanced one and balanced loading will be considered, it can be modelled on per phase basis for the reference phase a.

In a machine with more than two poles, the above defined structure repeats electrically for every pair of poles. The frequency of induced emf is given by

$$f = \frac{NP}{120} \text{ Hz}$$

where

---

[*] High-speed turbo-generators have cylindrical rotors and low speed hydro-generators have salient pole rotors.

N = rotor speed (synchronous speed) in rpm
P = number of poles

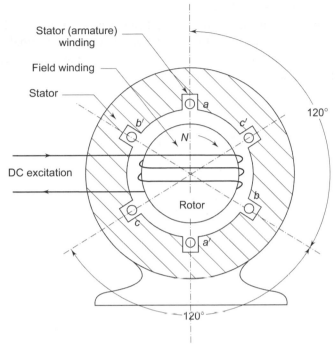

Fig. 4.13 Schematic diagram of a round rotor synchronous generator

On no load the voltage E_f induced in the reference phase a lags 90° behind ϕ_f which produces it and is proportional to ϕ_f if the magnetic circuit is assumed to be unsaturated. This phasor relationship is indicated in Fig. 4.14. Obviously the terminal voltage $V_t = E_f$.

Fig. 4.14 Phasor relationship between ϕ_f and E_f

As balanced steady load is drawn from the three-phase stator winding, the stator currents produce synchronously rotating flux ϕ_a/pole (in the direction of rotation of the rotor). This flux, called *armature reaction* flux, is therefore stationary with respect to field flux ϕ_f. It intuitively follows that ϕ_a is in phase with phase a current I_a which causes it. Since the magnetic circuit has been assumed to be unsaturated, the superposition principle is applicable so that the resultant air gap flux is given by the phasor sum

$$\phi_r = \phi_f + \phi_a \tag{4.23}$$

Further assuming that the armature leakage reactance and resistance are negligible, ϕ_r induces the armature emf which equals the terminal voltage V_t. Phasor diagram under loaded (balanced) conditions showing fluxes, currents and voltages as phasors is drawn in Fig. 4.15.

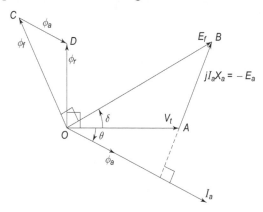

Fig. 4.15 Phasor diagram of synchronous generator

Here

θ = power factor angle

δ = angle by which E_f leads V_t called *load angle* or *torque angle*

We shall see in Sec. 5.10 that δ mainly determines the power delivered by the generator and the magnitude of E_f (i.e. excitation) determines the VARs delivered by it.

Because of the assumed linearity of the magnetic circuit, voltage phasor E_f, E_a and V_t are proportional to flux phasors ϕ_f, ϕ_a and ϕ_r, respectively; further, voltage phasors lag 90° behind flux phasors. It therefore easily follows from Fig. 4.15 that phasor $AB = - E_a$ is proportional to ϕ_a (and therefore I_a) and is 90° leading ϕ_a (or I_a). With the direction of phasor AB indicated on the diagram

$$AB = jI_a X_a$$

where X_a is the constant of propotionality.

In terms of the above definition of X_a, we can directly write the following expression for voltages without the need of invoking flux phasors.

$$V_t = E_f - jI_a X_a \tag{4.24}$$

where

E_f = voltage induced by field flux ϕ_f alone

= no load emf

The circuit model of Eq. (4.24) is drawn in Fig. 4.16 wherein X_a is interpreted as inductive reactance which accounts for the effect of armature reaction thereby avoiding the need of resorting to addition of fluxes [Eq.(4.23)].

The circuit of Fig. 4.16 can be easily modified to include the effect of armature leakage reactance and resistance (these are series effects) to give the complete circuit model of the synchronous generator as in Fig. 4.17. The total reactance $(X_a + X_l) = X_s$ is called the *synchronous reactance* of the machine. Equation (4.24) now becomes

$$V_t = E_f - jI_a X_s - I_a R_a \qquad (4.25)$$

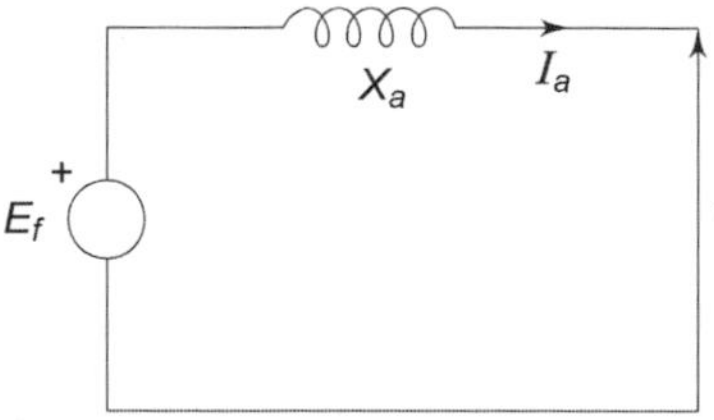

Fig. 4.16   Circuit model of round rotor synchronous generator (resistance and leakage reactance neglected)

This model of the synchronous machine can be further modified to account for the effect of magnetic saturation where the principle of super-position does not hold.

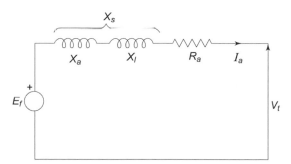

Fig. 4.17   Circuit model of round rotor synchronous generator

Armature resistance R_a is invariably neglected in power system studies. Therefore, in the place of the circuit model of Fig. 4.17, the simplified circuit model of Fig. 4.18 will be used throughout this book. The corresponding phasor diagram is given in Fig. 4.19. The field induced emf E_f leads the terminal voltage by the torque (load) angle δ. This, in fact, is the condition for active power to flow out of the generator. The magnitude of power delivered depends upon sin δ.

In the motoring operation of a synchronous machine, the current I_a reverses as shown in Fig. 4.20, so that Eq. (4.25) modifies to

$$E_f = V_t - jI_a X_s \qquad (4.26)$$

which is represented by the phasor diagram of Fig. 4.21. It may be noted that V_t now leads E_f by δ. This in fact is the condition for power to flow into motor terminals.

The flow of reactive power and terminal voltage of a synchronous machine is mainly controlled by means of its excitation. This is discussed in detail in Section 5.10. Voltage and reactive power flow are often automatically regulated by voltage regulators (see Section 8.6) acting on the field circuits of generators and by automatic tap changing devices on transformers.

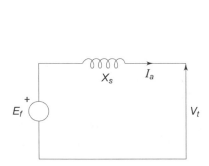

Fig. 4.18 Simplified circuit model of round rotor synchronous generator

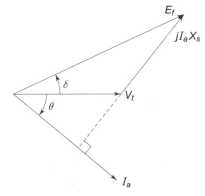

Fig. 4.19 Phasor diagram of synchronous generator

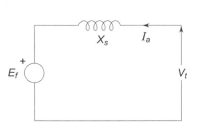

Fig. 4.20 Motoring operation of synchronous machine

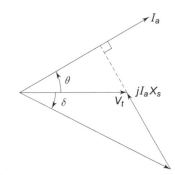

Fig. 4.21 Phasor diagram of motoring operation

Normally, a synchronous generator operates in parallel with other generators connected to the power system. For simplicity of operation we shall consider a generator connected to an *infinite bus* as shown in Fig. 4.22. As infinite bus means a large system whose voltage and frequency remain constant independent of the power exchange between the synchronous machine and the bus, and independent of the excitation of the synchronous machine.

Consider now a synchronous generator feeding constant active power into an infinite bus bar. As the machine excitation is varied, armature current I_a and its angle θ, i.e. power factor, change in such a manner as to keep

$$|V_t| \, |I_a| \cos \theta = \text{constant} = \text{active power output}$$

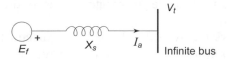

Fig. 4.22 Synchronous machine connected to infinite bus

It means that since $|V_t|$ is fixed, the projection $|I_a| \cos \theta$ of the phasor I_a on V_t remains constant, while the excitation is varied. Phasor diagrams corresponding to high, medium and low excitations are presented in Fig. 4.23. The phasor diagram of Fig. 4.23(b) corresponds to the unity power factor case. It is obvious from the phasor diagram that for this excitation

$$|E_f| \cos \delta = |V_t|$$

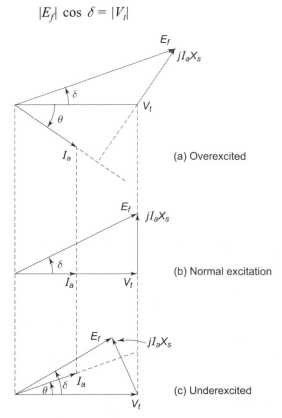

Fig. 4.23 Phasor diagrams of synchronous generator feeding constant power as excitation is varied

This is defined as *normal excitation*. For the *overexcited* case (Fig. 4.23a), i.e. $|E_f| \cos \delta > |V_t|$, I_a lags behind V_t so that the generator feeds positive reactive power into the bus (or draws negative reactive power from the bus). For the *underexcited* case (Fig. 4.23c), i.e. $|E_f|$ cos $\delta < |V_t|$, I_a leads V_t so that the generator feeds negative reactive power into the bus (or draws positive reactive power from the bus).

Figure 4.24 shows the overexcited and underexcited cases of synchronous motor (connected to infinite bus) with constant power drawn from the infinite bus. In the overexcited case, I_a leads V_t, i.e. the motor draws negative reactive power (or supplies positive reactive power); while in the underexcited case I_a lags V_t, i.e. the motor draws positive reactive power (or supplies negative reactive power).

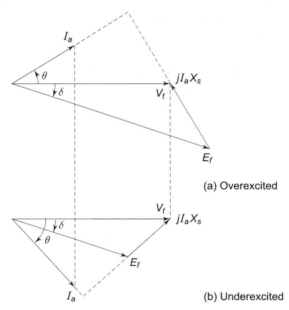

(a) Overexcited

(b) Underexcited

Fig. 4.24 Phasor diagrams of synchronous motor drawing constant power as excitation is varied

From the above discussion we can draw the general conclusion that a synchronous machine (generating or motoring) while operating at constant power supplies positive reactive power into the bus bar (or draws negative reactive power from the bus bar) when overexcited. An underexcited machine on the other hand, feeds negative reactive power into the bus bar (or draws positive reactive power from the bus bar).

Consider now the power delivered by a synchronous generator to an infinite bus. From Fig. 4.19 this power is

$$P = |V_t|\ |I_a|\ \cos\ \theta$$

The above expression can be written in a more useful form from the phasor geometry. From Fig. 4.19

$$\frac{|E_f|}{\sin\ (90° + \theta)} = \frac{|I_a|X_s}{\sin\ \delta}$$

or

$$|I_a| \cos \theta = \frac{|E_f|}{X_s} \sin \delta \qquad (4.27)$$

$$\therefore \qquad P = \frac{|E_f||V_t|}{X_s} \sin \delta \qquad (4.28)$$

The plot of P versus δ, shown in Fig. 4.25, is called the *power angle curve*. The maximum power that can be delivered occurs at $\delta = 90°$ and is given by

$$P_{max} = \frac{|E_f||V_t|}{X_s} \qquad (4.29)$$

For $P > P_{max}$ or for $\delta > 90°$ the generator falls out of step. This problem (the stability) will be discussed at length in Chapter 12.

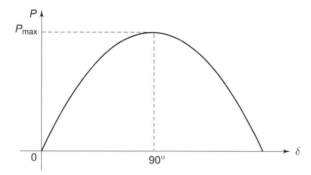

Fig. 4.25   Power angle curve of a synchronous generator

Power Factor and Power Control

While Figs 4.23 and 4.24 illustrate how a synchronous machine power factor changes with excitation for fixed power exchange, these do not give us a clue regarding the quantitative values of $|I_a|$ and δ. This can easily be accomplished by recognizing from Eq. (4.27) that

$$|E_f| \sin \delta = |I_a|X_s \cos \theta$$

$$= \frac{PX_s}{|V_t|} = \text{constant (for constant exchange of power to}$$
$$\text{infinite bus bar)} \qquad (4.30)$$

Figure 4.26 shows the phasor diagram for a generator delivering constant power to infinite bus but with varying excitation. As $|E_f| \sin \delta$ remains constant, the tip of phasor E_f moves along a line parallel to V_t as excitation is varied. The direction of phasor I_a is always 90° lagging jI_aX_s and its magnitude is obtained from $(|I_a|X_S)/X_S$. Figure 4.27 shows the case of limiting excitation with $\delta = 90°$. For excitation lower than this value the generator becomes unstable.

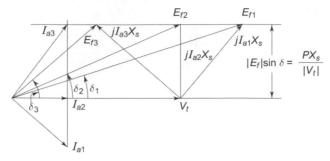

Fig. 4.26 Effect of varying excitation of generator delivering constant power to infinite bus bar

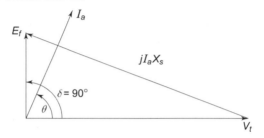

Fig. 4.27 Case of limiting excitation of generator delivering constant power to infinite bus bar

Similar phasor diagrams can be drawn for synchronous motor as well for constant input power (or constant load if copper and iron losses are neglected and mechanical loss is combined with load).

Another important operating condition is variable power and fixed excitation. In this case $|V_t|$ and $|E_f|$ are fixed, while δ and active power vary in accordance with Eq. (4.28). The corresponding phasor diagram for two values of δ is shown in Fig. 4.28. It is seen from this diagram that as δ increases, current magnitude increases and power factor improves. It will be shown in Section 5.10 that as δ changes, there is no significant change in the flow of reactive power.

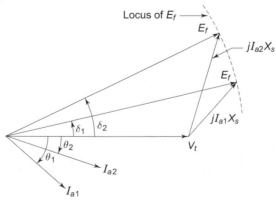

Fig. 4.28 Operation of synchronous generator with variable power and fixed excitation

Salient Pole Synchronous Generator

A salient pole synchronous machine, as shown in Fig. 4.29, is distinguished from a round rotor machine by constructional features of field poles which project with a large interpolar air gap. This type of construction is commonly employed in machines coupled to hydroelectric turbines which are inherently slow-speed ones so that the synchronous machine has multiple pole pairs as different from machines coupled to high-speed steam turbines (3,000/1,500 rpm) which have a two- or four-pole structure. Salient pole machine analysis is made through the *two-reaction theory* outlined below.

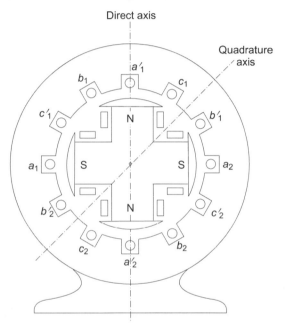

Fig. 4.29   Sallent pole synchronous machine (4-pole structure)

In a round rotor machine, armature current in phase with field induced emf E_f or in quadrature (at 90°) to E_f, produces the same flux linkages per ampere as the air gap is uniform so that the armature reaction reactance offered to in-phase or quadrature current is the same $(X_a + X_l = X_s)$. In a salient pole machine air gap is non-uniform along rotor periphery. It is the least along the axis of main poles (called *direct axis*) and is the largest along the axis of the interpolar region (called *quadrature axis*). Armature current in quadrature with E_f produces flux along the direct axis and the reluctance of flux path being low (because of small air gap), it produces larger flux linkages per ampere and hence the machine presents larger armature reaction reactance X_d (called direct axis reactance) to the flow of quadrature component I_d of armature current I_a. On the other hand, armature current in phase with E_f produces flux along the quadrature axis and the reluctance of the flux path being high (because of large interpolar air gap), it produces smaller flux linkages per ampere and hence the machine presents smaller armature reaction reactance X_q (quadrature axis reactance $< X_d$) to the flow of inphase component I_q of armature current I_a.

Since a salient pole machine offers different reactances to the flow of I_d and I_q components of armature current I_a, a circuit model cannot be drawn. The phasor diagram of a salient pole generator is shown in Fig. 4.30. It can be easily drawn by following the steps given below:

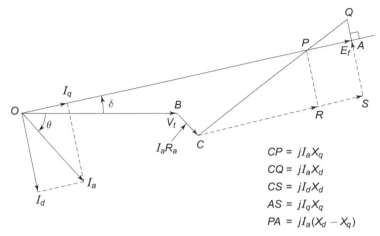

$CP = jI_aX_q$
$CQ = jI_aX_d$
$CS = jI_dX_d$
$AS = jI_qX_q$
$PA = jI_a(X_d - X_q)$

Fig. 4.30 Phasor diagram of salient pole synchronous generator

1. Draw V_t and I_a at angle θ
2. Draw I_aR_a. Draw $CQ = jI_a X_d (\perp$ to $I_a)$
3. Make $|CP| = |I_a| X_q$ and draw the line OP which gives the direction of E_f phasor
4. Draw a $\perp$ from Q to the extended line OP such that $OA = E_f$

It can be shown by the above theory that the power output of a salient pole generator is given by

$$P = \frac{|V_t||E_f|}{X_d} \sin \delta + \frac{|V_t|^2 (X_d - X_q)}{2X_dX_q} \sin 2\delta \qquad (4.31)$$

The first term is the same as for a round rotor machine with $X_s = X_d$ and constitutes the major part in power transfer. The second term is quite small (about 10–20%) compared to the first term and is known as *reluctance power*.

P versus δ is plotted in Fig. 4.31. It is noticed that the maximum power output occurs at $\delta < 90°$ (about 70°). Further $\dfrac{dP}{d\delta}$ (change in power per unit change in power angle for small changes in power angle), called the *synchronizing power coefficient*, in the operating region ($\delta < 70°$) is larger in a salient pole machine than in a round rotor machine.

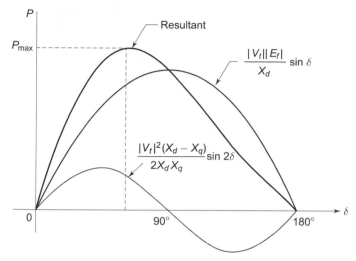

Fig. 4.31 Power angle curve for salient pole generator

In this book we shall neglect the effect of saliency and take

$$X_s = X_d$$

in all types of power system studies considered.

During a machine transient, the direct axis reactance changes with time acquiring the following distinct values during the complete transient.

$X_d''=$ subtransient direct axis reactance

$X_d' =$ transient direct axis reactance

$X_d =$ steady state direct axis reactance

The significance and use of these three values of direct axis reactance will be elaborated in Chapter 9.

Operating Chart of a Synchronous Generator

While selecting a large generator, besides rated MVA and power factor, the greatest allowable stator and rotor currents must also be considered as they influence mechanical stresses and temperature rise. Such limiting parameters in the operation are brought out by means of an *operating chart* or *performance chart*.

For simplicity of analysis, the saturation effects, saliency, and resistance are ignored and an unsaturated value of synchronous reactance is considered. Consider Fig. 4.32, the phasor diagram of a cylindrical rotor machine. The locus of constant $|I_a|X_s, |I_a|$ and hence MVA is a circle centered at M. The locus of constant $|E_f|$ (excitation) is also a circle centered at O. As MP is proportional to MVA, QP is proportional to MVAR and MQ to MW, all to the same scale which is obtained as follows.

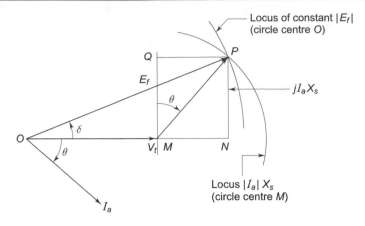

Fig. 4.32 Phasor diagram of synchronous generator

For zero excitation, i.e. $|E_f| = 0$

$$- jI_a X_s = V_t$$

or

$$I_a = jV_t/X_s$$

i.e. $|I_a| = |V_t|/X_s$ leading at 90° to OM which corresponds to VARs/phase.

Consider now the chart shown in Fig. 4.33 which is drawn for a synchronous machine having $X_3 = 1.43$ pu. For zero excitation, the current is $1.0/1.43 = 0.7$ pu, so that the length MO corresponds to reactive power of 0.7 pu, fixing both active and reactive power scales.

With centre at 0 a number of semicircles are drawn with radii equal to different pu MVA loadings. Circles of per unit excitation are drawn from centre M with 1.0 pu excitation corresponding to the fixed terminal voltage OM. Lines may also be drawn from 0 corresponding to various power factors but for clarity only 0.85 pf lagging line is shown. The operational limits are fixed as follows.

Taking 1.0 per unit active power as the maximum allowable power, a horizontal limit-line abc is drawn through b at 1.0 pu. It is assumed that the machine is rated to give 1.0 per unit active power at power factor 0.85 lagging and this fixes point c. Limitation of the stator current to the corresponding value requires the limit-line to become a circular arc cd about centre 0. At point d the rotor heating becomes more important and the arc de is fixed by the maximum excitation current allowable, in this case assumed to be $|E_f| = 2.40$ pu (i.e. 2.4 times $|V_t|$). The remaining limit is decided by loss of synchronism at leading power factors. The theoretical limit is the line perpendicular to MO at M (i.e. $\delta = 90°$), but in practice a safety margin is brought in to permit a further small increase in load before instability. In Fig. 4.33, a 0.1 pu margin is employed and is shown by the curve afg which is drawn in the following way.

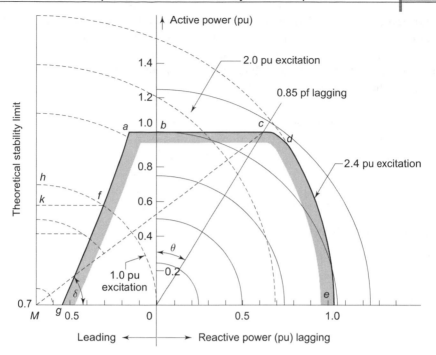

Fig. 4.33 Operating chart for large synchronous generator

Consider a point h on the theoretical limit on the $|E_f| = 1.0$ pu excitations arc, the power Mh is reduced by 0.1 pu to Mk; the operating point must, however, still be on the same $|E_f|$ arc and k is projected to f which is the required point on the desired limiting curve. This is repeated for other excitations giving the curve *afg*. The complete working area, shown shaded, is *gfabcde*. A working point placed within this area at once defines the MVA, MW, MVAR, current, power factor and excitation. The load angle δ can be measured as shown in the figure.

4.7 REPRESENTATION OF LOADS

Load drawn by consumers is the toughest parameter to assess scientifically. The magnitude of the load, in fact, changes continuously so that the load forecasting problem is truly a statistical one. A typical daily load curve is shown in Fig. 1.1. The loads are generally composed of industrial and domestic components. An industrial load consists mainly of large three-phase induction motors with sufficient load constancy and predictable duty cycle, whereas the domestic load mainly consists of lighting, heating and many single-phase devices used in a random way by householders. The design and operation of power systems both economically and electrically are greatly influenced by the nature and magnitude of loads.

In representation of loads for various system studies such as load flow and stability studies, it is essential to know the variation of real and reactive power with variation of voltage. Normally in such studies the load is of composite nature with both industrial and domestic components. A typical composition of load at a bus may be

Induction motors	55–75%
Synchronous motors	5–15%
Lighting and heating	20–30%

Though it is always better to consider the P-V and Q-V characteristics of each of these loads for simulation, the analytic treatment would be very cumbersome and complicated. In most of the analytical work one of the following three ways of load representation is used.

(i) Constant Power Representation

This is used in load flow studies. Both the specified MW and MVAR are taken to be constant.

(ii) Constant Current Representation

Here the load current is given by Eq. (4.17), i.e.

$$I = \frac{P - jQ}{V^*} = |I| \angle (\delta - \theta)$$

where $V = |V| \angle \delta$ and $\theta = \tan^{-1} Q/P$ is the power factor angle. It is known as constant current representation because the magnitude of current is regarded as constant in the study.

(iii) Constant Impedance Representation

This is quite often used in stability studies. The load specified in MW and MVAR at nominal voltage is used to compute the load impedance (Eq. (4.22b)). Thus

$$Z = \frac{V}{I} = \frac{VV^*}{P - jQ} = \frac{|V|^2}{P - jQ} = \frac{1}{Y}$$

which then is regarded as constant throughout the study.

Example 4.3

A synchronous generator is rated 645 MVA, 24 kV, 0.9 pf lagging. It has a synchronous reactance 1.2 Ω. The generator is feeding full load at 0.9 pf lagging at rated voltage. Calculate:

(a) Excitation emf (E_f) and power angle δ
(b) Reactive power drawn by the load
Carry out calculations in pu form and convert the result to actual values.

Solution

$$\text{Base MVA} = 645, \text{ 3-phase}$$

$$\text{Base kV} - 24, \text{ line-to-line}$$

$$\text{Load voltage} = \frac{24}{24} = 1 \text{ pu}$$

$$\text{Synchronous reactance } X_s = \frac{1.2 \times 645}{(24)^2} = 1.344 \text{ pu}$$

$$\text{Full load (MVA)} = 1 \text{ pu}, \, 0.9 \text{ pf lagging}$$

$$\text{Load current} = \text{generator current}$$

$$I_a = 1 \text{ pu}, \, 0.9 \text{ pf lagging}$$

$$= 0.9 - j\, 0.436 \text{ pu}$$

(a) Excitation emf (see Fig. 4.19)

$$E_f = V_t + j\, X_s I_a$$

$$= 1 \angle 0° + j\, 1.344 \, (0.9 - j\, 0.436)$$

$$= 1.586 \quad j\, 1.21 = 1.99 \angle 37.1°$$

$$E_f \text{ (actual)} = 1.99 \times 24 = 47.76 \text{ kV (line)}$$

$$\delta = 37.1° \text{ (leading)}$$

(b) Reactive power drawn by load

$$Q = V_t I_a \sin \phi$$

$$= 1 \times 1 \times 0.436 = 0.436 \text{ pu or } 0.436 \times 645$$

$$= 281 \text{ MVAR}$$

Example 4.4

The generator of Example 4.3 is carrying full load at rated voltage but its excitation emf is (i) increased by 20% and (ii) reduced by 20%. Calculate in each case
(a) load pf
(b) reactive power drawn by load
(c) load angle δ

Solution

$$\text{Full load, } P = 1 \times 0.9 = 0.9 \text{ pu}$$

$$E_f = 1.99$$

$$V_t = 1$$

(i) E_f is increased by 20% at same real load. Now

$$E_f = 1.99 \times 1.2 = 2.388$$

As per Eq. (4.28)

$$P = \frac{|E_f||V_f|}{X_s} \sin \delta \qquad\qquad \text{(i)}$$

$$0.9 = \left(\frac{2.388 x 1}{1.344}\right) \sin \delta$$

or $\sin \delta = 0.5065$

or $\delta = 30.4°$

$$I_a = \frac{E_f - V_i}{jX_s} = \frac{2.388\angle 30.4° - 1\angle 0°}{j1.344}$$

$$= 0.89 - j0.79 = 1.183 \angle -41.2°$$

(a)   pf = cos 41.2° = 0.75 lagging

(b)   Reactive power drawn by load

$$Q = |V_t||I_a| \sin \phi$$

$$= 1 \times 1.183 \times 0.659$$

$$= 0.78 \text{ pu or } 502.8 \text{ MVAR}$$

(ii) E_f decreased by 20% or

$$E_f = 1.99 \times 0.8 = 1.59$$

Substituting in Eq. (i)

$$0.9 = \left(\frac{1.59 \times 1}{1.344}\right) \sin \delta$$

which gives

$$\delta = 49.5°$$

$$I_a = \frac{1.59\angle 49.5° - 1\angle 0°}{j1.344}$$

$$= 0.9 - j\,0.024$$

$$= 0.9 \angle -1.5°$$

(a)   pf = cos 1.5° $\cong$ 1; unity pf

(b) $Q = 1 \times 0.9 \times \sin 1.5 = 0.024$

or $Q = 0.024 \times 645 = 15.2$ MVAR

Problems

4.1 Figure P-4.1 shows the schematic diagram of a radial transmission system. The ratings and reactances of the various components are shown therein. A load of 60 MW at 0.9 power factor lagging is tapped from the 66 kV substation which is to be maintained at 60 kV. Calculate the terminal voltage of the synchronous machine. Represent the transmission line and the transformers by series reactances only.

Fig. P-4.1

4.2 Draw the pu impedance diagram for the power system shown in Fig. P-4.2. Neglect resistance, and use a base of 100 MVA, 220 kV in 50 Ω line. The ratings of the generator, motor and transformers are

Generator 40 MVA, 25 kV, $X'' = 20\%$

Motor 50 MVA, 11 kV, $X'' = 30\%$

Y–Y transformer, 40 MVA, 33 Y-220 Y kV, $X = 15\%$

Y–Δ transformer, 30 MVA, 11 Δ-220 Y kV, $X = 15\%$

Fig. P-4.2

4.3 A synchronous generator is rated 60 MVA, 11 kV. It has a resistance $R_a = 0.1$ pu and $X_d = 1.65$ pu. It is feeding into an infinite bus bar at 11 kV delivering a current 3.15 kA at 0.9 pf lagging.
   (a) Determine E_f and angle δ.
   (b) Draw a phasor diagram for this operation.
   (c) Bus bar voltage falls to 10 kV while the mechanical power input to generator and its excitation remains unchanged. What is the value and pf of the current delivered to the bus. In this case assume the

generator resistance to be negligible.

4.4 A 250 MVA, 16 kV rated generator is feeding into an infinite bus bar at 15 kV. The generator has a synchronous reactance of 1.62 pu. It is found that the machine excitation and mechanical power input are adjusted to give $E_f = 24$ kV and power angle $\delta = 30°$.

   (a) Determine the line current and active and reactive powers fed to the bus bars.

   (b) The mechanical power input to the generator is increased by 20% from that in part (a) but its excitation is not changed. Find the new line current and power factor.

   (c) With reference to part (a) current is to be reduced by 20% at the same power factor by adjusting mechanical power input to the generator and its excitation. Determine E_f, δ and mechanical power input.

   (d) With the reduced current as in part (c), the power is to be delivered to bus bars at unity pf, what are the corresponding values of E_f and δ and also the mechanical power input to the generator.

4.5 The generator of Problem 4.4 is feeding 150 MVA at 0.85 pf lagging to infinite bus bar at 15 kV.

   (a) Determine E_f and δ for the above operation. What are P and Q fed to the bus bars?

   (b) Now E_f is reduced by 10% keeping mechanical input to generator same, find new δ and Q delivered.

   (c) E_f is now maintained as in part (a) but mechanical power input to generator is adjusted till $Q = 0$. Find new δ and P.

   (d) For the value of E_f in part (a) what is the maximum Q that can be delivered to bus bar. What is the corresponding δ and I_a? Sketch the phasor diagram for each part.

Answers

4.1 12 kV

4.3 (a) 26.8 kV (line), 42.3° leading

   (c) 1.13 $\angle$–28.8° kA; 0.876 lag

4.4 (a) 0.511°$\angle$– 25.6° kA; 108 MW, 51.75 MVAR

   (b) 6.14 kA, 0.908 lagging

   (c) 1.578, 13.5°, 53.3 MW

   (d) 18.37 kV, 35.5°, 96 MW

4.5 (a) 25.28 kV, 20.2°, 127.5 MW, 79.05 MVAR

   (b) 33.9°, 54.14 MVAR

   (c) 41.1°, 150.4 MW

   (d) 184.45 MVAR, 53.6°, $- j$ 0.787 pu

References

Books

1. Nagrath, I.J. and D.P. Kothari, *Electric Machines*, 2nd edn Tata McGraw-Hill, New Delhi, 1997.
2. Van E. Mablekos, *Electric Machine Theory for Power Engineers*, Harper and Raw, New York, 1980.
3. DelToro, V., *Electric Machines and Power Systems*, Prentice-Hall, Inc., New Jersey, 1985.
4. Kothari, D.P. and I.J. Nagrath, *Theory and Problems of Electric Machines*, 2nd Edn, Tata McGraw-Hill, New Delhi, 2002.
5. Kothari, D.P. and I.J. Nagrath, *Basic Electrical Engineering*, 2nd Edn., Tata McGraw-Hill, New Delhi, 2002.

Paper

6. IEEE Comittee Report, "The Effect of Frequency and Voltage on Power System Load", Presented at *IEEE Winter Power Meeting*, New York, 1966.

5

Characteristics and Performance of Power Transmission Lines

5.1 INTRODUCTION

This chapter deals primarily with the characteristics and performance of transmission lines. A problem of major importance in power systems is the flow of load over transmission lines such that the voltage at various nodes is maintained within specified limits. While this general interconnected system problem will be dealt with in Chapter 6, attention is presently focussed on performance of a single transmission line so as to give the reader a clear understanding of the principle involved.

Transmission lines are normally operated with a balanced three-phase load; the analysis can therefore proceed on a per phase basis. A transmission line on a per phase basis can be regarded as a two-port network, wherein the sending-end voltage V_S and current I_S are related to the receiving-end voltage V_R and current I_R through $ABCD$ constants[*] as

$$\begin{bmatrix} V_S \\ I_S \end{bmatrix} = \begin{bmatrix} A & B \\ C & D \end{bmatrix} \begin{bmatrix} V_R \\ I_R \end{bmatrix}$$
(5.1)

Also the following identity holds for $ABCD$ constants:

$$AD - BC = 1$$
(5.2)

These constants can be determined easily for short and medium-length lines by suitable approximations lumping the line impedance and shunt admittance. For long lines exact analysis has to be carried out by considering the distribution of resistance, inductance and capacitance parameters and the $ABCD$ constants of the line are determined therefrom. Equations for power flow on a line and receiving- and sending-end circle diagrams will also be developed in this chapter so that various types of end conditions can be handled.

---

[*]Refer to Appendix B.

The following nomenclature has been adopted in this chapter:

z = series impedance/unit length/phase

y = shunt admittance/unit length/phase to neutral

r = resistance/unit length/phase

L = inductance/unit length/phase

C = capacitance/unit length/phase to neutral

l = transmission line length

$Z = zl$ = total series impedance/phase

$Y = yl$ = total shunt admittance/phase to neutral

Note: Subscript S stands for a sending-end quantity and subscript R stands for a receiving-end quantity

5.2 SHORT TRANSMISSION LINE

For short lines of length 100 km or less, the total 50 Hz shunt admittance[*] ($j\omega Cl$) is small enough to be negligible resulting in the simple equivalent circuit of Fig. 5.1.

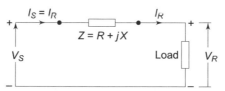

Fig. 5.1 Equivalent circuit of a short line

This being a simple series circuit, the relationship between sending-end receiving-end voltages and currents can be immediately written as:

$$\begin{bmatrix} V_S \\ I_S \end{bmatrix} = \begin{bmatrix} 1 & Z \\ 0 & 1 \end{bmatrix}\begin{bmatrix} V_R \\ I_R \end{bmatrix}$$ (5.3)

The phasor diagram for the short line is shown in Fig. 5.2 for the lagging current case. From this figure we can write

$$|V_S| = [(|V_R|\cos\phi_R + |I|R)^2 + (|V_R|\sin\phi_R + |I|X)^2]^{1/2}$$

$$|V_S| = [|V_R|^2 + |I|^2(R^2 + X^2) + 2|V_R||I|(R\cos\phi_R + X\sin\phi_R)]^{1/2}$$ (5.4)

$$= |V_R|\left[1 + \frac{2|I|R}{|V_R|}\cos\phi_R + \frac{2|I|X}{|V_R|}\sin\phi_R + \frac{|I|^2(R^2 + X^2)}{|V_R|^2}\right]^{1/2}$$

---

[*]For overhead transmission lines, shunt admittance is mainly capacitive susceptance ($j\omega Cl$) as the line conductance (also called *leakance*) is always negligible.

The last term is usually of negligible order.

$$\therefore \qquad |V_S| \simeq |V_R| \left[1 + \frac{2|I|R}{|V_R|} \cos \phi_R + \frac{2|I|X}{|V_R|} \sin \phi_R \right]^{1/2}$$

Expanding binomially and retaining first order terms, we get

$$|V_S| \simeq |V_R| \left[1 + \frac{|I|R}{|V_R|} \cos \phi_R + \frac{|I|X}{|V_R|} \sin \phi_R \right]^{1/2}$$

or

$$|V_S| \simeq |V_R| + |I| \, (R \cos \phi_R + X \sin \phi_R) \qquad (5.5)$$

The above equation is quite accurate for the normal load range.

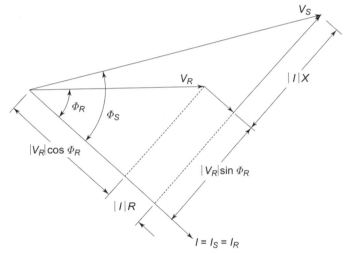

Fig. 5.2   Phasor diagram of a short line for lagging current

Voltage Regulation

Voltage regulation of a transmission line is defined as the rise in voltage at the receiving-end, expressed as percentage of full load voltage, when full load at a specified power factor is thrown off, i.e.

$$\text{Per cent regulation} = \frac{|V_{R0}| - |V_{RL}|}{|V_{RL}|} \times 100 \qquad (5.6)$$

where $|V_{R0}|$ = magnitude of no load receiving-end voltage

$|V_{RL}|$ = magnitude of full load receiving-end voltage
(at a specified power factor)

For short line, $|V_{R0}| = |V_S|$, $|V_{RL}| = |V_R|$

$$\therefore \quad \text{Per cent regulation} = \frac{|V_S| - |V_R|}{|V_R|}$$

$$= \frac{|I|R\cos\phi_R + |I|X\sin\phi_R}{|V_R|} \times 100 \qquad (5.7)$$

In the above derivation, ϕ_R has been considered positive for a lagging load. It will be negative for a leading load.

$$\text{Per cent regulation} = \frac{|I|R\cos\phi_R - |I|X\sin\phi_R}{|V_R|} \times 100 \qquad (5.8)$$
$$\text{(for leading load)}$$

Voltage regulation becomes negative (i.e. load voltage is more than no load voltage), when in Eq. (5.8)

$$X\sin\phi_R > R\cos\phi_R, \text{ or } \tan\phi_R \text{ (leading)} > \frac{R}{X}$$

It also follows from Eq. (5.8) that for zero voltage regulation

$$\tan\phi_R = \frac{R}{X} = \cot\theta$$

i.e., $$\phi_R \text{ (leading)} = \frac{\pi}{2} - \theta \qquad (5.9)$$

where θ is the angle of the transmission line impedance. This is, however, an approximate condition. The exact condition for zero regulation is determined as follows:

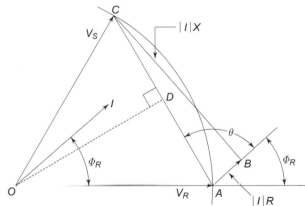

Fig. 5.3 Phasor diagram under zero regulation condition

Figure 5.3 shows the phasor diagram under conditions of zero voltage regulation, i.e.

$$|V_S| = |V_R|$$

or $$OC = OA$$

$$\sin\angle AOD = \frac{AD}{OA} = \frac{AC/2}{|V_R|} = \frac{|I||Z|}{2|V_R|}$$

or
$$\angle AOD = \sin^{-1} \frac{|I||Z|}{2|V_R|}$$

It follows from the geometry of angles at A, that for zero voltage regulation,

$$\phi_R \text{ (leading)} = \left(\frac{\pi}{2} - \theta + \sin^{-1} \frac{|I||Z|}{2|V_R|} \right) \tag{5.10}$$

From the above discussion it is seen that the voltage regulation of a line is heavily dependent upon load power factor. Voltage regulation improves (decreases) as the power factor of a lagging load is increased and it becomes zero at a leading power factor given by Eq. (5.10).

Example 5.1

A single-phase 50 Hz generator supplies an inductive load of 5,000 kW at a power factor of 0.707 lagging by means of an overhead transmission line 20 km long. The line resistance and inductance are 0.0195 ohm and 0.63 mH per km. The voltage at the receiving-end is required to be kept constant at 10 kV.

Find (a) the sending-end voltage and voltage regulation of the line; (b) the value of the capacitors to be placed in parallel with the load such that the regulation is reduced to 50% of that obtained in part (a); and (c) compare the transmission efficiency in parts (a) and (b).

Solution The line constants are

$$R = 0.0195 \times 20 = 0.39 \ \Omega$$

$$X = 314 \times 0.63 \times 10^{-3} \times 20 = 3.96 \ \Omega$$

(a) This is the case of a short line with $I = I_R = I_S$ given by

$$|I| = \frac{5000}{10 \times 0.707} = 707 \text{ A}$$

From Eq. (5.5),

$$|V_S| \simeq |V_R| + |I| (R \cos \phi_R + X \sin \phi_R)$$

$$= 10,000 + 707(0.39 \times 0.707 + 3.96 \times 0.707) \text{ V}$$

$$= 12.175 \text{ kV}$$

$$\text{Voltage regulation} = \frac{12.175 - 10}{10} \times 100 = 21.75\%$$

(b) Voltage regulation desired $= \dfrac{21.75}{2} = 10.9\%$

$$\therefore \qquad \frac{|V_S| - 10}{10} = 0.109$$

or   new value of $|V_S| = 11.09$ kV

Figure 5.4 shows the equivalent circuit of the line with a capacitive reactance placed in parallel with the load.

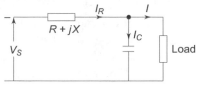

Fig. 5.4

Assuming cos ϕ_R now to be the power factor of load and capacitive reactance taken together, we can write

$$(11.09 - 10) \times 10^3 = |I_R| \, (R \cos \phi_R + X \sin \phi_R) \qquad \text{(i)}$$

Since the capacitance does not draw any real power, we have

$$|I_R| = \frac{5000}{10 \times \cos \phi_R} \qquad \text{(ii)}$$

Solving Eqs. (i) and (ii), we get

$$\cos \phi_R = 0.911 \text{ lagging}$$

and

$$|I_R| = 549 \text{ A}$$

Now

$$I_C = I_R - I$$

$$= 549(0.911 - j0.412) - 707(0.707 - j0.707)$$

$$= 0.29 + j273.7$$

Note that the real part of 0.29 appears due the approximation in (i) Ignoring it, we have

$$I_C = j273.7 \text{ A}$$

$$\therefore \qquad X_C = \frac{1}{314 \times C} = \left|\frac{V_R}{I_C}\right| = \frac{10 \times 1000}{273.7}$$

or $\qquad C = 87 \, \mu\text{F}$

(c)   Efficiency of transmission;

$$\eta = \frac{\text{output}}{\text{output} + \text{loss}}$$

Case (a)

$$\eta = \frac{5000}{5000 + (707)^2 \times 0.39 \times 10^{-3}} = 96.2\%$$

Case (b)

$$\eta = \frac{5000}{5000 + (549)^2 \times 0.39 \times 10^{-3}} = 97.7\%$$

It is to be noted that by placing a capacitor in parallel with the load, the receiving-end power factor improves (from 0.707 lag to 0.911 lag), the line current reduces (from 707 A to 549 A), the line voltage regulation decreases (one half the previous value) and the transmission efficiency improves (from 96.2 to 97.7%). Adding capacitors in parallel with load is a powerful method of improving the performance of a transmission system and will be discussed further towards the end of this chapter.

Example 5.2

A substation as shown in Fig. 5.5 receives 5 MVA at 6 kV, 0.85 lagging power factor on the low voltage side of a transformer from a power station through a cable having per phase resistance and reactance of 8 and 2.5 ohms, respectively. Identical 6.6/33 kV transformers are installed at each end of the line. The 6.6 kV side of the transformers is delta connected while the 33 kV side is star connected. The resistance and reactance of the star connected windings are 0.5 and 3.75 ohms, respectively and for the delta connected windings are 0.06 and 0.36 ohms. What is the voltage at the bus at the power station end?

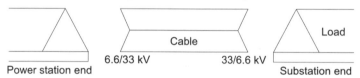

Fig. 5.5

Solution   It is convenient here to employ the per unit method. Let us choose,

   Base MVA = 5

      Base kV = 6.6 on low voltage side

         = 33 on high voltage side

Cable impedance = (8 + j2.5) Ω/phase

$$= \frac{(8 + j2.5) \times 5}{(33)^2} = (0.037 + j0.0115) \ pu$$

Equivalent star impedance of 6.6 kV winding of the transformer

$$= \frac{1}{3}(0.06 + j0.36) = (0.02 + j0.12) \ \Omega/\text{phase}$$

Per unit transformer impedance,

$$Z_T = \frac{(0.02 + j0.12) \times 5}{(6.6)^2} + \frac{(0.5 + j3.75) \times 5}{(33)^2}$$

$$= (0.0046 + j0.030) \text{ pu}$$

Total series impedance $= (0.037 + j0.0115) + 2(0.0046 + j0.030)$

$$= (0.046 + j0.072) \text{ pu}$$

Given: Load MVA $= 1$ pu

$$\text{Load voltage} = \frac{6}{6.6} = 0.91 \text{ pu}$$

$\therefore$ $\quad$ $\text{Load current} = \frac{1}{0.91} = 1.1 \text{ pu}$

Using Eq. (5.5), we get

$$|V_S| = 0.91 + 1.1(0.046 \times 0.85 + 0.072 \times 0.527)$$

$$= 0.995 \text{ pu}$$

$$= 0.995 \times 6.6 = 6.57 \text{ kV (line-to-line)}$$

Example 5.3

Input to a single-phase short line shown in Fig. 5.6 is 2,000 kW at 0.8 lagging power factor. The line has a series impedance of $(0.4 + j0.4)$ ohms. If the load voltage is 3 kV, find the load and receiving-end power factor. Also find the supply voltage.

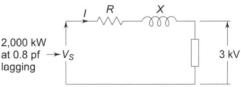

Fig. 5.6

Solution It is a problem with mixed-end conditions—load voltage and input power are specified. The exact solution is outlined below:

Sending-end active/reactive power = receiving-end active/reactive power + active/reactive line losses

For active power

$$|V_S| \, |I| \cos \phi_S = |V_R| \, |I| \cos \phi_R + |I|^2 R \tag{i}$$

For reactive power

$$|V_S| \, |I| \sin \phi_S = |V_R| \, |I| \sin \phi_R + |I|^2 X \tag{ii}$$

Squaring (i) and (ii), adding and simplifying, we get

$$|V_S|^2 \, |I|^2 = |V_R|^2 \, |I|^2 + 2|V_R| \, |I|^2 \, (|I|R \cos \phi_R$$
$$+ \; |I|X \sin \phi_R) + |I|^4 \, (R^2 + X^2) \qquad \text{(iii)}$$

Note: This, in fact, is the same as Eq. (5.4) if $|I|^2$ is cancelled throughout. For the numerical values given

$$|Z|^2 = (R^2 + X^2) = 0.32$$

$$|V_S| \, |I| = \frac{2{,}000 \times 10^3}{0.8} = 2{,}500 \times 10^3$$

$$|V_S| \, |I| \cos \phi_S = 2{,}000 \times 10^3$$

$$|V_S| \, |I| \sin \phi_S = 2{,}500 \times 10^3 \times 0.6 = 1{,}500 \times 10^3$$

From Eqs. (i) and (ii), we get

$$|I| \cos \phi_R = \frac{2000 \times 10^3 - 0.4 \, |I|^2}{3000} \qquad \text{(iv)}$$

$$|I| \sin \phi_R = \frac{1500 \times 10^3 - 0.4 \, |I|^2}{3000} \qquad \text{(v)}$$

Substituting all the known values in Eq. (iii), we have

$$(2{,}500 \times 10^3)^2 = (3{,}000)^2 \, |I|^2 + 2 \times 3{,}000 \, |I|^2 \left[0.4 \times \frac{2000 \times 10^3 - 0.4 |I|^2}{3000} \right.$$
$$\left. + 0.4 \times \frac{1500 \times 10^3 - 0.4|I|^2}{3000} \right] + 0.32 \, |I|^4$$

Simplifying, we get

$$0.32 \, |I|^4 - 11.8 \times 10^6 \, |I|^2 + 6.25 \times 10^{12} = 0$$

which upon solution yields

$$|I| = 725 \text{ A}$$

Substituting for $|I|$ in Eq. (iv), we get

$$\cos \phi_R = 0.82$$

$\therefore$ Load $P_R = |V_R| \, |I| \cos \phi_R = 3{,}000 \times 725 \times 0.82$

$$= 1{,}790 \text{ kW}$$

Now

$$|V_S| = |I| \cos \phi_S = 2{,}000$$

$\therefore$ $$|V_S| = \frac{2000}{725 \times 0.8} = 3.44 \text{ kV}$$

5.3 MEDIUM TRANSMISSION LINE

For lines more than 100 km long, charging currents due to shunt admittance cannot be neglected. For lines in range 100 km to 250 km length, it is sufficiently accurate to lump all the line admittance at the receiving-end resulting in the equivalent diagram shown in Fig. 5.7.

Starting from fundamental circuit equations, it is fairly straightforward to write the transmission line equations in the $ABCD$ constant form given below:

$$\begin{bmatrix} V_S \\ I_S \end{bmatrix} = \begin{bmatrix} 1+YZ & Z \\ Y & 1 \end{bmatrix} \begin{bmatrix} V_R \\ I_R \end{bmatrix}$$

(5.11)

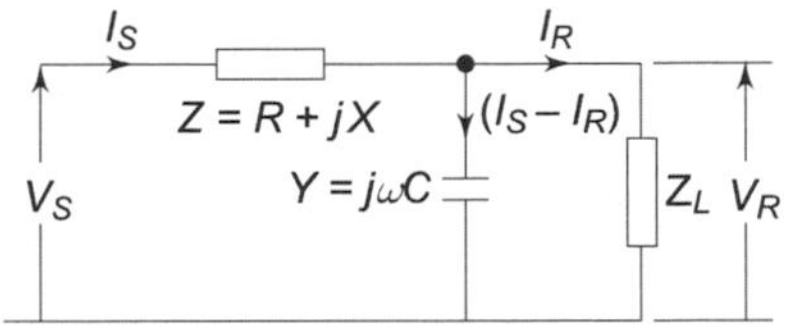

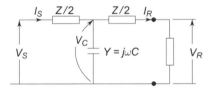

Fig. 5.7 Medium line, localized load-end capacitance

Nominal–T Representation

If all the shunt capacitance is lumped at the middle of the line, it leads to the nominal-T circuit shown in Fig. 5.8.

Fig. 5.8 Medium line, nominal-T representation

For the nominal-T circuit, the following circuit equations can be written:

$$V_C = V_R + I_R(Z/2)$$

$$I_S = I_R + V_C Y = I_R + YV_R + I_R(Z/2)Y$$

$$V_S = V_C + I_S(Z/2)$$

Substituting for V_C and I_S in the last equation, we get

$$V_S = V_R + I_R(Z/2) + (Z/2)\left[I_R\left(1+\frac{ZY}{2}\right)+YV_R\right]$$

$$= V_R\left(1+\frac{ZY}{2}\right)+I_RZ\left(1+\frac{YZ}{4}\right)$$

Rearranging the results, we get the following equations

$$\begin{bmatrix} V_S \\ I_S \end{bmatrix} = \begin{bmatrix} \left(1 + \dfrac{1}{2}ZY\right) & Z\left(1 + \dfrac{1}{4}YZ\right) \\ Y & \left(1 + \dfrac{1}{2}YZ\right) \end{bmatrix} \begin{bmatrix} V_R \\ I_R \end{bmatrix} \quad (5.12)$$

Nominal-π Representation

In this method the total line capacitance is divided into two equal parts which are lumped at the sending- and receiving-ends resulting in the nominal-π representation as shown in Fig. 5.9.

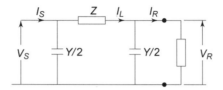

Fig. 5.9 Medium line, nominal-π representation

From Fig. 5.9, we have

$$I_S = I_R + \frac{1}{2}V_R Y + \frac{1}{2}V_S Y$$

$$V_S = V_R + (I_R + \frac{1}{2}V_R Y)Z = V_R\left(1 + \frac{1}{2}YZ\right) + I_R Z$$

$$\therefore \quad I_S = I_R + \frac{1}{2}V_R Y + \frac{1}{2}Y[V_R\left(1 + \frac{1}{2}YZ\right) + I_R Z\ ]$$

$$= V_R Y\left(1 + \frac{1}{4}YZ\right) + I_R\left(1 + \frac{1}{2}YZ\right)$$

Finally, we have

$$\begin{bmatrix} V_S \\ I_S \end{bmatrix} = \begin{bmatrix} \left(1 + \dfrac{1}{2}YZ\right) & Z \\ Y\left(1 + \dfrac{1}{4}YZ\right) & \left(1 + \dfrac{1}{2}YZ\right) \end{bmatrix} \begin{bmatrix} V_R \\ I_R \end{bmatrix} \quad (5.13)$$

It should be noted that nominal-T and nominal-π with the above constants are not equivalent to each other. The reader should verify this fact by applying star-delta transformation to either one.

Example 5.4

Using the nominal-π method, find the sending-end voltage and voltage regulation of a 250 km, three-phase, 50 Hz, transmission line delivering 25

MVA at 0.8 lagging power factor to a balanced load at 132 kV. The line conductors are spaced equilaterally 3 m apart. The conductor resistance is 0.11 ohm/km and its effective diameter is 1.6 cm. Neglect leakance.

Solution   Now, $L = 0.461 \log\dfrac{D}{r'} = 0.461 \log\dfrac{300}{0.7788 \times 0.8} = 1.24$ mH/km

$$C = \frac{0.0242}{\log D/r} = \frac{0.0242}{\log\dfrac{300}{0.8}} = 0.0094 \ \mu F/km$$

$$R = 0.11 \times 250 = 27.5 \ \Omega$$

$$X = 2\pi f L = 2\pi \times 50 \times 1.24 \times 10^{-3} \times 250 = 97.4 \ \Omega$$

$$Z = R + jX = 27.5 + j97.4 = 101.2 \ \angle 74.2° \ \Omega$$

$$Y = j\omega Cl = 314 \times 0.0094 \times 10^{-6} \times 250 \ \angle 90°$$

$$= 7.38 \times 10^{-4} \ \angle 90° \ \mho$$

$$I_R = \frac{25 \times 1000}{\sqrt{3} \times 132} \angle -36.9° \ \Omega = 109.3 \ \angle -36.9° \ A$$

V_R (per phase) = $(132/(\sqrt{3}) \ \angle 0° = 76.2 \ \angle 0°$ kV

$$V_S = \left(1 + \frac{1}{2} YZ\right) V_R + ZI_R$$

$$= \left(1 + \frac{1}{2} \times 7.38 \times 10^{-4} \ \angle 90° \times 101.2 \angle 74.2°\right) 76.2$$

$$+ \ 101.2 \ \angle 74.2° \times 109.3 \times 10^{-3} \ \angle -36.9°$$

$$= 76.2 + 2.85 \ \angle 164.2° + 11.06 \ \angle 37.3°$$

$$= 82.26 + j7.48 = 82.5 \ \angle 5.2°$$

∴ $|V_S|$ (line) = $82.6 \times \sqrt{3} = 143$ kV

$$1 + \frac{1}{2} YZ = 1 + 0.0374 \ \angle 164.2° = 0.964 + j0.01$$

$|V_{R0}|$ (line no load) = $\dfrac{143}{\left|1 + \dfrac{1}{2} YZ\right|} = \dfrac{143}{0.964} = 148.3$ kV

∴   Voltage regulation = $\dfrac{148.3 - 132}{132} \times 100 = 12.3\%$

5.4   THE LONG TRANSMISSION LINE—RIGOROUS SOLUTION

For lines over 250 km, the fact that the parameters of a line are not lumped but distributed uniformly throughout its length, must be considered.

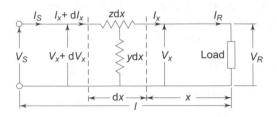

Fig. 5.10   Schematic diagram of a long line

Figure 5.10 shows one phase and the neutral return (of zero impedance) of a transmission line. Let dx be an elemental section of the line at a distance x from the receiving-end having a series impedance zdx and a shunt admittance ydx. The rise in voltage[*] to neutral over the elemental section in the direction of increasing x is dV_x. We can write the following differential relationships across the elemental section:

$$dVx = I_x\, zdx \quad \text{or} \quad \frac{dV_x}{dx} = zI_x \tag{5.14}$$

$$dIx = V_x ydx \quad \text{or} \quad \frac{dI_x}{dx} = yVx \tag{5.15}$$

It may be noticed that the kind of connection (e.g. T or π) assumed for the elemental section, does not affect these first order differential relations.

Differentiating Eq. (5.14) with respect to x, we obtain

$$\frac{d^2V_x}{dx^2} = \frac{dI_x}{dx} z$$

Substituting the value of $\dfrac{dI_x}{dx}$ from Eq. (5.15), we get

$$\frac{d^2V_x}{dx^2} = yzV_x \tag{5.16}$$

This is a linear differential equation whose general solution can be written as follows:

$$V_x = C_1 e^{\gamma x} + C_2 e^{-\gamma x} \tag{5.17}$$

where

$$\gamma = \sqrt{yx} \tag{5.18}$$

and C_1 and C_2 are arbitrary constants to be evaluated.

Differentiating Eq. (5.17) with respect to x:

---

[*] Here V_x is the complex expression of the rms voltage, whose magnitude and phase vary with distance along the line.

$$\frac{dV_x}{dx} = C_1\gamma e^{\gamma x} - C_2\gamma e^{-\gamma x} = zI_x$$

$$\therefore \qquad I_x = \frac{C_1}{Z_c}e^{\gamma x} - \frac{C_2}{Z_c}e^{-\gamma x} \qquad\qquad (5.19)$$

where

$$Z_c = \left(\frac{z}{y}\right)^{1/2} \qquad\qquad (5.20)$$

The constants C_1 and C_2 may be evaluated by using the end conditions, i.e. when $x = 0$, $V_x = V_R$ and $I_x = I_R$. Substituting these values in Eqs. (5.17) and (5.19) gives

$$V_R = C_1 + C_2$$

$$I_R = \frac{1}{Z_c}\,(C_1 - C_2)$$

which upon solving yield

$$C_1 = \frac{1}{2}\,(V_R + Z_c I_R)$$

$$C_2 = \frac{1}{2}(V_R - Z_c I_R)$$

With C_1 and C_2 as determined above, Eqs. (5.17) and (5.19) yield the solution for V_x and I_x as

$$V_x = \left(\frac{V_R + Z_c I_R}{2}\right)e^{\gamma x} + \left(\frac{V_R - Z_c I_R}{2}\right)e^{-\gamma x}$$

$$I_x = \left(\frac{V_R/Z_c + I_R}{2}\right)e^{\gamma x} - \left(\frac{V_R/Z_c - I_R}{2}\right)e^{-\gamma x} \qquad (5.21)$$

Here Z_c is called the *characteristic impedance* of the line and γ is called the *propagation constant*.

Knowing V_R, I_R and the parameters of the line, using Eq. (5.21) complex number rms values of V_x and I_x at any distance x along the line can be easily found out.

A more convenient form of expression for voltage and current is obtained by introducing hyperbolic functions. Rearranging Eq. (5.21), we get

$$V_x = V_R\left(\frac{e^{\gamma x} + e^{-\gamma x}}{2}\right) + I_R Z_c\left(\frac{e^{\gamma x} - e^{-\gamma x}}{2}\right)$$

$$I_x = V_R\frac{1}{Z_c}\left(\frac{e^{\gamma x} - e^{-\gamma x}}{2}\right) + I_R\left(\frac{e^{\gamma x} + e^{-\gamma x}}{2}\right)$$

These can be rewritten after introducing hyperbolic functions, as

$$V_x = V_R \cosh \gamma x + I_R Z_c \sinh \gamma x \tag{5.22}$$

$$I_x = I_R \cosh \gamma x + V_R \frac{1}{Z_c} \sinh \gamma x$$

when $x = l$, $V_x = V_s$, $I_x = I_S$

$$\therefore \qquad \begin{bmatrix} V_S \\ I_S \end{bmatrix} = \begin{bmatrix} \cosh \gamma l & Z_c \sinh \gamma l \\ \frac{1}{Z_c} \sinh \gamma l & \cosh \gamma l \end{bmatrix} \begin{bmatrix} V_R \\ I_R \end{bmatrix} \tag{5.23}$$

Here

$$A = D = \cosh \gamma l$$

$$B = Z_c \sinh \gamma l \tag{5.24}$$

$$C = \frac{1}{Z_c} \sinh \gamma l$$

In case $[V_S I_S]$ is known, $[V_R I_R]$ can be easily found by inverting Eq. (5.23). Thus

$$\begin{bmatrix} V_R \\ I_R \end{bmatrix} = \begin{bmatrix} D & -B \\ -C & A \end{bmatrix} \begin{bmatrix} V_S \\ I_S \end{bmatrix} \tag{5.25}$$

Evaluation of ABCD Constants

The *ABCD* constants of a long line can be evaluated from the results given in Eq. (5.24). It must be noted that $\gamma = \sqrt{yz}$ is in general a complex number and can be expressed as

$$\gamma = \alpha + j\beta \tag{5.26}$$

The hyperbolic function of complex numbers involved in evaluating *ABCD* constants can be computed by any one of the three methods given below.

Method 1

$$\cosh (\alpha l + j\beta l) = \cosh \alpha l \cos \beta l + j \sinh \alpha l \sin \beta l \tag{5.27}$$

$$\sinh (\alpha l + j\beta l) = \sinh \alpha l \cos \beta l + j \cosh \alpha l \sin \beta l$$

Note that sinh, cosh, sin and cos of real numbers as in Eq. (5.27) can be looked up *in standard tables.*

Method 2

$$\cosh \gamma l = 1 + \frac{\gamma^2 l^2}{2!} + \frac{\gamma^4 l^4}{4!} + \dots \approx \left(1 + \frac{YZ}{2}\right)$$

$$\sinh \gamma l = \gamma l + \frac{\gamma^3 l^3}{3!} + \frac{\gamma^5 l^5}{5!} + \ldots \approx \sqrt{YZ}\left(1 + \frac{YZ}{6}\right) \qquad (5.28a)$$

This series converges rapidly for values of γl usually encountered for power lines and can be conveniently approximated as above. The corresponding expressions for *ABCD* constants are

$$A = D \approx 1 + \frac{YZ}{2}$$

$$B \approx Z\left(1 + \frac{YZ}{6}\right) \qquad (5.28b)$$

$$C \approx Y\left(1 + \frac{YZ}{6}\right)$$

The above approximation is computationally convenient and quite accurate for lines up to 400/500 km.

Method 3

$$\cosh(\alpha l + j\beta l) = \frac{e^{\alpha l} e^{j\beta l} + e^{-\alpha l} e^{-j\beta l}}{2} = \frac{1}{2}(e^{\alpha}\angle\beta l + e^{-\alpha l}\angle-\beta l)$$

$$(5.29)$$

$$\sinh(\alpha l + j\beta l) = \frac{e^{\alpha l} e^{j\beta l} - e^{-\alpha l} e^{-j\beta l}}{2} = \frac{1}{2}(e^{\alpha l}\angle\beta l - e^{-\alpha l}\angle-\beta l)$$

5.5 INTERPRETATION OF THE LONG LINE EQUATIONS

As already said in Eq. (5.26), γ is a complex number which can be expressed as

$$\gamma = \alpha + j\beta$$

The real part α is called the *attenuation constant* and the imaginary part β is called the *phase constant*. Now V_x of Eq. (5.21) can be written as

$$V_x = \left|\frac{V_R + Z_c I_R}{2}\right| e^{\alpha x} e^{j(\beta x + \phi_1)} + \left|\frac{V_R - Z_c I_R}{2}\right| e^{-\alpha x} e^{-j(\beta x - \phi_2)}$$

where

$$\phi_1 = \angle(V_R + I_R Z_c) \qquad (5.30)$$

$$\phi_2 = \angle(V_R - I_R Z_c)$$

The instantaneous voltage $v_x(t)$ can be written from Eq. (5.30) as

$$v_x(t) = \mathrm{Re}\left[\sqrt{2}\left|\frac{V_R + Z_c I_R}{2}\right| e^{\alpha x} e^{j(\omega t + \beta x + \phi_1)}\right.$$

$$+\sqrt{2}\left|\frac{V_R - Z_c I_R}{2}\right| e^{-\alpha x} e^{j(\omega t - \beta x + \phi_2)}\right] \tag{5.31}$$

The instantaneous voltage consists of two terms each of which is a function of two variables—time and distance. Thus they represent two travelling waves, i.e.

$$v_x = v_{x1} + v_{x2} \tag{5.32}$$

Now

$$v_{x1} = \sqrt{2}\left|\frac{V_R + Z_c I_R}{2}\right| e^{\alpha x} \cos(\omega t + \beta x + \phi_1) \tag{5.33}$$

At any instant of time t, v_{x1} is sinusoidally distributed along the distance from the receiving-end with amplitude increasing exponentially with distance, as shown in Fig. 5.11 ($\alpha > 0$ for a line having resistance).

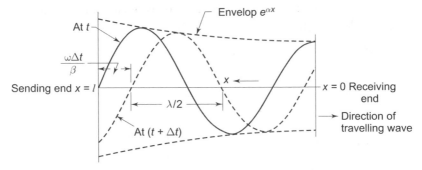

Fig. 5.11 Incident wave

After time Δt, the distribution advances in distance phase by $(\omega \Delta t / \beta)$. Thus this wave is travelling towards the receiving-end and is the *incident wave*. Line losses cause its amplitude to decrease exponentially in going from the sending to the receiving-end.

Now

$$v_{x2} = \sqrt{2}\left|\frac{V_R - Z_c I_R}{2}\right| e^{-\alpha x} \cos(\omega t - \beta x + \phi_2) \tag{5.34}$$

After time Δt the voltage distribution retards in distance phase by $(\omega \Delta t / \beta)$. This is the *reflected wave* travelling from the receiving-end to the sending-end with amplitude decreasing exponentially in going from the receiving-end to the sending-end, as shown in Fig. 5.12.

At any point along the line, the voltage is the sum of incident and reflected voltage waves present at the point [Eq. (5.32)]. The same is true of current waves. Expressions for incident and reflected current waves can be similarly written down by proceeding from Eq. (5.21). If Z_c is pure resistance, current waves can be simply obtained from voltage waves by dividing by Z_c.

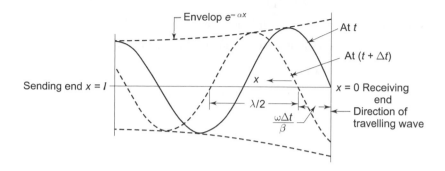

Fig. 5.12 Reflected wave

If the load impedance $Z_L = \dfrac{V_R}{I_R} = Z_c$, i.e. the line is terminated in its *characteristic impedance*, the reflected voltage wave is zero ($V_R - Z_c I_R = 0$).

A line terminated in its characteristic impedance is called the *infinite line*. The incident wave under this condition cannot distinguish between a termination and an infinite continuation of the line.

Power system engineers normally call Z_c the *surge impedance*. It has a value of about 400 ohms for an overhead line and its phase angle normally varies from $0°$ to $-15°$. For underground cables Z_c is roughly one-tenth of the value for overhead lines. The term surge impedance is, however, used in connection with surges (due to lightning or switching) or transmission lines, where the lines loss can be neglected such that

$$Z_c = Z_s = \left(\frac{j\omega L}{-j\omega C} \right)^{1/2} = \left(\frac{L}{C} \right)^{1/2}, \text{ a pure resistance.}$$

Surge Impedance Loading (SIL) of a transmission line is defined as the power delivered by a line to purely resistive load equal in value to the surge impedance of the line. Thus for a line having 400 ohms surge impedance,

$$\text{SIL} = \sqrt{3} \frac{|V_R|}{\sqrt{3} \times 400} \; |V_R| \times 1000 \text{ kW}$$

$$= 2.5 \; |V_R|^2 \text{ kW} \tag{5.35}$$

where $|V_R|$ is the line-to-line receiving-end voltage in kV. Sometimes, it is found convenient to express line loading in per unit of SIL, i.e. as the ratio of the power transmitted to surge impedance loading.

At any time the voltage and current vary harmonically along the line with respect to x, the space coordinate. A complete voltage or current cycle along the line corresponds to a change of 2π rad in the angular argument βx. The corresponding line length is defined as the *wavelength*.

If β is expressed in rad/m,

$$\lambda = 2\pi/\beta \text{ m} \tag{5.36}$$

Now for a typical power transmission line

$$g \text{ (shunt conductance/unit length)} \simeq 0$$

$$r \ll \omega L$$

$$\therefore \qquad \gamma = (yz)^{1/2} = (j\omega C(r + j\omega L))^{1/2}$$

$$= j\omega (LC)^{1/2} \left(1 - j\frac{r}{\omega L}\right)^{1/2}$$

or $$\gamma = \alpha + j\beta \simeq j\omega(LC)^{1/2}\left(1 - j\frac{r}{2\omega L}\right)$$

$$\therefore \qquad \alpha \simeq \frac{r}{2}\left(\frac{C}{L}\right)^{1/2} \tag{5.37}$$

$$\beta \approx \omega (LC)^{1/2} \tag{5.38}$$

Now time for a phase change of 2π is $1/f$ s, where $f = \omega/2\pi$ is the frequency in cycles/s. During this time the wave travels a distance equal to λ, i.e. one wavelength.

$$\therefore \quad \text{Velocity of propagation of wave, } v = \frac{\lambda}{1/f} = f\lambda \text{ m/s} \tag{5.39}$$

which is a well known result.

For a lossless transmission line ($R = 0$, $G = 0$),

$$\gamma = (yz)^{1/2} = j\omega(LC)^{1/2}$$

such that $\alpha = 0$, $\beta = \omega (LC)^{1/2}$

$$\therefore \qquad \lambda = 2\pi/\beta = \frac{2\pi}{\omega(LC)^{1/2}} = \frac{1}{f(LC)^{1/2}} \text{ m} \tag{5.40}$$

and

$$v = f\lambda = 1/(LC)^{1/2} \text{ m/s} \tag{5.41}$$

For a single-phase transmission line

$$L = \frac{\mu_0}{2\pi} \ln \frac{D}{r'}$$

$$C = \frac{2\pi k_0}{\ln D/r}$$

$$\therefore \qquad v = \frac{1}{\left(\dfrac{\mu_0}{2\pi} \ln \dfrac{D}{r'} \dfrac{2\pi k_0}{\ln D/r}\right)^{1/2}}$$

Since r and r' are quite close to each other, when log is taken, it is sufficiently accurate to assume that $\ln \dfrac{D}{r'} \simeq \ln D/r$.

$$\therefore \qquad v \simeq \frac{1}{(\mu_0 k_0)^{1/2}} = \text{velocity of light} \qquad (5.42)$$

The actual velocity of the propagation of wave along the line would be somewhat less than the velocity of light.
Wavelength of a 50 Hz power transmission is approximately given by

$$\lambda \simeq \frac{3 \times 10^8}{50} = 6{,}000 \text{ km}$$

Practical transmission lines are much shorter than this (usually a few hundred kilometres). *It needs to be pointed out here that the waves drawn in Figs. 5.11 and 5.12 are for illustration only and do not pertain to a real power transmission line.*

Example 5.5

A three-phase 50 Hz transmission line is 400 km long. The voltage at the sending-end is 220 kV. The line parameters are $r = 0.125$ ohm/km, $x = 0.4$ ohm/km and $y = 2.8 \times 10^{-6}$ mho/km.
   Find the following:
   (i) The sending-end current and receiving-end voltage when there is no-load on the line.
  (ii) The maximum permissible line length if the receiving-end no-load voltage is not to exceed 235 kV.
 (iii) For part (i), the maximum permissible line frequency, if the no-load voltage is not to exceed 250 kV.

Solution   The total line parameters are:

$$R = 0.125 \times 400 = 50.0 \ \Omega$$

$$X = 0.4 \times 400 = 160.0 \ \Omega$$

$$Y = 2.8 \times 10^{-6} \times 400 \ \angle 90° = 1.12 \times 10^{-3} \ \angle 90° \ \mho$$

$$Z = R + jX = (50.0 + j160.0) = 168.0 \ \angle \ 72.6° \ \Omega$$

$$YZ = 1.12 \times 10^{-3} \ \angle 90° \times 168 \ \angle 72.6°$$

$$= 0.188 \ \angle 162.6°$$

(i)   At no-load

$$V_S = AV_R; \ I_S = CV_R$$

A and C are computed as follows:

$$A \simeq 1 + \frac{1}{2} YZ = 1 + \frac{1}{2} \times 0.188 \ \angle 162.6°$$

$$= 0.91 + j0.028$$

$$|A| = 0.91$$

$$C = Y(1 + YZ/6) = 1.12 \times 10^{-3} \angle 90° \left(1 + \frac{0.188}{6} \angle 162.6°\right)$$

$$= 1.09 \times 10^{-3} \angle 90.55°$$

Now

$$|V_R|_{line} = \frac{220}{|A|} = \frac{220}{0.91} = 242 \text{ kV}$$

$$|I_S| = |C| \, |V_R| = 1.09 \times 10^{-3} \times \frac{242}{\sqrt{3}} \times 10^3 = 152 \text{ A}$$

It is to be noted that under no-load conditions, the receiving-end voltage (242 kV) is more than the sending-end voltage. This phenomenon is known as the Ferranti effect and is discussed at length in Sec. 5.6.

(ii) Maximum permissible no-load receiving-end voltage = 235 kV.

$$|A| = \left|\frac{V_s}{V_R}\right| = \frac{220}{235} = 0.936$$

Now

$$A \approx 1 + \frac{1}{2}YZ$$

$$= 1 + \frac{1}{2}l^2 \times j2.8 \times 10^{-6} \times (0.125 + j0.4)$$

$$= (1 - 0.56 \times 10^{-6}l^2) + j0.175 \times 10^{-6}l^2$$

Since the imaginary part will be less than $\frac{1}{10}$ th of the real part, $|A|$ can be approximated as

$$|A| = 1 - 0.56 \times 10^{-6}l^2 = 0.936$$

$$\therefore \qquad l^2 = \frac{1 - 0.936}{0.56 \times 10^{-6}}$$

or $\qquad l = 338 \text{ km}$

(iii) $\quad |A| = \dfrac{220}{250} = 0.88$

$$A \simeq 1 + \frac{1}{2} \times j1.12 \times 10^{-3} \times \frac{f}{50}\left(50 + j160 \times \frac{f}{50}\right)$$

Neglecting the imaginary part, we can write

$$|A| = 1 - \frac{1}{2} \times 1.12 \times 10^{-3} \times 160 \times \frac{f^2}{(50)^2} = 0.88$$

Simplifying, we obtain the maximum permissible frequency as

$$f = 57.9 \text{ Hz}$$

Example 5.6

If in Example 5.5 the line is open circuited with a receiving-end voltage of 220 kV, find the rms value and phase angle of the following:

(a) The incident and reflected voltages to neutral at the receiving-end.

(b) The incident and reflected voltages to neutral at 200 km from the receiving-end.

(c) The resultant voltage at 200 km from the receiving-end.

Note: Use the receiving-end line to neutral voltage as reference.

Solution From Example 5.5, we have following line parameters:

$$r = 0.125 \ \Omega/\text{km}; \ x = 0.4 \ \Omega/\text{km}; \ y = j2.8 \times 10^{-6} \ \mho/\text{km}$$

$$\therefore \qquad z = (0.125 + j0.4) \ \Omega/\text{km} = 0.42 \ \angle 72.6° \ \Omega/\text{km}$$

$$\gamma = (yz)^{1/2} = (2.8 \times 10^{-6} \times 0.42 \ \angle(90° + 72.6°))^{1/2}$$

$$= 1.08 \times 10^{-3} \ \angle 81.3°$$

$$= (0.163 + j1.068) \times 10^{-3} = \alpha + j\beta$$

$$\therefore \qquad \alpha = 0.163 \times 10^{-3}; \ \beta = 1.068 \times 10^{-3}$$

(a)   At the receiving-end;

For open circuit $I_R = 0$

$$\text{Incident voltage} = \frac{V_R + Z_C I_R}{2} = \frac{V_R}{2}$$

$$= \frac{220/\sqrt{3}}{2} = 63.51 \ \angle 0° \text{ kV (to neutral)}$$

$$\text{Reflected voltage} = \frac{V_R - Z_C I_R}{2} = \frac{V_R}{2}$$

$$= 63.51 \ \angle 0° \text{ kV (to neutral)}$$

(b)   At 200 km from the receiving-end:

$$\text{Incident voltage} = \left. \frac{V_R}{2} e^{+\alpha x} e^{j\beta x} \right|_{x = 200 \text{ km}}$$

$$= 63.51 \ \exp (0.163 \times 10^{-3} \times 200)$$
$$\times \exp (j1.068 \times 10^{-3} \times 200)$$

$$= 65.62 \ \angle 12.2° \ \text{kV (to neutral)}$$

$$\text{Reflected voltage} = \left. \frac{V_R}{2} e^{-\alpha x} e^{-j\beta x} \right|_{x=200 \text{ km}}$$

$$= 63.51 \ e^{-0.0326} \ e^{-j0.2135}$$

$$= 61.47 \ \angle -12.2° \ \text{kV (to neutral)}$$

(c) Resultant voltage at 200 km from the receiving-end

$$= 65.62 \ \angle 12.2° + 61.47 \ \angle -12.2°$$

$$= 124.2 + j0.877 = 124.2 \ \angle 0.4°$$

Resultant line-to-line voltage at 200 km

$$= 124.2 \times \sqrt{3} = 215.1 \ \text{kV}$$

5.6 FERRANTI EFFECT

As has been illustrated in Example 5.5, the effect of the line capacitance is to cause the no-load receiving-end voltage to be more than the sending-end voltage. The effect becomes more pronounced as the line length increases. This phenomenon is known as the *Ferranti effect*. A general explanation of this effect is advanced below:

Substituting $x = l$ and $I_R = 0$ (no-load) in Eq. (5.21), we have

$$V_S = \frac{V_R}{2} e^{\alpha l} e^{j\beta l} + \frac{V_R}{2} e^{-\alpha l} \ e^{-j\beta l} \qquad (5.43)$$

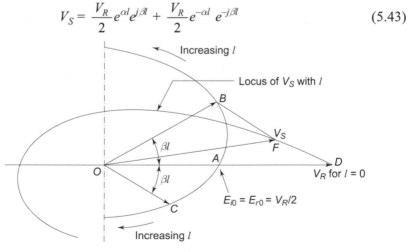

Fig. 5.13

The above equation shows that at $l = 0$, the incident (E_{i0}) and reflected (E_{r0}) voltage waves are both equal to $V_R/2$. With reference to Fig. 5.13, as l increases, the incident voltage wave increases exponentially in magnitude

$\left(\dfrac{V_R}{2}e^{\alpha l}\right)$ and turns through a positive angle βl (represented by phasor OB); while the reflected voltage wave decreases in magnitude exponentially $\left(\dfrac{V_R}{2}e^{-\alpha l}\right)$ and turns through a negative angle βl (represented by phasor OC). It is apparent from the geometry of this figure that the resultant phasor voltage V_S (OF) is such that $|V_R| > |V_S|$.

A simple explanation of the Ferranti effect on an approximate basis can be advanced by lumping the inductance and capacitance parameters of the line. As shown in Fig. 5.14 the capacitance is lumped at the receiving-end of the line.

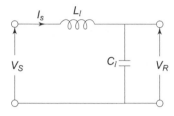

Fig. 5.14

Here
$$I_S = \dfrac{V_S}{\left(\dfrac{1}{j\omega Cl}+j\omega Ll\right)}$$

Since C is small compared to L, ωLl can be neglected in comparison to $1/\omega Cl$. Thus

$$I_S \simeq jV_S\omega Cl$$

Now
$$V_R = V_S - I_S\,(j\omega Ll) = V_S + V_S\omega^2\,CLl^2$$
$$= V_S\,(1 + \omega^2 CLl^2) \tag{5.44}$$

Magnitude of voltage rise $= |V_S|\omega^2\,CLl^2$

$$= |V_S|\dfrac{\omega^2 l^2}{v^2} \tag{5.45}$$

where $v = 1/\sqrt{LC}$ is the velocity of propagation of the electromagnetic wave along the line, which is nearly equal to the velocity of light.

5.7 TUNED POWER LINES

Equation (5.23) characterizes the performance of a long line. For an overhead line shunt conductance G is always negligible and it is sufficiently accurate to neglect line resistance R as well. With this approximation

$$\gamma = \sqrt{yz} = j\omega\sqrt{LC}$$

$$\cosh \gamma l = \cosh j\omega l \sqrt{LC} = \cos \omega l \sqrt{LC}$$

$$\sinh \gamma l = \sinh j\omega l \sqrt{LC} = j \sin \omega l \sqrt{LC}$$

Hence Eq. (5.23) simplifies to

$$\begin{bmatrix} V_S \\ I_S \end{bmatrix} = \begin{bmatrix} \cos \omega l \sqrt{LC} & jZ_c \sin \omega l \sqrt{LC} \\ \dfrac{j}{Z_c} \sin \omega l \sqrt{LC} & \cos \omega l \sqrt{LC} \end{bmatrix} \begin{bmatrix} V_R \\ I_R \end{bmatrix} \qquad (5.46)$$

Now if $\omega l \sqrt{LC} = n\pi$, $n = 1, 2, 3, \ldots$

$$|V_S| = |V_R|$$

$$|I_S| = |I_R|$$

i.e. the receiving-end voltage and current are numerically equal to the corresponding sending-end values, so that there is no voltage drop on load. Such a line is called a *tuned line*.

For 50 Hz, the length of line for tuning is

$$l = \frac{n\pi}{2 \pi f \sqrt{LC}}$$

Since $1/\sqrt{LC} \simeq v$, the velocity of light

$$l = \frac{1}{2}(n\lambda) = \frac{1}{2}\lambda, \lambda, \frac{3}{2}\lambda, \ldots \qquad (5.47)$$

$$= 3{,}000 \text{ km}, 6{,}000 \text{ km}, \ldots$$

It is too long a distance of transmission from the point of view of cost and efficiency (note that line resistance was neglected in the above analysis). For a given line, length and frequency tuning can be achieved by increasing L or C, i.e. by adding series inductances or shunt capacitances at several places along the line length. The method is impractical and uneconomical for power frequency lines and is adopted for telephony where higher frequencies are employed.

A method of tuning power lines which is being presently experimented with, uses series capacitors to cancel the effect of the line inductance and shunt inductors to neutralize line capacitance. A long line is divided into several sections which are individually tuned. However, so far the practical method of improving line regulation and power transfer capacity is to add series capacitors to reduce line inductance; shunt capacitors under heavy load conditions; and shunt inductors under light or no-load conditions.

5.8 THE EQUIVALENT CIRCUIT OF A LONG LINE

So far as the end conditions are concerned, the exact equivalent circuit of a transmission line can be established in the form of a T- or π-network.
The parameters of the equivalent network are easily obtained by comparing the performance equations of a π-network and a transmission line in terms of end quantities.

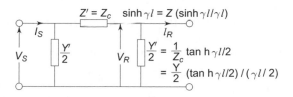

Fig. 5.15 Equivalent-π network of a transmission line

For a π-network shown in Fig. 5.15 [refer to Eq. (5.13)].

$$\begin{bmatrix} V_S \\ I_S \end{bmatrix} = \begin{bmatrix} \left(1 + \frac{1}{2} Y' Z'\right) & Z' \\ Y'\left(1 + \frac{1}{4} Y' Z'\right) & \left(1 + \frac{1}{2} Y' Z'\right) \end{bmatrix} \begin{bmatrix} V_R \\ I_R \end{bmatrix} \tag{5.48}$$

According to exact solution of a long line [refer to Eq. (5.23)].

$$\begin{bmatrix} V_S \\ I_S \end{bmatrix} = \begin{bmatrix} \cosh \gamma l & Z_c \sinh \gamma l \\ \frac{1}{Z_c} \sinh \gamma l & \cosh \gamma l \end{bmatrix} \begin{bmatrix} V_R \\ I_R \end{bmatrix} \tag{5.49}$$

For exact equivalence, we must have

$$Z' = Z_c \sin h \, \gamma l \tag{5.50}$$

$$1 + \frac{1}{2} Y'Z' = \cosh \gamma l \tag{5.51}$$

From Eq. (5.50)

$$Z' = \sqrt{\frac{z}{y}} \sinh \gamma l = Zl \frac{\sinh \gamma l}{l\sqrt{yz}} = Z\left(\frac{\sinh \gamma l}{\gamma l}\right) \tag{5.52}$$

Thus $\dfrac{\sinh \gamma l}{\gamma l}$ is the factor by which the series impedance of the nominal-π

must be multiplied to obtain the Z' parameter of the equivalent-π, Substituting Z' from Eq. (5.50) in Eq. (5.51), we get

$$1 + \frac{1}{2} Y'Z_c \sinh \gamma l = \cosh \gamma l$$

$$\therefore \qquad \frac{1}{2} Y' = \frac{1}{Z_c} \left(\frac{\cosh \gamma l - 1}{\sinh \gamma l}\right)$$

$$= \frac{1}{Z_c} \tanh \frac{\gamma l}{2} = \sqrt{\frac{y}{z}} \tanh \frac{\gamma l}{2}$$

$$= \frac{yl}{2} \left(\frac{\tanh \gamma l/2}{\gamma l/2}\right)$$

or $\qquad \dfrac{1}{2} Y' = \dfrac{Y}{2} \left(\dfrac{\tanh \gamma l/2}{\gamma l/2}\right) \tag{5.53}$

Thus $\left(\dfrac{\tanh \gamma l/2}{\gamma l / 2}\right)$ is the factor by which the shunt admittance arm of the nominal-π must be multiplied to obtain the shunt parameter $(Y'/2)$ of the equivalent-π.

Note that $Y'\left(1+\dfrac{1}{4}Y'Z'\right)=\dfrac{1}{Z_c}\sinh \gamma l$ is a consistent equation in terms of the above values of Y' and Z'.

For a line of medium length $\dfrac{\tanh \gamma l/2}{\gamma l/2}\simeq 1$ and $\dfrac{\sinh \gamma l}{\gamma l}\simeq 1$ so that the equivalent-π network reduces to that of nominal-π.

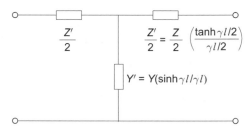

$$\frac{Z'}{2} \qquad \frac{Z'}{2}=\frac{Z}{2}\left(\frac{\tanh \gamma l/2}{\gamma l/2}\right)$$

$$Y' = Y(\sinh \gamma l/\gamma l)$$

Fig. 5.16 Equivalent-T network of a transmission line

Equivalent-T network parameters of a transmission line are obtained on similar lines. The equivalent-T network is shown in Fig. 5.16.

As we shall see in Chapter 6 equivalent-π (or nominal-π) network is easily adopted to load flow studies and is, therefore, universally employed.

Example 5.7

A 50 Hz transmission line 300 km long has a total series impedance of 40 + $j125$ ohms and a total shunt admittance of 10^{-3} mho. The receiving-end load is 50 MW at 220 kV with 0.8 lagging power factor. Find the sending-end voltage, current, power and power factor using

(a) short line approximation,
(b) nominal-π method,
(c) exact transmission line equation [Eq. (5.27)],
(d) approximation [Eq. (5.28b)].

Compare the results and comment.

Solution $Z = 40 + j125 = 131.2 \angle 72.3° \ \Omega$

$$Y = 10^{-3} \angle 90° \ \mho$$

The receiving-end load is 50 MW at 220 kV, 0.8 pf lagging.

$\therefore \qquad\qquad I_R = \dfrac{50}{\sqrt{3}\times 220\times 0.8} \ \angle -36.9° = 0.164 \ \angle -36.9° \text{ kA}$

$$V_R = \frac{220}{\sqrt{3}} \angle 0° = 127 \angle 0° \text{ kV}$$

(a) Short line approximation:

From Eq. (5.3)

$$V_S = 127 + 0.164 \angle -36.9° \times 131.2 \angle 72.3°$$

$$= 145 \angle 4.9°$$

$$|V_S|_{\text{line}} = 251.2 \text{ kV}$$

$$I_S = I_R = 0.164 \angle -36.9° \text{ kA}$$

Sending-end power factor = $\cos (4.9° + 36.9° = 41.8°)$

$$= 0.745 \text{ lagging}$$

Sending-end power = $\sqrt{3} \times 251.2 \times 0.164 \times 0.745$

$$= 53.2 \text{ MW}$$

(b) Nominal-π method:

$$A = D = 1 + \frac{1}{2} YZ = 1 + \frac{1}{2} \times 10^{-3} \angle 90° \times 131.2 \angle 72.3°$$

$$= 1 + 0.0656 \angle 162.3° = 0.938 \angle 1.2°$$

$$B = Z = 131.2 \angle 72.3°$$

$$C = Y\left(1 + \frac{1}{4} YZ\right) = Y + \frac{1}{4} Y^2 Z$$

$$= 0.001 \angle 90° + \frac{1}{4} \times 10^{-6} \angle 180° \times 131.2 \angle 72.3°$$

$$= 0.001 \angle 90°$$

$$V_S = 0.938 \angle 1.2° \times 127 + 131.2 \angle 72.3° \times 0.164 \angle -36.9°$$

$$= 119.1 \angle 1.2° + 21.5 \angle 35.4° = 137.4 \angle 6.2°$$

$$|V_S|_{\text{line}} = 238 \text{ kV}$$

$$I_S = 0.001 \angle 90° \times 127 + 0.938 \angle 1.2° \times 0.164 \angle -36.9°$$

$$= 0.127 \angle 90° + 0.154 \angle -35.7° = 0.13 \angle 16.5°$$

Sending-end pf = $\cos (16.5° - 6.2°) = 0.984$ leading

Sending-end power = $\sqrt{3} \times 238 \times 0.13 \times 0.984$

$$= 52.7 \text{ MW}$$

(c) Exact transmission line equations (Eq. (5.29)).

$$\gamma l = \alpha l + j\beta l = \sqrt{YZ}$$

$$= \sqrt{10^{-3} \angle 90° \times 131.2 \angle 72.3°} = 0.0554 + j0.3577$$

$$= 0.362 \angle 81.2°$$

$$\cosh(\alpha l + j\beta l) = \frac{1}{2}(e^{\alpha l} \angle \beta l + e^{-\alpha l} \angle -\beta l)$$

$$\beta l = 0.3577 \text{ (radians)} = \angle 20.49°$$

$$e^{0.0554} \angle(20.49°) = 1.057 \angle 20.49° = 0.99 + j0.37$$

$$e^{-0.0554} \angle -20.49° = 0.946 \angle - 20.49° = 0.886 - j0.331$$

$\therefore$

$$\cosh \gamma l = 0.938 + j0.02 = 0.938 \angle 1.2°$$

$$\sinh \gamma l = 0.052 + j0.35 = 0.354 \angle 81.5°$$

$$Z_c = \sqrt{\frac{Z}{Y}} = \sqrt{\frac{131.2 \angle 72.3°}{10^{-3} \angle 90°}} = 362.21 \angle - 8.85°$$

$$A = D = \cosh \gamma l = 0.938 \angle 1.2°$$

$$B = Z_c \sinh \gamma l = 362.21 \angle - 8.85° \times 0.354 \angle 81.5°$$

$$= 128.2 \angle 72.65°$$

Now

$$V_S = 0.938 \angle 1.2° \times 127 \angle 0° + 128.2 \angle 72.65° \times 0.164 \angle -36.9°$$

$$= 119.13 \angle 1.2° + 21.03 \angle 35.75°$$

$$= 136.97 \angle 6.2° \text{ kV}$$

$$|V_S|_{\text{line}} = 237.23 \text{ kV}$$

$$C = \frac{1}{Z_c} \sinh \gamma l = \frac{1}{362.21 \angle - 8.85°} \times 0.354 \angle 81.5°$$

$$= 9.77 \times 10^{-4} \angle 90.4°$$

$$I_S = 9.77 \times 10^{-4} \angle 90.4° \times 127 + 0.938 \angle 1.2° \times 0.164 \angle -36.9°$$

$$= 0.124 \angle 90.4° + 0.154° - 35.7°$$

$$= 0.1286 \angle 15.3° \text{ kA}$$

Sending-end pf $= \cos(15.3° - 6.2° = 9.1°) = 0.987$ leading

Sending-end power $= \sqrt{3} \times 237.23 \times 0.1286 \times 0.987$

$$= \textbf{52.15 MW}$$

(d) Approximation (5.28b):

$$A = D = 1 + \frac{1}{2}YZ$$

= 0.938 $\angle 1.2°$ (already calculated in part (b))

$$B = Z\left(1+\frac{YZ}{6}\right) = Z + \frac{1}{6}YZ^2$$

$$= 131.2\ \angle 72.3° + \frac{1}{6} \times 10^{-3}\ \angle 90° \times (131.2)^2\ \angle 144.6°$$

$$= 131.2\ \angle 72.3° + 2.87\ \angle -125.4°$$

$$= 128.5\ \angle 72.7°$$

$$C = Y\left(1+\frac{YZ}{6}\right) = 0.001\ \angle 90° + \frac{1}{6} \times 10^{-6}\ \angle 180° \times 131.2\ \angle 72.3°$$

$$= 0.001\ \angle 90°$$

$$V_S = 0.938\ \angle 1.2° \times 127\ \angle 0° + 128.5\ \angle 72.7° \times 0.164\ \angle -36.9°$$

$$= 119.13\ \angle 1.2° + 21.07\ \angle 35.8° = 136.2 + j14.82$$

$$= 137\ \angle 6.2°\ \text{kV}$$

$$|V_S|_{\text{line}} = 237.3\ \text{kV}$$

$$I_S = 0.13\ \angle 16.5°\ \text{(same as calculated in part (b))}$$

Sending-end pf = cos (16.5° – 6.2° = 10.3°) = 0.984 leading

Sending-end power = $\sqrt{3} \times 237.3 \times 0.13 \times 0.984$

$$= 52.58\ \text{MW}$$

The results are tabulated below:

	Short line approximation	Nominal-π	Exact	Approximation (5.28b)
$\|V_S\|_{\text{line}}$	251.2 kV	238 kV	237.23 kV	237.3 kV
I_s	0.164 $\angle -36.9°$ kA	0.13 $\angle 16.5°$ kA	0.1286 $\angle 15.3°$ kA	0.13 $\angle 16.5°$ kA
pf_s	0.745 lagging	0.984 leading	0.987 leading	0.984 leading
P_g	53.2 MW	52.7 MW	52.15 MW	52.58 MW

Comments

We find from the above example that the results obtained by the nominal-π method and the approximation (5.28b) are practically the same and are very close to those obtained by exact calculations (part (c)). On the other hand the results obtained by the short line approximation are in considerable error. Therefore, for a line of this length (about 300 km), it is sufficiently accurate to use the nominal-π (or approximation (5.28b)) which results in considerable saving in computational effort.

5.9 POWER FLOW THROUGH A TRANSMISSION LINE

So far the transmission line performance equation was presented in the form of voltage and current relationships between sending-and receiving-ends. Since loads are more often expressed in terms of real (watts/kW) and reactive (VARs/kVAR) power, it is convenient to deal with transmission line equations in the form of sending- and receiving-end complex power and voltages. While the problem of flow of power in a general network will be treated in the next chapter, the principles involved are illustrated here through a single transmission line (2-node/2-bus system) as shown in Fig. 5.17.

Fig. 5.17 A two-bus system

Let us take the receiving-end voltage as a reference phasor ($V_R = |V_R| \angle 0°$) and let the sending-end voltage lead it by an angle δ ($V_S = |V_S| \angle \delta$). The angle δ is known as the torque angle whose significance has been explained in Chapter 4 and will further be taken up in Chapter 12 while dealing with the problem of stability.

The complex power leaving the receiving-end and entering the sending-end of the transmission line can be expressed as (on per phase basis)

$$S_R = P_R + jQ_R = V_R I_R^* \qquad (5.54)$$

$$S_S = P_S + jQ_S = V_S I_S^* \qquad (5.55)$$

Receiving- and sending-end currents can, however, be expressed in terms of receiving- and sending-end voltages [see Eq. (5.1)] as

$$I_R = \frac{1}{B} V_S - \frac{A}{B} V_R \qquad (5.56)$$

$$I_S = \frac{D}{B} V_S - \frac{1}{B} V_R \qquad (5.57)$$

Let A, B, D, the transmission line constants, be written as

$$A = |A| \angle \alpha, \; B = |B| \angle \beta, \; D = |D| \angle \alpha \text{ (since } A = D)$$

Therefore, we can write

$$I_R = \left|\frac{1}{B}\right| |V_S| \angle(\delta - \beta) - \left|\frac{A}{B}\right| |V_R| \angle(\alpha - \beta)$$

$$I_S = \left|\frac{D}{B}\right| |V_S| \angle(\alpha + \delta - \beta) - \left|\frac{1}{B}\right| |V_R| \angle -\beta$$

Substituting for I_R in Eq. (5.54) we get

$$S_R = |V_R| \angle 0 \left[\left| \frac{1}{B} \right| |V_S| \angle(\beta - \delta) - \left| \frac{A}{B} \right| |V_R| \angle(\beta - \alpha) \right]$$

$$= \frac{|V_S||V_R|}{|B|} \angle(\beta - \delta) - \left| \frac{A}{B} \right| |V_R|^2 \angle(\beta - \alpha) \qquad (5.58)$$

Similarly,
$$S_S = \left| \frac{D}{B} \right| |V_S|^2 \angle(\beta - \alpha) - \frac{|V_S||V_R|}{|B|} \angle(\beta + \delta) \qquad (5.59)$$

In the above equations S_R and S_S are per phase complex voltamperes, while V_R and V_S are expressed in per phase volts. If V_R and V_S are expressed in kV line, then the three-phase receiving-end complex power is given by

$$S_R \text{ (3-phase VA)} = 3 \left\{ \frac{|V_S||V_R| \times 10^6}{\sqrt{3} \times \sqrt{3} |B|} \angle(\beta - \delta) - \left| \frac{A}{B} \right| \frac{|V_R|^2 \times 10^6}{3} \angle(\beta - \alpha) \right\}$$

$$S_R \text{ (3-phase MVA)} = \frac{|V_S||V_R|}{|B|} \angle(\beta - \delta) - \left| \frac{A}{B} \right| |V_R|^2 \angle(\beta - \alpha) \qquad (5.60)$$

This indeed is the same as Eq. (5.58). The same result holds for S_S. Thus we see that Eqs. (5.58) and (5.59) give the three-phase MVA if V_S and V_R are expressed in kV line.

If Eq. (5.58) is expressed in real and imaginary parts, we can write the real and reactive powers at the receiving-end as

$$P_R = \frac{|V_S||V_R|}{|B|} \cos(\beta - \delta) - \left| \frac{A}{B} \right| |V_R|^2 \cos(\beta - \alpha) \qquad (5.61)$$

$$Q_R = \frac{|V_S||V_R|}{|B|} \sin(\beta - \delta) - \left| \frac{A}{B} \right| |V_R|^2 \sin(\beta - \alpha) \qquad (5.62)$$

Similarly, the real and reactive powers at sending-end are

$$P_S = \left| \frac{D}{B} \right| |V_S|^2 \cos(\beta - \alpha) - \frac{|V_S||V_R|}{|B|} \cos(\beta + \delta) \qquad (5.63)$$

$$Q_S = \left| \frac{D}{B} \right| |V_S|^2 \sin(\beta - \alpha) - \frac{|V_S||V_R|}{|B|} \sin(\beta + \delta) \qquad (5.64)$$

It is easy to see from Eq. (5.61) that the received power P_R will be maximum at

$$\delta = \beta$$

such that

$$P_R \text{ (max)} = \frac{|V_S||V_R|}{|B|} - \frac{|A||V_R|^2}{|B|} \cos(\beta - \alpha) \qquad (5.65)$$

The corresponding Q_R (at max P_R) is

$$Q_R = -\frac{|A||V_R|^2}{|B|} \sin(\beta - \alpha)$$

Thus the load must draw this much leading MVAR in order to receive the maximum real power.

Consider now the special case of a short line with a series impedance Z. Now

$$A = D = 1 \angle 0; \ B = Z = |Z| \ \angle \theta$$

Substituting these in Eqs. (5.61) to (5.64), we get the simplified results for the short line as

$$P_R = \frac{|V_S||V_R|}{|Z|} \cos(\theta - \delta) - \frac{|V_R|^2}{|Z|} \cos \theta \tag{5.66}$$

$$Q_R = \frac{|V_S||V_R|}{|Z|} \sin(\theta - \delta) - \frac{|V_R|^2}{|Z|} \sin \theta \tag{5.67}$$

for the receiving-end and for the sending-end

$$P_S = \frac{|V_S|^2}{|Z|} \cos \theta - \frac{|V_S||V_R|}{|Z|} \cos(\theta + \delta) \tag{5.68}$$

$$Q_S = \frac{|V_S|^2}{|Z|} \sin \theta - \frac{|V_S||V_R|}{|Z|} \sin(\theta + \delta) \tag{5.69}$$

The above short line equation will also apply for a long line when the line is replaced by its equivalent-π (or nominal-π) and the shunt admittances are lumped with the receiving-end load and sending-end generation. In fact, this technique is always used in the load flow problem to be treated in the next chapter.

From Eq.(5.66), the maximum receiving-end power is received, when $\delta = \theta$

so that P_R (max) $= \dfrac{|V_S||V_R|}{|Z|} - \dfrac{|V_R|^2}{|Z|} \cos \theta$

Now $\cos \theta = R/|Z|$,

$$\therefore \qquad P_R \text{ (max)} = \frac{|V_S||V_R|}{|Z|} - \frac{|V_R|^2}{|Z|^2} R \tag{5.70}$$

Normally the resistance of a transmission line is small compared to its reactance (since it is necessary to maintain a high efficiency of transmission), so that $\theta = \tan^{-1} X/R \simeq 90°$; where $Z = R + jX$. The receiving-end Eqs. (5.66) and (5.67) can then be approximated as

$$P_R = \frac{|V_S||V_R|}{X} \sin \delta \tag{5.71}$$

$$Q_R = \frac{|V_S||V_R|}{X} \cos \delta - \frac{|V_R|^2}{X} \tag{5.72}$$

Equation (5.72) can be further simplified by assuming cos $\delta \simeq 1$, since δ is normally small[*]. Thus

$$Q_R = \frac{|V_R|}{X}(|V_S| - |V_R|) \tag{5.73}$$

Let $|V_S| - |V_R| = |\Delta V|$, the magnitude of voltage drop across the transmission line.

$$\therefore \qquad Q_R = \frac{|V_R|}{X}|\Delta V| \tag{5.74}$$

Several important conclusions that easily follow from Eqs. (5.71) to (5.74) are enumerated below:

1. For $R \simeq 0$ (which is a valid approximation for a transmission line) the real power transferred to the receiving-end is proportional to sin $\delta (\simeq \delta$ for small values of δ), while the reactive power is proportional to the magnitude of the voltage drop across the line.

2. The real power received is maximum for $\delta = 90°$ and has a value $|V_S||V_R|/X$. Of course, δ is restricted to values well below $90°$ from considerations of stability to be discussed in Chapter 12.

3. Maximum real power transferred for a given line (fixed X) can be increased by raising its voltage level. It is from this consideration that voltage levels are being progressively pushed up to transmit larger chunks of power over longer distances warranted by large size generating stations.

   For very long lines voltage level cannot be raised beyond the limits placed by present-day high voltage technology. To increase power transmitted in such cases, the only choice is to reduce the line reactance. This is accomplished by adding series capacitors in the line. This idea will be pursued further in Chapter 12. Series capacitors would of course increase the severity of line over voltages under switching conditions.

4. As said in 1 above, the VARs (lagging reactive power) delivered by a line is proportional to the line voltage drop and is independent of δ. Therefore, in a transmission system if the VARs demand of the load is large, the voltage profile at that point tends to sag rather sharply. To maintain a desired voltage profile, the VARs demand of the load must be met locally by employing positive VAR generators (condensers). This will be discussed at length in Sec. 5.10.

A somewhat more accurate yet approximate result expressing line voltage drop in terms of active and reactive powers can be written directly from Eq. (5.5), i.e.

$$|\Delta V| = |I_R| R \cos \phi + |I_R| X \sin \phi$$
$$= \frac{|V_R||I_R| R \cos \phi + |V_R||I_R| X \sin \phi}{|V_R|}$$

---

[*]Small δ is necessary from considerations of system stability which will be discussed at length in Chapter 12.

$$= \frac{RP_R + XQ_R}{|V_R|} \tag{5.75}$$

This result reduces to that of Eq. (5.74) if $R = 0$.

Example 5.8

An interconnector cable links generating stations 1 and 2 as shown in Fig. 5.18. The desired voltage profile is flat, i.e. $|V_1| = |V_2| = 1$ pu. The total demands at the two buses are

$$S_{D1} = 15 + j5 \text{ pu}$$
$$S_{D2} = 25 + j15 \text{ pu}$$

The station loads are equalized by the flow of power in the cable. Estimate the torque angle and the station power factors: (a) for cable $Z = 0 + j0.05$ pu, and (b) for cable $Z = 0.005 + j0.05$ pu. It is given that generator G_1 can generate a maximum of 20.0 pu real power.

Solution The powers at the various points in the fundamental (two-bus) system are defined in Fig. 5.18(a).

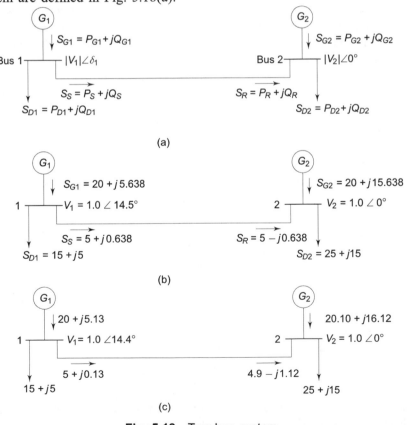

(a)

(b)

(c)

Fig. 5.18 Two-bus system

Case (a):   Cable impedance = $j0.05$ pu.

Since cable resistance is zero, there is no real power loss in the cable. Hence

$$P_{G1} + P_{G2} = P_{D1} + P_{D2} = 40 \text{ pu}$$

For equalization of station loads,

$$P_{G1} = P_{G2} = 20 \text{ pu}$$

The voltage of bus 2 is taken as reference, i.e. $V_2 \angle 0°$ and voltage of bus 1 is $V_1 \angle \delta_1$. Further, for flat voltage profile $|V_1| = |V_2| = 1$.

Real power flow from bus 1 to bus 2 is obtained from Eq. (5.68) by recognizing that since $R = 0$, $\theta = 90°$.
Hence

$$P_S = P_R = \frac{|V_1||V_2|}{X} \sin \delta_1$$

$$5 = \frac{1 \times 1}{0.05} \sin \delta_1$$

or

$$\delta_1 = 14.5°$$

∴

$$V_1 = 1 \angle 14.5°$$

From Eq. (5.69)

$$Q_S = \frac{|V_1|^2}{X} - \frac{|V_1||V_2|}{X} \cos \delta_1$$

$$= \frac{1}{0.05} - \frac{1}{0.05} \times 0.968 = 0.638 \text{ pu}$$

From Eq. (5.67)

$$Q_R = \frac{|V_1||V_2|}{X} \cos \delta_1 - \frac{|V_1|^2}{X} = -Q_S = -0.638 \text{ pu}$$

Reactive power loss[*] in the cable is

$$Q_L = Q_S - Q_R = 2Q_S = 1.276 \text{ pu}$$

Total load on station 1 = $(15 + j5) + (5 + j0.638)$

$$= 20 + j5.638$$

Power factor at station 1 = $\cos \left(\tan^{-1} \dfrac{5.638}{20} \right)$ = 0.963 lagging

Total load on station 2 = $(25 + j15) - (5 - j0.638)$

$$= 20 + j15.638$$

---

[*]Reactive power loss can also be computed as $|I|^2 X = \dfrac{5^2 + (0.638)^2}{1} \times 0.05 = 1.27 \text{ pu.}$

Power factor at station $2 = \cos\left(\tan^{-1}\dfrac{15.638}{20}\right) = 0.788$ lagging

The station loads, load demands, and line flows are shown in Fig. 5.18(b). It may be noted that to maintain a flat voltage profile, the generators are required to supply reactive powers $Q_{G1} = 5.638$ and $Q_{G2}=15.638$, respectively. *Case (b)*: Cable impedance $= 0.005 + j0.05 = 0.0502 \angle 84.3°$ pu. In this case the cable resistance causes real power loss which is not known a priori. The real load flow is thus not obvious as was in the case of $R = 0$. We specify the generation at station 1 as

$$P_{G1} = 20 \text{ pu}$$

The consideration for fixing this generation is economic as we shall see in Chapter 7.

The generation at station 2 will be 20 pu plus the cable loss. The unknown variables in the problem are

$$P_{G2},\ \delta_1,\ Q_{G1},\ Q_{G2}$$

Let us now examine as to how many system equations can be formed.

From Eqs. (5.68) and (5.69)

$$P_{G1} - P_{D1} = P_S = \frac{|V_1|^2}{|Z|}\cos\theta - \frac{|V_1||V_2|}{|Z|}\cos(\theta + \delta_1)$$

$$5 = \frac{1}{0.0502}\cos 84.3° - \frac{1}{0.0502}\cos(84.3° + \delta_1) \qquad \text{(i)}$$

$$Q_{G1} - Q_{D1} = Q_S = \frac{|V_1|^2}{|Z|}\sin\theta - \frac{|V_1||V_2|}{|Z|}\sin(\theta + \delta_1)$$

$$Q_{G1} - 5 = \frac{1}{0.0502}\sin 84.3° - \frac{1}{0.0502}\sin(84.3° + \delta_1) \qquad \text{(ii)}$$

From Eqs. (5.66) and (5.67)

$$P_{D2} - P_{G2} = P_R = \frac{|V_1||V_2|}{|Z|}\cos(\theta - \delta_1) - \frac{|V_1|^2}{|Z|}\cos\theta$$

$$25 - P_{G2} = \frac{1}{0.0502}\cos(84.3° - \delta_1) - \frac{1}{0.0502}\cos 84.3° \qquad \text{(iii)}$$

$$Q_{D2} - Q_{G2} = Q_R = \frac{|V_1||V_2|}{|Z|}\sin(\theta - \delta_1) - \frac{|V_1|^2}{|Z|}\sin\theta$$

$$15 - Q_{G2} = \frac{1}{0.0502}\sin(84.3° - \delta_1) - \frac{1}{0.0502}\sin 84.3° \qquad \text{(v)}$$

Thus we have four equations, Eqs. (i) to (iv), in four unknowns P_{G2}, δ_1, Q_{G1}, Q_{G2}. Even though these are non-linear algebraic equations, solution is possible in this case. Solving Eq. (i) for δ_1, we have

$$\delta_1 = 14.4°$$

Substituting δ_1 in Eqs. (ii), (iii) and (iv), we get

$$Q_{G1} = 5.13, \ Q_{G2} = 16.12, \ P_{G2} = 20.10$$

The flow of real and reactive powers for this case is shown in Fig. 5.18 (c).

It may be noted that the real power loss of 0.1 pu is supplied by $G_2(P_{G2} = 20.10)$.

The above presented problem is a two-bus load flow problem. Explicit solution is always possible in a two-bus case. The reader should try the case when

$$Q_{G2} = j10 \text{ and } |V_2| = ?$$

The general load flow problem will be taken up in Chapter 6. It will be seen that explicit solution is not possible in the general case and iterative techniques have to be resorted to.

Example 5.9

A 275 kV transmission line has the following line constants:

$$A = 0.85 \ \angle 5°; \ B = 200 \ \angle 75°$$

(a) Determine the power at unity power factor that can be received if the voltage profile at each end is to be maintained at 275 kV.

(b) What type and rating of compensation equipment would be required if the load is 150 MW at unity power factor with the same voltage profile as in part (a).

(c) With the load as in part (b), what would be the receiving-end voltage if the compensation equipment is not installed?

Solution (a) Given $|V_S| = |V_R| = 275$ kV; $\alpha = 5°$, $\beta = 75°$. Since the power is received at unity power factor,

$$Q_R = 0$$

Substituting these values in Eq. (5.62), we can write

$$0 = \frac{275 \times 275}{200} \sin (75° - \delta) - \frac{0.85}{200} \times (275)^2 \sin (75° - 5°)$$

$$0 = 378 \sin (75° - \delta) - 302$$

which gives

$$\delta = 22°$$

From Eq. (5.61)

$$P_R = \frac{275 \times 275}{200} \cos (75° - 22°) - \frac{0.85}{200} \times (275)^2 \cos 70°$$

$$= 227.6 - 109.9 = \textbf{117.7 MW}$$

(b)   Now $|V_S| = |V_R| = 275$ kV

Power demanded by load $= 150$ MW at UPF

$\therefore$ $\qquad P_D = P_R = 150$ MW; $Q_D = 0$

From Eq. (5.61)

$$150 = \frac{275 \times 275}{200} \cos (75° - \delta) - \frac{0.85}{200} \times (275)^2 \cos 70°$$

$$150 = 378 \cos (75° - \delta) - 110$$

or $\qquad \qquad \delta = 28.46°$

From Eq. (5.62)

$$Q_R = \frac{275 \times 275}{200} \sin (75° - 28.46°) - \frac{0.85}{200} \times (275)^2 \sin 70°$$

$$= 274.46 - 302 = -27.56 \text{ MVAR}$$

Thus in order to maintain 275 kV at a receiving-end, $Q_R = -27.56$ MVAR must be drawn along with the real power of $P_R = 150$ MW. The load being 150 MW at unity power factor, i.e. $Q_D = 0$, compensation equipment must be installed at the receiving-end. With reference to Fig. 5.19, we have

$\qquad -27.56 + Q_C = 0$

or $\qquad \qquad Q_C = +27.56$ MVAR

i.e. the compensation equipment must feed positive VARs into the line. See subsection 5.10 for a more detailed explanation.

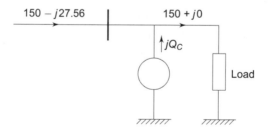

Fig. 5.19

(c)   Since no compensation equipment is provided

$\qquad \qquad P_R = 150$ MW, $Q_R = 0$

Now,

$\qquad \qquad |V_S| = 275$ kV, $|V_R| = ?$

Substituting this data in Eqs. (5.61) and (5.62), we have

$$150 = \frac{275|V_R|}{200} \cos (75° - \delta) - \frac{0.85}{200}|V_R|^2 \cos 70° \qquad \qquad \text{(i)}$$

$$0 = \frac{275|V_R|}{200} \sin (75° - \delta) - \frac{0.85}{200}|V_R|^2 \sin 70° \qquad \text{(ii)}$$

From Eq. (ii), we get

$$\sin (75° - \delta) = 0.0029|V_R|$$

$$\therefore \quad \cos (75° - \delta) = (1 - (0.0029)^2|V_R|^2)^{1/2}$$

Substituting in Eq. (i), we obtain

$$150 = 1.375 \, |V_R| \, (1 - (0.0029)^2 \, |V_R|^2)^{1/2} - 0.00145|V_R|^2$$

Solving the quadratic and retaining the higher value of $|V_R|$, we obtain

$$|V_R| = 244.9 \text{ kV}$$

Note: The second and lower value solution of $|V_R|$ though feasible, is impractical as it corresponds to abnormally low voltage and efficiency.

It is to be observed from the results of this problem that larger power can be transmitted over a line with a fixed voltage profile by installing compensation equipment at the receiving-end capable of feeding positive VARs into the line.

Circle Diagrams

It has been shown above that the flow of active and reactive power over a transmission line can be handled computationally. It will now be shown that the locus of complex sending- and receiving-end power is a circle. Since circles are convenient to draw, the circle diagrams are a useful aid to visualize the load flow problem over a single transmission.

The expressions for complex number receiving- and sending-end powers are reproduced below from Eqs. (5.58) and (5.59).

$$S_R = - \left|\frac{A}{B}\right| \, |V_R|^2 \, \angle(\beta - \alpha) + \frac{|V_S||V_R|}{|B|} \angle(\beta - \delta) \qquad (5.58)$$

$$S_S = \left|\frac{D}{B}\right| \, |V_S|^2 \, \angle(\beta - \alpha) - \frac{|V_S||V_R|}{|B|} \angle(\beta + \delta) \qquad (5.59)$$

The units for S_R and S_S are MVA (three-phase) with voltages in KV line. As per the above equations, S_R and S_S are each composed of two phasor components—one a constant phasor and the other a phasor of fixed magnitude but variable angle. The loci for S_R and S_S would, therefore, be circles drawn from the tip of constant phasors as centres.

It follows from Eq. (5.58) that the centre of receiving-end circle is located at the tip of the phasor.

$$- \left|\frac{A}{B}\right| |V_R|^2 \, \angle(\beta - \alpha) \qquad (5.76)$$

in polar coordinates or in terms of rectangular coordinates,
Horizontal coordinate of the centre

$$= - \left|\frac{A}{B}\right| |V_R|^2 \, \cos (\beta - \alpha) \text{ MW} \qquad (5.77)$$

Vertical coordinate of the centre

$$= -\left|\frac{A}{B}\right||V_R|^2 \sin (\beta - \alpha) \text{ MVAR}$$

The radius of the receiving-end circle is

$$\frac{|V_S||V_R|}{|B|}\text{MVA} \tag{5.78}$$

The receiving-end circle diagram is drawn in Fig. 5.20. The centre is located by drawing OC_R at an angle $(\beta - \alpha)$ in the positive direction from the negative MW-axis. From the centre C_R the receiving-end circle is drawn with the radius $|V_S| |V_R|/|B|$. The operating point M is located on the circle by means of the received real power P_R. The corresponding Q_R (or θ_R) can be immediately read from the circle diagram. The torque angle δ can be read in accordance with the positive direction indicated from the reference line.

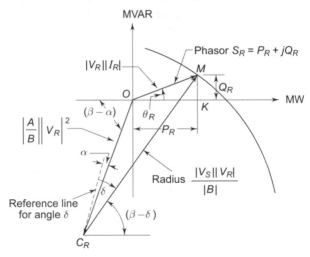

Fig. 5.20 Receiving-end circle diagram

For constant $|V_R|$, the centre C_R remains fixed and concentric circles result for varying $|V_S|$. However, for the case of constant $|V_S|$ and varying $|V_R|$ the centres of circles move along the line OC_R and have radii in accordance to $|V_S| |V_R|/|B|$.

Similarly, it follows from Eq. (5.59) that the centre of the sending-end circle is located at the tip of the phasor

$$\left|\frac{D}{B}\right||V_S|^2 \angle(\beta - \alpha) \tag{5.79}$$

in the polar coordinates or in terms of rectangular coordinates.

Horizontal coordinate of the centre

$$= \left|\frac{D}{B}\right||V_S|^2 \cos (\beta - \alpha) \text{ MW} \tag{5.80}$$

Vertical coordinate of the centre

$$= \left|\frac{D}{B}\right| |V_S|^2 \sin (\beta - \alpha) \text{ MVAR}$$

The radius of the sending-end circle is

$$\frac{|V_S||V_R|}{|B|} \tag{5.81}$$

The sending-end circle diagram is shown in Fig. 5.21. The centre is located by drawing OC_S at angle $(\beta - \alpha)$ from the positive MW-axis. From the centre the

sending-end circle is drawn with a radius $\dfrac{|V_S||V_R|}{|B|}$ (same as in the case of

receiving-end). The operating point N is located by measuring the torque angle δ (as read from the receiving-end circle diagram) in the direction indicated from the reference line.

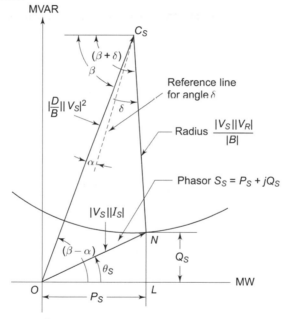

Fig. 5.21 Sending-end circle diagram

For constant $|V_S|$ the centre C_S remains fixed and concentric circles result for varying $|V_R|$. However, if $|V_R|$ is fixed and $|V_S|$ varies, the centres of the circles move along the line OC_S and have radii in accordance to $|V_S|\,|V_R|/|B|$.

For the case of a short line with a series impedance $|Z| \angle \theta$, the simplified circle diagrams can be easily drawn by recognizing

$$|A| = |D| = 1, \ \alpha = 0$$
$$|B| = |Z|, \ \beta = \theta$$

The corresponding receiving- and sending-end circle diagrams have been drawn in Figs 5.22 and 5.23.

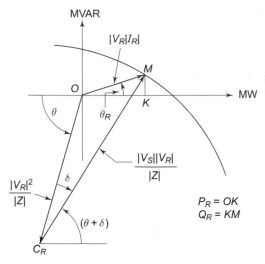

Fig. 5.22 Receiving-end circle diagram for a short line

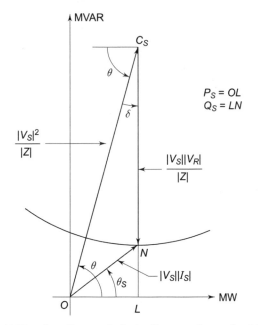

Fig. 5.23 Sending-end circle diagram for a short line

The use of circle diagrams is illustrated by means of the two examples given below:

Example 5.10

A 50 Hz, three-phase, 275 kV, 400 km transmission line has the following parameters:

Resistance = 0.035 Ω/km per phase
Inductance = 1.1 mH/km per phase
Capacitance = 0.012 μF/km per phase

If the line is supplied at 275 kV, determine the MVA rating of a shunt reactor having negligible losses that would be required to maintain 275 kV at the receiving-end when the line is delivering no load. Use nominal-π method.

Solution

$$R = 0.035 \times 400 = 14\ \Omega$$

$$X = 314 \times 1.1 \times 10^{-3} \times 400 = 138.2\ \Omega$$

$$Z = 14 + j138 = 138.7\ \angle 84.2°\ \Omega$$

$$Y = 314 \times 0.012 \times 10^{-6} \times 400\ \angle 90° - 1.507 \times 10^{-3}\ \angle 90°\ \mho$$

$$A = \left(1 + \frac{1}{2}YZ\right) = 1 + \frac{1}{2} \times 1.507 \times 10^{-3} \times 138.7\ \angle 174.2°$$

$$= (0.896 + j0.0106) = 0.896\ \angle 0.7°$$

$$B = Z = 138.7\ \angle 84.2°$$

$$|V_S| = 275\ \text{kV},\ |V_R| = 275\ \text{kV}$$

Radius of receiving-end circle $= \dfrac{|V_S||V_R|}{|B|} = \dfrac{275 \times 275}{138.7} = 545.2$ MVA

Location of the centre of receiving-end circle,

$$\left|\frac{A}{B}\right||V_R|^2 = \frac{275 \times 275 \times 0.896}{138.7} = 488.5\ \text{MVA}$$

$$\angle(\beta - \alpha) = 84.2° - 0.7° = 83.5°$$

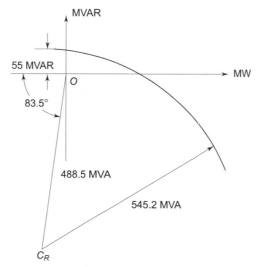

Fig. 5.24 Circle diagram for Example 5.10

From the circle diagram of Fig. 5.24, + 55 MVAR must be drawn from the receiving-end of the line in order to maintain a voltage of 275 kV. Thus rating of shunt reactor needed = 55 MVA.

Example 5.11

A 275 kV, three-phase line has the following line parameters:

$$A = 0.93 \ \angle 1.5°, \ B = 115 \ \angle 77°$$

If the receiving-end voltage is 275 kV, determine:

(a) The sending-end voltage required if a load of 250 MW at 0.85 lagging pf is being delivered at the receiving-end.

(b) The maximum power that can be delivered if the sending-end voltage is held at 295 kV.

(c) The additional MVA that has to be provided at the receiving-end when delivering 400 MVA at 0.8 lagging pf, the supply voltage being maintained at 295 kV.

Solution In Fig. 5.25 the centre of the receiving-end circle is located at

$$\left|\frac{A}{B}\right| |V_R|^2 = \frac{275 \times 275 \times 0.93}{115} = 611.6 \text{ MVA}$$

$$\cos^{-1} 0.85 = 31.8°$$

$$\angle(\beta - \alpha) = 77° - 1.5° = 75.5°$$

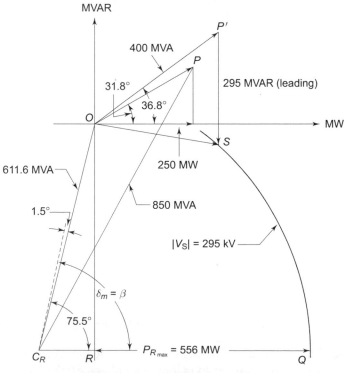

Fig. 5.25 Circle diagram for Example 5.11

(a) Locate OP corresponding to the receiving-end load of 250 MW at 0.85 lagging pf (+ 31.8°). Then

$$C_R P = 850 = \frac{|V_S||V_R|}{|B|} = \frac{275|V_S|}{115}$$

∴ $|V_S| = 355.5$ kV

(b)   Given $|V_S| = 295$ kV.

Radius of circle diagram $= \dfrac{295 \times 275}{115} = 705.4$ MVA

Drawing the receiving-end circle (see Fig. 5.25) and the line $C_R Q$ parallel to the MW-axis, we read

$$P_{R\ max} = RQ = 556 \text{ MW}$$

(c)   Locate OP' corresponding to 400 MVA at 0.8 lagging pf (+ 36.8°). Draw $P'S$ parallel to MVAR-axis to cut the circle drawn in part (b) at S. For the specified voltage profile, the line load should be OS. Therefore, additional MVA to be drawn from the line is

$$P'S = 295 \text{ MVAR or } 295 \text{ MVA leading}$$

5.10 METHODS OF VOLTAGE CONTROL

Practically each equipment used in power system are rated for a certain voltage with a permissible band of voltage variations. Voltage at various buses must, therefore, be controlled within a specified regulation figure. This article will discuss the two methods by means of which voltage at a bus can be controlled.

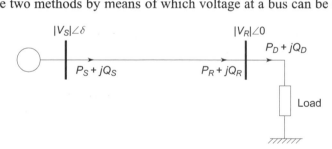

Fig. 5.26 A two-bus system

Consider the two-bus system shown in Fig. 5.26 (already exemplified in Sec. 5.9). For the sake of simplicity let the line be characterized by a series reactance (i.e. it has negligible resistance). Further, since the torque angle δ is small under practical conditions, real and reactive powers delivered by the line for fixed sending-end voltage $|V_S|$ and a specified receiving-end voltage $|V_R^S|$ can be written as below from Eqs. (5.71) and (5.73).

$$P_R = \frac{|V_S||V_R^S|}{X} \sin \delta \qquad\qquad (5.82)$$

$$Q_R^S = \frac{|V_R^S|}{X}(|V_S| - |V_R^S|) \tag{5.83}$$

Equation (5.83) upon quadratic solution[*] can also be written as

$$|V_R^S| = \frac{1}{2}|V_S| + \frac{1}{2}|V_S|\,(1 - 4XQ_R^S/|V_S|^2)^{1/2} \tag{5.84}$$

Since the real power demanded by the load must be delivered by the line,

$$P_R = P_D$$

Varying real power demand P_D is met by consequent changes in the torque angle δ.

It is, however, to be noted that the received reactive power of the line must remain fixed at Q_R^S as given by Eq. (5.83) for fixed $|V_S|$ and specified $|V_R^S|$. The line would, therefore, operate with specified receiving-end voltage for only one value of Q_D given by

$$Q_D = Q_R^S$$

Practical loads are generally lagging in nature and are such that the VAR demand Q_D may exceed Q_R^S. It easily follows from Eq. (5.83) that for $Q_D > Q_R^S$ the receiving-end voltage must change from the specified value $|V_R^S|$ to some value $|V_R|$ to meet the demanded VARs. Thus

$$Q_D = Q_R = \frac{|V_R|}{X}\,(|V_S| - |V_R|) \text{ for } (Q_D > Q_R^S)$$

The modified $|V_R|$ is then given by

$$|V_R| = \frac{1}{2}|V_S| + \frac{1}{2}|V_S|\,(1 - 4XQ_R/|V_S|^2)^{1/2} \tag{5.85}$$

Comparison of Eqs. (5.84) and (5.85) reveals that for $Q_D = Q_R = Q_R^S$, the receiving-end voltage is $|V_R^S|$, but for $Q_D = Q_R > Q_R^S$,

$$|V_R| < |V_R^S|$$

Thus a VAR demand larger than Q_R^S is met by a consequent fall in receiving-end voltage from the specified value. Similarly, if the VAR demand is less than Q_R^S, it follows that

$$|V_R| > |V_R^S|$$

Indeed, under light load conditions, the charging capacitance of the line may cause the VAR demand to become negative resulting in the receiving-end voltage exceeding the sending-end voltage (this is the Ferranti effect already illustrated in Section 5.6).

In order to regulate the line voltage under varying demands of VARs, the two methods discussed below are employed.

---

[*]Negative sign in the quadratic solution is rejected because otherwise the solution would not match the specified receiving-end voltage which is only slightly less than the sending-end voltage (the difference is less than 12%).

Reactive Power Injection

It follows from the above discussion that in order to keep the receiving-end voltage at a specified value $|V_R^S|$, a fixed amount of VARs (Q_R^S) must be drawn from the line[*]. To accomplish this under conditions of a varying VAR demand Q_D, a local VAR generator (controlled reactive power source/compensating equipment) must be used as shown in Fig. 5.27. The VAR balance equation at the receiving-end is now

$$Q_R^S + Q_C = Q_D$$

Fluctuations in Q_D are absorbed by the *local VAR generator* Q_C such that the VARs drawn from the line remain fixed at Q_R^S. The receiving-end voltage would thus remain fixed at $|V_R^S|$ (this of course assumes a fixed sending-end voltage $|V_S|$). Local VAR compensation can, in fact, be made automatic by using the signal from the VAR meter installed at the receiving-end of the line.

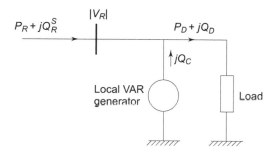

Fig. 5.27 Use of local VAR generator at the load bus

Two types of VAR generators are employed in practice—*static type* and *rotating type*. These are discussed below.

Static VAR generator

It is nothing but a bank of three-phase static capacitors and/or inductors. With reference to Fig. 5.28, if $|V_R|$ is in line kV, and X_C is the per phase capacitive reactance of the capacitor bank on an equivalent star basis, the expression for the VARs fed into the line can be derived as under.

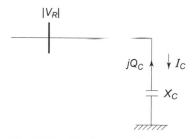

Fig. 5.28 Static capacitor bank

$$I_C = j \frac{|V_R|}{\sqrt{3}\, X_C} \quad \text{kA}$$

---

[*]Of course, since $|V_R^S|$ is specified within a band, Q_R^S may vary within a corresponding band.

$$jQ_C \text{ (3-phase)} = 3 \frac{|V_R|}{\sqrt{3}} (- I_C^*)$$

$$= j3 \times \frac{|V_R|}{\sqrt{3}} \times \frac{|V_R|}{\sqrt{3} X_C} \text{ MVA}$$

$$\therefore \quad Q_C \text{ (3-phase)} = \frac{|V_R|^2}{X_C} \text{ MVAR} \qquad (5.86)$$

If inductors are employed instead, VARs fed into the line are

$$Q_L(\text{3-phase}) = - \frac{|V_R|^2}{X_L} \text{MVAR} \qquad (5.87)$$

Under heavy load conditions, when positive VARs are needed, capacitor banks are employed; while under light load conditions, when negative VARs are needed, inductor banks are switched on.

The following observations can be made for static VAR generators.

(i) Capacitor and inductor banks can be switched on in steps. However, stepless (smooth) VAR control can now be achieved using SCR (Silicon Controlled Rectifier) circuitry.

(ii) Since Q_C is proportional to the square of terminal voltage, for a given capacitor bank, their effectiveness tends to decrease as the voltage sags under full load conditions.

(iii) If the system voltage contains appreciable harmonics, the fifth being the most troublesome, the capacitors may be overloaded considerably.

(iv) Capacitors act as short circuit when switched on.

(v) There is a possibility of series resonance with the line inductance particularly at harmonic frequencies.

Rotating VAR generator

It is nothing but a synchronous motor running at no-load and having excitation adjustable over a wide range. It feeds positive VARs into the line under overexcited conditions and feeds negative VARs when underexcited. A machine thus running is called a *synchronous condenser*.

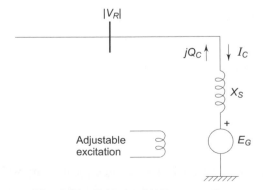

Fig. 5.29 Rotating VAR generation

Figure 5.29 shows a synchronous motor connected to the receiving-end bus bars and running at no load. Since the motor draws negligible real power from the bus bars, E_G and V_R are nearly in phase. X_S is the synchronous reactance of the motor which is assumed to have negligible resistance. If $|E_G|$ and $|V_R|$ are in line kV, we have

$$I_C = \frac{(|V_R| - |E_G|)\angle 0°}{\sqrt{3} \times jXs} \text{ kA}$$

$$jQ_C = 3\frac{|V_R|\angle 0°}{\sqrt{3}} (-I_C^*)$$

$$= 3\frac{|V_R|}{\sqrt{3}}\left(-\frac{|V_R| - |E_G|}{-jX_S\sqrt{3}}\right)$$

$$= j|V_R|(|E_G| - |V_R|)/X_S \text{ MVA}$$

$$\therefore \qquad Q_C = |V_R| (|E_G| - |V_R|)/X_S \text{ MVAR} \qquad (5.88)$$

It immediately follows from the above relationship that the machine feeds positive VARs into the line when $|E_G| > |V_R|$ (overexcited case) and injects negative VARs if $|E_G| < |V_R|$ (underexcited case). VARs are easily and continuously adjustable by adjusting machine excitation which controls $|E_G|$.

In contrast to static VAR generators, the following observations are made in respect of rotating VAR generators.

(i) These can provide both positive and negative VARs which are continuously adjustable.

(ii) VAR injection at a given excitation is less sensitive to changes in bus voltage. As $|V_R|$ decreases and $(|E_G| - |V_R|)$ increases with consequent smaller reduction in Q_C compared to the case of static capacitors.

From the observations made above in respect of static and rotating VAR generators, it seems that rotating VAR generators would be preferred. However, economic considerations, installation and maintenance problems limit their practical use to such buses in the system where a large amount of VAR injection is needed.

Control by Transformers

The VAR injection method discussed above lacks the flexibility and economy of voltage control by transformer tap changing. The transformer tap changing is obviously limited to a narrow range of voltage control. If the voltage correction needed exceeds this range, tap changing is used in conjunction with the VAR injection method.

Receiving-end voltage which tends to sag owing to VARs demanded by the load, can be raised by simultaneously changing the taps of sending-and receiving-end transformers. Such tap changes must be made 'on-load' and can be done either manually or automatically, the transformer being called a Tap Changing Under Load (TCUL) transformer.

Consider the operation of a transmission line with a tap changing transformer at each end as shown in Fig. 5.30. Let t_S and t_R be the fractions of the nominal transformation ratios, i.e. the tap ratio/nominal ratio. For example, a transformer with nominal ratio 3.3 kV/11 kV when tapped to give 12 kV with 3.3 kV input has $t_S = 12/11 = 1.09$.

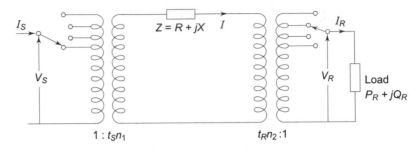

Fig. 5.30 Transmission line with tap changing transformer at each end

With reference to Fig. 5.30 let the impedances of the transformer be lumped in Z along with the line impedance. To compensate for voltage in the line and transformers, let the transformer taps be set at off nominal values, t_S and t_R. With reference to the circuit shown, we have

$$t_S n_1 V_S = t_R n_2 V_R + IZ \qquad (5.89)$$

From Eq. (5.75) the voltage drop referred to the high voltage side is given by

$$|\Delta V| = \frac{RP_R + XQ_R}{t_R n_2 |V_R|} \qquad (5.90)$$

Now
$$|\Delta V| = t_S n_1 |V_S| - t_R n_2 |V_R|$$

$\therefore$
$$t_S n_1 |V_S| = t_R n_2 |V_R| + \frac{RP_R + XQ_R}{t_R n_2 |V_R|} \qquad (5.91)$$

In order that the voltage on the HV side of the two transformers be of the same order and the tap setting of each transformer be the minimum, we choose

$$t_S\, t_R = 1 \qquad (5.92)$$

Substituting $t_R = 1/t_S$ in Eq. (5.91) and reorganising, we obtain

$$t_S^2 \left(1 - \frac{RP_R + XQ_R}{n_1 n_2 |V_S||V_R|}\right) = \frac{n_2}{n_1}\frac{|V_R|}{|V_S|} \qquad (5.93)$$

For complete voltage drop compensation, the right hand side of Eq. (5.93) should be unity.

It is obvious from Fig. 5.30 that $t_S > 1$ and $t_R < 1$ for voltage drop compensation. Equation (5.90) indicates that t_R tends to increase[*] the voltage

---

[*]This is so because $t_R < 1$ increases the line current I and hence voltage drop.

$|\Delta V|$ which is to be compensated. Thus merely tap setting as a method of voltage drop compensation would give rise to excessively large tap setting if compensation exceeds certain limits. Thus, if the tap setting dictated by Eq. (5.93), to achieve a desired receiving-end voltage exceeds the normal tap setting range (usually not more than + 20%), it would be necessary to simultaneously inject VARs at the receiving-end in order to maintain the desired voltage level.

Compensation of Transmission Lines

The performance of long EHV AC transmission systems can be improved by reactive compensation of series or shunt (parallel) type. Series capacitors and shunt reactors are used to reduce artificially the series reactance and shunt susceptance of lines and thus they act as the line compensators. Compensation of lines results in improving the system stability (Ch. 12) and voltage control, in increasing the efficiency of power transmission, facilitating line energization and reducing temporary and transient overvoltages.

Series compensation reduces the series impedance of the line which causes voltage drop and is the most important factor in finding the maximum power transmission capability of a line (Eq. (5.70)). A, C and D constants are functions of Z and therefore the also affected by change in the value of Z, but these changes are small in comparison to the change in B as $B = Z$ for the nominal $-\pi$ and equals Z (sinh $\gamma 1/\gamma 1$) for the equivalent π.

The voltage drop ΔV due to series compensation is given by

$$\Delta V \approx IR \cos \phi_r + I(X_L - X_C) \sin \phi_r \qquad (5.94)$$

Here X_C = capacitive reactance of the series capacitor bank per phase and X_L is the total inductive reactance of the line/phase. In practice, X_C may be so selected that the factor $(X_L - X_C) \sin \phi_r$ becomes negative and equals (in magnitude) $R \cos \phi_r$ so that ΔV becomes zero. The ratio X_C/X_L is called "compensation factor" and when expressed as a percentage is known as the "percentage compensation".

The extent of effect of compensation depends on the number, location and circuit arrangements of series capacitor and shunt reactor stations. While planning long-distance lines, besides the average degree of compensation required, it is required to find out the most appropriate location of the reactors and capacitor banks, the optimum connection scheme and the number of intermediate stations. For finding the operating conditions along the line, the *ABCD* constants of the portions of line on each side of the capacitor bank, and *ABCD* constants of the bank may be first found out and then equivalent constants of the series combination of line-capacitor-line can then be arrived at by using the formulae given in Appendix B.

In India, in states like UP, series compensation is quite important since super thermal plants are located (east) several hundred kilometers from load centres (west) and large chunks of power must be transmitted over long distances. Series capacitors also help in balancing the voltage drop of two parallel lines.

When series compensation is used, there are chances of sustained overvoltage to the ground at the series capacitor terminals. This overvoltage can be the power limiting criterion at high degree of compensation. A spark gap with a high speed contactor is used to protect the capacitors under overvoltage conditions.

Under light load or no-load conditions, charging current should be kept less than the rated full-load current of the line. The charging current is approximately given by $B_C|V|$ where B_C is the total capacitive susceptance of the line and $|V|$ is the rated voltage to neutral. If the total inductive susceptance is B_I due to several inductors connected (shunt compensation) from line to neutral at appropriate places along the line, then the charging current would be

$$I_{chg} = (B_C - B_I)\ |V| = B_C|\ V\ |\left(1 - \frac{B_I}{B_C}\right) \qquad (5.95)$$

Reduction of the charging current is by the factor of $(1 - B_I/B_C)$ and B_I/B_C is the shunt compensation factor. Shunt compensation at no-load also keeps the receiving end voltage within limits which would otherwise be quite high because of the Ferranti Effect. Thus reactors should be introduced as load is removed for proper voltage control.

As mentioned earlier, the shunt capacitors are used across an inductive load so as to provide part of the reactive VARs required by the load to keep the voltage within desirable limits. Similarly, the shunt reactors are kept across capacitive loads or in light load conditions, as discussed above, to absorb some of the leading VARs for achieving voltage control. Capacitors are connected either directly to a bus or through tertiary winding of the main transformer and are placed along the line to minimise losses and the voltage drop.

It may be noted that for the same voltage boost, the reactive power capacity of a shunt capacitor is greater than that of a series capacitor. The shunt capacitor improves the *pf* of the load while the series capacitor has hardly any impact on the *pf*. Series capacitors are more effective for long lines for improvement of system stability.

Thus, we see that in both series and shunt compensation of long transmission lines it is possible to transmit large amounts of power efficiently with a flat voltage profile. Proper type of compensation should be provided in proper quantity at appropriate places to achieve the desired voltage control. The reader is encouraged to read the details about the Static Var Systems (SVS) in References 7, 8 and 16. For complete treatment on 'compensation', the reader may refer to Chapter 15.

PROBLEMS

5.1 A three-phase voltage of 11 kV is applied to a line having $R = 10\ \Omega$ and $X = 12\ \Omega$ per conductor. At the end of the line is a balanced load of P

kW at a leading power factor. At what value of P is the voltage regulation zero when the power factor of the load is (a) 0.707, (b) 0.85?

5.2 A long line with $A = D = 0.9 \angle 1.5°$ and $B = 150 \angle 65°$ Ω has at the load end a transformer having a series impedance $Z_T = 100 \angle 67°$ Ω. The load voltage and current are V_L and I_L. Obtain expressions for V_S and I_S in form of

$$\begin{bmatrix} V_S \\ I_S \end{bmatrix} = \begin{bmatrix} A' & B' \\ C' & D' \end{bmatrix} \begin{bmatrix} V_L \\ I_L \end{bmatrix}$$

and evaluate these constants.

5.3 A three-phase overhead line 200 km long has resistance = 0.16 Ω/km and conductor diameter of 2 cm with spacing 4 m, 5 m and 6 m transposed. Find: (a) the *ABCD* constants using Eq. (5.28b), (b) the V_s, I_s, pf_s, P_s when the line is delivering full load of 50 MW at 132 kV and 0.8 lagging *pf*, (c) efficiency of transmission, and (d) the receiving-end voltage regulation.

5.4 A short 230 kV transmission line with a reactance of 18 Ω/phase supplies a load at 0.85 lagging power factor. For a line current of 1,000 A the receiving- and sending-end voltages are to be maintained at 230 kV. Calculate (a) rating of synchronous capacitor required, (b) the load current, (c) the load MVA. Power drawn by the synchronous capacitor may be neglected.

5.5 A 40 MVA generating station is connected to a 'three-phase line having
$Z = 300 \angle 75°$ Ω $Y = 0.0025 \angle 90°$ $\mho$.

The power at the generating station is 40 MVA at unity power factor at a voltage of 120 kV. There is a load of 10 MW at unity power factor at the mid point of the line. Calculate the voltage and load at the distant end of the line. Use nominal-T circuit for the line.

5.6 The generalized circuit constants of a transmission line are
$A = 0.93 + j0.016$
$B = 20 + j140$
The load at the receiving-end is 60 MVA, 50 Hz, 0.8 power factor lagging. The voltage at the supply end is 220 kV. Calculate the load voltage.

5.7 Find the incident and reflected currents for the line of Problem 5.3 at the receiving-end and 200 km from the receiving-end.

5.8 If the line of Problem 5.6 is 200 km long and delivers 50 MW at 220 kV and 0.8 power factor lagging, determine the sending-end voltage, current, power factor and power. Compute the efficiency of transmission, characteristic impedance, wavelength, and velocity of propagation.

5.9 For Example 5.7 find the parameters of the equivalent-π circuit for the line.

5.10 An interconnector cable having a reactance of 6 Ω links generating stations 1 and 2 as shown in Fig. 5.18a. The desired voltage profile is $|V_1|$ = $|V_2|$ = 22 kV. The loads at the two-bus bars are 40 MW at 0.8 lagging power factor and 20 MW at 0.6 lagging power factor, respectively. The station loads are equalized by the flow of power in the cable. Estimate the torque angle and the station power factors.

5.11 A 50 Hz, three-phase, 275 kV, 400 km transmission line has following parameters (per phase).

> Resistance = 0.035 Ω/km
>
> Inductance = 1 mH/km
>
> Capacitance = 0.01 μF/km

If the line is supplied at 275 kV, determine the MVA rating of a shunt reactor having negligible losses that would be required to maintain 275 kV at the receiving-end, when the line is delivering no-load. Use nominal-π method.

5.12 A three-phase feeder having a resistance of 3 Ω and a reactance of 10 Ω supplies a load of 2.0 MW at 0.85 lagging power factor. The receiving-end voltage is maintained at 11 kV by means of a static condenser drawing 2.1 MVAR from the line. Calculate the sending-end voltage and power factor. What is the voltage regulation and efficiency of the feeder?

5.13 A three-phase overhead line has resistance and reactance of 5 and 20 Ω, respectively. The load at the receiving-end is 30 MW, 0.85 power factor lagging at 33 kV. Find the voltage at the sending-end. What will be the kVAR rating of the compensating equipment inserted at the receiving-end so as to maintain a voltage of 33 kV at each end? Find also the maximum load that can be transmitted.

5.14 Construct a receiving-end power circle diagram for the line of Example 5.7. Locate the point corresponding to the load of 50 MW at 220 kV with 0.8 lagging power factor. Draw the circle passing through the load point. Measure the radius and determine therefrom $|V_S|$. Also draw the sending-end circle and determine therefrom the sending-end power and power factor.

5.15 A three-phase overhead line has resistance and reactance per phase of 5 and 25 Ω, respectively. The load at the receiving-end is 15 MW, 33 kV, 0.8 power factor lagging. Find the capacity of the compensation equipment needed to deliver this load with a sending-end voltage of 33 kV.

Calculate the extra load of 0.8 lagging power factor which can be delivered with the compensating equipment (of capacity as calculated above) installed, if the receiving-end voltage is permitted to drop to 28 kV.

REFERENCES

Books

1. *Transmission Line Reference Book-345 kV and Above*, Electric Power Research Institute, Palo Alto Calif, 1975.
2. McCombe, J. and F.J. Haigh, *Overhead-line Practice*, Macdonalel, London, 1966.
3. Stevenson, W.D., *Elements of Power System Analysis*, 4th edn, McGraw-Hill, New York, 1982.
4. Arrillaga, J., *High Voltage Direct Current Transmission*, IEE Power Engineering Series 6, Peter Peregrinus Ltd., London, 1983.
5. Kimbark, E.W., *Direct Current Transmission*, Vol. 1, Wiley, New York, 1971.
6. Uhlmann, E., *Power Transmission by Direct Current*, Springer-Verlag, Berlin-Heidelberg, 1975.
7. Miller, T.J.E., *Reactive Power Control in Electric Systems*, Wiley, New York 1982.
8. Mathur, R.M. (Ed.), *Static Compensators for Reactive Power Control*, Context Pub., Winnipeg, 1984.
9. Desphande, M.V., *Electrical Power System Design*, Tata McGraw-Hill, New Delhi, 1984.

Papers

10. Dunlop, R.D., R. Gutman and D.P. Marchenko, "Analytical Development of Loadability Characteristics for EHV and UHV Transmission Lines", *IEEE Trans, PAS*, 1979, 98: 606.
11. "EHV Transmission", (Special Issue), *IEEE Trans*, June 1966, No. 6, PAS-85.
12. Goodrich, R.D., "A Universal Power Circle Diagram", *AIEE Trans.*, 1951, 70: 2042.
13. Indulkar, C.S. Parmod Kumar and D.P. Kothari, "Sensitivity Analysis of a Multiconductor Transmission Line", *Proc. IEEE*, March 1982, 70: 299.
14. Indulkar, C.S., Parmod Kumar and D.P.Kothari, "Some Studies on Carrier Propagation in Overhead Transmission Lines", *IEEE Trans. on PAS*, No. 4, 1983, 102: 942.
15. Bijwe, P.R., D.P. Kothari, J. Nanda and K.S. Lingamurthy, "Optimal Voltage Control Using Constant Sensitivity Matrix", *Electric Power System Research*, Oct. 1986, 3: 195.
16. Kothari, D.P., et al. "Microprocessors Controlled Static Var Systems", *Proc. Int. conf. Modelling & Simulation*, Gorakhpur, Dec. 1985, 2: 139.

6

Load Flow Studies

6.1 INTRODUCTION

With the background of the previous chapters, we are now ready to study the operational features of a composite power system. Symmetrical steady state is, in fact, the most important mode of operation of a power system. Three major problems encountered in this mode of operation are listed below in their hierarchical order.

1. Load flow problem
2. Optimal load scheduling problem
3. Systems control problem

This chapter is devoted to the load flow problem, while the other two problems will be treated in later chapters. Load flow study in power system parlance is the steady state solution of the power system network. The main information obtained from this study comprises the magnitudes and phase angles of load bus voltages, reactive powers at generator buses, real and reactive power flow on transmission lines, other variables being specified. This information is essential for the continuous monitoring of the current state of the system and for analyzing the effectiveness of alternative plans for future system expansion to meet increased load demand.

Before the advent of digital computers, the AC calculating board was the only means of carrying out load flow studies. These studies were, therefore, tedious and time consuming. With the availability of fast and large size digital computers, all kinds of power system studies, including load flow, can now be carried out conveniently. In fact, some of the advanced level sophisticated studies which were almost impossible to carry out on the AC calculating board have now become possible. The AC calculating board has been rendered obsolete for all practical purposes.

6.2 NETWORK MODEL FORMULATION

The load flow problem has, in fact, been already introduced in Chapter 5 with the help of a fundamental system, i.e. a two-bus problem (*see* Example 5.8). For a load flow study of a real life power system comprising a large number of buses, it is necessary to proceed systematically by first formulating the network model of the system.

A power system comprises several buses which are interconnected by means of transmission lines. Power is injected into a bus from generators, while the loads are tapped from it. Of course, there may be buses with only generators and no-loads, and there may be others with only loads and no generators. Further, VAR generators may also be connected to some buses. The surplus power at some of the buses is transported via transmission lines to buses deficient in power. Figure 6.1a shows the one-line diagram of a four-bus system with generators and loads at each bus. To arrive at the network model of a power system, it is sufficiently accurate to represent a short line by a series impedance and a long line by a nominal-π model[*] (equivalent-π may be used for very long lines). Often, line resistance may be neglected with a small loss in accuracy but a great deal of saving in computation time.

For systematic analysis, it is convenient to regard loads as negative generators and lump together the generator and load powers at the buses. Thus at the ith bus, the net complex power injected into the bus is given by

$$S_i = P_i + jQ_i = (P_{Gi} - P_{Di}) + j(Q_{Gi} - Q_{Di})$$

where the complex power supplied by the generators is

$$S_{Gi} = P_{Gi} + jQ_{Gi}$$

and the complex power drawn by the loads is

$$S_{Di} = P_{Di} + jQ_{Di}$$

The real and reactive powers injected into the ith bus are then

$$P_i = P_{Gi} - P_{Di} \quad i = 1, 2, ..., n \tag{6.1}$$

$$Q_i = Q_{Gi} - Q_{Di}$$

Figure 6.1b shows the network model of the sample power system prepared on the above lines. The equivalent power source at each bus is represented by a shaded circle. The equivalent power source at the ith bus injects current J_i into the bus. It may be observed that the structure of a power system is such that all the sources are always connected to a *common ground node*.

The network model of Fig. 6.1b has been redrawn in Fig. 6.1c after lumping the shunt admittances at the buses. Besides the ground node, it has four other nodes (buses) at which the current from the sources is injected into the network. The line admittance between nodes i and k is depicted by $y_{ik} = y_{ki}$. Further, the mutual admittance between lines is assumed to be zero.

---

[*]Line transformers are represented by a series impedance (or for accurate representation by series and shunt impedances, i.e. inverted L-network).

Applying Kirchhoff's current law (KCL) at nodes 1, 2, 3 and 4, respectively, we get the following four equations:

$$J_1 = V_1 y_{10} + (V_1 - V_2) y_{12} + (V_1 - V_3) y_{13}$$
$$J_2 = V_2 y_{20} + (V_2 - V_1) y_{12} + (V_2 - V_3) y_{23} + (V_2 - V_4) y_{24}$$
$$J_3 = V_3 y_{30} + (V_3 - V_1) y_{13} + (V_3 - V_2) y_{23} + (V_3 - V_4) y_{34} \qquad (6.2)$$
$$J_4 = V_4 y_{40} + (V_4 - V_2) y_{24} + (V_4 - V_3) y_{34}$$

Rearranging and writing in matrix form, we get

$$
\begin{bmatrix} J_1 \\ J_2 \\ J_3 \\ J_4 \end{bmatrix} =
\begin{bmatrix}
(y_{10} + y_{12} + y_{13}) & -y_{12} & -y_{13} & 0 \\
-y_{12} & (y_{20} + y_{12} + y_{23} + y_{24}) & -y_{23} & -y_{24} \\
-y_{13} & -y_{23} & (y_{30} + y_{13} + y_{23} + y_{34}) & -y_{34} \\
0 & -y_{24} & -y_{34} & (y_{40} + y_{24} + y_{34})
\end{bmatrix}
\begin{bmatrix} V_1 \\ V_2 \\ V_3 \\ V_4 \end{bmatrix} \qquad (6.3)
$$

Equation (6.3) can be recognized to be of the standard form

$$
\begin{bmatrix} J_1 \\ J_2 \\ J_3 \\ J_4 \end{bmatrix} =
\begin{bmatrix}
Y_{11} & Y_{12} & Y_{13} & Y_{14} \\
Y_{21} & Y_{22} & Y_{23} & Y_{24} \\
Y_{31} & Y_{32} & Y_{33} & Y_{34} \\
Y_{41} & Y_{42} & Y_{43} & Y_{44}
\end{bmatrix}
\begin{bmatrix} V_1 \\ V_2 \\ V_3 \\ V_4 \end{bmatrix} \qquad (6.4)
$$

Comparing Eqs. (6.3) and (6.4), we can write

$$Y_{11} = y_{10} + y_{12} + y_{13}$$
$$Y_{22} = y_{20} + y_{12} + y_{23} + y_{24}$$
$$Y_{33} = y_{30} + y_{13} + y_{23} + y_{34}$$
$$Y_{44} = y_{40} + y_{24} + y_{34}$$
$$Y_{12} = Y_{21} = -y_{12}; \quad Y_{23} = Y_{32} = -y_{23}$$

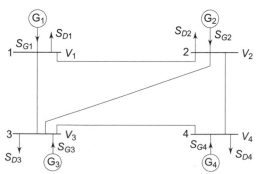

(a) One-line diagram

Fig. 6.1 Sample four bus system

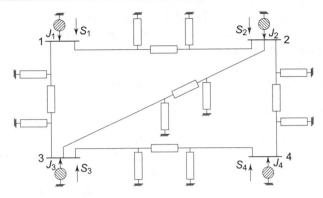

(b) Equivalent circuit

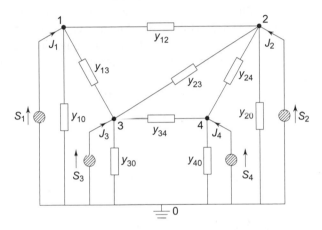

(c) Power network of Fig. 6.1 (b) lumped and redrawn

Fig. 6.1   Sample four-bus system

$$Y_{31} = Y_{13} = - y_{13}; \ Y_{14} = Y_{41} = - y_{14} = 0$$

$$Y_{24} = Y_{42} = - y_{24}; \ Y_{34} = Y_{43} = - y_{34}$$

Each admittance Y_{ii} (i = 1, 2, 3, 4) is called the *self admittance (or driving point admittance)* of node i and equals the algebraic sum of all the admittances terminating on the node. Each off-diagonal term Y_{ik} (i, k = 1, 2, 3, 4) is the *mutual admittance (transfer admittance)* between nodes i and k and equals the negative of the sum of all admittances connected directly between these nodes. Further, $Y_{ik} = Y_{ki}$.

Using index notation, Eq. (6.4) can be written in compact form as

$$J_i = \sum_{k=1}^{n} Y_{ik} V_k; \ i = 1, 2, ..., n \tag{6.5}$$

or, in matrix form

$$J_{BUS} = Y_{BUS} \ V_{BUS} \tag{6.6}$$

where Y_{BUS} denotes the matrix of bus admittance and is known as *bus admittance matrix*. The dimension of the Y_{BUS} matrix is $(n \times n)$ where n is the number of buses. [The total number of nodes are $m = n + 1$ including the ground (reference) node.]

As seen above, Y_{BUS} is a symmetric matrix, except when phase shifting transformers are involved, so that only $\dfrac{n(n+1)}{2}$ terms are to be stored for an n-bus system. Furthermore, $Y_{ik} = 0$ if buses i and k are not connected (e.g. $Y_{14} = 0$). Since in a power network each bus is connected only to a few other buses (usually to two or three buses), the Y_{BUS} of a large network is very sparse, i.e. it has a large number of zero elements. Though this property is not evident in a small system like the sample system under consideration, in a system containing hundreds of buses, the sparsity may be as high as 90%. Tinney and associates [22] at Bonnevile Power Authority were the first to exploit the sparsity feature of Y_{BUS} in greatly reducing numerical computations in load flow studies and in minimizing the memory required as only non-zero terms need be stored.

Equation (6.6) can also be written in the form

$$V_{\text{BUS}} = Z_{\text{BUS}} J_{\text{BUS}} \tag{6.7}$$

where

$$Z_{\text{BUS}} \text{ (bus impedance matrix)} = Y_{\text{BUS}}^{-1} \tag{6.8}$$

for a network of four buses (four independent nodes)

$$Z_{\text{BUS}} = \begin{bmatrix} Z_{11} & Z_{12} & Z_{13} & Z_{14} \\ Z_{21} & Z_{22} & Z_{23} & Z_{24} \\ Z_{31} & Z_{32} & Z_{33} & Z_{34} \\ Z_{41} & Z_{42} & Z_{43} & Z_{44} \end{bmatrix}$$

Symmetric Y_{BUS} yields symmetric Z_{BUS}. The diagonal elements of Z_{BUS} are called *driving point impedances* of the nodes, and the off-diagonal elements are called *transfer impedances* of the nodes. Z_{BUS} need not be obtained by inverting Y_{BUS}. While Y_{BUS} is a sparse matrix, Z_{BUS} is a full matrix. i.e., zero elements of Y_{BUS} become non-zero in the corresponding Z_{BUS} elements.

It is to be stressed here that $Y_{\text{BUS}}/Z_{\text{BUS}}$ constitute models of the passive portions of the power network.

Bus admittance matrix is often used in solving load flow problem. It has gained widespread application owing to its simplicity of data preparation and the ease with which the bus admittance matrix can be formed and modified for network changes—addition of lines, regulating transformers, etc. (*see* Examples 6.2 and 6.7). Of course, sparsity is one of its greatest advantages as it heavily reduces computer memory and time requirements. In contrast to this, the formation of a bus impedance matrix requires either matrix inversion[*] or use of involved algorithms. Furthermore, the impedance matrix is a full matrix.[**]

The bus impedance matrix, however, is most useful for short circuit studies as will be seen in Chapters 10 and 11.

Note: In the sample system of Fig. 6.1, the buses are numbered in an arbitrary manner, although in more sophisticated studies of large power systems, it has been shown that certain ordering of nodes produces faster convergence and solutions. Appendix C deals with the topics of *sparsity and optimal ordering*.

6.3   FORMATION OF Y_{BUS} BY SINGULAR TRANSFORMATION

The matrix pair Y_{BUS} and Z_{BUS} form the network models for load flow studies. The Y_{BUS} can be alternatively assembled by use of singular transformations given by a graph theoretical approach. This alternative approach is of great theoretical and practical significance and is therefore discussed here.

To start with, the graph theory is briefly reviewed.

Graph

To describe the geometrical features of a network, it is replaced by single line segments called *elements* whose terminals are called *nodes*. A *linear graph* depicts the geometrical interconnection of the elements of a network. A graph is *connected* if, and only if, there is a path between every pair of nodes. If each element of a connected graph is assigned a direction,[***] it is called an *oriented graph*.

Power networks are so structured that out of the m total nodes, one node (normally described by 0) is always at ground potential and the remaining $n = m - 1$ nodes are the buses at which the source power is injected. Figure 6.2 shows the graph of the power network of Fig. 6.1c. It may be noted

---

[*]Z_{BUS} can be referred to ground or slack bus. In the former case, it is usually necessary to create at least one strong artificial tie to ground to avoid numerical difficulties when obtaining Z_{BUS}, because in absence of this, Y_{BUS} is ill-conditioned or even singular. A large shunt admittance inserted at the slack bus most simply achieves the desired result [20].

[**]The disadvantages of the conventional impedance matrix may be overcome by making use of LU factors of the admittance matrix and by employing compact storage scheme. Piecewise methods or tearing techniques (dialoptics) have recently been applied to overcome the disadvantages of excessive storage requirements [18].

[***]For convenience, direction is so assigned as to coincide with the assumed positive direction of the element current.

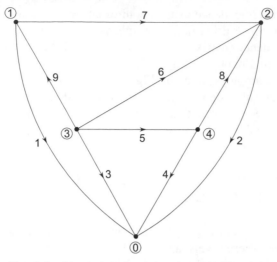

Fig. 6.2   Linear graph of the circuit in Fig. 6.1c

here that each source and the shunt admittance connected across it are represented by a single element. In fact, this combination represents the most general network element and is described under the subheading "Primitive Network".

A connected subgraph containing all the nodes of a graph but having no closed paths is called a *tree*. The elements of a tree are called *branches* or *tree branches*. The number of branches b that form a tree are given by

$$b = m - 1 = n \text{ (number of buses)} \qquad (6.9)$$

Those elements of the graph that are not included in the tree are called *links* (or *link branches*) and they form a subgraph, not necessarily connected, called

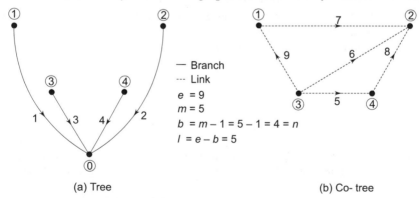

(a) Tree   (b) Co- tree

Fig. 6.3 Tree and cotree of the oriented connected graph of Fig. 6.2

cotree. The number of links l of a connected graph with e elements is

$$l = e - b = e - m + 1 \qquad (6.10)$$

Note that a tree (and therefore, cotree) of a graph is not unique.

A tree and the corresponding cotree of the graph of Fig. 6.2 are shown in Fig. 6.3. The reader should try and find some other tree and cotree pairs.

If a link is added to the tree, the corresponding graph contains one closed path called a *loop*. Thus a graph has as many loops as the number of links.

Primitive Network

A network element may in general contain active and passive components. Figures 6.4a and b show, respectively, the alternative impedance and admittance form representation of a general network element. The impedance form is a voltage source e_{rs} in series with an impedance z_{rs}; while the admittance form is a current source j_{rs} in parallel with an admittance y_{rs}. The element current is i_{rs} and the element voltage is

$$v_{rs} = E_r - E_s$$

where E_r and E_s are the voltages of the element nodes r and s, respectively.

It may be remembered here that for steady state AC performance, all element variables (v_{rs}, E_r, E_s, i_{rs}, j_{rs}) are phasors and element parameters (z_{rs}, y_{rs}) are complex numbers.

The voltage relation for Fig. 6.4a can be written as

$$v_{rs} + e_{rs} = z_{rs}i_{rs} \tag{6.11}$$

Similarly, the current relation for Fig. 6.4b is

$$i_{rs} + j_{rs} = y_{rs}v_{rs} \tag{6.12}$$

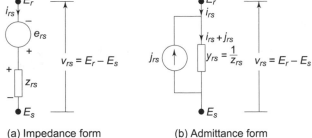

(a) Impedance form   (b) Admittance form

Fig. 6.4 Representation of a network element

The forms of Figs. 6.4a and b are equivalent wherein the parallel source current in admittance form is related to the series voltage in impedance form by

$$j_{rs} = - y_{rs}e_{rs}$$

Also

$$y_{rs} = 1/z_{rs}$$

A set of unconnected elements is defined as a *primitive network*. The performance equations of a primitive network are given below:

In impedance form

$$V + E = ZI \tag{6.13}$$

In admittance form

$$I + J = YV \tag{6.14}$$

Here V and I are the element voltage and current vectors respectively, and J and E are the source vectors. Z and Y are referred to as the primitive impedance and admittance matrices, respectively. These are related as $Z = Y^{-1}$. If there is no mutual coupling between elements, Z and Y are diagonal where the diagonal entries are the impedances/admittances of the network elements and are reciprocal.

Network Variables in Bus Frame of Reference

Linear network graph helps in the systematic assembly of a network model. The main problem in deriving mathematical models for large and complex power networks is to select a minimum or zero redundancy (linearly independent) set of current or voltage variables which is sufficient to give the information about all element voltages and currents. One set of such variables is the b tree voltages.* It may easily be seen by using topological reasoning that these variables constitute a non-redundant set. The knowledge of b tree voltages allows us to compute all element voltages and therefore, all bus currents assuming all element admittances being known.

Consider a tree graph shown in Fig. 6.3a where the ground node is chosen as the reference node. This is the most appropriate tree choice for a power network. With this choice, the b tree branch voltages become identical with the bus voltages as the tree branches are incident to the ground node.

Bus Incidence Matrix

For the specific system of Fig. 6.3a, we obtain the following relations between the nine element voltages and the four bus (i.e. tree branch) voltages V_1, V_2, V_3 and V_4.

$$
\begin{aligned}
V_{b1} &= V_1 \\
V_{b2} &= V_2 \\
V_{b3} &= V_3 \\
V_{b4} &= V_4 \\
V_{l5} &= V_3 - V_4 \\
V_{l6} &= V_3 - V_2 \\
V_{l7} &= V_1 - V_2 \\
V_{l8} &= V_4 - V_2 \\
V_{l9} &= V_3 - V_1
\end{aligned}
\tag{6.15}
$$

or, in matrix form

---

*Another useful set of network variables are the l link (loop) currents which constitute a zero redundancy set of network variables [6].

$$V = AV_{\text{BUS}} \tag{6.16}$$

where the *bus incidence matrix A* is

bus	1	2	3	4
e				
1	1	0	0	0
2	0	1	0	0
3	0	0	1	0
4	0	0	0	1
5	0	0	1	-1
6	0	-1	1	0
7	1	-1	0	0
8	0	-1	0	1
9	-1	0	1	0

$$= \quad \begin{matrix} \text{buses} \\ \text{elements} \\ \text{branches} \\ \text{links} \end{matrix} \quad \begin{bmatrix} A_b \\ \hline A_l \end{bmatrix} = \begin{bmatrix} I \\ \hline A_l \end{bmatrix} \tag{6.17}$$

This matrix is rectangular and therefore singular. Its elements a_{ik} are found as per the following rules:

$a_{ik} = 1$ if ith element is incident to and oriented away from the kth node (bus)

$= -1$ if ith element is incident to but oriented towards the kth node

$= 0$ if the ith element is not incident to the kth node

Substituting Eq. (6.16) into Eq. (6.14), we get

$$I + J = YAV_{\text{BUS}} \tag{6.18}$$

Premultiplying by A^{T},

$$A^{\text{T}}I + A^{\text{T}}J = A^{\text{T}}YAV_{\text{BUS}} \tag{6.19}$$

Each component of the n-dimensional vector $A^{\text{T}}I$ is the algebraic sum of the element currents leaving the nodes 1, 2, ..., n.

Therefore, the application[*] of the KCL must result in

$$A^{\text{T}}I = 0 \tag{6.20}$$

Similarly, each component of the vector $A^{\text{T}}J$ can be recognized as the algebraic sum of all source currents injected into nodes 1, 2, ..., n. These components are therefore the bus currents[**]. Hence we can write

[*]For node 1, $A^{\text{T}}I$ gives
$i_{10} + i_{12} - i_{31} = 0$
The reader should verify this for another node.
[**]For node 1, $A^{\text{T}}J$ gives
j_{01} = current injected into bus 1
because other elements connected to bus 1 have no sources.

$$A^{\mathrm{T}}J = J_{\mathrm{BUS}} \tag{6.21}$$

Equation (6.19) then is simplified to

$$J_{\mathrm{BUS}} = A^{T}YAV_{\mathrm{BUS}} \tag{6.22}$$

Thus, following an alternative systematic approach, we have in fact, obtained the same nodal current equation as (6.6). The bus admittance matrix can then be obtained from the singular transformation of the primitive Y, i.e.

$$Y_{\mathrm{BUS}} = A^{T}YA \tag{6.23}$$

A computer programme can be developed to write the bus incidence matrix A from the interconnection data of the directed elements of the power system. Standard matrix transpose and multiplication subroutines can then be used to compute Y_{BUS} from Eq. (6.23).

Example 6.1

Find the Y_{BUS} using singular transformation for the system of Fig. 6.2.

Solution

$$Y = \begin{bmatrix} y_{10} & & & & & & & & 0 \\ & y_{20} & & & & & & & \\ & & y_{30} & & & & & & \\ & & & y_{40} & & & & & \\ & & & & y_{34} & & & & \\ & & & & & y_{23} & & & \\ & & & & & & y_{12} & & \\ & & & & & & & y_{24} & \\ 0 & & & & & & & & y_{13} \end{bmatrix}$$

Using A from Eq. (6.17), we get

$$YA = \begin{bmatrix} y_{10} & 0 & 0 & 0 \\ 0 & y_{20} & 0 & 0 \\ 0 & 0 & y_{30} & 0 \\ 0 & 0 & 0 & y_{40} \\ 0 & 0 & y_{34} & -y_{34} \\ 0 & -y_{23} & y_{23} & 0 \\ y_{12} & -y_{12} & 0 & 0 \\ 0 & -y_{24} & 0 & y_{24} \\ -y_{13} & 0 & y_{13} & 0 \end{bmatrix}$$

Finally,

$$Y_{BUS} = A^{T}YA$$

$$= \begin{bmatrix} (y_{10} + y_{12} \\ \quad + y_{13}) & -y_{12} & -y_{13} & 0 \\ -y_{12} & (y_{20} + y_{12} \\ \quad + y_{23} + y_{24}) & -y_{23} & -y_{24} \\ -y_{13} & -y_{23} & (y_{30} + y_{13} \\ \quad + y_{23} + y_{34}) & -y_{34} \\ 0 & -y_{24} & -y_{34} & (y_{40} + y_{24} \\ \quad + y_{34}) \end{bmatrix}$$

The elements of this matrix, of course, agree with those previously calculated in Eq. (6.3)

Example 6.2

Figure 6.5 shows the one-line diagram of a simple four-bus system. Table 6.1 gives the line impedances identified by the buses on which these terminate. The shunt admittance at all the buses is assumed negligible.

(a) Find Y_{BUS} assuming that the line shown dotted is not connected.
(b) What modifications need to be carried out in Y_{BUS} if the line shown dotted is connected.

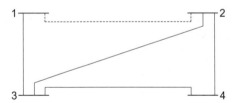

Fig. 6.5 Sample system for Example 6.2

Table 6.1

Line, bus to bus	R, pu	X, pu
1–2	0.05	0.15
1–3	0.10	0.30
2–3	0.15	0.45
2–4	0.10	0.30
3–4	0.05	0.15

Table 6.2

Line	G, pu	B, pu
1–2	2.0	– 6.0
1–3	1.0	– 3.0
2–3	0.666	– 2.0
2–4	1.0	– 3.0
3–4	2.0	– 6.0

Solution   (a) From Table 6.1, Table 6.2 is obtained from which Y_{BUS} for the system can be written as

$$Y_{BUS} = \begin{bmatrix} Y_{11} & Y_{12} & Y_{13} & Y_{14} \\ Y_{21} & Y_{22} & Y_{23} & Y_{24} \\ Y_{31} & Y_{32} & Y_{33} & Y_{34} \\ Y_{41} & Y_{42} & Y_{43} & Y_{44} \end{bmatrix} \tag{i}$$

$$= \begin{bmatrix} y_{13} & 0 & -y_{13} & 0 \\ 0 & y_{23}+y_{24} & -y_{23} & -y_{24} \\ -y_{13} & -y_{23} & y_{31}+y_{32}+y_{34} & y_{34} \\ 0 & -y_{24} & -y_{34} & y_{43}+y_{42} \end{bmatrix} \tag{ii}$$

$$Y_{BUS} = \begin{bmatrix} 1-j3 & 0 & -1+j3 & 0 \\ 0 & 1.666-j5 & -0.666+j2 & -1+j3 \\ -1+j3 & -0.666+j2 & 3.666-j11 & -2+j6 \\ 0 & -1+j3 & -2+j6 & 3-j9 \end{bmatrix} \tag{iii}$$

(b) The following elements of Y_{BUS} of part (a) are modified when a line is added between buses 1 and 2.

$$Y_{12,\,new} = Y_{12,\,old} - (2-j6) = Y_{21,\,new}$$
$$Y_{11,\,new} = Y_{11,\,old} + (2-j6) \tag{iv}$$
$$Y_{22,\,new} = Y_{22,\,old} + (2-j6)$$

Modified Y_{BUS} is written below

$$Y_{BUS} = \begin{bmatrix} 3-j9 & -2+j6 & -1+j3 & 0 \\ -2+j6 & 3.666-j11 & -0.666+j2 & -1+j3 \\ -1+j3 & -0.666+j2 & 3.666-j11 & -2+j6 \\ 0 & -1+j3 & -2+j6 & 3-j9 \end{bmatrix} \tag{v}$$

6.4   LOAD FLOW PROBLEM

The complex power injected by the source into the *i*th bus of a power system is

$$S_i = P_i + jQ_i = V_i J_i^*; \quad i = 1, 2, ..., n \tag{6.24}$$

where V_i is the voltage at the ith bus with respect to ground and J_i is the source current injected into the bus.

The load flow problem is handled more conveniently by use of J_i rather than J_i^*. Therefore, taking the complex conjugate of Eq. (6.24), we have

$$P_i - jQ_i = V_i^* J_i; \quad i = 1, 2, ..., n \tag{6.25a}$$

Substituting for $J_i = \sum_{k=1}^{n} Y_{ik} V_k$ from Eq. (6.5), we can write

$$P_i - jQ_i = V_i^* \sum_{k=1}^{n} Y_{ik} V_k; \quad i = 1, 2, ..., n \tag{6.25b}$$

Equating real and imaginary parts

$$P_i \text{ (real power)} = \text{Re} \left\{ V_i^* \sum_{k=1}^{n} Y_{ik} V_k \right\} \tag{6.26a}$$

$$Q_i \text{ (reactive power)} = -\text{Im} \left\{ V_i^* \sum_{k=1}^{n} Y_{ik} V_k \right\} \tag{6.26b}$$

In polar form

$$V_i = |V_i| e^{j\delta_i}$$

$$Y_{ik} = |Y_{ik}| e^{j\theta_{ik}}$$

Real and reactive powers can now be expressed as

$$P_i \text{ (real power)} = |V_i| \sum_{k=1}^{n} |V_k| |Y_{ik}| \cos (\theta_{ik} + \delta_k - \delta_i);$$

$$i = 1, 2, ..., n \tag{6.27}$$

$$Q_i \text{ (reactive power)} = - |V_i| \sum_{k=1}^{n} |V_k| |Y_{ik}| \sin (\theta_{ik} + \delta_k - \delta_i);$$

$$i = 1, 2, ..., n \tag{6.28}$$

Equations (6.27) and (6.28) represent $2n$ power flow equations at n buses of a power system (n real power flow equations and n reactive power flow equations). Each bus is characterized by four variables; P_i, Q_i, $|V_i|$ and δ_i resulting in a total of $4n$ variables. Equations (6.27) and (6.28) can be solved for $2n$ variables if the remaining $2n$ variables are specified. Practical considerations allow a power system analyst to fix *a priori* two variables at each bus. The solution for the remaining $2n$ bus variables is rendered difficult by the fact that Eqs. (6.27) and (6.28) are non-linear algebraic equations (bus voltages are involved in product form and sine and cosine terms are present) and therefore, explicit solution is not possible. Solution can only be obtained by iterative numerical techniques.

Depending upon which two variables are specified *a priori*, the buses are classified into three categories.

(1) PQ Bus

At this type of bus, the net powers P_i and Q_i are known (P_{Di} and Q_{Di} are known from load forecasting and P_{Gi} and Q_{Gi} are specified). The unknowns are $|V_i|$ and δ_i. A pure load bus (no generating facility at the bus, i.e., $P_{Gi} = Q_{Gi} = 0$) is a *PQ* bus.

(2) PV Bus/Generator Bus/Voltage Controlled Bus

At this type of bus P_{Di} and Q_{Di} are known *a priori* and $|V_i|$ and P_i (hence P_{Gi}) are specified. The unknowns are Q_i (hence Q_{Gi}) and δ_i.

(3) Slack Bus/Swing Bus/Reference Bus

This bus is distinguished from the other two types by the fact that real and reactive powers at this bus are not specified. Instead, voltage magnitude and phase angle (normally set equal to zero) are specified. Normally there is only one bus of this type in a given power system. The need of such a bus for a load flow study is explained below.

In a load flow study real and reactive powers (i.e. complex power) cannot be fixed *a priori* at all the buses as the net complex power flow into the network is not known in advance, the system power loss being unknown till the load flow study is complete. It is, therefore, necessary to have one bus (i.e. the slack bus) at which complex power is unspecified so that it supplies the difference in the total system load plus losses and the sum of the complex powers specified at the remaining buses. By the same reasoning the slack bus must be a generator bus. The complex power allocated to this bus is determined as part of the solution. In order that the variations in real and reactive powers of the slack bus during the iterative process be a small percentage of its generating capacity, the bus connected to the largest generating station is normally selected as the slack bus. Further, for convenience the slack bus is numbered as bus 1.

Equations (6.27) and (6.28) are referred to as *static load flow equations* (SLFE). By transposing all the variables on one side, these equations can be written in the vector form

$$f(x, y) = 0 \qquad (6.29)$$

where

        f = vector function of dimension $2n$

        x = dependent or state vector of dimension $2n$
            ($2n$ unspecified variables)

        y = vector of independent variables of dimension 2n
            ($2n$ independent variables which are specified *a priori*)

Some of the independent variables in y can be used to manipulate some of the state variables. These adjustable independent variables are called control parameters. Vector y can then be partitioned into a vector u of control parameters and a vector p of fixed parameters.

$$y = \begin{bmatrix} u \\ p \end{bmatrix} \tag{6.30}$$

Control parameters may be voltage magnitudes at PV buses, real powers P_i, etc. The vector p includes all the remaining parameters which are uncontrollable.

For SLFE solution to have practical significance, all the state and control variables must lie within specified practical limits. These limits, which are dictated by specifications of power system hardware and operating constraints, are described below:

(i) Voltage magnitude $|V_i|$ must satisfy the inequality

$$|V_i|_{min} \leq |V_i| \leq |V_i|_{max} \tag{6.31}$$

The power system equipment is designed to operate at fixed voltages with allowable variations of $\pm (5-10)\%$ of the rated values.

(ii) Certain of the δ_is (state variables) must satisfy the inequality constraint

$$|\delta_i - \delta_k| \leq |\delta_i - \delta_k|_{max} \tag{6.32}$$

This constraint limits the maximum permissible power angle of transmission line connecting buses i and k and is imposed by considerations of system stability (see Chapter 12).

(iii) Owing to physical limitations of P and/or Q generation sources, P_{Gi} and Q_{Gi} are constrained as follows:

$$P_{Gi,\ min} \leq P_{Gi} \leq P_{Gi,\ max} \tag{6.33}$$

$$Q_{Gi,\ min} \leq Q_{Gi} \leq Q_{Gi,\ max}{}^{*} \tag{6.34}$$

It is, of course, obvious that the total generation of real and reactive power must equal the total load demand plus losses, i.e.

$$\sum_i P_{Gi} = \sum_i P_{Di} + P_L \tag{6.35}$$

$$\sum_i Q_{Gi} = \sum_i Q_{Di} + Q_L \tag{6.36}$$

where P_L and Q_L are system real and reactive power loss, respectively. Optimal sharing of active and reactive power generation between sources will be discussed in Chapter 7.

---

[*]Voltage at a PV bus can be maintained constant only if controllable Q source is available at the bus and the reactive generation required is within prescribed limits.

The *load flow problem* can now be fully defined as follows:

Assume a certain nominal bus load configuration. Specify $P_{Gi} + jQ_{Gi}$ at all the PQ buses (this specifies $P_i + jQ_i$ at these buses); specify P_{Gi} (this specifies P_i) and $|V_i|$ at all the PV buses. Also specify $|V_1|$ and δ_1 (= 0) at the slack bus. Thus $2n$ variables of the vector y are specified. The $2n$ SLFE can now be solved (iteratively) to determine the values of the $2n$ variables of the vector x comprising voltages and angles at the PQ buses, reactive powers and angles at the PV buses and active and reactive powers at the slack bus. The next logical step is to compute line flows.

So far we have presented the methods of assembling a Y_{BUS} matrix and load flow equations and have defined the load flow problem in its general form with definitions of various types of buses. It has been demonstrated that load flow equations, being essentially non-linear algebraic equations, have to be solved through iterative numerical techniques. Section 6.5 presents some of the algorithms which are used for load flow solutions of acceptable accuracy for systems of practical size.

At the cost of solution accuracy, it is possible to linearize load flow equations by making suitable assumptions and approximations so that fast and explicit solutions become possible. Such techniques have value particularly for planning studies, where load flow solutions have to be carried out repeatedly but a high degree of accuracy is not needed.

An Approximate Load Flow Solution

Let us make the following assumptions and approximations in the load flow Eqs. (6.27) and (6.28).

(i) Line resistances being small are neglected (shunt conductance of overhead lines is always negligible), i.e. P_L, the active power loss of the system is zero. Thus in Eqs. (6.27) and (6.28) $\theta_{ik} \simeq 90°$ and $\theta_{ii} \simeq -90°$.

(ii) $(\delta_i - \delta_k)$ is small ($< \pi/6$) so that $\sin(\delta_i - \delta_k) \simeq (\delta_i - \delta_k)$. This is justified from considerations of stability (*see* Chapter 12).

(iii) All buses other than the slack bus (numbered as bus 1) are PV buses, i.e. voltage magnitudes at all the buses including the slack bus are specified.

Equations (6.27) and (6.28) then reduce to

$$P_i = |V_i| \sum_{k=1}^{n} |V_k| |Y_{ik}| (\delta_i - \delta_k); \quad i = 2, 3, ..., n \tag{6.37}$$

$$Q_i = -|V_i| \sum_{\substack{k=1 \\ k \neq i}}^{n} |V_k| |Y_{ik}| \cos(\delta_i - \delta_k) + |V_i|^2 |Y_{ii}|; \quad i = 1, 2, ..., n \tag{6.38}$$

Since $|V_i|$s are specified, Eq. (6.37) represents a set of linear algebraic equations in δ_is which are $(n-1)$ in number as δ_1 is specified at the slack bus ($\delta_1 = 0$). The nth equation corresponding to slack bus ($n = 1$) is redundant as the real power injected at this bus is now fully specified as

$$P_1 = \sum_{i=2}^{n} P_{Di} - \sum_{i=2}^{n} P_{Gi}; \ (P_L = 0).$$ Equations (6.37) can be solved explicitly (non-iteratively) for $\delta_2, \delta_3, ..., \delta_n$, which, when substituted in Eq. (6.38), yields Q_is, the reactive power bus injections. It may be noted that the assumptions made have decoupled Eqs. (6.37) and (6.38) so that these need not be solved simultaneously but can be solved sequentially [solution of Eq. (6.38) follows immediately upon simultaneous solution of Eq. (6.37)]. Since the solution is non-iterative and the dimension is reduced to $(n-1)$ from $2n$, it is computationally highly economical.

Example 6.3

Consider the four-bus sample system of Fig. 6.6 wherein line reactances are indicated in pu. Line resistances are considered negligible. The magnitude of all the four bus voltages are specified to be 1.0 pu. The bus powers are specified in the table below:

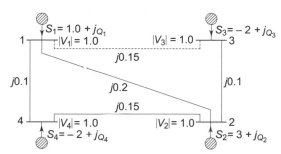

Fig. 6.6 Four-bus lossless sample system

Bus	Real demand	Reactive demand	Real generation	Reactive generation
1	$P_{D1} = 1.0$	$Q_{D1} = 0.5$	$P_{G1} = ?$	Q_{G1} (unspecified)
2	$P_{D2} = 1.0$	$Q_{D2} = 0.4$	$P_{G2} = 4.0$	Q_{G2} (unspecified)
3	$P_{D3} = 2.0$	$Q_{D3} = 1.0$	$P_{G3} = 0$	Q_{G3} (unspecified)
4	$P_{D4} = 2.0$	$Q_{D4} = 1.0$	$P_{G4} = 0$	Q_{G4} (unspecified)

Figure 6.6 indicates bus injections for the data specified in the table. As bus voltages are specified, all the buses must have controllable Q sources. It is also obvious from the data that buses 3 and 4 have only Q sources. Further, since the system is assumed lossless, the real power generation at bus 1 is known *a priori* to be

$$P_{G1} = P_{D1} + P_{D2} + P_{D3} + P_{D4} - P_{G2} = 2.0 \text{ pu}$$

Therefore, we have 7 unknowns instead of $2 \times 4 = 8$ unknowns. In the present problem the unknown state and control variables are $\delta_2, \delta_3, \delta_4, Q_{G1}, Q_{G2}, Q_{G3}$ and Q_{G4}.

Though the real losses are zero, the presence of the reactive losses requires that the total reactive generation must be more than the total reactive demand (2.9 pu).

From the data given, Y_{BUS} can be written as follows:

$$Y_{BUS} = \begin{array}{c} \\ 1 \\ 2 \\ 3 \\ 4 \end{array} \begin{array}{cccc} 1 & 2 & 3 & 4 \\ \left[\begin{array}{cccc} -j21.667 & j5.000 & j6.667 & j10.000 \\ j5.000 & -j21.667 & j10.000 & j6.667 \\ j6.667 & j10.000 & -j16.667 & j0.000 \\ j10.000 & j6.667 & j0.000 & -j16.667 \end{array}\right] \end{array} \quad (i)$$

Using the above Y_{BUS} and bus powers as shown in Fig. 6.6, approximate load flow Eqs. (6.37) are expressed as (all voltage magnitudes are equal to 1.0 pu)

$$P_2 = 3 = 5 \ (\delta_2 - \delta_1) + 10 \ (\delta_2 - \delta_3) + 6.667 \ (\delta_2 - \delta_4) \quad (ii)$$

$$P_3 = -2 = 6.667 \ (\delta_3 - \delta_1) + 10 \ (\delta_3 - \delta_2) \quad (iii)$$

$$P_4 = -2 = 10 \ (\delta_4 - \delta_1) + 6.667 \ (\delta_4 - \delta_2) \quad (iv)$$

Taking bus 1 as a reference bus, i.e. $\delta_1 = 0$, and solving (ii), (iii) and (iv), we get

$$\delta_2 = 0.077 \text{ rad} = 4.41°$$

$$\delta_3 = -0.074 \text{ rad} = -4.23° \quad (v)$$

$$\delta_4 = -0.089 \text{ rad} = -5.11°$$

Substituting δs in Eqs. (6.38), we have

$$Q_1 = -5 \cos 4.41° - 6.667 \cos 4.23° - 10 \cos 5.11° + 21.667$$

$$Q_2 = -5 \cos 4.41° - 10 \cos 8.64° - 6.667 \cos 9.52° + 21.667$$

$$Q_3 = -6.667 \cos 4.23° - 10 \cos 8.64° + 16.667$$

$$Q_4 = -10 \cos 5.11° - 6.667 \cos 9.52° + 16.667$$

or,

$$Q_1 = 0.07 \text{ pu}$$

$$Q_2 = 0.22 \text{ pu} \quad (vi)$$

$$Q_3 = 0.132 \text{ pu}$$

$$Q_4 = 0.132 \text{ pu}$$

Reactive power generation at the four buses are

$$Q_{G1} = Q_1 + 0.5 = 0.57 \text{ pu}$$

$$Q_{G2} = Q_2 + 0.4 = 0.62 \text{ pu} \quad (vii)$$

$$Q_{G3} = Q_3 + 1.0 = 1.132 \text{ pu}$$

$$Q_{G4} = Q_4 + 1.0 = 1.132 \text{ pu}$$

Reactive line losses are

$$Q_L = \sum_{i=1}^{4} Q_{Gi} - \sum_{i=1}^{4} Q_{Di}$$

$$= 3.454 - 2.9 = 0.554 \text{ pu} \tag{viii}$$

Now, let us find the line flows. Equation (5.68) can be written in the form ($|Z| = X$, $\theta = 90°$)

$$P_{ik} = -P_{ki} = \frac{|V_i| |V_k|}{X_{ik}} \sin(\delta_i - \delta_k)$$

where P_{ik} is the real power flow from bus i to bus k.

$$P_{13} = -P_{31} = \frac{1}{0.15} \sin(\delta_1 - \delta_3) = \frac{\sin 4.23°}{0.15} = 0.492 \text{ pu}$$

$$P_{12} = -P_{21} = \frac{1}{0.2} \sin(\delta_1 - \delta_2) = -\frac{\sin 4.41°}{0.2} = -0.385 \text{ pu} \quad \text{(ix)}$$

$$P_{14} = -P_{41} = \frac{1}{0.1} \sin(\delta_1 - \delta_4) = 10 \sin 5.11° = 0.891 \text{ pu}$$

Real power flows on other lines can be similarly calculated.

For reactive power flow, Eq. (5.69) can be written in the general form ($|Z| = X$, $\theta = 90°$)

$$Q_{ik} = \frac{|V_i|^2}{X_{ik}} - \frac{|V_i| |V_k|}{X_{ik}} \cos(\delta_i - \delta_k)$$

where Q_{ik} is the reactive power flow from bus i to bus k.

$$Q_{12} = Q_{21} = \frac{1}{0.2} - \frac{1}{0.2} \cos(\delta_1 - \delta_2) = 0.015 \text{ pu}$$

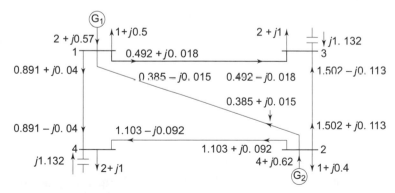

Fig. 6.7 Load flow solution for the four-bus system

$$Q_{13} = Q_{31} = \frac{1}{0.15} - \frac{1}{0.15} \cos(\delta_1 - \delta_3) = 0.018 \text{ pu}$$

$$Q_{14} = Q_{41} = \frac{1}{0.1} - \frac{1}{0.1} \cos (\delta_1 - \delta_4) = 0.04 \text{ pu}$$

Reactive power flows on other lines can be similarly calculated.

Generations and load demands at all the buses and all the line flows are indicated in Fig. 6.7.

6.5 GAUSS-SEIDEL METHOD

The Gauss-Seidel (GS) method is an iterative algorithm for solving a set of non-linear algebraic equations. To start with, a solution vector is assumed, based on guidance from practical experience in a physical situation. One of the equations is then used to obtain the revised value of a particular variable by substituting in it the present values of the remaining variables. The solution vector is immediately updated in respect of this variable. The process is then repeated for all the variables thereby completing one *iteration*. The iterative process is then repeated till the solution vector converges within prescribed accuracy. The convergence is quite sensitive to the starting values assumed. Fortunately, in a load flow study a starting vector close to the final solution can be easily identified with previous experience.

To explain how the GS method is applied to obtain the load flow solution, let it be assumed that all buses other than the slack bus are *PQ* buses. We shall see later that the method can be easily adopted to include *PV* buses as well. The slack bus voltage being specified, there are $(n - 1)$ bus voltages starting values of whose magnitudes and angles are assumed. These values are then updated through an iterative process. During the course of any iteration, the revised voltage at the ith bus is obtained as follows:

$$J_i = (P_i - jQ_i)/V_i^* \quad \text{[from Eq. (6.25a)]} \tag{6.39}$$

From Eq. (6.5)

$$V_i = \frac{1}{Y_{ii}} \left[J_i - \sum_{\substack{k=1 \\ k \neq i}}^{n} Y_{ik} V_k \right] \tag{6.40}$$

Substituting for J_i from Eq. (6.39) into (6.40)

$$V_i = \frac{1}{Y_{ii}} \left[\frac{P_i - jQ_i}{V_i^*} - \sum_{\substack{k=1 \\ k \neq i}}^{n} Y_{ik} V_k \right] ; i = 2, 3, ..., n \tag{6.41}$$

The voltages substituted in the right hand side of Eq. (6.41) are the most recently calculated (updated) values for the corresponding buses. During each iteration voltages at buses $i = 2, 3, ..., n$ are sequentially updated through use of Eq. (6.41). V_1, the slack bus voltage being fixed is not required to be updated. Iterations are repeated till no bus voltage magnitude changes by more than a prescribed value during an iteration. The computation process is then said to *converge* to a solution.

If instead of updating voltages at every step of an iteration, updating is carried out at the end of a complete iteration, the process is known as the *Gauss iterative method*. It is much slower to converge and may sometimes fail to do so.

Algorithm for Load Flow Solution

Presently we shall continue to consider the case where all buses other than the slack are *PQ* buses. The steps of a computational algorithm are given below:

1. With the load profile known at each bus (i.e. P_{Di} and Q_{Di} known), allocate[*] P_{Gi} and Q_{Gi} to all generating stations.

    While active and reactive generations are allocated to the slack bus, these are permitted to vary during iterative computation. This is necessary as voltage magnitude and angle are specified at this bus (only two variables can be specified at any bus).

    With this step, bus injections $(P_i + jQ_i)$ are known at all buses other than the slack bus.

2. *Assembly of bus admittance matrix* Y_{BUS}: With the line and shunt admittance data stored in the computer, Y_{BUS} is assembled by using the rule for self and mutual admittances (Sec. 6.2). Alternatively Y_{BUS} is assembled using Eq. (6.23) where input data are in the form of primitive matrix Y and singular connection matrix A.

3. *Iterative computation of bus voltages* $(V_i; i = 2, 3,..., n)$: To start the iterations a set of initial voltage values is assumed. Since, in a power system the voltage spread is not too wide, it is normal practice to use a *flat voltage start*,[**] i.e. initially all voltages are set equal to $(1 + j0)$ except the voltage of the slack bus which is fixed. It should be noted that $(n - 1)$ equations (6.41) in complex numbers are to be solved iteratively for finding $(n - 1)$ complex voltages $V_2, V_3, ..., V_n$. If complex number operations are not available in a computer, Eqs (6.41) can be converted into $2(n - 1)$ equations in real unknowns $(e_i, f_i$ or $|V_i|, \delta_i)$ by writing

$$V_i = e_i + jf_i = |V_i|\ e^{j\delta_i} \tag{6.42}$$

A significant reduction in the computer time can be achieved by performing in advance all the arithmetic operations that do not change with the iterations.

Define

$$A_i = \frac{P_i - jQ_i}{Y_{ii}}\quad i = 2, 3, ..., n \tag{6.43}$$

---

[*]Active and reactive generation allocations are made on economic considerations discussed in Chapter 7.

[**]A flat voltage start means that to start the iteration set the voltage magnitudes and angles at all buses other than the *PV* buses equal to $(i + j0)$. The slack bus angle is conveniently taken as zero. The voltage magnitudes at the *PV* buses and slack bus are set equal to the specified values.

$$B_{ik} = \frac{Y_{ik}}{Y_{ii}} \quad i = 2, 3, \dots, n;$$ (6.44)

$$k = 1, 2, \dots, n;$$

$$k \neq i$$

Now for the $(r + 1)$th iteration, the voltage Eq. (6.41) becomes

$$V_i^{(r+1)} = \frac{A_i}{\left(V_i^{(r)}\right)^*} - \sum_{k=1}^{i-1} B_{ik} V_k^{(r+1)} - \sum_{k=i+1}^{n} B_{ik} V_k^{(r)} \quad i = 2, 3, \dots, n$$ (6.45)

The iterative process is continued till the change in magnitude of bus voltage, $|\Delta V_i^{(r+1)}|$, between two consecutive iterations is less than a certain tolerance for all bus voltages, i.e.

$$|\Delta V_i^{(r+1)}| = |V_i^{(r+1)} - V_i^{(r)}| < \varepsilon \; ; \; i = 2, 3, \dots, n$$ (6.46)

4. *Computation of slack bus power*: Substitution of all bus voltages computed in step 3 along with V_1 in Eq. (6.25b) yields $S_1^* = P_1 - jQ_1$.

5. *Computation of line flows*: This is the last step in the load flow analysis wherein the power flows on the various lines of the network are computed. Consider the line connecting buses i and k. The line and transformers at each end can be represented by a circuit with series admittance y_{ik} and two shunt admittances y_{ik0} and y_{ki0} as shown in Fig. 6.8.

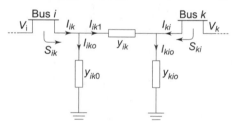

Fig. 6.8 π-representation of a line and transformers connected between two buses

The current fed by bus i into the line can be expressed as

$$I_{ik} = I_{ik1} + I_{ik0} = (V_i - V_k) y_{ik} + V_i y_{ik0}$$ (6.47)

The power fed into the line from bus i is.

$$S_{ik} = P_{ik} + jQ_{ik} = V_i I_{ik}^* = V_i(V_i^* - V_k^*) y_{ik}^* + V_i V_i^* y_{ik0}^*$$ (6.48)

Similarly, the power fed into the line from bus k is

$$S_{ki} = V_k (V_k^* - V_i^*) y_{ik}^* + V_k V_k^* y_{ki0}^*$$ (6.49)

The power loss in the $(i-k)$th line is the sum of the power flows determined from Eqs. (6.48) and (6.49). Total transmission loss can be computed by summing all the line flows (i.e. $S_{ik} + S_{ki}$ for all i, k).

It may be noted that the slack bus power can also be found by summing the flows on the lines terminating at the slack bus.

Acceleration of convergence

Convergence in the GS method can sometimes be speeded up by the use of the acceleration factor. For the ith bus, the accelerated value of voltage at the $(r + 1)$th iteration is given by

$$V_i^{(r+1)} \text{ (accelerated)} = V_i^{(r)} + \alpha(V_i^{(r+1)} - V_i^{(r)})$$

where α is a real number called the *acceleration factor*. A suitable value of α for any system can be obtained by trial load flow studies. A generally recommended value is $\alpha = 1.6$. A wrong choice of α may indeed slow down convergence or even cause the method to diverge.

This concludes the load flow analysis for the case of PQ buses only.

Algorithm Modification when PV Buses are also Present

At the PV buses, P and $|V|$ are specified and Q and δ are the unknowns to be determined. Therefore, the values of Q and δ are to be updated in every GS iteration through appropriate bus equations. This is accomplished in the following steps for the ith PV bus.

1. From Eq. (6.26b)

$$Q_i = - \text{Im} \left\{ V_i^* \sum_{k=1}^{n} Y_{ik} V_k \right\}$$

The revised value of Q_i is obtained from the above equation by substituting most updated values of voltages on the right hand side. In fact, for the $(r + 1)$th iteration one can write from the above equation

$$Q_i^{(r+1)} = - \text{Im} \left\{ (V_i^{(r)})^* \sum_{k=1}^{i-1} Y_{ik} V_k^{(r+1)} + (V_i^{(r)})^* \sum_{k=i}^{n} Y_{ik} V_k^{(r)} \right\} \quad (6.50)$$

2. The revised value of δ_i is obtained from Eq. (6.45) immediately following step 1. Thus

$$\delta_i^{(r+1)} = \angle V_i^{(r+1)}$$

$$= \text{Angle of} \left[\frac{A_i^{(r+1)}}{(V_i^{(r)})^*} - \sum_{k=1}^{i-1} B_{ik} V_k^{(r+1)} - \sum_{k=i+1}^{n} B_{ik} V_k^{(r)} \right] \quad (6.51)$$

where

$$A_i^{(r+1)} = \frac{P_i - jQ_i^{(r+1)}}{Y_{ii}} \quad (6.52)$$

The algorithm for PQ buses remains unchanged.

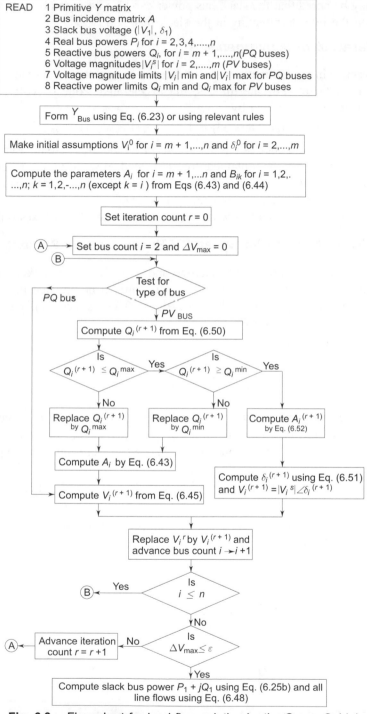

Fig. 6.9 Flow chart for load flow solution by the Gauss-Seidel iterative method using Y_{BUS}

As explained already, physical limitations of Q generation require that Q demand at any bus must be in the range $Q_{min} \rightarrow Q_{max}$. If at any stage during the computation, Q at any bus goes outside these limits, it is fixed at Q_{min} or Q_{max} as the case may be, and the bus voltage specification is dropped, i.e. the bus is now treated like a PQ bus. Thus step 1 above branches out to step 3 below.

3. If $Q_i^{(r+1)} < Q_{i,\ min}$, set $Q_i^{(r+1)} = Q_{i,\ min}$ and treat bus i as a PQ bus. Compute $A_i^{(r+1)}$ and $V_i^{(r+1)}$ from Eqs. (6.52) and (6.45), respectively. If $Q_i^{(r+1)} > Q_{i,\ max}$, set $Q_i^{(r+1)} = Q_{i,\ max}$ and treat bus i as a PQ bus. Compute $A_i^{(r+1)}$ and $V_i^{(r+1)}$ from Eqs. (6.52) and (6.45), respectively.

Now all the computational steps are summarized in the detailed flow chart of Fig. 6.9 which serves as a basis for the reader to write his own computer programme. It is assumed that out of n buses, the first is slack as usual, then 2, 3, ..., m are PV buses and the remaining $m + 1$, ..., n are PQ buses.

Example 6.4

For the sample system of Fig. 6.5 the generators are connected at all the four buses, while loads are at buses 2 and 3. Values of real and reactive powers are listed in Table 6.3. All buses other than the slack are PQ type.

Assuming a flat voltage start, find the voltages and bus angles at the three buses at the end of the first GS iteration.

Solution

Table 6.3 Input data

Bus	P_i, pu	Q_i, pu	V_i, pu	Remarks
1	–	–	1.04 $\angle 0°$	Slack bus
2	0.5	– 0.2	–	PQ bus
3	– 1.0	0.5	–	PQ bus
4	0.3	0.1	–	PQ bus

The Y_{BUS} for the sample system has been calculated earlier in Example 6.2b (i.e. the dotted line is assumed to be connected). In order to approach the accuracy of a digital computer, the computations given below have been performed on an electronic calculator.

Bus voltages at the end of the first iteration are calculated using Eq. (6.45).

$$V_2^1 = \frac{1}{Y_{22}} \left\{ \frac{P_2 - jQ_2}{(V_2^0)^*} - Y_{21} V_1 - Y_{23} V_3^0 - Y_{24} V_4^0 \right\}$$

$$= \frac{1}{Y_{22}} \left\{ \frac{0.5 + j0.2}{1 - j0} - 1.04(-2 + j6) - (-0.666 + j2) - (-1 + j3) \right\}$$

$$= \frac{4.246 - j11.04}{3.666 - j11} = 1.019 + j0.046 \text{ pu}$$

$$V_3^1 = \frac{1}{Y_{33}} \left\{ \frac{P_3 - jQ_3}{(V_3^0)^*} - Y_{31} V_1 - Y_{32} V_3^1 - Y_{34} V_4^0 \right\}$$

$$= \frac{1}{Y_{33}} \left\{ \frac{-1 - j0.5}{1 - j0} - 1.04 \, (-1 + j3) \right.$$

$$\left. - (-0.666 + j2)(1.019 + j0.046) - (-2 + j6) \right\}$$

$$= \frac{2.81 - j11.627}{3.666 - j11} = 1.028 - j0.087 \text{ pu}$$

$$V_4^1 = \frac{1}{Y_{44}} \left\{ \frac{P_4 - jQ_4}{(V_4^0)^*} - Y_{41} V_1 - Y_{42} V_2^1 - Y_{43} V_3^1 \right\}$$

$$= \frac{1}{Y_{44}} \left\{ \frac{0.3 + j0.1}{1 - j0} - (-1 + j3)(1.019 + j0.046) \right.$$

$$\left. - (-2 + j6)(1.028 - j0.087) \right\}$$

$$= \frac{2.991 - j9.253}{3 - j9} = 1.025 - j0.0093 \text{ pu}$$

Example 6.5

In Example 6.4, let bus 2 be a PV bus now with $|V_2| = 1.04$ pu. Once again assuming a flat voltage start, find Q_2, δ_2, V_3, V_4 at the end of the first GS iteration.

Given: $0.2 \leq Q_2 \leq 1$.

From Eq. (6.5), we get (*Note* $\delta_2^0 = 0$, i.e. $V_2^0 = 1.04 + j0$)

$$Q_2^1 = - \text{Im} \left\{ (V_2^0)^* Y_{21} V_1 + (V_2^0)^* \left[Y_{22} V_2^0 + Y_{23} V_3^0 + Y_{24} V_4^0 \right] \right\}$$

$$= - \text{Im} \left\{ 1.04 \, (-2 + j6) \, 1.04 + 1.04 \, [(3.666 - j11) \, 1.04 \right.$$

$$\left. + (-0.666 + j2) + (-1 + j3)] \right\}$$

$$= - \text{Im} \left\{ -0.0693 - j0.2079 \right\} = 0.2079 \text{ pu}$$

$$\therefore \; Q_2^1 = 0.2079 \text{ pu}$$

From Eq. (6.51)

$$\delta_2^1 = \angle\left\{\frac{1}{Y_{22}}\left[\frac{P_2 - jQ_2^1}{(V_2^0)^*} - Y_{21}V_1 - Y_{23}V_3^0 - Y_{24}V_4^0\right]\right\}$$

$$= \angle\left\{\frac{1}{3.666 - j11}\left[\frac{0.5 - j0.2079}{1.04 - j0} - (-2 + j6)(1.04 + j0)\right.\right.$$

$$\left.\left. - (-0.666 + j2)(1 + j0) - (-1 + j3)(1 + j0)\right]\right\}$$

$$= \angle\left(\frac{4.2267 - j11.439}{3.666 - j11}\right) = \angle(1.0512 + j0.0339)$$

or $\delta_2^1 = 1.84658° = 0.032$ rad

$\therefore V_2^1 = 1.04 (\cos \delta_2^1 + j \sin \delta_2^1)$

$\qquad = 1.04 (0.99948 + j0.0322)$

$\qquad = 1.03946 + j0.03351$

$$V_3^1 = \frac{1}{Y_{33}}\left\{\frac{P_3 - jQ_3}{(V_3^0)^*} - Y_{31}V_1 - Y_{32}V_2^1 - Y_{34}V_4^0\right\}$$

$$= \frac{1}{3.666 - j11}\left[\frac{-1 - j0.5}{(1 - j0)} - (-1 + j3)1.04\right.$$

$$\left. - (-0.666 + j2)(1.03946 + j0.03351) - (-2 + j6)\right]$$

$$= \frac{2.7992 - j11.6766}{3.666 - j11} = 1.0317 - j0.08937$$

$$V_4^1 = \frac{1}{Y_{44}}\left\{\frac{P_4 - jQ_4}{(V_4^0)^*} - Y_{41}V_1 - Y_{42}V_2^1 - Y_{43}V_3^1\right\}$$

$$= \frac{1}{3 - j9}\left[\frac{0.3 + j0.1}{1 - j0} - (-1 + j3)(1.0394 + j0.0335)\right.$$

$$\left. - (-2 + j6)(1.0317 - j0.08937)\right]$$

$$= \frac{2.9671 - j8.9962}{3 - j9} = 0.9985 - j0.0031$$

Now, suppose the permissible limits on Q_2 (reactive power injection) are revised as follows:

$$0.25 \le Q_2 \le 1.0 \text{ pu}$$

It is clear, that other data remaining the same, the calculated Q_2 (= 0.2079) is now less than the $Q_{2,\,min}$. Hence Q_2 is set equal to $Q_{2,\,min}$, i.e.

$$Q_2 = 0.25 \text{ pu}$$

Bus 2, therefore, becomes a *PQ* bus from a *PV* bus. Therefore, $|V_2|$ can no longer remain fixed at 1.04 pu. The value of V_2 at the end of the first iteration is calculated as follows. (*Note* $V_2^0 = 1 + j0$ by virtue of a flat start.)

$$V_2^1 = \frac{1}{Y_{22}}\left(\frac{P_2 - jQ_2}{(V_2^0)^*} - Y_{21}V_1 - Y_{23}V_3^0 - Y_{24}V_4^0\right)$$

$$= \frac{1}{3.666 - j11}\left[\frac{0.5 - j0.25}{1 - j0}\right.$$

$$\left. - (-2 + j6)1.04 - (-0.666 + j2) - (-1 + j3)\right]$$

$$= \frac{4.246 - j11.49}{3.666 - j11} = 1.0559 + j0.0341$$

$$V_3^1 = \frac{1}{Y_{33}}\left(\frac{P_3 - jQ_3}{(V_3^0)^*} - Y_{31}V_1 - Y_{32}V_2^1 - Y_{34}V_4^0\right)$$

$$= \frac{1}{3.666 - j11}\left[\frac{-1 - j0.5}{1 - j0} - (-1 + j3)\,1.04\right.$$

$$\left. - (-0.666 + j2)(1.0559 + j0.0341) - (-2 + j6)\right]$$

$$= \frac{2.8112 - j11.709}{3.666 - j11} = 1.0347 - j0.0893 \text{ pu}$$

$$V_4^1 = \frac{1}{Y_{44}}\left(\frac{P_4 - jQ_4}{(V_4^0)^*} - Y_{41}V_1 - Y_{42}V_2^1 - Y_{43}V_3^1\right)$$

$$= \frac{1}{3 - j9}\left[\frac{0.3 + j0.1}{1 - j0} - (-1 + j3)\,(1.0509 + j0.0341)\right.$$

$$\left. - (-2 + j6)\,(1.0347 - j0.0893)\right]$$

$$= \frac{4.0630 - j9.4204}{3 - j9} = 1.0775 + j0.0923 \text{ pu}$$

6.6 NEWTON-RAPHSON (NR) METHOD

The Newton-Raphson method is a powerful method of solving non-linear algebraic equations. It works faster and is sure to converge in most cases as compared to the GS method. It is indeed the practical method of load flow solution of large power networks. Its only drawback is the large requirement of computer memory which has been overcome through a compact storage scheme (*see* Appendix C). Convergence can be considerably speeded up by performing the first iteration through the GS method and using the values so obtained for starting the NR iterations. Before explaining how the NR method is applied to solve the load flow problem, it is useful to review the method in its general form.

Consider a set of n non-linear algebraic equations

$$f_i(x_1, x_2, \ldots, x_n) = 0; \quad i = 1, 2, \ldots, n \tag{6.53}$$

Assume initial values of unknowns as $x_1^0, x_2^0, \ldots, x_n^0$. Let $\Delta x_1^0, \Delta x_2^0, \ldots, \Delta x_n^0$ be the corrections, which on being added to the initial guess, give the actual solution. Therefore

$$f_i(x_1^0 + \Delta x_1^0, x_2^0 + \Delta x_2^0, \ldots, x_n^0 + \Delta x_n^0) = 0; \quad i = 1, 2, \ldots, n \tag{6.54}$$

Expanding these equations in Taylor series around the initial guess, we have

$$f_i(x_1^0, x_2^0, \ldots, x_n^0) + \left[\left(\frac{\partial f_i}{\partial x_1} \right)^0 \Delta x_1^0 + \left(\frac{\partial f_i}{\partial x_2} \right)^0 \Delta x_2^0 + \ldots \right.$$

$$\left. + \left(\frac{\partial f_i}{\partial x_n} \right)^0 \Delta x_n^0 \right] + \text{higher order terms} = 0 \tag{6.55}$$

where $\left(\dfrac{\partial f_i}{\partial x_1} \right)^0, \left(\dfrac{\partial f_i}{\partial x_2} \right)^0, \ldots, \left(\dfrac{\partial f_i}{\partial x_n} \right)^0$ are the derivatives of f_i with respect to $x_1, x_2, \ldots, x_n$ evaluated at $(x_1^0, x_2^0, \ldots, x_n^0)$.

Neglecting higher order terms we can write Eq. (6.55) in matrix form

$$\begin{bmatrix} f_1^0 \\ f_2^0 \\ \vdots \\ f_n^0 \end{bmatrix} + \begin{bmatrix} \left(\dfrac{\partial f_1}{\partial x_1} \right)^0 & \left(\dfrac{\partial f_1}{\partial x_2} \right)^0 & \cdots & \left(\dfrac{\partial f_1}{\partial x_n} \right)^0 \\ \left(\dfrac{\partial f_2}{\partial x_1} \right)^0 & \left(\dfrac{\partial f_2}{\partial x_2} \right)^0 & \cdots & \left(\dfrac{\partial f_2}{\partial x_n} \right)^0 \\ \vdots & & & \vdots \\ \left(\dfrac{\partial f_n}{\partial x_1} \right)^0 & \left(\dfrac{\partial f_n}{\partial x_2} \right)^0 & \cdots & \left(\dfrac{\partial f_n}{\partial x_n} \right)^0 \end{bmatrix} \begin{bmatrix} \Delta x_1^0 \\ \Delta x_2^0 \\ \vdots \\ \Delta x_n^0 \end{bmatrix} \simeq \begin{bmatrix} 0 \\ 0 \\ \vdots \\ 0 \end{bmatrix} \tag{6.56a}$$

or in vector matrix form

$$f^0 + J^0 \Delta x^0 \simeq 0 \qquad (6.56b)$$

J^0 is known as the *Jacobian matrix* (obtained by differentiating the function vector f with respect to x and evaluating it at x^0). Equation (6.56b) can be written as

$$f^0 \simeq [- J^0] \, \Delta x^0 \qquad (6.57)$$

Approximate values of corrections Δx^0 can be obtained from Eq (6.57). These being a set of linear algebraic equations can be solved efficiently by *triangularization and back substitution* (*see* Appendix C).

Updated values of x are then

$$x^1 = x^0 + \Delta x^0$$

or, in general, for the $(r + 1)$th iteration

$$x^{(r+1)} = x^{(r)} + \Delta x^{(r)} \qquad (6.58)$$

Iterations are continued till Eq. (6.53) is satisfied to any desired accuracy, i.e.

$$|f_i(x^{(r)})| < \varepsilon \text{ (a specified value)}; \quad i = 1, 2, \ldots, n \qquad (6.59)$$

NR Algorithm for Load flow Solution

First, assume that all buses are PQ buses. At any PQ bus the load flow solution must satisfy the following non-linear algebraic equations

$$f_{iP} \, (|V|, \, \delta) = P_i \text{ (specified)} - P_i = 0 \qquad (6.60a)$$
$$f_{iQ} \, (|V|, \, \delta) = Q_i \text{ (specified)} - Q_i = 0 \qquad (6.60b)$$

where expressions for P_i and Q_i are given in Eqs. (6.27) and (6.28). For a trial set of variables $|V_i|$, δ_i, the vector of residuals f^0 of Eq. (6.57) corresponds to

$$f_{iP} = P_i \text{ (specified)} - P_i \text{ (calculated)} = \Delta P_i \qquad (6.61a)$$
$$f_{iQ} = Q_i \text{ (specified)} - Q_i \text{ (calculated)} = \Delta Q_i \qquad (6.61b)$$

while the vector of corrections Δx^0 corresponds to

$$\Delta|V_i|, \; \Delta\delta_i$$

Equation (6.57) for obtaining the approximate corrections vector can be written for the load flow case as

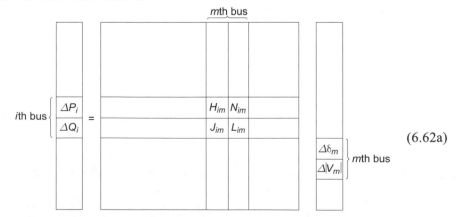

where

$$H_{im} = \frac{\partial P_i}{\partial \delta_m}$$

$$N_{im} = \frac{\partial P_i}{\partial |V_m|}$$

$$J_{im} = \frac{\partial Q_i}{\partial \delta_m}$$ (6.63a)

$$L_{im} = \frac{\partial Q_i}{\partial |V_m|}$$

It is to be immediately observed that the Jacobian elements corresponding to the ith bus residuals and mth bus corrections are a 2×2 matrix enclosed in the box in Eq. (6.62a) where i and m are both PQ buses.

Since at the slack bus (bus number 1), P_1 and Q_1 are unspecified and $|V_1|$, δ_1 are fixed, there are no equations corresponding to Eq. (6.60) at this bus. Hence the slack bus does not enter the Jacobian in Eq. (6.62a).

Consider now the presence of PV buses. If the ith bus is a PV bus, Q_i is unspecified so that there is no equation corresponding to Eq. (6.60b) for this bus. Therefore, the Jacobian elements of the ith bus become a single row pertaining to ΔP_i, i.e.

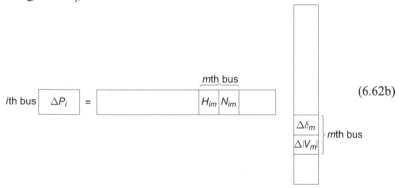

(6.62b)

If the mth bus is also a PV bus, $|V_m|$ becomes fixed so that $\Delta|V_m| = 0$. We can now write

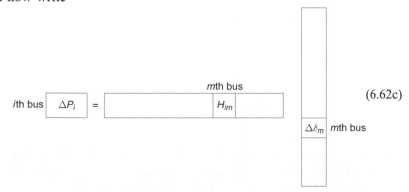

(6.62c)

Also if the ith bus is a PQ bus while the mth bus is a PV bus, we can then write

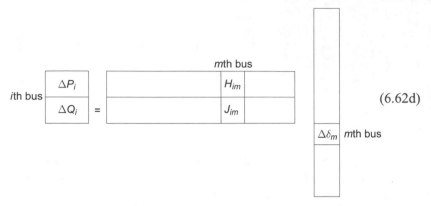

$$(6.62d)$$

It is convenient for numerical solution to normalize the voltage corrections as

$$\frac{\Delta|V_m|}{|V_m|}$$

as a consequence of which, the corresponding Jacobian elements become

$$N_{im} = \frac{\partial P_i}{\partial|V_m|}|V_m|$$

$$L_{im} = \frac{\partial Q_i}{\partial|V_m|}|V_m| \qquad (6.63b)$$

Expressions for elements of the Jacobian (in normalized form) of the load flow Eqs. (6.60a and b) are derived in Appendix D and are given below:

Case 1 $m \neq i$

$$H_{im} = L_{im} = a_m f_i - b_m e_i \qquad (6.64)$$

$$N_{im} = -J_{im} = a_m e_i + b_m f_i$$

where

$$Y_{im} = G_{im} + jB_{im}$$

$$V_i = e_i + jf_i$$

$$(a_m + jb_m) = (G_{im} + jB_{im})(e_m + jf_m)$$

Case 2 $m = i$

$$H_{ii} = -Q_i - B_{ii}|V_i|^2$$

$$N_{ii} = P_i + G_{ii}|V_i|^2 \qquad (6.65)$$

$$J_{ii} = P_i - G_{ii}|V_i|^2$$

$$L_{ii} = Q_i - B_{ii}|V_i|^2$$

An important observation can be made in respect of the Jacobian by examination of the Y_{BUS} matrix. If buses i and m are not connected, $Y_{im} = 0$ ($G_{im} = B_{im} = 0$). Hence from Eqs. (6.63) and (6.64), we can write

$$H_{im} = H_{mi} = 0$$
$$N_{im} = N_{mi} = 0$$
$$J_{im} = J_{mi} = 0 \tag{6.66}$$
$$L_{im} = L_{mi} = 0$$

Thus the Jacobian is as sparse as the Y_{BUS} matrix.

Formation of Eq. (6.62) of the NR method is best illustrated by a problem. Figure 6.10 shows a five-bus power network with bus types indicated therein. The matrix equation for determining the vector of corrections from the vector of residuals is given below.

Corresponding to a particular vector of variables $[\delta_2|V_2|\delta_3\delta_4|V_4|\delta_5]^T$, the vector of residuals $[\varDelta P_2 \; \varDelta Q_2 \; \varDelta P_3 \; \varDelta P_4 \; \varDelta Q_4 \; \varDelta P_5]^T$ and the Jacobian (6 × 6 in this example) are computed. Equation (6.67) is then solved by triangularization and back substitution procedure to obtain the vector of

corrections $\left[\varDelta\delta_2 \; \dfrac{\varDelta|V_2|}{|V_2|} \; \varDelta\delta_3\varDelta\delta_4 \; \dfrac{\varDelta|V_4|}{|V_4|} \; \varDelta\delta_5 \right]^T$. Corrections are then added to

update the vector of variables.

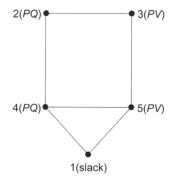

Fig. 6.10 Sample five-bus network

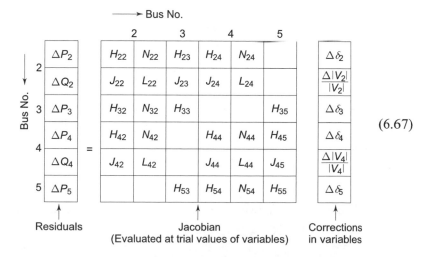

		Residuals	2	3	4	5		Corrections in variables
2		$\varDelta P_2$	H_{22} N_{22}	H_{23}	H_{24} N_{24}			$\varDelta\delta_2$
		$\varDelta Q_2$	J_{22} L_{22}	J_{23}	J_{24} L_{24}			$\dfrac{\varDelta|V_2|}{|V_2|}$
3		$\varDelta P_3$	H_{32} N_{32}	H_{33}		H_{35}		$\varDelta\delta_3$
4	=	$\varDelta P_4$	H_{42} N_{42}		H_{44} N_{44}	H_{45}		$\varDelta\delta_4$
		$\varDelta Q_4$	J_{42} L_{42}		J_{44} L_{44}	J_{45}		$\dfrac{\varDelta|V_4|}{|V_4|}$
5		$\varDelta P_5$		H_{53}	H_{54} N_{54}	H_{55}		$\varDelta\delta_5$

$$\tag{6.67}$$

Iterative Algorithm

Omitting programming details, the iterative algorithm for the solution of the load flow problem by the NR method is as follows:

1. With voltage and angle (usually $\delta = 0$) at slack bus fixed, assume $|V|$, δ at all PQ buses and δ at all PV buses. In the absence of any other information flat voltage start is recommended.

2. Compute ΔP_i (for PV and PQ buses) and ΔQ_i (for all PQ buses) from Eqs. (6.60a and b). If all the values are less than the prescribed tolerance, stop the iterations, calculate P_1 and Q_1 and print the entire solution including line flows.

3. If the convergence criterion is not satisfied, evaluate elements of the Jacobian using Eqs. (6.64) and (6.65).

4. Solve Eq. (6.67) for corrections of voltage angles and magnitudes.

5. Update voltage angles and magnitudes by adding the corresponding changes to the previous values and return to step 2.

Note: 1. In step 2, if there are limits on the controllable Q sources at PV buses, Q is computed each time and if it violates the limits, it is made equal to the limiting value and the corresponding PV bus is made a PQ bus in that iteration. If in the subsequent computation, Q does come within the prescribed limits, the bus is switched back to a PV bus.

2. If there are limits on the voltage of a PQ bus and if any of these limits is violated, the corresponding PQ bus is made a PV bus in that iteration with voltage fixed at the limiting value.

Example 6.6

Consider the three-bus system of Fig. 6.11. Each of the three lines has a series impedance of $0.02 + j0.08$ pu and a total shunt admittance of $j0.02$ pu. The specified quantities at the buses are tabulated below:

Bus	Real load demand P_D	Reactive load demand Q_D	Real power generation P_G	Reactive power generation Q_G	Voltage specification
1	2.0	1.0	Unspecified	Unspecified	$V_1 = 1.04 + j0$ (slack bus)
2	0.0	0.0	0.5	1.0	Unspecified (PQ bus)
3	1.5	0.6	0.0	$Q_{G3} = ?$	$|V_3| = 1.04$ (PV bus)

Controllable reactive power source is available at bus 3 with the constraint

$$0 \leq Q_{G3} \leq 1.5 \text{ pu}$$

Find the load flow solution using the NR method. Use a tolerance of 0.01 for power mismatch.

Solution Using the nominal-π model for transmission lines, Y_{BUS} for the given system is obtained as follows:

For each line

$$y_{series} = \frac{1}{0.02 + j0.08} = 2.941 - j11.764 = 12.13 \angle -75.96°$$

Each off-diagonal term $= -2.941 + j11.764$
Each self term $= 2[(2.941 - j11.764) + j0.01]$
$$= 5.882 - j23.528 = 24.23 \angle -75.95°$$

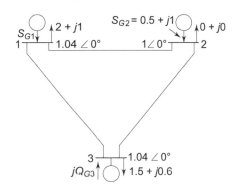

Fig. 6.11 Three-bus system for Example 6.6

$$\therefore \quad Y_{BUS} = \begin{bmatrix} 24.23\angle -75.95° & 12.13\angle 104.04° & 12.13\angle 104.04° \\ 12.13\angle 104.04° & 24.23\angle -75.95° & 12.13\angle 104.04° \\ 12.13\angle 104.04° & 12.13\angle 104.04° & 24.23\angle -75.95° \end{bmatrix}$$

To start iteration choose $V_2^0 = 1 + j0$ and $\delta_3^0 = 0$. From Eqs. (6.27) and (6.28), we get

$$P_2 = |V_2| |V_1| |Y_{21}| \cos(\theta_{21} + \delta_1 - \delta_2) + |V_2|^2 |Y_{22}| \cos\theta_{22} + |V_2| |V_3|$$
$$|Y_{23}| \cos(\theta_{23} + \delta_3 - \delta_2)$$

$$P_3 = |V_3| |V_1| |Y_{31}| \cos(\theta_{31} + \delta_1 - \delta_3) + |V_3| |V_2| |Y_{32}| \times \cos(\theta_{32} + \delta_2$$
$$- \delta_3) + |V_3|^2 |Y_{33}| \cos\theta_{33}$$

$$Q_2 = -|V_2| |V_1| |Y_{21}| \sin(\theta_{21} + \delta_1 - \delta_2) - |V_2|^2 |Y_{22}| \times \sin\theta_{22} - |V_2|$$
$$|V_3| |Y_{23}| \sin(\theta_{23} + \delta_2 - \delta_3)$$

Substituting given and assumed values of different quantities, we get the values of powers as

$$P_2^0 = -0.23 \text{ pu}$$

$$P_3^0 = 0.12 \text{ pu}$$

$$Q_2^0 = -0.96 \text{ pu}$$

Power residuals as per Eq. (6.61) are

$$\Delta P_2^0 = P_2 \text{ (specified)} - P_2^0 \text{ (calculated)}$$

$$= 0.5 - (-0.23) = 0.73$$

$$\Delta P_3^0 = -1.5 - (0.12) = -1.62$$

$$\Delta Q_2^0 = 1 - (-0.96) = 1.96$$

The changes in variables at the end of the first iteration are obtained as follows:

$$
\begin{bmatrix} \Delta P_2 \\ \Delta P_3 \\ \Delta Q_2 \end{bmatrix} =
\begin{bmatrix}
\dfrac{\partial P_2}{\partial \delta_2} & \dfrac{\partial P_2}{\partial \delta_3} & \dfrac{\partial P_2}{\partial |V_2|} \\[2mm]
\dfrac{\partial P_3}{\partial \delta_2} & \dfrac{\Delta P_3}{\partial \delta_3} & \dfrac{\partial P_3}{\partial |V_2|} \\[2mm]
\dfrac{\partial Q_2}{\partial \delta_2} & \dfrac{\partial Q_2}{\partial \delta_3} & \dfrac{\partial Q_2}{\partial |V_2|}
\end{bmatrix}
\begin{bmatrix} \Delta \delta_2 \\ \Delta \delta_3 \\ \Delta |V_2| \end{bmatrix}
$$

Jacobian elements can be evaluated by differentiating the expressions given above for P_2, P_3, Q_2 with respect to δ_2, δ_3 and $|V_2|$ and substituting the given and assumed values at the start of iteration. The changes in variables are obtained as

$$
\begin{bmatrix} \Delta \delta_2^1 \\ \Delta \delta_3^1 \\ \Delta |V_2|^1 \end{bmatrix} =
\begin{bmatrix} 24.47 & -12.23 & 5.64 \\ -12.23 & 24.95 & -3.05 \\ -6.11 & 3.05 & 22.54 \end{bmatrix}^{-1}
\begin{bmatrix} 0.73 \\ -1.62 \\ 1.96 \end{bmatrix} =
\begin{bmatrix} -0.023 \\ -0.0654 \\ 0.089 \end{bmatrix}
$$

$$
\begin{bmatrix} \delta_2^1 \\ \delta_3^1 \\ |V_2|^1 \end{bmatrix} =
\begin{bmatrix} \delta_2^0 \\ \delta_3^0 \\ |V_2|^0 \end{bmatrix} +
\begin{bmatrix} \Delta \delta_2^1 \\ \Delta \delta_3^1 \\ \Delta |V_2|^1 \end{bmatrix} =
\begin{bmatrix} 0 \\ 0 \\ 1 \end{bmatrix} +
\begin{bmatrix} -0.023 \\ -0.0654 \\ 0.089 \end{bmatrix} =
\begin{bmatrix} -0.023 \\ -0.0654 \\ 1.089 \end{bmatrix}
$$

We can now calculate [using Eq. (6.28)]

$$Q_3^1 = 0.4677$$

$$Q_{G3}^1 = Q_3^1 + Q_{D3} = 0.4677 + 0.6 = 1.0677$$

which is within limits.

If the same problem is solved using a digital computer, the solution converges in three iterations. The final results are given below:

$$V_2 = 1.081 \angle -0.024 \text{ rad}$$

$$V_3 = 1.04 \angle -0.0655 \text{ rad}$$

$$Q_{G3} = -0.15 + 0.6 = 0.45 \text{ (within limits)}$$

$$S_1 = 1.031 + j(-0.791)$$

$$S_2 = 0.5 + j1.00$$

$$S_3 = -1.5 - j0.15$$

Transmission loss = 0.031 pu

Line flows

The following matrix shows the real part of line flows

$$\begin{bmatrix} 0.0 & 0.191312E00 & 0.839861E00 \\ -0.184229E00 & 0.0 & 0.684697E00 \\ -0.826213E00 & -0.673847E00 & 0.0 \end{bmatrix}$$

The following matrix shows the imaginary part of line flows

$$\begin{bmatrix} 0.0 & -0.599464E00 & -0.191782E00 \\ 0.605274E00 & 0.0 & 0.396045E00 \\ 0.224742E00 & -0.375165E00 & 0.0 \end{bmatrix}$$

Rectangular Power-Mismatch Version

This version uses e_i and f_i the real and imaginary parts of the voltages respectively, as variables. The number of equations and variables is greater than that for Eq. (6.67), by the number of PV buses. Since at PV buses e_i and f_i can vary but $e_i^2 + f_i^2 = |V_i|^2$, a voltage-magnitude squared mismatch equation is required for each PV bus. With sparsity programming, this increase in order is of hardly any significance. Indeed each iteration is marginally faster than for Eq. (6.67) since there are no time-consuming sine and cosine terms. It may, however, be noted that even the polar version avoids these as far as possible by using rectangular arithemetic in constructing Eqs (6.64) and (6.65). However, the rectangular version seems to be slightly less reliable but faster in convergence than the polar version.

The total number of non-linear power flow equations considered in this case are fixed and equal 2 (n–1). These follow from Eqs. (6.26a) and (6.26b) and are

$$P_i \text{ (specified)} - \sum_{k=1}^{n} \{e_i(e_k G_{ik} - f_i B_{ik}) + f_i(f_k G_{ik} + e_k B_{ik})\} = 0$$

$$i = 1, 2, ..., n \tag{6.68}$$

$$i \neq \text{slack(s)}$$

$$Q_i \text{ (specified)} - \sum_{k=1}^{n} \{f_i(e_k G_{ik} - f_k B_{ik}) - e_i(f_k G_{ik} + e_k B_{ik})\} = 0$$

$$\text{(for each } PQ \text{ bus)} \tag{6.69}$$

$$|V_i| \text{ (specified)}^2 - (e_i^2 + f_i^2) = 0 \qquad \text{(for each } PV \text{ bus)} \qquad (6.70)$$

Using the NR method, the linearised equations in the rth iteration of iterative process can be written as

$$\begin{bmatrix} \Delta P \\ \Delta Q \\ \Delta |V|^2 \end{bmatrix}^{(r)} = \begin{bmatrix} J_1 & J_2 \\ J_3 & J_4 \\ J_5 & J_6 \end{bmatrix}^{(r)} \begin{bmatrix} \Delta e \\ \Delta f \end{bmatrix}^{(r)} \qquad (6.71)$$

For $i \neq j$

$$\begin{aligned} J_{1ij} &= -J_{4ij} = G_{ij} \, e_i \, f_i \\ J_{2ij} &= J_{3ij} = -B_{ij} e_i + G_{ij} f_i \\ J_{5ij} &= J_{6ij} = 0 \end{aligned} \qquad (6.72)$$

For $i = j$

$$\begin{aligned} J_{1ii} &= a_i + G_{ii} e_i + B_{ii} f_i \\ J_{4ii} &= a_i - G_{ii} e_i - B_{ii} f_i \\ J_{2ii} &= b_i - B_i \, e_i + G_{ii} \, f_i \\ J_{3ii} &= -b_i - B_{ii} e_i + G_{ii} f_i \\ J_{5ii} &= 2_{ei} \quad J6_{ii} = 2f_i \end{aligned} \qquad (6.73)$$

a_i and b_i are the components of the current flowing into node i, i.e.,

$$a_i + jb_i = \sum_{k=1}^{n} (G_{ik} + jB_{ik})(e_k + jf_k) \qquad (6.74)$$

Steps in solution procedure are similar to the polar coordinates case, except that the initial estimates of real and imaginary parts of the voltages at the PQ buses are made and the corrections required are obtained in each iteration using

$$\begin{bmatrix} \Delta e^{(r)} \\ \Delta f^{(r)} \end{bmatrix} = \begin{bmatrix} J_1^{(r)} & J_2^{(r)} \\ J_3^{(r)} & J_4^{(r)} \\ J_5^{(r)} & J_6^{(r)} \end{bmatrix}^{-1} \begin{bmatrix} \Delta P^{(r)} \\ \Delta Q^{(r)} \\ \Delta |V|^{2(r)} \end{bmatrix} \qquad (6.75)$$

The corrections are then applied to e and f and the calculations are repeated till convergence is achieved. A detailed flow chart describing the procedure for load flow analysis using NR method is given in Fig. 6.12.

6.7 DECOUPLED LOAD FLOW METHODS

An important characteristic of any practical electric power transmission system operating in steady state is the strong interdependence between real powers and bus voltages angles and between reactive powers and voltage magnitudes. This interesting property of weak coupling between P-δ and Q-V variables gave the necessary motivation in developing the decoupled load flow (DLF) method, in which P-δ and Q-V problems are solved separately.

Decoupled Newton Methods

In any conventional Newton method, half of the elements of the Jacobean matrix represent the weak coupling referred to above, and therefore may be ignored. Any such approximation reduces the true quadratic convergence to geometric one, but there are compensating computational benefits. A large number of decoupled algorithms have been developed in the literature. However, only the most popular decoupled Newton version is presented here [19].

In Eq. (6.67), the elements to be neglected are submatrices [N] and [J]. The resulting decoupled linear Newton equations become

$$[\Delta P] = [H]\,[\Delta \delta] \tag{6.76}$$

$$[\Delta Q] = [L]\left[\frac{\Delta |V|}{|V|}\right] \tag{6.77}$$

where it can be shown that

$$H_{ij} = L_{ij} = |V_i|\,|V_j|\,(C_{ij}\sin\delta_{ij} - B_{ij}\cos\delta_{ij}) \tag{6.78}$$
$$i \neq j$$

and
$$H_{ii} = -B_{ii}|V_i|^2 - Q_i \qquad \text{(Eq. (6.65))} \tag{6.79}$$

$$L_{ii} = -B_{ii}|V_i|^2 + Q_i \qquad \text{(Eq. (6.65))} \tag{6.80}$$

Equations (6.76) and (6.77) can be constructed and solved simultaneously with each other at each iteration, updating the [H] and [L] matrices in each iteration using Eqs (6.78) to (6.80). A better approach is to conduct each iteration by first solving Eq. (6.76) for $\Delta\delta$, and use the updated δ in constructing and then solving Eq. (6.77) for $\Delta|V|$. This will result in faster convergence than in the simultaneous mode.

The main advantage of the Decoupled Load Flow (DLF) as compared to the NR method is its reduced memory requirements in storing the Jacobean. There is not much of an advantage from the point of view of speed since the time per iteration of the DLF is almost the same as that of NR method and it always takes more number of iterations to converge because of the approximation.

Fast Decoupled Load Flow (FDLF)

Further physically justifiable simplifications may be carried out to achieve some speed advantage without much loss in accuracy of solution using the DLF model described in the previous subsection. This effort culminated in the development of the Fast Decoupled Load Flow (FDLF) method by B. Stott in 1974 [21]. The assumptions which are valid in normal power system operation are made as follows:

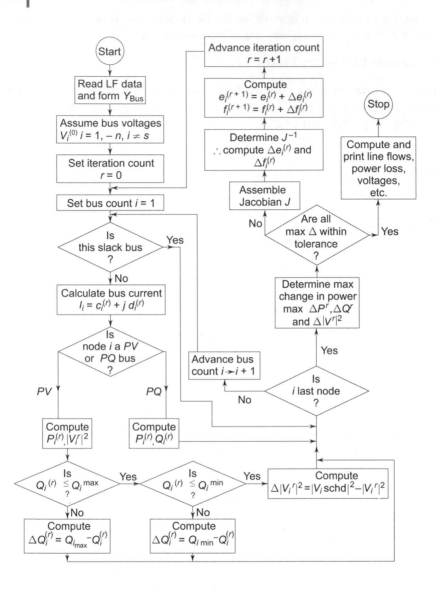

Fig. 6.12

$$\cos \delta_{ij} \approx 1;$$

$$\sin \delta_{ij} \approx 0$$

$$G_{ij} \sin \delta_{ij} \ll B_{ij}; \qquad (6.81)$$

and $$Q_i \ll B_{ii} |V_i|^2$$

With these assumptions, the entries of the $[H]$ and $[L]$ submatrices will become considerably simplified and are given by

$$H_{ij} = L_{ij} = -|V_i|\,|V_j|\,B_{ij} \qquad \text{for} \qquad i \neq j \qquad (6.82)$$

and $\qquad\qquad H_{ii} = L_{ii} = -B_{ii}\,|V_i|^2 \qquad \text{for} \qquad i = j \qquad (6.83)$

Matrices $[H]$ and $[L]$ are square matrics with dimension $(n_{PQ} + n_{PV})$ and n_{PQ} respectively.

Equations (6.76) and (6.77) can now be written as

$$[\Delta P] = [\,|V_i|\,|V_j|\,B_{ij}'\,]\,[\Delta\delta] \qquad (6.84)$$

$$[\Delta Q] = [\,|V_i|\,|V_j|\,B''_{ij}\,]\left[\frac{\Delta|V|}{|V|}\right] \qquad (6.85)$$

where B'_{ij}, B''_{ij} are elements of $[-B]$ matrix.

Further decoupling and logical simplification of the FDLF algorithm is achieved by:

1. Omitting from $[B']$ the representation of those network elements that predominantly affect reactive power flows, i.e., shunt reactances and transformer off-nominal in-phase taps;

2. Neglecting from $[B'']$ the angle shifting effects of phase shifters;

3. Dividing each of the Eqs. (6.84) and (6.85) by $|V_i|$ and setting $|V_j| = 1$ pu in the equations;

4. Ignoring series resistance in calculating the elements of $[B']$ which then becomes the dc approximation power flow matrix.

With the above modifications, the resultant simplified FDLF equations become

$$[\Delta P/\,|V|\,] = [B']\,[\Delta\delta] \qquad (6.86)$$

$$[\Delta Q/\,|V|\,] = [B'']\,[\Delta|V|] \qquad (6.87)$$

In Eqs. (6.86) and (6.87), both $[B']$ and $[B'']$ are real, sparse and have the structures of $[H]$ and $[L]$, respectively. Since they contain only admittances, they are constant and need to be inverted only once at the beginning of the study. If phase shifters are not present, both $[B']$ and $[B'']$ are always symmetrical, and their constant sparse upper triangular factors are calculated and stored only once at the beginning of the solution.

Equations (6.86) and (6.87) are solved alternatively always employing the most recent voltage values. One iteration implies one solution for $[\Delta\delta]$, to update $[\delta]$ and then one solution for $[\Delta\,|V|]$ to update $[|V|]$ to be called $1-\delta$ and $1-V$ iteration. Separate convergence tests are applied for the real and reactive power mismatches as follows:

$$\max\,[\Delta P] \leq \varepsilon_p;\ \max\,|\Delta Q| \leq \varepsilon_Q \qquad (6.88)$$

where ε_p and ε_Q are the tolerances.

A flow chart giving FDLF algorithm is presented in Fig. 6.13.

Example 6.8

Consider the three-bus system of Example 6.6. Use (a) Decoupled NR method and (b) FDLF method to obtain one iteration of the load flow solution.
Solution:
(a) *Decoupled NR method:*
Equations to be solved are (see Eqs. (6.76) and (6.77)).
Substituting relevant values in Eqs. (6.78) – (6.80) we get

$$H_{22} = 0.96 + 23.508 = 24.47$$

$$H_{23} = H_{32} = 1.04 \, (-B_{23}) = -1.04 \times 11.764 = -12.23$$

$$H_{33} = -Q_3 - B_{33} \, (1.04)^2$$

$$= [-B_{31} \, |V_3|^2 - B_{32} \, |V_3| - B_{33} \, |V_3|^2] - B_{33} \, (1.04)^2$$

$$= -11.764 \times (1.04)^2 - 11.764 \times 1.04 + (1.04)^2$$

$$\times 2 \times 23.508 = 25.89$$

$$L_{22} = Q_2 - B_{22} = 1 + 23.508 = 24.508$$

$$\begin{bmatrix} 0.73 \\ -1.62 \end{bmatrix} = \begin{bmatrix} 24.47 & -12.23 \\ -12.23 & 25.89 \end{bmatrix} \begin{bmatrix} \Delta\delta_2^{(1)} \\ \Delta\delta_3^{(1)} \end{bmatrix} \tag{i}$$

$$[\Delta Q_2] = [24.51] \begin{bmatrix} \dfrac{\Delta|V_2^{(1)}|}{|V_2^{(1)}|} \end{bmatrix} \tag{ii}$$

Solving Eq. (i) we get

$$\Delta\delta_2^{(1)} = -0.0082 - 0.0401 = -0.002$$

$$\Delta\delta_3^{(1)} = -0.018 - 0.08 = -0.062$$

$$Q_2 = -|V_2| \, |V_1| \, |Y_{21}| \sin (\theta_{21} + \delta_1 - \delta_2) - |V_2|^2 \, |Y_{22}| \sin \theta_{22}$$
$$- |V_2| \, |V_3| \, |Y_{23}| \sin (\theta_{23} + \delta_2 - \delta_3)$$

$$= -1.04 \times 12.13 \sin (104.04 + 0 - 0.115) - 24.23$$
$$\sin(-75.95°) - 1.04 \times 12.13 \sin (104.04 + .115 - 3.55°)$$

$$= -12.24 + 23.505 - 12.39$$

$$Q_2 = -1.125$$

∴ $$\Delta Q_2 = 1 - (-1.125) = 2.125$$

Substituting in Eq. (ii)

∴ $$[2.125] = [24.51] \begin{bmatrix} \dfrac{\Delta|V_2^{(1)}|}{|V_2^{(1)}|} \end{bmatrix}$$

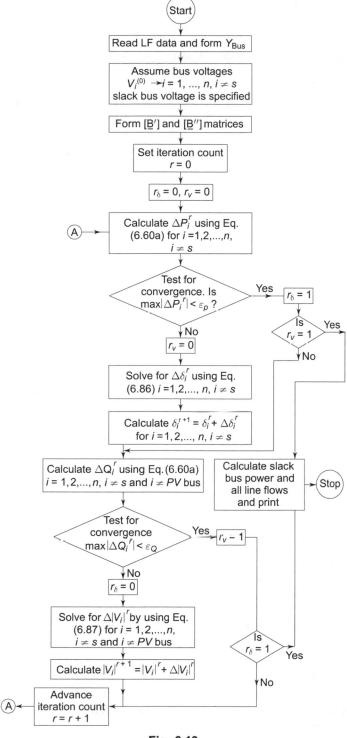

Fig. 6.13

$$\therefore \qquad \Delta|V_2^{(1)}| = 0.086$$

$$\therefore \qquad |V_2^{(1)}| = |V_2^{(0)}| + \Delta|V_2^{(1)}| = 1.086 \text{ pu}$$

$Q_3^{(1)}$ can be similarly calculated using Eq. (6.28).

(b) *FDLF Method*

The matrix equations for the solution of load flow by FDLF method are [see Eqs. (6.86) and (6.87)]

$$\begin{bmatrix} \dfrac{\Delta P_2^{(0)}}{|V_2^0|} \\ \dfrac{\Delta P_3^{(0)}}{|V_3^0|} \end{bmatrix} = \begin{bmatrix} -B_{22} & -B_{23} \\ -B_{32} & -B_{33} \end{bmatrix} \begin{bmatrix} \Delta\delta_2^{(1)} \\ \Delta\delta_3^{(1)} \end{bmatrix} \tag{iii}$$

and

$$\begin{bmatrix} \dfrac{\Delta Q_2^{(0)}}{|V_2^{(0)}|} \end{bmatrix} = [-B_{22}] \, [\Delta|V^{(i)}|] \tag{iv}$$

$$\therefore \qquad \begin{bmatrix} 0.73 \\ -1.62 \\ 1.04 \\ =-1.557 \end{bmatrix} = \begin{bmatrix} 23.508 & -11.764 \\ -11.764 & 23.508 \end{bmatrix} \begin{bmatrix} \Delta\delta_2^{(1)} \\ \Delta\delta_3^{(1)} \end{bmatrix} \tag{v}$$

Solving Eq. (v) we get

$$\Delta\delta_2^{(1)} = -0.003$$

$$\Delta\delta_3^{(1)} = -0.068$$

$$\therefore \qquad \delta_2^{(1)} = -0.003 \text{ rad}; \ \delta_3^{(1)} = -0.068 \text{ rad}$$

$$[2.125] = [23.508] \, [\Delta|V_2'|]$$

$$\therefore \qquad \Delta|V_2'| = 0.09$$

$$|V_2^1| = 1.09 \text{ pu}$$

Now Q_3 can be calculated.

These values are used to compute bus power mismatches for the next iteration. Using the values of $[\Delta P/|V|]$ and $[\Delta Q/|V|]$ the above equations are solved alternatively, using the most recent values, till the solution converges within the specified limits.

6.8 COMPARISON OF LOAD FLOW METHODS

In this section, GS and NR methods are compared when both use Y_{BUS} as the network model. It is experienced that the GS method works well when programmed using rectangular coordinates, whereas NR requires more memory when rectangular coordinates are used. Hence, polar coordinates are preferred for the NR method.

The GS method requires the fewest number of arithmetic operations to complete an iteration. This is because of the sparsity of the network matrix and the simplicity of the solution techniques. Consequently, this method requires less time per iteration. With the NR method, the elements of the Jacobian are to be computed in each iteration, so the time is considerably longer. For typical large systems, the time per iteration in the NR method is roughly equivalent to 7 times that of the GS method [20]. The time per iteration in both these methods increases almost directly as the number of buses of the network.

The rate of convergence of the GS method is slow (linear convergence characteristic), requiring a considerably greater number of iterations to obtain a solution than the NR method which has quadratic convergence characteristics and is the best among all methods from the standpoint of convergence. In addition, the number of iterations for the GS method increases directly as the number of buses of the network, whereas the number of iterations for the NR method remains practically constant, independent of system size. The NR method needs 3 to 5 iterations to reach an acceptable solution for a large system. In the GS method and other methods, convergence is affected by the choice of slack bus and the presence of series capacitor, but the sensitivity of the NR method is minimal to these factors which cause poor convergence.

Therefore, for large systems the NR method is faster, more accurate and more reliable than the GS method or any other known method. In fact, it works for any size and kind of problem and is able to solve a wider variety of ill-conditioned problems [23]. Its programming logic is considerably more complex and it has the disadvantage of requiring a large computer memory even when a compact storage scheme is used for the Jacobian and admittance matrices. In fact, it can be made even faster by adopting the scheme of optimally renumbered buses. The method is probably best suited for optimal load flow studies (Chapter 7) because of its high accuracy which is restricted only by round-off errors.

The chief advantage of the GS method is the ease of programming and most efficient utilization of core memory. It is, however, restricted in use of small size system because of its doubtful convergence and longer time needed for solution of large power networks.

Thus the NR method is decidedly more suitable than the GS method for all but very small systems.

For FDLF, the convergence is geometric, two to five iterations are normally required for practical accuracies, and it is more reliable than the formal NR method. This is due to the fact that the elements of $[B']$ and $[B'']$ are fixed approximation to the tangents of the defining functions $\Delta P/|V|$ and $\Delta Q/|V|$, and are not sensitive to any 'humps' in the defining functions.

If $\Delta P/|V|$ and $\Delta Q/|V|$ are calculated efficiently, then the speed for iterations of the FDLF is nearly five times that of the formal NR or about two-thirds that of the GS method. Storage requirements are around 60 percent of the formal NR, but slightly more than the decoupled NR method.

Changes in system configurations can be easily taken into account and though adjusted solutions take many more iterations, each one of them takes less time and hence the overall solution time is still low.

The FDLF can be employed in optimization studies and is specially used for accurate information of both real and reactive power for multiple load flow studies, as in contingency evaluation for system security assessment and enhancement analysis.

Note: When a series of load flow calculations are performed, the final values of bus voltages in each case are normally used as the initial voltages of the next case. This reduces the number of iterations, particularly when there are minor changes in system conditions.

6.9 CONTROL OF VOLTAGE PROFILE

Control by Generators

Control of voltage at the receiving bus in the fundamental two-bus system was discussed in Section 5.10. Though the same general conclusions hold for an interconnected system, it is important to discuss this problem in greater detail.

At a bus with generation, voltage can be conveniently controlled by adjusting generator excitation. This is illustrated by means of Fig. 6.14 where the equivalent generator at the ith bus is modelled by a synchronous reactance (resistance is assumed negligible) and voltage behind synchronous reactance. It immediately follows upon application of Eqs. (5.71) and (5.73) that

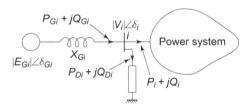

Fig. 6.14

$$P_{Gi} = \frac{|V_i||E_{Gi}|}{X_{Gi}} \sin(\delta_{Gi} - \delta_i) \qquad (6.89)$$

$$Q_{Gi} = \frac{|V_i|}{X_{Gi}}(-|V_i| + |E_{Gi}|) \qquad (6.90)$$

With $(P_{Gi} + jQ_{Gi})$ and $|V_i| \angle \delta_i$ given by the load flow solution, these values can be achieved at the bus by adjusting generator excitation to give $|E_{Gi}|$ as required by Eq. (6.90) and by adjusting the governor setting so that power input to generator from turbine is P_{Gi} plus losses, resulting in load angle of $(\delta_{Gi} - \delta_i)$ corresponding to Eq. (6.89). If Q_{Gi} demand exceeds the capacity of generators, VAR generators (synchronous or static capacitor) have to be used to modify the local load.

It follows from above that to control the voltage profile of an inter-connected system, buses with generators are usually made PV (i.e. voltage control) buses. Load flow solution then gives the voltage levels at the load buses. If some of the load bus voltages work out to be less than the specified lower voltage limit, it is indicative of the fact that the reactive power flow capacity of transmission lines for specified voltage limits cannot meet the reactive load demand (reactive line flow from bus i to bus k is proportional to $|\Delta V| = |V_i| - |V_k|$). This situation can be remedied by installing VAR generators at some of the load buses. These buses in the load flow analysis are then regarded as PV buses with the resulting solution giving the requisite values of VAR (jQ_C) injection at these buses.

The fact that positive VAR injection at any bus of an interconnected system would help to raise the voltage at the bus is easily demonstrated below: Figure 6.15a shows the Thevenin equivalent circuit of the power system as seen from the ith bus. Obviously, $E_{th} = V_i$. If now jQ_C from VAR generator is injected into this bus as shown in Fig. 6.15b, we have from Eq. (5.73)

$$|\Delta V| = |E_{th}| - |V_i'| = -\frac{X_{th}}{|V_i'|}\,Q_C$$

or

$$|V_i'| = |E_{th}| + \frac{X_{th}}{|V_i'|}\,Q_C$$

$$= |V_i| + \frac{X_{th}}{|V_i'|}\,Q_C$$

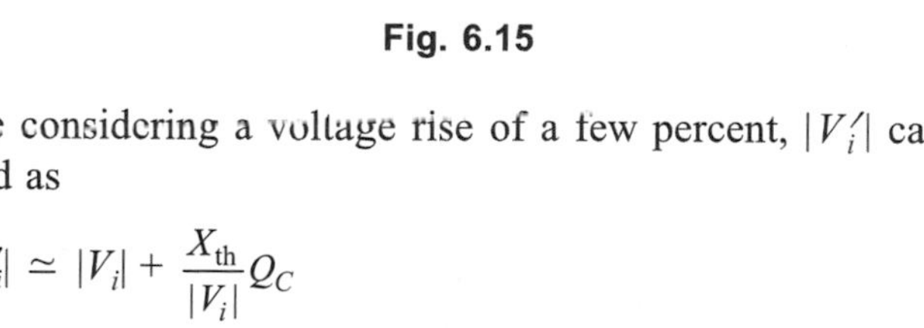

Fig. 6.15

Since we are considering a voltage rise of a few percent, $|V_i'|$ can be further approximated as

$$|V_i'| \simeq |V_i| + \frac{X_{th}}{|V_i|}\,Q_C \tag{6.91}$$

Thus the VAR injection of $+jQc$ causes the voltage at the ith bus to rise approximately by $(X_{th}/|V_i|)Q_C$. The voltages at other load buses will also rise owing to this injection to a varying but smaller extent.

Control by Transformers

Apart from being VAR generators, transformers provide a convenient means of controlling real power, and reactive power flow along a transmission line. As has already been clarified, real power is controlled by means of shifting the phase of voltage, and reactive power by changing its magnitude. Voltage magnitude can be changed by transformers provided with tap changing under load (TCUL) gear. Transformers specially designed to adjust voltage magnitude or phase angle through small values are called *regulating transformers*.

Figure 6.16 shows a regulating transformer for control of voltage magnitude, which is achieved by adding in-phase boosting voltage in the line. Figure 6.17a shows a regulating transformer which shifts voltage phase angle with no appreciable change in its magnitude. This is achieved by adding a voltage in series with the line at 90° phase angle to the corresponding line to neutral voltage as illustrated by means of the phasor diagram of Fig. 6.17b. Here

$$V'_{an} = (V_{an} + tV_{bc}) = (1 - j\sqrt{3}t) \, V_{an} = \alpha V_{an} \qquad (6.92)$$

where $\alpha = (1 - j\sqrt{3}t) \simeq 1\angle - \tan^{-1}\sqrt{3}t$
since t is small.

The presence of regulating transformers in lines modifies the Y_{BUS} matrix thereby modifying the load flow solution. Consider a line, connecting two buses, having a regulating transformer with off-nominal turns (tap) ratio α included at one end as shown in Fig. 6.18a. It is quite accurate to neglect the small impedance of the regulating transformer, i.e. it is regarded as an ideal device. Figure 6.18b gives the corresponding circuit representation with line represented by a series admittance.

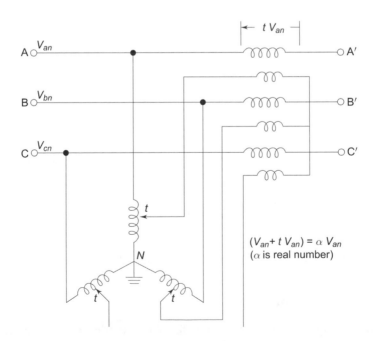

Fig. 6.16 Regulating transformer for control of voltage magnitude

Since the transformer is assumed to be ideal, complex power output from it equals complex power input, i.e.

$$S_1 = V_1 I_1^* = \alpha V_1 I_1^{\prime*}$$

or

$$I_1 = \alpha^* I_1' \tag{6.93}$$

For the transmission line

$$I_1' = y \, (\alpha V_1 - V_2)$$

or

$$I_1 = \alpha^* I_1' = |\alpha|^2 y V_1 - \alpha^* y V_2 \tag{6.94}$$

Also

$$I_2 = y(V_2 - \alpha V_1) = -\alpha y V_1 + y V_2 \tag{6.95}$$

Equations (6.94) and (6.95) cannot be represented by a bilateral network. The Y matrix representation can be written down as follows from Eqs. (6.94) and (6.95).

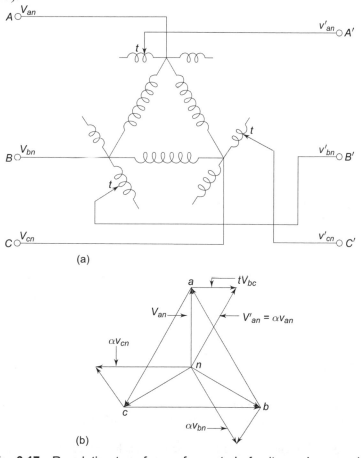

Fig. 6.17 Regulating transformer for control of voltage phase angle

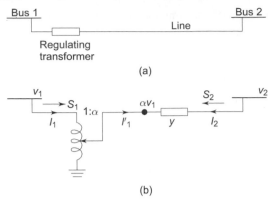

(a)

(b)

Fig. 6.18 Line with regulating transformer and its circuit representation

$$Y = \begin{bmatrix} |\alpha|^2 y & -\alpha^* y \\ -\alpha y & y \end{bmatrix} \tag{6.96}$$

The entries of Y matrix of Eq. (6.96) would then be used in writing, the Y_{BUS} matrix of the complete power network.

For a voltage regulating transformer α is real, i.e. $\alpha^* = \alpha$, therefore, Eqs. (6.94) and (6.95) can be represented by the π-network of Fig. 6.19.

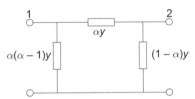

Fig. 6.19 Circuit representation of a line with off-nominal tap setting or voltage regulating transformer

If the line shown in Fig. 6.18a is represented by a π-network with shunt admittance y_0 at each end, additional shunt admittance $|\alpha|^2 y_0$ appears at bus 1 and y_0 at bus 2.

The above derivations also apply for a transformer with off-nominal tap setting[*] where $\alpha = (kV)_{Base}/(kV)_{Tap}$, a real value.

Example 6.7

The four-bus system of Fig. 6.5 is now modified to include a regulating transformer in line 3-4 near bus 3. Find the modified Y_{BUS} of the system for

---

[*]With off-nominal tap setting transformers at each end of the line as shown in Fig. 6.20, we can write.

(i) $V_3/V_3' = 1.04$ or $\alpha = 1/1.04$

(ii) $V_3/V_3' = e^{j3}$ or $\alpha = e^{-j3\circ}$

Solution (i) With regulating transformer in line 3-4, the elements of the corresponding submatrix in (v) of Example 6.2 are modified as under.

$$
\begin{array}{cc}
\;3 & 4 \\
\begin{array}{c} 3 \\ \\ 4 \end{array}
\left[
\begin{array}{c|c}
\begin{array}{c}(1-j3)+(0.666-j2) \\ +\dfrac{1}{(1.04)^2}(2-j6)\end{array} & \dfrac{1}{1.04}(-2+j6) \\ \hline
\dfrac{1}{1.04}(-2+j6) & 3-j9
\end{array}
\right]
\end{array}
=
\begin{array}{c} 3 \\ \\ 4 \end{array}
\left[
\begin{array}{c|c}
3.516 & -1.92 \\
-j10.547 & +j5.777 \\ \hline
-1.92 & \\
+j5.777 & 3-j9
\end{array}
\right]
$$

(ii) Modified submatrix in (v) of Example 6.2 is

$$
\begin{array}{c} 3 \\ \\ 4 \end{array}
\left[
\begin{array}{c|c}
\begin{array}{c}(1-j3) \\ +(0.666-j2) \\ +(2-j6)\end{array} & e^{j3^\circ}(-2+j6) \\ \hline
e^{-j3^\circ}(-2+j6) & 3-j9
\end{array}
\right]
=
\begin{array}{c} 3 \\ \\ 4 \end{array}
\left[
\begin{array}{c|c}
3.666-j11 & -2.3113 \\
 & +j5.8871 \\ \hline
-1.683 & \\
+j6.0965 & 3-j9
\end{array}
\right]
$$

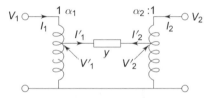

Fig. 6.20 Off-nominal transformers at both line ends (α_1, α_2 real)

$$
\begin{bmatrix} y & -y \\ -y & y \end{bmatrix}
\begin{bmatrix} V_1' \\ V_2' \end{bmatrix}
=
\begin{bmatrix} I_1' \\ I_2' \end{bmatrix}
\tag{i}
$$

$$
\begin{bmatrix} V_1' \\ V_2' \end{bmatrix}
=
\begin{bmatrix} \alpha_1 & 0 \\ 0 & \alpha_2 \end{bmatrix}
\begin{bmatrix} V_1 \\ V_2 \end{bmatrix};
\quad
\begin{bmatrix} I_1' \\ I_2' \end{bmatrix}
=
\begin{bmatrix} 1/\alpha_1 & 0 \\ 0 & 1/\alpha_2 \end{bmatrix}
\begin{bmatrix} I_1 \\ I_2 \end{bmatrix}
\tag{ii}
$$

Substituting (ii) in (i) and solving we get

$$
\begin{bmatrix} \alpha_1^2 y & -\alpha_1\alpha_2 y \\ -\alpha_1\alpha_2 y & \alpha_2^2 y \end{bmatrix}
\begin{bmatrix} V_1 \\ V_2 \end{bmatrix}
=
\begin{bmatrix} I_1 \\ I_2 \end{bmatrix}
\tag{iii}
$$

Thus

$$
Y = \begin{bmatrix} \alpha_1^2 y & -\alpha_1\alpha_2 y \\ -\alpha_1\alpha_2 y & \alpha_2^2 y \end{bmatrix}
\tag{iv}
$$

Note: Solve it for the case when α_1, α_2 are complex.

6.10 CONCLUSION

In this chapter, perhaps the most important power system study, viz. load flow has been introduced and discussed in detail. Important methods available have been briefly described. It is almost impossible to say which one of the existing methods is the best, because the behaviour of different load flow methods is dictated by the types and sizes of the problems to be solved as well as the precise details to implementation. Choice of a particular method in any given situation is normally a compromise between the various criteria of goodness of the load flow methods. It would not be incorrect to say that among the existing methods no single load flow method meets all the desirable requirements of an ideal load flow method; high speed, low storage, reliability for ill-conditioned systems, versatility in handling various adjustments and simplicity in programming. Fortunately, not all the desirable features of a load flow method are needed in all situations.

Inspite of a large number of load flow methods available, it is easy to see that only the NR and the FDLF load flow methods are the most important ones for general purpose load flow analysis. The FDLF method is clearly superior to the NR method from the point of view of speed as well as storage. Yet, the NR method is still in use because of its high versatility, accuracy and reliability and as such is widely being used for a variety of system optimization calculations; it gives sensitivity analyses and can be used in modern dynamic-response and outage-assessment calculations. Of course newer methods would continue to be developed which would either reduce the computation requirements for large systems or which are more amenable to on-line implementation.

Problems

6.1 For the power system shown in Fig. P-6.1, obtain the bus incidence matrix *A*. Take ground as reference. Is this matrix unique? Explain.

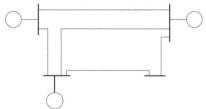

Fig. P-6.1

6.2 For the network shown in Fig. P-6.2, obtain the complex bus bar voltage at bus 2 at the end of the first iteration. Use the GS method. Line impedances shown in Fig. P-6.2 are in pu. Given:

Bus 1 is slack bus with $V_1 = 1.0 \angle 0°$

$P_2 + jQ_2 = -5.96 + j1.46$

$|V_3| = 1.02$

Assume:

$V_3^0 = 1.02\angle 0°$ and $V_2^0 = 1\angle 0°$

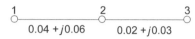

Fig. P-6.2

6.3 For the system of Fig. P-6.3 find the voltage at the receiving bus at the end of the first iteration. Load is $2 + j0.8$ pu. Voltage at the sending end (slack) is $1 + j0$ pu. Line admittance is $1.0 - j4.0$ pu. Transformer reactance is $j0.4$ pu. Off-nominal turns ratio is $1/1.04$. Use the GS technique. Assume $V_R = 1\angle 0°$.

Fig. P-6.3

6.4 (a) Find the bus incidence matrix A for the four-bus system in Fig. P-6.4. Take ground as a reference.

(b) Find the primitive admittance matrix for the system. It is given that all the lines are characterized by a series impedance of $0.1 + j0.7$ Ω/km and a shunt admittance of $j0.35 \times 10^{-5}$ $\mho$/km. Lines are rated at 220 kV.

(c) Find the bus admittance matrix for the system. Use the base values 220 kV and 100 MVA. Express all impedances and admittances in per unit.

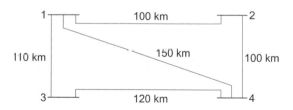

Fig. P-6.4

6.5 Consider the three-bus system of Fig. P-6.5. The pu line reactances are indicated on the figure; the line resistances are negligible. The magnitude of all the three-bus voltages are specified to be 1.0 pu. The bus powers are specified in the following table.

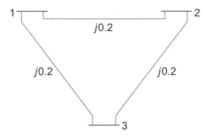

Fig. P-6.5

Bus	Real demand	Reactive demand	Real generation	Reactive generation
1	$P_{D1} = 1.0$	$Q_{D1} = 0.6$	$P_{G1} = ?$	Q_{G1} (unspecified)
2	$P_{D2} = 0$	$Q_{D2} = 0$	$P_{G2} = 1.4$	Q_{G2} (unspecified)
3	$P_{D3} = 1.0$	$Q_{D3} = 1.0$	$P_{G3} = 0$	Q_{G3} (unspecified)

Carry out the complete approximate load flow solution. Mark generations, load demands and line flows on the one-line diagram.

6.6 (a) Repeat Problem 6.5 with bus voltage specifications changed as below:

$|V_1| = 1.00$ pu

$|V_2| = 1.04$ pu

$|V_3| = 0.96$ pu

Your results should show that no significant change occurs in real power flows, but the reactive flows change appreciably as Q is sensitive to voltage.

(b) Resolve Problem 6.5 assuming that the real generation is scheduled as follows:

$P_{G1} = 1.0$ pu, $P_{G2} = 1.0$ pu, $P_{G3} = 0$

The real demand remains unchanged and the desired voltage profile is flat, i.e. $|V_1| = |V_2| = |V_3| = 1.0$ pu. In this case the results will show that the reactive flows are essentially unchanged, but the real flows are changed.

6.7 Consider the three-bus system of Problem 6.5. As shown in Fig. P-6.7 where a regulating transformer (RT) is now introduced in the line 1–2 near bus 1. Other system data remain as that of Problem 6.5. Consider two cases:

(i) RT is a magnitude regulator with a ratio $= V_1/V_1' = 0.99$,

(ii) RT is a phase angle regulator having a ratio $= V_1/V_1' = e^{j3°}$

(a) Find out the modified Y_{BUS} matrix.

(b) Solve the load flow equations in cases (i) and (ii). Compare the load flow picture with the one in Problem 6.5. The reader should verify that in case (i) only the reactive flow will change; whereas in case (ii) the changes will occur in the real power flow.

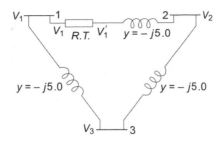

Fig. P-6.7 Three-bus sample system containing a
regulating transformer

6.8 Calculate V_3 for the system of Fig. 6.5 for the first iteration, using the data of Example 6.4. Start the algorithm with calculations at bus 3 rather than at bus 2.

6.9 For the sample system of Example 6.4 with bus 1 as slack, use the following methods to obtain a load flow solution.

(a) Gauss-Seidel using Y_{BUS}, with acceleration factor of 1.6 and tolerances of 0.0001 for the real and imaginary components of voltage.

(b) Newton-Raphson using Y_{BUS}, with tolerances of 0.01 pu for changes in the real and reactive bus powers.

Note: This problem requires the use of the digital computer.

6.10 Perform a load flow study for the system of Problem 6.4. The bus power and voltage specifications are given in Table P-6.10.

Table P-6.10

Bus	Bus power, pu		Voltage magnitude, pu	Bus type
	Real	Reactive		
1	Unspecified	Unspecified	1.02	Slack
2	0.95	Unspecified	1.01	PV
3	− 2.0	− 1.0	Unspecified	PQ
4	− 1.0	− 0.2	Unspecified	PQ

Compute the unspecified bus voltages, all bus powers and all line flows. Assume unlimited Q sources. Use the NR method.

References

Books

1. Mahalanabis, A.K., D.P. Kothari and S.I. Ahson, *Computer Aided Power System Analysis and Control*, Tata McGraw-Hill, New Delhi, 1988.

2. Weedy, B.M. and B.J. Cory, *Electrical Power Systems*, 4th Ed., John Wiley, New York, 1998.
3. Gross, C.A., *Power System Analysis*, 2nd Ed., John Wiley, New York, 1986.
4. Sterling, M.J.H., *Power System Control*, IEE, England, 1978.
5. Elgerd, O.I., *Electric Energy System Theory: An Introduction*, 2nd Ed, McGraw-Hill, New York, 1982.
6. Stagg, G.W. and A.H. El-Abiad, *Computer Methods in Power System Analysis*, McGraw-Hill, New York, 1968.
7. Rose, D.J. and R.A Willough (Eds), *Sparse Matrices and their Applications*, Plenum, New York, 1972.
8. Anderson, P.M., *Analysis of Faulted Power Systems*, The Iowa State University Press, Ames, Iowa, 1973.
9. Brown, H.E., *Solution of Large Networks by Matrix Methods*, John Willey, New York, 1975.
10. Knight, U.G., *Power System Engineering and Mathematics*, Pergamon Press, New York 1972.
11. Shipley, R.B., *Introduction to Matrices and Power Systems*, John Wiley, New York, 1976.
12. Happ, H.H., *Diakoptics and Networks, Academic Press*, New York, 1971.
13. Arrillaga, J. and N.R. Watson, *Computer Modelling of Electrical Power Systems*, 2nd edn, Wiley, New York, 2001.
14. Nagrath, I.J. and D.P. Kothari, *Power System Engineering*, Tata McGraw-Hill, New Delhi, 1994.
15. Bergen, A.R., *Power System Analysis*, Prentice-Hall, Englewood Cliffs, N.J. 1986.
16. Arrillaga, J and N.R. Watson, *Computer Modelling of Electrical Power Systems*, 2nd edn. Wiley, N.Y. 2001.

Papers

17. Happ, H.H., 'Diakoptics—The solution of System Problems by Tearing', *Proc. of the IEEE*, July 1974, 930.
18. Laughton, M.A., 'Decomposition Techniques in Power System Network Load Flow Analysis Using the Nodal Impedance Matrix', *Proc. IEE*, 1968, 115, 539
19. Stott, B., 'Decoupled Newton Load Flow', *IEEE Trans.*, 1972, PAS-91, 1955.
20. Stott, B., 'Review of Load-Flow Calculation Method', *Proc. IEEE*, July 1974, 916.
21. Stott, B., and O. Alsac, 'Fast Decoupled Load Flow', *IEEE Trans.*, 1974, PAS-93: 859.
22. Tinney, W.F., and J.W. Walker, 'Direct Solutions of Sparse Network Equations by Optimally Ordered Triangular Factorizations', *Proc. IEEE*, November 1967, 55:1801.
23. Tinney, W.F. and C.E. Hart, 'Power Flow Solution by Newton's Method', *IEEE Trans.*, November 1967, No. 11, PAS-86: 1449.
24. Ward, J.B. and H.W. Hale, 'Digital Computer Solution of Power Problems', *AIEE Trans.*, June 1956, Pt III, 75: 398.

25. Murthy, P.G., D.L. Shenoy, J. Nanda and D.P. Kothari, 'Performance of Typical Power Flow Algorithms with Reference to Indian Power Systems', *Proc. of II Symp. on Power Plant Dynamics and Control*, Hyderabad, 14-16 Feb. 1979, 219.

26. Tinney, W.F. and W.S. Mayer, 'Solution of Large Sparse Systems by Ordered Triangular Factorization', *IEEE Trans. on Auto Control*, August 1973, Vol. AC-18, 333.

27. Sato, N. and W.F Tinney, 'Techniques for Exploiting Sparsity of Network Admittance Matrix', *IEEE Trans.* PAS, December 1963, 82, 44.

28. Sachdev, M.S. and T.K.P. Medicherla, 'A Second Order Load Flow Technique', *IEEE Trans.*, PAS, Jan/Feb 1977, 96, 189.

29. Iwamoto, S.and Y. Tamura, 'A Fast Load Flow Method Retaining Non linearity', *IEEE Trans.*, PAS. Sept/Oct. 1978, 97, 1586.

30. Roy, L., 'Exact Second Order Load Flow', *Proc. of the Sixth PSCC Conf. Dermstadt*, Vol. 2, August 1978, 711.

31. Iwamoto, S. and Y. Tamura, 'A Load Flow Calculation Method for III-Conditioned Power Systems', *IEEE Trans PAS*, April 1981, 100, 1736.

32. Happ, H.H. and C.C. Young, 'Tearing Algorithms for Large Scale Network Problems', *IEEE Trans PAS*, Nov/Dec 1971, 90, 2639.

33. Dopazo, J.F.O.A. Kiltin and A.M. Sarson, 'Stochastic Load Flows', *IEEE Trans. PAS*, 1975, 94, 299.

34. Nanda, J.D.P. Kothari and S.C. Srivastava, "Some Important Observations on FDLF Algorithm", *Proc. of the IEEE*, May 1987, pp. 732-33.

35. Nanda, J., P.R., Bijwe, D.P. Kothari and D.L. Shenoy, "Second Order Decoupled Load Flow', *Electric Machines and Power Systems*, Vol 12, No. 5, 1987, pp. 301–312.

36. Nanda J., D.P. Kothari and S.C. Srivastva, "A Novel Second Order Fast Decoupled Load Flow Method in Polar Coordinates", *Electric Machines and Power Systems*, Vol. 14, No. 5, 1989, pp 339–351.

37. Das, D., H.S. Nagi and D.P. Kothari, "A Novel Method for Solving Radial Distribution Networks", *Proc. IEE*, ptc, vol. 141, no. 4, July 1994, pp. 291–298.

38. Das, D., D.P. Kothari and A. Kalam. "A Simple and Efficient Method for Load Flow Solution of Radial Distribution Networks", *Int. J. of EPES*, 1995, pp. 335-346.

7

Optimal System Operation

7.1 INTRODUCTION

The optimal system operation, in general, involved the consideration of economy of operation, system security, emissions at certain fossil-fuel plants, optimal releases of water at hydro generation, etc. All these considerations may make for conflicting requirements and usually a compromise has to be made for optimal system operation. In this chapter we consider the economy of operation only, also called the *ecomonic dispatch problem.*

The main aim in the economic dispatch problem is to minimize the total cost of generating real power (production cost) at various stations while satisfying the loads and the losses in the transmission links. For simplicity we consider the presence of thermal plants only in the beginning. In the later part of this chapter we will consider the presence of hydro plants which operate in conjunction with thermal plants. While there is negligible operating cost at a hydro plant, there is a limitation of availability of water over a period of time which must be used to save maximum fuel at the thermal plants.

In the load flow problem as detailed in Chapter 6, two variables are specified at each bus and the solution is then obtained for the remaining variables. The specified variables are real and reactive powers at PQ buses, real powers and voltage magnitudes at PV buses, and voltage magnitude and angle at the slack bus. The additional variables to be specified for load flow solution are the tap settings of regulating transformers. If the specified variables are allowed to vary in a region constrained by practical considerations (upper and lower limits on active and reactive generations, bus voltage limits, and range of transformer tap settings), there results an infinite number of load flow solutions, each pertaining to one set of values of specified variables. The 'best' choice in some sense of the values of specified variables leads to the 'best' load flow solution. Economy of operation is naturally predominant in determining allocation of generation to each station for various system load levels. The first problem in power system parlance is called the 'unit commitment' (UC) problem and the second is called the 'load scheduling' (LS) problem. One must first solve the UC problem before proceeding with the LS problem.

Throughout this chapter we shall concern ourselves with an existing installation, so that the economic considerations are that of operating (running) cost and not the capital outlay.

7.2 OPTIMAL OPERATION OF GENERATORS ON A BUS BAR

Before we tackle the unit commitment problem, we shall consider the optimal operation of generators on a bus bar.

Generator Operating Cost

The major component of generator operating cost is the fuel input/hour, while maintenance contributes only to a small extent. The fuel cost is meaningful in case of thermal and nuclear stations, but for hydro stations where the energy storage is 'apparently free', the operating cost as such is not meaningful. A suitable meaning will be attached to the cost of hydro stored energy in Section 7.7 of this chapter. Presently we shall concentrate on fuel fired stations.

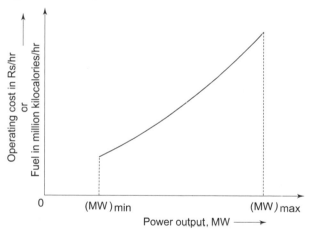

Fig. 7.1 Input-output curve of a generating unit

The input-output curve of a unit[*] can be expressed in a million kilocalories per hour or directly in terms of rupees per hour versus output in megawatts. The cost curve can be determined experimentally. A typical curve is shown in Fig. 7.1 where $(MW)_{min}$ is the minimum loading limit below which it is uneconomical (or may be technically infeasible) to operate the unit and $(MW)_{max}$ is the maximum output limit. The input-output curve has discontinuities at steam valve openings which have not been indicated in the figure. By fitting a suitable degree polynomial, an analytical expression for operating cost can be written as

---

[*]A unit consists of a boiler, turbine and generator.

$$C_i(P_{Gi}) \quad \text{Rs/hour at output } P_{Gi}$$

where the suffix i stands for the unit number. It generally suffices to fit a second degree polynomial, i.e.

$$C_i = \frac{1}{2} a_i P_{Gi}^2 + b_i P_{Gi} + d_i \text{ Rs/hour} \qquad (7.1)$$

The slope of the cost curve, i.e. $\dfrac{dC_i}{dP_{Gi}}$ is called the *incremental fuel cost (IC)*,

and is expressed in units of rupees per megawatt hour (Rs/MWh). A typical plot of incremental fuel cost versus power output is shown in Fig. 7.2. If the cost curve is approximated as a quadratic as in Eq. (7.1), we have

$$(IC)_i = a_i P_{Gi} + b_i \qquad (7.2)$$

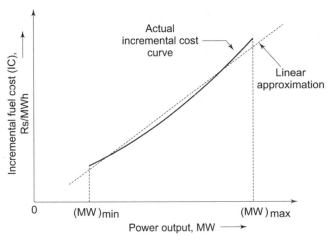

Fig. 7.2   Incremental fuel cost versus power output for the unit whose input-output curve is shown in Fig. 7.1

i.e. a linear relationship. For better accuracy incremental fuel cost may be expressed by a number of short line segments (piecewise linearization). Alternatively, we can fit a polynomial of suitable degree to represent IC curve in the inverse form

$$P_{Gi} = \alpha_i + \beta_i (IC)_i + \gamma_i (IC)^2_i + \ldots \qquad (7.3)$$

Optimal Operation

Let us assume that it is known a *priori* which generators are to run to meet a particular load demand on the station. Obviously

$$\Sigma P_{Gi, \text{ max}} \geq P_D \qquad (7.4)$$

where $P_{Gi, \text{ max}}$ is the rated real power capacity of the ith generator and P_D is the total power demand on the station. Further, the load on each generator is to be constrained within lower and upper limits, i.e.

$$P_{Gi, \text{ min}} \leq P_{Gi} \leq P_{Gi, \text{ max}}, \ i = 1, 2, \ldots, k \qquad (7.5)$$

Considerations of *spinning reserve*, to be explained later in this section, require that

$$\sum_i P_{Gi,\ max} > P_D \qquad (7.6)$$

by a proper margin, i.e. Eq. (7.6) must be a strict inequality.

Since the operating cost is insensitive to reactive loading of a generator, the manner in which the reactive load of the station is shared among various on-line generators does not affect the operating economy[*].

The question that has now to be answered is: 'What is the optimal manner in which the load demand P_D must be shared by the generators on the bus?' This is answered by minimizing the operating cost

$$C = \sum_{t=1}^{k} C_i(P_{Gi}) \qquad (7.7)$$

under the equality constraint of meeting the load demand, i.e.

$$\sum_{i=1}^{k} P_{Gi} - P_D = 0 \qquad (7.8)$$

where k = the number of generators on the bus.

Further, the loading of each generator is constrained by the inequality constraint of Eq. (7.5).

Since $C_i(P_{Gi})$ is non-linear and C_i is independent of P_{Gj} ($j \neq i$), this is a separable non-linear programming problem.

If it is assumed at present, that the inequality constraint of Eq. (7.4) is not effective, the problem can be solved by the method of Lagrange multipliers. Define the Lagrangian as

$$\mathrm{L} = \sum_{i=1}^{k} C_i(P_{Gi}) - \lambda \left[\sum_{i=1}^{k} P_{Gi} - P_D \right] \qquad (7.9)$$

where λ is the Lagrange multiplier.
Minimization is achieved by the condition

$$\frac{\partial \mathrm{L}}{\partial P_{Gi}} = 0$$

or
$$\frac{dC_i}{dP_{Gi}} = \lambda; \ i = 1, 2, ..., k \qquad (7.10)$$

where $\dfrac{dC_i}{dP_{Gi}}$ is the incremental cost of the ith generator (units: Rs/MWh),

a function of generator loading P_{Gi}. Equation (7.10) can be written as

$$\frac{dC_1}{dP_{G1}} = \frac{dC_2}{dP_{G2}} = ... = \frac{dC_k}{dP_{Gk}} = \lambda \qquad (7.11)$$

---

[*]The effect of reactive loading on generator losses is of negligible order.

i.e. the optimal loading of generators corresponds to the equal incremental cost point of all the generators. Equation (7.11), called the *coordination equations* numbering k are solved simultaneously with the load demand equation (7.8) to yield a solution for the Lagrange multiplier λ and the optimal loading of k generators. This is illustrated by means of Example 7.1 at the end of this section.

Computer solution for optimal loading of generators can be obtained iteratively as follows:

1. Choose a trial value of λ, i.e. $IC = (IC)^0$.
2. Solve for P_{Gi} ($i = 1, 2, ..., k$) from Eq. (7.3).
3. If $|\sum_i P_{Gi} - P_D| < \varepsilon$ (a specified value), the optimal solution is reached. Otherwise,
4. Increment (IC) by $\Delta (I_C)$, if $\left[\sum_i P_{Gi} - P_D\right] < 0$ or decrement (I_C) by $\Delta(I_C)$ if $[\sum P_{Gi} - P_D] > 0$ and repeat from step 2. This step is possible because P_{Gi} is monotonically increasing function of (I_C).

Consider now the effect of the inequality constraint (7.5). As (IC) is increased or decreased in the iterative process, if a particular generator loading P_{Gj} reaches the limit $P_{Gj, \, max}$ or P_{Gj}, min, its loading from now on is held fixed at this value and the balance load is then shared between the remaining generators on equal incremental cost basis. The fact that this operation is optimal can be shown by the Kuhn-Tucker theory (*see* Appendix E).

Example 7.1

Incremental fuel costs in rupees per MWh for a plant consisting of two units are:

$$\frac{dC_1}{dP_{G1}} = 0.20P_{G1} + 40.0$$

$$\frac{dC_2}{dP_{G2}} = 0.25P_{G2} + 30.0$$

Assume that both units are operating at all times, and total load varies from 40 MW to 250 MW, and the maximum and minimum loads on each unit are to be 125 and 20 MW, respectively. How will the load be shared between the two units as the system load varies over the full range? What are the corresponding values of the plant incremental costs?

Solution At light loads, unit 1 has the higher incremental fuel cost and will, therefore, operate at its lower limit of 20 MW, for which dC_1/dP_{G1} is Rs 44 per MWh. When the output of unit 2 is 20 MW, $dC_2/dP_{G2} = $ Rs 35 per MWh. Thus, with an increase in the plant output, the additional load should be borne by unit

2 until dC_2/dP_{G2}= Rs 44/MWh. Until this point is reached, the incremental fuel cost of the plant corresponds to that of unit 2 alone. When the plant load is 40 MW, each unit operates at its minimum bound, i.e. 20 MW with plant λ = Rs 35/MWh.

When $\qquad dC_2/dP_{G2}$ = Rs 44/MWh,

$$0.25P_{G2} + 30 = 44$$

or $\qquad\qquad P_{G2} = \dfrac{14}{0.25} = 56 \text{ MW}$

The total plant output is then (56 + 20) = 76 MW. From this point onwards, the values of plant load shared by the two units are found by assuming various values of λ. The results are displayed in Table 7.1.

Table 7.1 Output of each unit and plant output for various values of λ for Example 7.1

Plant λ, Rs/MWh	Unit 1 P_{G1}, MW	Unit 2 P_{G2}, MW	Plant Output $(P_{G1} + P_{G2})$, MW
35	20.0	20.0	40.0
44	20.0	56.0	76.0
50	50.0	80.0	130.0
55	75.0	100.0	175.0
60	100.0	120.0	220.0
61.25	106.25	125.0	231.25
65	125.0	125.0	250.0

Figure 7.3 shows the plot of the plant λ versus plant output. It is seen from Table 7.1 that at λ = 61.25, unit 2 is operating at its upper limit and therefore, the additional load must now be taken by unit 1, which then determines the plant λ.

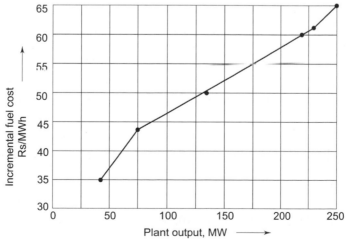

Fig. 7.3 Incremental fuel cost versus plant output, as found in Example 7.1

To find the load sharing between the units for a plant output of say 150 MW, we find from the curve of Fig. 7.3, that the corresponding plant λ is Rs 52.22 per MWh. Optimum schedules for each unit for 150 MW plant load can now be found as

$$0.2P_{G1} + 40 = 52.22; \qquad \therefore \qquad P_{G1} = 61.11 \text{ MW}$$

$$0.25P_{G2} + 30 = 52.22; \qquad \therefore \qquad P_{G2} = 88.89 \text{ MW}$$

$$P_{G1} + P_{G2} = 150 \text{ MW}$$

Proceeding on the above lines, unit outputs for various plant outputs are computed and have been plotted in Fig. 7.4. Optimum load sharing for any plant load can be directly read from this figure.

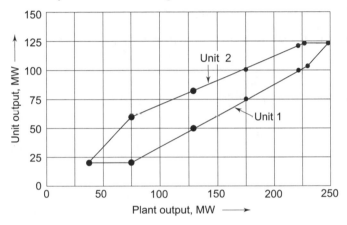

Fig. 7.4 Output of each unit versus plant output for Example 7.1

Example 7.2

For the plant described in Example 7.1 find the saving in fuel cost in rupees per hour for the optimal scheduling of a total load of 130 MW as compared to equal distribution of the same load between the two units.

Solution Example 7.1 reveals that unit 1 should take up a load of 50 MW and unit 2 should supply 80 MW. If each unit supplies 65 MW, the increase in cost for unit 1 is

$$\int_{50}^{65} (0.2P_{G1} + 40)dP_{G1} = (0.1P_{G1}^2 + 40P_{G1})\Big|_{50}^{65} = 772.5 \text{ Rs/hr}$$

Similarly, for unit 2,

$$\int_{80}^{65} (0.25P_{G2} + 30)\, dP_{G2} = (0.125P_{G2}^2 + 30P_{G2})\Big|_{80}^{65}$$

$$= - \ 721.875 \text{ Rs/hr}$$

Net saving caused by optimum scheduling is

$$772.5 - 721.875 = 50.625 \text{ Rs/hr}$$

Total yearly saving assuming continuous operation

$$= \text{Rs } 4,43,475.00$$

This saving justifies the need for optimal load sharing and the devices to be installed for controlling the unit loadings automatically.

Example 7.3

Let the two units of the system studied in Example 7.1 have the following cost curves.

$$C_1 = 0.1P_{G1}^2 + 40P_{G1} + 120 \text{ Rs/hr}$$

$$C_2 = 0.125P_{G2}^2 + 30P_{G2} + 100 \text{ Rs/hr}$$

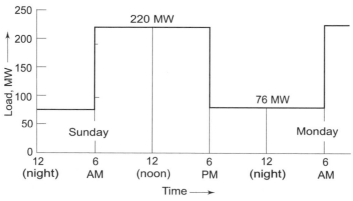

Fig. 7.5 Daily load cycle

Let us assume a daily load cycle as given in Fig. 7.5. Also assume that a cost of Rs 400 is incurred in taking either unit off the line and returning it to service after 12 hours. Consider the 24 hour period from 6 a.m. one morning to 6 a.m. the next morning. Now, we want to find out whether it would be more economical to keep both the units in service for this 24 hour period or to remove one of the units from service for the 12 hours of light load.

For the twelve-hour period when the load is 220 MW, referring to Table 7.1 of Example 7.1, we get the optimum schedule as

$$P_{G1} = 100 \text{ MW}, \quad P_{G2} = 120 \text{ MW}$$

Total fuel cost for this period is

$$[0.1 \times 100^2 + 40 \times 100 + 120 + 0.125 \times 120^2 + 30 \times 120 + 100] \times 12$$

$$= \text{Rs. } 1,27,440$$

If both units operate in the light load period (76 MW from 6 p.m. to 6 a.m.) also, then from the same table, we get the optimal schedule as

$$P_{G1} = 20 \text{ MW}, \quad P_{G2} = 56 \text{ MW}$$

Total fuel cost for this period is then

$$(0.1 \times 20^2 + 40 \times 20 + 120 + 0.125 \times 56^2 + 30 \times 56 + 100) \times 12$$
$$= \text{Rs } 37,584$$

Thus the total fuel cost when the units are operating throughout the 24 hour period is Rs 1,65,024.

If only one of the units is run during the light load period, it is easily verified that it is economical to run unit 2 and to put off unit 1. Then the total fuel cost during this period will be

$$= (0.125 \times 76^2 + 30 \times 76 + 100) \times 12 = \text{Rs } 37,224$$

∴ Total fuel cost for this case = 1,27,440 + 37,224

$$= \text{Rs } 1,64,664$$

Total operating cost for this case will be the total fuel cost plus the start-up cost of unit 1, i.e.

1,64,664 + 400 = Rs 1,65,064

Comparing this with the earlier case, it is clear that it is economical to run both the units.

It is easy to see that if the start-up cost is Rs 200, then it is economical to run only 2 in the light load period and to put off unit 1.

7.3 OPTIMAL UNIT COMMITMENT (UC)

As is evident, it is not economical to run all the units available all the time. To determine the units of a plant that should operate for a particular load is the problem of unit commitment (UC). This problem is of importance for thermal plants as for other types of generation such as hydro; their operating cost and start-up times are negligible so that their on-off status is not important.

A simple but sub-optimal approach to the problem is to impose priority ordering, wherein the most efficient unit is loaded first to be followed by the less efficient units in order as the load increases.

A straightforward but highly time-consuming way of finding the most economical combination of units to meet a particular load demand, is to try all possible combinations of units that can supply this load; to divide the load optimally among the units of each combination by use of the coordination equations, so as to find the most economical operating cost of the combination; then, to determine the combination which has the least operating cost among all these. Considerable computational saving can be achieved by using branch and bound or a dynamic programming method for comparing the economics of combinations as certain combinations neet not be tried at all.

Dynamic Programming Method

In a practical problem, the UC table is to be arrived at for the complete load cycle. If the load is assumed to increase in small but finite size steps, dynamic programming (DP) can be used to advantage for computing the UC table, wherein it is not necessary to solve the coordination equations; while at the same time the unit combinations to be tried are much reduced in number. For these reasons, only the DP approach will be advanced here.

The total number of units available, their individual cost characteristics and the load cycle on the station are assumed to be known *a priori*. Further, it shall be assumed that the load on each unit of combination of units changes in suitably small but uniform steps of size ΔMW (e.g. 1 MW).

Starting arbitrarily with any two units, the most economical combination is determined for all the discrete load levels of the combined output of the two units. At each load level the most economic answer may be to run either unit or both units with a certain load sharing between the two. The most economical cost curve in discrete form for the two units thus obtained, can be viewed as the cost curve of a single equivalent unit. The third unit is now added and the procedure repeated to find the cost curve of the three combined units. It may be noted that in this procedure the operating combinations of third and first, also third and second are not required to be worked out resulting in considerable saving in computational effort. The process is repeated, till all available units are exhausted. The advantage of this approach is that having obtained the optimal way of loading k units, it is quite easy to determine the optimal manner of loading $(k + 1)$ units.

Let a cost function $F_N(x)$ be defined as follows:

$F_N(x) = $ the minimum cost in Rs/hr of generating x MW by N units,

$f_N(y) = $ cost of generating y MW by the Nth unit

$F_{N-1}(x - y) = $ the minimum cost of generating $(x - y)$ MW by the remaining $(N - 1)$ units

Now the application of DP results in the following recursive relation

$$F_N(x) = \overset{\min}{y} \{f_N(y) + F_{N-1}(x - y)\} \tag{7.12}$$

Using the above recursive relation, we can easily determine the combination of units, yielding minimum operating costs for loads ranging in convenient steps from the minimum permissible load of the smallest unit to the sum of the capacities of all available units. In this process the total minimum operating cost and the load shared by each unit of the optimal combination are automatically determined for each load level.

The use of DP for solving the UC problem is best illustrated by means of an example. Consider a sample system having four thermal generating units with parameters listed in Table 7.2. It is required to determine the most economical units to be committed for a load of 9 MW. Let the load changes be in steps of 1 MW.

Table 7.2 Generating unit parameters for the sample system

Unit No.	Capacity (MW)		Cost curve parameters (d = 0)	
	Min	Max	a (Rs/MW^2)	b (Rs/MW)
1	1.0	12.0	0.77	23.5
2	1.0	12.0	1.60	26.5
3	1.0	12.0	2.00	30.0
4	1.0	12.0	2.50	32.0

Now

$$F_1(x) = f_1(x)$$

$$\therefore \quad F_1(9) = f_1(9) = \frac{1}{2}a_1 P_{G1}^2 + b_1 P_{G1}$$

$$= 0.385 \times 9^2 + 23.5 \times 9 = \text{Rs } 242.685/\text{hour}$$

From the recursive relation (7.12), computation is made for $F_2(0)$, $F_2(1)$, $F_2(2)$, ..., $F_2(9)$. Of these

$$F_2(9) = \min \ \{[f_2(0) + F_1(9)], \ [f_2(1) + F_1(8)],$$

$$[f_2(2) + F_1(7)], \ [f_2(3) + F_1(6)], \ [f_2(4) + F_1(5)],$$

$$[f_2(5) + F_1(4)], \ [f_2(6) + F_1(3)], \ [f_2(7) + F_1(2)],$$

$$[f_2(8) + F_1(1)], \ [f_2(9) + F_1(0)]\}$$

On computing term-by-term and comparing, we get

$$F_2(9) = [f_2(2) + F_1(7)] = \text{Rs } 239.565/\text{hour}$$

Similarly, we can calculate $F_2(8)$, $F_2(7)$, ..., $F_2(1)$, $F_2(0)$.

Using the recursive relation (7.12), we now compute $F_3(0)$, $F_3(1)$, ..., $F_3(9)$. Of these

$$F_3(9) = \min \ \{[f_3(0) + F_2(9)], \ [f_3(1) + F_2(8)], \ ..., \ [f_3(9) + F_2(0)]\}$$

$$= [f_3(0) + F_2(9)] = \text{Rs } 239.565/\text{hour}$$

Proceeding similarly, we get

$$F_4(9) = [f_4(0) + F_3(9)] = \text{Rs } 239.565/\text{hour}$$

Examination of $F_1(9)$, $F_2(9)$, $F_3(9)$ and $F_4(9)$ leads to the conclusion that optimum units to be committed for a 9 MW load are 1 and 2 sharing the load as 7 MW and 2 MW, respectively with a minimum operating cost of Rs 239.565/hour.

It must be pointed out here, that the optimal UC table is independent of the numbering of units, which could be completely arbitrary. To verify, the reader may solve the above problem once again by choosing a different unit numbering scheme.

If a higher accuracy is desired, the step size could be reduced (e.g. $\frac{1}{2}$ MW), with a considerable increase in computation time and required storage capacity.

The effect of step size could be altogether eliminated, if the branch and bound technique [30] is employed. The answer to the above problem using branch and bound is the same in terms of units to be committed, i.e. units 1 and 2, but with a load sharing of 7.34 MW and 1.66 MW, respectively and a total operating cost of Rs 239.2175/hour.

In fact the best scheme is to restrict the use of the DP method to obtain the UC table for various discrete load levels; while the load sharing among committed units is then decided by use of the coordination Eq. (7.10).

For the example under consideration, the UC table is prepared in steps of 1 MW. By combining the load range over which the unit commitment does not change, the overall result can be telescoped in the form of Table 7.3.

Table 7.3 Status* of units for minimum operating cost (Unit commitment table for the sample system)

Load range	Unit number			
	1	2	3	4
1–5	1	0	0	0
6–13	1	1	0	0
14–18	1	1	1	0
19–48	1	1	1	1

*1 = unit running; 0 = unit not running.

The UC table is prepared once and for all for a given set of units. As the load cycle on the station changes, it would only mean changes in starting and stopping of units with the basic UC table remaining unchanged.

Using the UC table and increasing load in steps, the most economical station operating cost is calculated for the complete range of station capacity by using the coordination equations. The result is the overall station cost characteristic in the form of a set of data points. A quadratic equation (or higher order equation, if necessary) can then be fitted to this data for later use in economic load sharing among generating stations.

7.4 RELIABILITY CONSIDERATIONS

With the increasing dependence of industry, agriculture and day-to-day household comfort upon the continuity of electric supply, the reliability of power systems has assumed great importance. Every electric utility is normally under obligation to provide to its consumers a certain degree of continuity and quality of service (e.g. voltage and frequency in a specified range). Therefore, economy and reliability (security) must be properly coordinated in arriving at the operational unit commitment decision. In this section, we will see how the purely economic UC decision must be modified through considerations of reliability.

In order to meet the load demand under contingency of failure (forced outage) of a generator or its derating caused by a minor defect, *static reserve capacity* is always provided at a generating station so that the total installed capacity exceeds the yearly peak load by a certain margin. This is a planning problem and is beyond the scope of this book.

In arriving at the economic UC decision at any particular time, the constraint taken into account was merely the fact that the total capacity on line was at least equal to the load. The margin, if any, between the capacity of units committed and load was incidental. If under actual operation one or more of the units were to fail perchance (random outage), it may not be possible to meet the load requirements. To start a spare (standby) thermal unit[*] and to bring it on steam to take up the load will take several hours (2–8 hours), so that the load cannot be met for intolerably long periods of time. Therefore, to meet contingencies, the capacity of units on line (running) must have a definite margin over the load requirements at all times. This margin which is known as the *spinning reserve* ensures continuity by meeting the load demand up to a certain extent of probable loss of generation capacity. While rules of thumb have been used, based on past experience to determine the system's spinning reserve at any time, Patton's analytical approach to this problem is the most promising.

Since the probability of unit outage increases with operating time and since a unit which is to provide the spinning reserve at a particular time has to be started several hours ahead, the problem of security of supply has to be treated in totality over a period of one day. Furthermore, the loads are never known with complete certainty. Also, the spinning reserve has to be provided at suitable generating stations of the system and not necessarily at every generating station. This indeed is a complex problem. A simplified treatment of the problem is presented below:

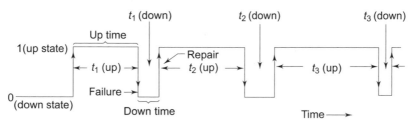

Fig. 7.6 Random unit performance record neglecting scheduled outages

A unit during its useful life span undergoes alternate periods of operation and repair as shown in Fig. 7.6. The lengths of individual operating and repair

---

[*] If hydro generation is available in the system, it could be brought on line in a matter of minutes to take up load.

periods are a random phenomenon with operating periods being much longer than repair periods. When a unit has been operating for a long time, the random phenomenon can be described by the following parameters.

Mean time to failure (mean 'up' time),

$$\overline{T}(\text{up}) = \frac{\sum_j t_j(\text{up})}{\text{No. of cycles}} \tag{7.13}$$

Mean time to repair (mean 'down' time),

$$\overline{T}(\text{down}) = \frac{\sum_j t_j(\text{down})}{\text{No. of cycles}} \tag{7.14}$$

Mean cycle time $= \overline{T}(\text{up}) + \overline{T}(\text{down})$

Inverse of these times can be defined as rates [1], i.e.

Failure rate, $\lambda = 1/\overline{T}$ (up) (failures/year)

Repair rate, $\mu = 1/\overline{T}$ (down) (repairs/year)

Failure and repair rates are to be estimated from the past data of units (or other similar units elsewhere) by use of Eqs. (7.13) and (7.14). Sound engineering judgement must be exercised in arriving at these estimates. The failure rates are affected by preventive maintenance and the repair rates are sensitive to size, composition and skill of repair teams.

By ratio definition of probability, we can write the probability of a unit being in 'up' or 'down' states at any time as

$$p(\text{up}) = \frac{\overline{T}(\text{up})}{\overline{T}(\text{up}) + \overline{T}(\text{down})} = \frac{\mu}{\mu + \lambda} \tag{7.15}$$

$$p(\text{down}) = \frac{\overline{T}(\text{down})}{\overline{T}(\text{up}) + \overline{T}(\text{down})} = \frac{\lambda}{\mu + \lambda} \tag{7.16}$$

Obviously,

$$p(\text{up}) + p(\text{down}) = 1$$

$p(\text{up})$ and $p(\text{down})$ in Eqs. (7.15) and (7.16) are also termed as *availability* and *unavailability*, respectively.

When k units are operating, the system state changes because of random outages. Failure of a unit can be regarded as an event independent of the state of other units. If a particular system state i is defined as X_i units in 'down' state and Y_i in 'up' state ($k = X_i + Y_i$), the probability of the system being in this state is

$$p_i = \prod_{j \in Y_i} p_j(\text{up}) \prod_{l \in X_i} p_l(\text{down}) \tag{7.17}$$

Patton's Security Function

A breach of system security is defined as some intolerable or undesirable condition. The only breach of security considered here is insufficient generation capacity. The Patton's security function, which quantitatively estimates the probability that the available generation capacity (sum of capacities of units committed) at a particular hour is less than the system load at that time, is defined as [25]

$$S = \Sigma p_i r_i \qquad (7.18)$$

where

p_i = probability of system being in state i [see Eq. (7.17)]

r_i = probability that system state i causes breach of system security.

When system load is deterministic (i.e. known with complete certainty), $r_i = 1$ if available capacity is less than load and 0 otherwise. S indeed, is a quantitative estimate of system insecurity.

Though theoretically Eq. (7.18) must be summed over all possible system states (this in fact can be very large), from a practical point of view the sum needs to be carried out over states reflecting a relatively small number of units on forced outage, e.g. states with more than two units out may be neglected as the probability of their occurrence will be too low.

Security Constrained Optimal Unit Commitment

Once the units to be committed at a particular load level are known from purely economic considerations, the security function S is computed as per Eq. (7.18). This figure should not exceed a certain maximum tolerable insecurity level (MTIL). MTIL for a given system is a management decision which is guided by past experience. If the value of S exceeds MTIL, the economic unit commitment schedule is modified by bringing in the next most economical unit as per the UC table. S is then recalculated and checked. The process is continued till $S \leq$ MTIL. As the economic UC table has some inherent spinning reserve, rarely more than one iteration is found to be necessary.

For illustration, reconsider the four unit example of Sec. 7.3. Let the daily load curve for the system be as indicated in Fig. 7.7. The economically optimal UC for this load curve is immediately obtained by use of the previously prepared UC table (*see* Table 7.3) and is given in Table 7.4.

Let us now check if the above optimal UC table is secure in every period of the load curve.

For the minimum load of 5 MW (period E of Fig. 7.7) according to optimal UC Table 7.4., only unit 1 is to be operated. Assuming identical failure rate λ of 1/year and repair rate μ of 99/year for all the four units, let us check if the system is secure for the period E. Further assume the system MTIL to be 0.005. Unit 1 can be only in two possible states—operating or on forced outage.

Table 7.4 Economically optimal UC table for the sample system for the load curve of Fig. 7.7

Period	Unit number			
	1	2	3	4
A	1	1	1	1
B	1	1	1	0
C	1	1	0	0
D	1	1	1	0
E	1	0	0	0
F	1	1	0	0

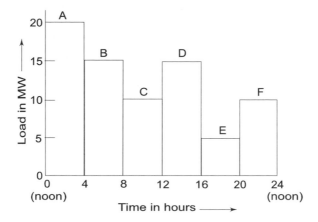

Fig. 7.7 Daily load curve

Therefore,

$$S = \sum_{i=1}^{2} p_i r_i = p_1 r_1 + p_2 r_2$$

where

$$p_1 = p(\text{up}) = \frac{\mu}{\mu + \lambda} = 0.99, \quad r_1 = 0 \ (\text{unit } 1 = 12 \text{ MW} > 5 \text{ MW})$$

$$p_2 = p(\text{down}) = \frac{\lambda}{\mu + \lambda} = 0.01, \quad r_2 = 1 \quad \text{(with unit 1 down load demand cannot be met)}$$

Hence

$$S = 0.99 \times 0 + 0.01 \times 1 = 0.01 > 0.005 \text{ (MTIL)}$$

Thus unit 1 alone supplying 5 MW load fails to satisfy the prescribed security criterion. In order to obtain optimal and yet secure UC, it is necessary to run the next most economical unit, i.e. unit 2 (Table 7.3) along with unit 1.

With both units 1 and 2 operating, security function is contributed only by the state when both the units are on forced outage. The states with both units operating or either one failed can meet the load demand of 5 MW and so do not contribute to the security function. Therefore,

$$S = p \text{ (down)} \times p \text{ (down)} \times 1 = 0.0001$$

This combination (units 1 and 2 both committed) does meet the prescribed MTIL of 0.005, i.e. $S <$ MTIL.

Proceeding similarly and checking security functions for periods A, B, C, D and F, we obtain the following optimal and secure UC table for the sample system for the load curve given in Fig. 7.7.

Table 7.5 Optimal and secure UC table

Period	Unit number			
	1	2	3	4
A	1	1	1	1
B	1	1	1	0
C	1	1	0	0
D	1	1	1	0
E	1	1*	0	0
F	1	1	0	0

* Unit was started due to security considerations.

Start-up Considerations

The UC table as obtained above is secure and economically optimal over each individual period of the load curve. Such a table may require that certain units have to be started and stopped more than once. Therefore, start-up cost must be taken into consideration from the point of view of overall economy. For example, unit 3 has to be stopped and restarted twice during the cycle. We must, therefore, examine whether or not it will be more economical to avoid one restarting by continuing to run the unit in period C.

Case a When unit 3 is not operating in period C.
Total fuel cost for periods B, C and D as obtained by most economic load sharing are as under (detailed computation is avoided)

$$= 1,690.756 + 1,075.356 + 1.690.756 = \text{Rs } 4,456.868$$

$$\text{Start-up cost of unit } 3 = \text{Rs} \quad 50.000 \text{ (say)}$$

$$\text{Total operating cost} = \text{Rs } 4,506.868$$

Case b When all three units are running in period C, i.e. unit 3 is not stopped at the end of period B.

$$\text{Total operating costs} = 1,690.756 + 1,081.704 + 1,690.756$$

$$= \text{Rs } 4{,}463.216 \text{ (start-up cost} = 0)$$

Clearly Case b results in overall economy. Therefore, the optimal and secure UC table for this load cycle is modified as under, with due consideration to the overall cost.

Table 7.6 Overall optimal and secure UC table

Period	Unit number			
	1	2	3	4
A	1	1	1	1
B	1	1	1	0
C	1	1	1*	0
D	1	1	1	0
E	1	1	0	0
F	1	1	0	0

*Unit was started due to start-up considerations.

7.5 OPTIMUM GENERATION SCHEDULING

From the unit commitment table of a given plant, the fuel cost curve of the plant can be determined in the form of a polynomial of suitable degree by the method of least squares fit. If the transmission losses are neglected, the total system load can be optimally divided among the various generating plants using the equal incremental cost criterion of Eq. (7.10). It is, however, unrealistic to neglect transmission losses particularly when long distance transmission of power is involved.

A modern electric utility serves over a vast area of relatively low load density. The transmission losses may vary from 5 to 15% of the total load, and therefore, it is essential to account for losses while developing an economic load dispatch policy. It is obvious that when losses are present, we can no longer use the simple 'equal incremental cost' criterion. To illustrate the point, consider a two-bus system with identical generators at each bus (i.e. the same IC curves). Assume that the load is located near plant 1 and plant 2 has to deliver power via a lossy line. Equal incremental cost criterion would dictate that each plant should carry half the total load; while it is obvious in this case that the plant 1 should carry a greater share of the load demand thereby reducing transmission losses.

In this section, we shall investigate how the load should be shared among various plants, when line losses are accounted for. The objective is to minimize the overall cost of generation

$$C = \sum_{i=1}^{k} C_i(P_{Gi}) \tag{7.7}$$

at any time under equality constraint of meeting the load demand with transmission loss, i.e.

$$\sum_{i=1}^{k} P_{Gi} - P_D - P_L = 0 \qquad (7.19)$$

where

k = total number of generating plants

P_{Gi} = generation of ith plant

P_D = sum of load demand at all buses (system load demand)

P_L = total system transmission loss

To solve the problem, we write the Lagrangian as

$$L = \sum_{i=1}^{k} C_i(P_{Gi}) - \lambda \left[\sum_{i=1}^{k} P_{Gi} - P_D - P_L \right] \qquad (7.20)$$

It will be shown later in this section that, if the power factor of load at each bus is assumed to remain constant, the system loss P_L can be shown to be a function of active power generation at each plant, i.e.

$$P_L = P_L (P_{G1}, P_{G2}, ..., P_{Gk}) \qquad (7.21)$$

Thus in the optimization problem posed above, P_{Gi} ($i = 1, 2, ..., k$) are the only control variables.

For optimum real power dispatch,

$$\frac{\partial L}{\partial P_{Gi}} = \frac{dC_i}{dP_{Gi}} - \lambda + \lambda \frac{\partial P_L}{\partial P_{Gi}} = 0, \quad i = 1, 2, ..., k \qquad (7.22)$$

Rearranging Eq. (7.22) and recognizing that changing the output of only one plant can affect the cost at only that plant, we have

$$\frac{\dfrac{dC_i}{dP_{Gi}}}{\left(1 - \dfrac{\partial P_L}{\partial P_{Gi}}\right)} = \lambda \text{ or } \frac{dC_i}{dP_{Gi}} L_i = \lambda, \ i = 1, 2, ..., k \qquad (7.23)$$

where

$$L_i = \frac{1}{(1 - \partial P_L / \partial P_{Gi})} \qquad (7.24)$$

is called the *penalty factor* of the ith plant.

The Lagrangian multiplier λ is in rupees per megawatt-hour, when fuel cost is in rupees per hour. Equation (7.23) implies that minimum fuel cost is obtained, when the incremental fuel cost of each plant multiplied by its penalty factor is the same for all the plants.

The ($k + 1$) variables ($P_{G1}, P_{G2}, ..., P_{Gk}, \lambda$) can be obtained from k optimal dispatch Eq. (7.23) together with the power balance Eq. (7.19). The partial derivative $\partial P_L / \partial P_{Gi}$ is referred to as the *incremental transmission loss* (ITL)$_i$, associated with the ith generating plant.

Equation (7.23) can also be written in the alternative form

$$(IC)_i = \lambda[1 - (ITL)_i] \quad i = 1, 2, ..., k \tag{7.25}$$

This equation is referred to as the *exact coordination equation*.

Thus it is clear that to solve the optimum load scheduling problem, it is necessary to compute ITL for each plant, and therefore we must determine the functional dependence of transmission loss on real powers of generating plants. There are several methods, approximate and exact, for developing a transmission loss model. A full treatment of these is beyond the scope of this book. One of the most important, simple but approximate, methods of expressing transmission loss as a function of generator powers is through B-coefficients. This method is reasonably adequate for treatment of loss coordination in economic scheduling of load between plants. The general form of the loss formula (derived later in this section) using B-coefficients is

$$P_L = \sum_{m=1}^{k}\sum_{n=1}^{k} P_{Gm} B_{mn} P_{Gn} \tag{7.26}$$

where

P_{Gm}, P_{Gn} = real power generation at m, nth plants

B_{mn} = loss coefficients which are constants under certain assumed operating conditions

If P_Gs are in megawatts, B_{mn} are in reciprocal of megawatts[*]. Computations, of course, may be carried out in per unit. Also, $B_{mn} = B_{nm}$.

Equation (7.26) for transmission loss may be written in the matrix form as

$$P_L = P_G^T B P_G \tag{7.27}$$

where

$$P_G = \begin{bmatrix} P_{G1} \\ P_{G2} \\ \vdots \\ P_{Gk} \end{bmatrix} \quad \text{and} \quad B = \begin{bmatrix} B_{11} & B_{12} & \cdots & B_{1k} \\ B_{21} & B_{22} & \cdots & B_{2k} \\ \vdots & & & \vdots \\ B_{k} & B_{k2} & \cdots & B_{kk} \end{bmatrix}$$

It may be noted that B is a symmetric matrix.

For a three plant system, we can write the expression for loss as

$$P_L = B_{11}P_{G1}^2 + B_{22}P_{G2}^2 + B_{33}P_{G3}^2 + 2B_{12}P_{G1}P_{G2} + 2B_{23}P_{G2}P_{G3} + 2B_{31}P_{G3}P_{G1} \tag{7.28}$$

With the system power loss model as per Eq. (7.26), we can now write

$$\frac{\partial P_L}{\partial P_{Gi}} = \frac{\partial}{\partial P_{Gi}}\left[\sum_{m=1}^{k}\sum_{n=1}^{k} P_{Gm} B_{mn} P_{Gn}\right]$$

---

[*]B_{mn} (in pu) = B_{mn} (in MW^{-1}) × Base MVA

$$= \frac{\partial}{\partial P_{Gi}} \left[\sum_{\substack{n=1 \\ n \neq i}}^{k} P_{Gi} B_{in} P_{Gn} + \sum_{\substack{m=1 \\ m \neq i}}^{k} P_{Gm} B_{mi} P_{Gi} + P_{Gi} B_{ii} P_{Gi} \right] \qquad (7.29)$$

It may be noted that in the above expression other terms are independent of P_{Gi} and are, therefore, left out.

Simplifying Eq. (7.29) and recognizing that $B_{ij} = B_{ji}$, we can write

$$\frac{\partial P_L}{\partial P_{Gi}} = \sum_{j=1}^{k} 2 B_{ij} P_{Gj} \qquad (7.30a)$$

Assuming quadratic plant cost curves as

$$C_i(P_{Gi}) = \frac{1}{2} a_i P_{Gi}^2 + b_i P_{Gi} + d_i$$

We obtain the incremental cost as

$$\frac{dC_i}{dP_{Gi}} = a_i P_{Gi} + b_i \qquad (7.30b)$$

Substituting $\partial P_L / \partial P_{Gi}$ and dC_i/dP_{Gi} from above in the coordination Eq. (7.22), we have

$$a_i P_{Gi} + b_i + \lambda \sum_{j=1}^{k} 2 B_{ij} P_{G_j} = \lambda \qquad (7.31)$$

Collecting all terms of P_{Gi} and solving for P_{Gi}, we obtain

$$(a_i + 2\lambda B_{ii}) P_{Gi} = - \lambda \sum_{\substack{j=1 \\ j \neq i}}^{k} 2 B_{ij} P_{Gj} - b_i + \lambda$$

$$P_{Gi} = \frac{1 - \dfrac{b_i}{\lambda} - \sum_{\substack{j=1 \\ j \neq i}}^{k} 2 B_{ij} P_{Gj}}{\dfrac{a_i}{\lambda} + 2 B_{ii}}; \quad i = 1,2,\ldots,k \qquad (7.32)$$

For any particular value of λ, Eq. (7.32) can be solved iteratively by assuming initial values of P_{Gi}s (a convenient choice is $P_{Gi} = 0$; $i = 1, 2, \ldots, k$). Iterations are stopped when P_{Gi}s converge within specified accuracy.

Equation (7.32) along with the power balance Eq. (7.19) for a particular load demand P_D are solved iteratively on the following lines:

1. Initially choose $\lambda = \lambda_0$.
2. Assume $P_{Gi} = 0$; $i = 1, 2, \ldots, k$.
3. Solve Eq. (7.32) iteratively for P_{Gi}s.

4. Calculate $P_L = \sum\limits_{i=1}^{k} \sum\limits_{j=1}^{k} P_{Gi} B_{ij} P_{Gj}$.

5. Check if power balance equation (7.19) is satisfied, i.e.

$$\left| \sum\limits_{n \neq 1}^{k} P_{Gi} - P_D - P_L \right| < \varepsilon \quad \text{(a specified value)}$$

If yes, stop. Otherwise, go to step 6.

6. Increase λ by $\Delta\lambda$ (a suitable step size); if $\left(\sum\limits_{i=1}^{k} P_{Gi} - P_D - P_L \right) < 0$ or

decrease λ by $\Delta\lambda$ (a suitable step size); if $\left(\sum\limits_{i=1}^{k} P_{Gi} - P_D - P_L \right) > 0$,

repeat from step 3.

Example 7.4

A two-bus system is shown in Fig. 7.8. If 100 MW is transmitted from plant 1 to the load, a transmission loss of 10 MW is incurred. Find the required generation for each plant and the power received by load when the system λ is Rs 25/MWh.

The incremental fuel costs of the two plants are given below:

$$\frac{dC_1}{dP_{G1}} = 0.02 P_{G1} + 16.0 \text{ Rs/MWh}$$

$$\frac{dC_2}{dP_{G2}} = 0.04 P_{G2} + 20.0 \text{ Rs/MWh}$$

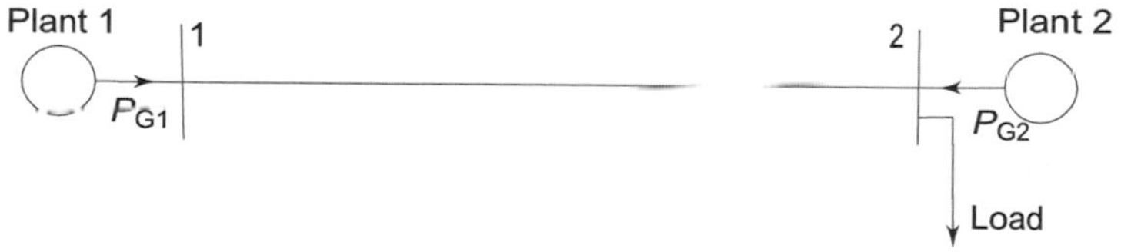

Fig. 7.8 A two-bus system for Example 7.4

Solution Since the load is at bus 2 alone, P_{G2} will not have any effect on P_L. Therefore

$$B_{22} = 0 \text{ and } B_{12} = 0 = B_{21}$$

Hence

$$P_L = B_{11} P_{G1}^2 \qquad \text{(i)}$$

For $P_{G1} = 100$ MW, $P_L = 10$ MW, i.e.

$$10 = B_{11} (100)^2$$

$$B_{11} = 0.001 \text{ MW}^{-1}$$

Equation (7.31) for plant 1 becomes

$$0.02P_{G1} + 2\lambda B_{11}P_{G1} + 2\lambda B_{12}P_{G2} = \lambda - 16 \qquad \text{(ii)}$$

and for plant 2

$$0.04P_{G2} + 2\lambda B_{22}P_{G2} + 2\lambda B_{21}P_{G1} = \lambda - 20 \qquad \text{(iii)}$$

Substituting the values of B-coefficients and $\lambda = 25$, we get

$$P_{G1} = 128.57 \text{ MW}$$

$$P_{G2} = 125 \text{ MW}$$

The transmission power loss is

$$P_L = 0.001 \times (128.57)^2 = 16.53 \text{ MW}$$

and the load is

$$P_D = P_{G1} + P_{G2} - P_L = 128.57 + 125 - 16.53 = 237.04 \text{ MW}$$

Example 7.5

Consider the system of Example 7.4 with a load of 237.04 MW at bus 2. Find the optimum load distribution between the two plants for (a) when losses are included but not coordinated, and (b) when losses are also coordinated. Also find the savings in rupees per hour when losses are coordinated.

Solution   *Case a*   If the transmission loss is not coordinated, the optimum schedules are obtained by equating the incremental fuel costs at the two plants. Thus

$$0.02P_{G1} + 16 = 0.04P_{G2} + 20 \qquad \text{(i)}$$

The power delivered to the load is

$$P_{G1} + P_{G2} = 0.001P_{G1}^2 + 237.04 \qquad \text{(ii)}$$

Solving Eqs. (i) and (ii) for P_{G1} and P_{G2}, we get

$$P_{G1} = 275.18 \text{ MW; and } P_{G2} = 37.59 \text{ MW}$$

Case b   This case is already solved in Example 7.4. Optimum plant loadings with loss coordination are

$$P_{G1} = 128.57 \text{ MW; } P_{G2} = 125 \text{ MW}$$

Loss coordination causes the load on plant 1 to reduce from 275.18 MW to 128.57 MW. Therefore, saving at plant 1 due to loss coordination is

$$\int_{128.57}^{275.18}(0.02P_{G1} + 16)dP_{G1} = 0.01P_{G1}^2 + 16P_{G1}\Big|_{128.57}^{275.18}$$

$$= \text{Rs } 2,937.69/\text{hr}$$

At plant 2 the load increases from 37.59 MW to 125 MW due to loss coordination. The saving at plant 2 is

$$\int_{125}^{37.59} (0.04 P_{G2} + 20) \, dP_{G2} = 0.02 P_{G2}^2 + 20 P_{G2} \Big|_{125}^{37.59}$$

$$= - \text{ Rs } 2,032.43/\text{hr}$$

The net saving achieved by coordinating losses while scheduling the received load of 237.04 MW is

$$2,937.69 - 2,032.43 = \text{Rs } 905.26/\text{hr}$$

Derivation of Transmission Loss Formula

An accurate method of obtaining a general formula for transmission loss has been given by Kron [4]. This, however, is quite complicated. The aim of this article is to give a simpler derivation by making certain assumptions.

Figure 7.9 (c) depicts the case of two generating plants connected to an arbitrary number of loads through a transmission network. One line within the network is designated as branch p.

Imagine that the total load current I_D is supplied by plant 1 only, as in Fig. 7.9a. Let the current in line p be I_{p1}. Define

$$M_{p1} = \frac{I_{p1}}{I_D} \tag{7.33}$$

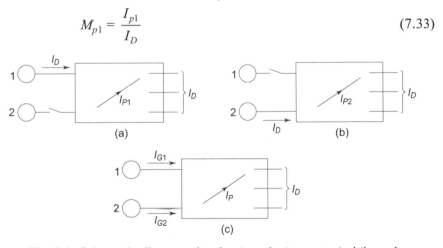

Fig. 7.9 Schematic diagram showing two plants connected through a power network to a number of loads

Similarly, with plant 2 alone supplying the total load current (Fig. 7.9b), we can define

$$M_{p2} = \frac{I_{p2}}{I_D} \tag{7.34}$$

M_{p1} and M_{p2} are called *current distribution factors*. The values of current distribution factors depend upon the impedances of the lines and their interconnection and are independent of the current I_D.

When both generators 1 and 2 are supplying current into the network as in Fig. 7.9(c), applying the principle of superposition the current in the line p can be expressed as

$$I_p = M_{p1}I_{G1} + M_{p2}I_{G2} \tag{7.35}$$

where I_{G1} and I_{G2} are the currents supplied by plants 1 and 2, respectively.

At this stage let us make certain simplifying assumptions outlined below:

(1) All load currents have the same phase angle with respect to a common reference. To understand the implication of this assumption consider the load current at the ith bus. It can be written as

$$|I_{Di}| \angle (\delta_i - \phi_i) = |I_{Di}| \angle \theta_i$$

where δ_i is the phase angle of the bus voltage and ϕ_i is the lagging phase angle of the load. Since δ_i and ϕ_i vary only through a narrow range at various buses, it is reasonable to assume that θ_i is the same for all load currents at all times.

(2) Ratio X/R is the same for all network branches.

These two assumptions lead us to the conclusion that I_{p1} and I_D [Fig. 7.9(a)] have the same phase angle and so have I_{p2} and I_D [Fig. 7.9(b)], such that the current distribution factors M_{p1} and M_{p2} are real rather than complex.

Let, $$I_{G1} = |I_{G1}| \angle \sigma_1 \text{ and } I_{G2} = |I_{G2}| \angle \sigma_2$$

where σ_1 and σ_2 are phase angles of I_{G1} and I_{G2}, respectively with respect to the common reference.

From Eq. (7.35), we can write

$$|I_p|^2 = (M_{p1}|I_{G1}| \cos \sigma_1 + M_{p2}|I_{G2}| \cos \sigma_2)^2 + (M_{p1}|I_{G1}| \sin \sigma_1 + M_{p2}|I_{G2}| \sin \sigma_2)^2 \tag{7.36}$$

Expanding the simplifying the above equation, we get

$$|I_p|^2 = M_{p1}^2|I_{G1}|^2 + M_{p2}^2|I_{G2}|^2 + 2M_{p1}M_{p2}|I_{G1}| \, |I_{G2}|\cos (\sigma_1 - \sigma_2) \tag{7.37}$$

Now $$|I_{G1}| = \frac{P_{G1}}{\sqrt{3}|V_1|\cos \phi_1}; \; |I_{G2}| = \frac{P_{G2}}{\sqrt{3}|V_2| \cos \phi_2} \tag{7.38}$$

where P_{G1} and P_{G2} are the three-phase real power outputs of plants 1 and 2 at power factors of $\cos \phi_1$, and $\cos \phi_2$, and V_1 and V_2 are the bus voltages at the plants.

If R_p is the resistance of branch p, the total transmission loss is given by[*]

---

[*]The general expression for the power system with k plants is expressed as

$$P_L = \frac{P_{G1}^2}{|V_1|^2 (\cos \phi_1)^2} \sum_v M_{p1}^2 R_p + \dots + \frac{P_{Gk}^2}{|V_k|^2 (\cos \phi_k)^2} \sum_p M_{pk}^2 R_p$$

$$+ 2 \sum_{\substack{m,n=1 \\ m \neq n}}^{k} \left\{ \frac{P_{Gm}P_{Gn} \cos(\sigma_m - \sigma_n)}{|V_m||V_n|\cos \phi_m \cos \phi_n} \sum_p M_{pm} M_{pn} R_p \right\}$$

$$P_L = \sum_p 3|I_p|^2 R_p$$

Substituting for $|I_p|^2$ from Eq. (7.37), and $|I_{G1}|$ and $|I_{G2}|$ from Eq. (7.38), we obtain

$$P_L = \frac{P_{G1}^2}{|V_1|^2 (\cos\phi_1)^2} \sum_p M_{p1}^2 R_p$$

$$+ \frac{2 P_{G1} P_{G2} \cos(\sigma_1 - \sigma_2)}{|V_1||V_2|\cos\phi_1 \cos\phi_2} \sum_p M_{p1} M_{p2} R_p$$

$$+ \frac{P_{G2}^2}{|V_2|^2 (\cos\phi_2)^2} \sum_p M_{p2}^2 R_p \qquad (7.39)$$

Equation (7.39) can be recognized as

$$P_L = P_{G1}^2 B_{11} + 2 P_{G1} P_{G2} B_{12} + P_{G2}^2 B_{22}$$

$$B_{11} = \frac{1}{|V_1|^2 (\cos\phi_1)^2} \sum_p M_{p1}^2 R_p$$

$$B_{12} = \frac{\cos(\sigma_1 - \sigma_2)}{|V_1||V_2|\cos\phi_1 \cos\phi_2} \sum_p M_{p1} M_{p2} R_p \qquad (7.40)$$

$$B_{22} = \frac{1}{|V_2|^2 (\cos\phi_2)^2} \sum_p M_{p2}^2 R_p$$

The terms B_{11}, B_{12} and B_{22} are called *loss coefficients* or *B-coefficients*. If voltages are line to line kV with resistances in ohms, the units of B-coefficients are in MW^{-1}. Further, with P_{G1} and P_{G2} expressed in MW, P_L will also be in MW.

The above results can be extended to the general case of k plants with transmission loss expressed as

$$P_L = \sum_{m=1}^{k}\sum_{n=1}^{k} P_{Gm} B_{mn} P_{Gn} \qquad (7.41)$$

where

$$B_{mn} = \frac{\cos(\sigma_m - \sigma_n)}{|V_m||V_n|\cos\phi_m \cos\phi_n} \sum_p M_{pm} M_{pn} R_p \qquad (7.42)$$

It can be recognized as

$$P_L = P_{G1}^2 B_{11} + \dots + P^2{}_{Gk} B_{kk} + 2 \sum_{\substack{m,n=1 \\ m \neq n}}^{k} P_{Gm} B_{mn} P_{Gn}$$

The following assumptions including those mentioned already are necessary, if B-coefficients are to be treated as constants as total load and load sharing between plants vary. These assumptions are:

1. All load currents maintain a constant ratio to the total current.
2. Voltage magnitudes at all plants remain constant.
3. Ratio of reactive to real power, i.e. power factor at each plant remains constant.
4. Voltage phase angles at plant buses remain fixed. This is equivalent to assuming that the plant currents maintain constant phase angle with respect to the common reference, since source power factors are assumed constant as per assumption 3 above.

In spite of the number of assumptions made, it is fortunate that treating B-coefficients as constants, yields reasonably accurate results, when the coefficients are calculated for some average operating conditions. Major system changes require recalculation of the coefficients.

Losses as a function of plant outputs can be expressed by other methods[*], but the simplicity of loss equations is the chief advantage of the B-coefficients method.

Accounting for transmission losses results in considerable operating economy. Furthermore, this consideration is equally important in future system planning and, in particular, with regard to the location of plants and building of new transmission lines:

Example 7.6

Figure 7.10 shows a system having two plants 1 and 2 connected to buses 1 and 2, respectively. There are two loads and a network of four branches. The reference bus with a voltage of $1.0\angle 0°$ pu is shown on the diagram. The branch currents and impedances are:

$$I_a = 2 - j0.5 \text{ pu} \qquad I_c = 1 - j0.25 \text{ pu}$$

$$I_b = 1.6 - j0.4 \text{ pu} \qquad I_d = 3.6 - j0.9 \text{ pu}$$

$$Z_a = 0.015 + j0.06 \text{ pu} \qquad Z_c = 0.01 + j0.04 \text{ pu}$$

$$Z_b = 0.015 + j0.06 \text{ pu} \qquad Z_d = 0.01 + j0.04 \text{ pu}$$

Calculate the loss formula coefficients of the system in pu and in reciprocal megawatts, if the base is 100 MVA.

---

[*]For more accurate methods and exact expression for $\partial P_L / \partial P_{Gi}$, references [22, 23] may be consulted.

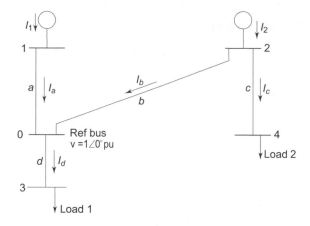

Fig. 7.10 Sample system of Example 7.6

Solution As all load currents maintain a constant ratio to the total current, we have

$$\frac{I_d}{I_c + I_d} = \frac{3.6 - j0.9}{4.6 - j1.15} = 0.7826$$

$$\frac{I_c}{I_c + I_d} = \frac{1 - j0.25}{4.6 - j1.15} = 0.2174$$

∴ $M_{a1} = 1$, $M_{b1} = -0.2174$, $M_{c1} = 0.2174$, $M_{d1} = 0.7826$

$M_{a2} = 0$, $M_{b2} = 0.7826$, $M_{c2} = 0.2174$, $M_{d2} = 0.7826$

Since the source currents are known, the voltages at the source buses can be calculated. However, in a practical size network a load flow study has to be made to find power factors at the buses, bus voltages and phase angles. The bus voltages at the plants are

$$V_1 = 1.0 + (2 - j0.5)(0.015 + j0.06)$$

$$= 1.06 + j0.1125 - 1.066 \angle 6.05° \text{ pu}$$

$$V_2 = 1 + (1.6 - j0.4)(0.015 + j0.06)$$

$$= 1.048 + j0.09 = 1.051 \angle 4.9° \text{ pu}$$

The current phase angles at the plants are ($I_1 = I_a$, $I_2 = I_b + I_c$)

$$\sigma_1 = \tan^{-1}\frac{-0.5}{2} = -14°; \quad \sigma_2 = \tan^{-1}\frac{-0.65}{2.6} = -14°$$

$$\cos(\sigma_2 - \sigma_1) = \cos 0° = 1$$

The plant power factors are

$$pf_1 = \cos(6.05° + 14°) = 0.9393$$

$$pf_2 = \cos(4.9° + 14°) = 0.946$$

The loss coefficients are [Eq. (7.42)]

$$B_{11} = \frac{0.015 \times 1^2 + 0.015 \times (0.2174)^2 + 0.01 \times (0.2174)^2 + 0.01 \times (0.7826)^2}{(1.066)^2 \times (0.9393)^2}$$

$$= 0.02224 \text{ pu}$$

$$B_{22} = \frac{0.015 \times (0.7826)^2 + 0.01 \times (0.2174)^2 + 0.01 \times (0.7826)^2}{(1.051)^2 \times (0.946)^2}$$

$$= 0.01597 \text{ pu}$$

$$B_{12} = \frac{(-0.2174)(0.7826)(0.015) + 0.01 \times (0.2174)^2 + 0.01 \times (0.7826)^2}{1.066 \times 1.051 \times 0.9393 \times 0.946}$$

$$= 0.00406 \text{ pu}$$

For a base of 100 MVA, these loss coefficients must be divided by 100 to obtain their values in units of reciprocal megawatts, i.e.

$$B_{11} = \frac{0.02224}{100} = 0.02224 \times 10^{-2} \text{ MW}^{-1}$$

$$B_{22} = \frac{0.01597}{100} = 0.01597 \times 10^{-2} \text{ MW}^{-1}$$

$$B_{12} = \frac{0.00406}{100} = 0.00406 \times 10^{-2} \text{ MW}^{-1}$$

7.6 OPTIMAL LOAD FLOW SOLUTION

The problem of optimal real power dispatch has been treated in the earlier section using the approximate loss formula. This section presents the more general problem of real and reactive power flow so as to minimize the instantaneous operating costs. It is a static optimization problem with a scalar objective function (also called cost function).

The solution technique given here was first given by Dommel and Tinney [34]. It is based on load flow solution by the NR method, a first order gradient adjustment algorithm for minimizing the objective function and use of penalty functions to account for inequality constraints on dependent variables. The problem of unconstrained optimal load flow is first tackled. Later the inequality constraints are introduced, first on control variables and then on dependent variables.

Optimal Power Flow without Inequality Constraints

The objective function to be minimized is the operating cost

$$C = \sum_{i}^{k} C_i(P_{Gi}) \tag{7.7}$$

subject to the load flow equations [see Eqs. (6.27) and (6.28)

$$P_i - \sum_{j=1}^{n} |V_i||V_j||Y_{ij}| \cos(\theta_{ij} + \delta_j - \delta_i) = 0 \tag{7.43}$$

for each

PQ bus

$$Q_i + \sum_{j=1}^{n} |V_i||V_j||Y_{ij}| \sin(\theta_{ij} + \delta_j - \delta_i) = 0 \tag{7.44}$$

and

$$P_i - \sum_{j=1}^{n} |V_i||V_j||Y_{ij}| \cos(\theta_{ij} + \delta_j - \delta_i) = 0 \text{ for each } PV \text{ bus} \tag{7.45}$$

It is to be noted that at the ith bus

$$P_i = P_{Gi} - P_{Di}$$

$$Q_i = Q_{Gi} - Q_{Di} \tag{7.46}$$

where P_{Di} and Q_{Di} are load demands at bus i.

Equations (7.43), (7.44) and (7.45) can be expressed in vector form

$$f(x, y) = \begin{bmatrix} \text{Eq. (7.43)} \\ \text{Eq. (7.44)} \\ \text{Eq. (7.45)} \end{bmatrix} \begin{array}{l} \text{for each } PQ \text{ bus} \\ \\ \text{for each } PV \text{ bus} \end{array} \tag{7.47}$$

where the vector of dependent variables is

$$\mathbf{x} = \begin{bmatrix} |V_i| \\ \delta_i \\ \delta_i \end{bmatrix} \begin{array}{l} \text{for each } PQ \text{ bus} \\ \\ \text{for each } PV \text{ bus} \end{array} \tag{7.48a}$$

and the vector of independent variables is

$$\mathbf{y} = \begin{bmatrix} |V_1| \\ \delta_1 \\ P_i \\ Q_i \\ P_i \\ |V_i| \end{bmatrix} \begin{array}{l} \text{for slack bus} \\ \\ \text{for each } PQ \text{ bus} \\ \\ \text{for each } PV \text{ bus} \end{array} = \begin{bmatrix} \mathbf{u} \\ \mathbf{p} \end{bmatrix} \tag{7.48b}$$

In the above formulation, the objective function must include the slack bus power.

The vector of independent variables $\mathbf{y}$ can be partitioned into two parts—a vector $\mathbf{u}$ of control variables which are to be varied to achieve optimum value of the objective function and a vector $\mathbf{p}$ of fixed or disturbance or uncontrollable parameters. Control parameters[*] may be voltage magnitudes on PV buses, P_{Gi} at buses with controllable power, etc.

The optimization problem[**] can now be restated as

$$\min_{u} C(\mathbf{x}, \mathbf{u}) \tag{7.49}$$

subject to equality constraints

$$f(\mathbf{x}, \mathbf{u}, \mathbf{p}) = 0 \tag{7.50}$$

To solve the optimization problem, define the Lagrangian function as

$$\mathbf{L}(x, u, p) = C(x, u) + \lambda^T f(x, u, p) \tag{7.51}$$

where λ is the vector of Lagrange multipliers of same dimension as $f(x, u, p)$

The necessary conditions to minimize the unconstrained Lagrangian function are (see Appendix A for differentiation of matrix functions).

$$\frac{\partial \mathbf{L}}{\partial x} = \frac{\partial C}{\partial x} + \left[\frac{\partial f}{\partial x}\right]^T \lambda = 0 \tag{7.52}$$

$$\frac{\partial \mathbf{L}}{\partial u} = \frac{\partial C}{\partial u} + \left[\frac{\partial f}{du}\right]^T \lambda = 0 \tag{7.53}$$

$$\frac{\partial \mathbf{L}}{\partial \lambda} = f(x, u, p) = 0 \tag{7.54}$$

Equation (7.54) is obviously the same as the equality constraints. The expressions for $\dfrac{\partial C}{\partial x}$ and $\dfrac{\partial f}{\partial u}$ as needed in Eqs. (7.52) and (7.53) are rather involved[***]. It may however be observed by comparison with Eq. (6.56a) that

$\dfrac{\partial f}{\partial x}$ = Jacobian matrix [same as employed in the NR method of load flow solution; the expressions for the elements of Jacobian are given in Eqs. (6.64) and (6.65)].

Equations (7.52), (7.53) and (7.54) are non-linear algebraic equations and can only be solved iteratively. A simple yet efficient iteration scheme, that can by employed, is the *steepest descent method* (also called *gradient method*).

---

[*]Slack bus voltage and regulating transformer tap setting may be employed as additional control variables. Dopazo *et al* [26] use Q_{Gi} as control variable on buses with reactive power control.

[**]If the system real power loss is to be minimized, the objective function is

$$C = P_1(|V|, \delta)$$

Since in this case the net injected real powers are fixed, the minimization of the real injected power P_1 at the slack bus is equivalent to minimization of total system loss. This is known as *optimal reactive power flow problem.*

[***]The original paper of Dommel and Tinney [34] may be consulted for details.

The basic technique is to adjust the control vector u, so as to move from one feasible solution point (a set of values of x which satisfies Eq. (7.54) for given u and p; it indeed is the load flow solution) in the direction of steepest descent (negative gradient) to a new feasible solution point with a lower value of objective function. By repeating these moves in the direction of the negative gradient, the minimum will finally be reached.

The computational procedure for the gradient method with relevant details is given below:

Step 1   Make an initial guess for u, the control variables.

Step 2   Find a feasible load flow solution from Eq. (7.54) by the NR iterative method. The method successively improves the solution x as follows.

$$x^{(r+1)} = x^{(r)} + \Delta x$$

where Δx is obtained by solving the set of linear equations (6.56b) reproduced below:

$$\left[\frac{\partial f}{\partial x}(x^{(r)}, y) \right] \Delta x = -f(x^{(r)}, y)$$

or

$$\Delta x = - (J^{(r)})^{-1} f(x^{(r)}, y)$$

The end results of Step 2 are a feasible solution of x and the Jacobian matrix.

Step 3   Solve Eq. (7.52) for

$$\lambda = -\left[\left(\frac{\partial f}{\partial x} \right)^T \right]^{-1} \frac{\partial C}{\partial x} \tag{7.55}$$

Step 4   Insert λ from Eq. (7.55) into Eq. (7.53), and compute the gradient

$$\nabla L = \frac{\partial C}{\partial u} + \left[\frac{\partial f}{\partial u} \right]^T \lambda \tag{7.56}$$

It may be noted that for computing the gradient, the Jacobian $J = \dfrac{\partial f}{\partial x}$ is already known from the load flow solution (step 2 above).

Step 5   If ∇L equals zero within prescribed tolerance, the minimum has been reached. Otherwise,

Step 6   Find a new set of control variables

$$u_{\text{new}} = u_{\text{old}} + \Delta u \tag{7.57}$$

where

$$\Delta u = - \alpha \nabla L \tag{7.58}$$

Here Δu is a step in the negative direction of the gradient. The step size is adjusted by the positive scalar α.

Steps 1 through 5 are straightforward and pose no computational problems. Step 6 is the critical part of the algorithm, where the choice of α is very important. Too small a value of α guarantees the convergence but slows down the rate of convergence; too high a value causes oscillations around the minimum. Several methods are available for optimum choice of step size.

Inequality Constraints on Control Variables

Though in the earlier discussion, the control variables are assumed to be unconstrained, the permissible values are, in fact, always contrained,

$$u_{\min} \leq u \leq u_{\max} \tag{7.59}$$

e.g. $P_{Gi,\,\min} \leq P_{Gi} \leq P_{Gi,\,\max}$

These inequality constraints on control variables can be easily handled. If the correction Δu_i in Eq. (7.57) causes u_i to exceed one of the limits, u_i is set equal to the corresponding limit, i.e.

$$u_{i,\text{ new}} = \begin{cases} u_{i,\max} & \text{if } u_{i,\text{old}} + \Delta u_i > u_{i,\max} \\ u_{i,\min} & \text{if } u_{i,\text{old}} + \Delta u_i < u_{i,\min} \\ u_{i,\text{old}} + \Delta u_i & \text{otherwise} \end{cases} \tag{7.60}$$

After a control variable reaches any of the limits, its component in the gradient should continue to be computed in later iterations, as the variable may come within limits at some later stage.

In accordance with the Kuhn-Tucker theorem (see Appendix E), the necessary conditions for minimization of L under constraint (7.59) are:

$$\left. \begin{aligned} \frac{\partial \text{L}}{\partial u_i} &= 0 & \text{if } u_{i,\,\min} < u_i < u_{i,\max} \\ \frac{\partial \text{L}}{\partial u_i} &\leq 0 & \text{if } u_i = u_{i,\max} \\ \frac{\partial \text{L}}{\partial u_i} &\geq 0 & u_i = u_{i,\max} \end{aligned} \right\} \tag{7.61}$$

Therefore, now, in step 5 of the computational algorithm, the gradient vector has to satisfy the optimality condition (7.61).

Inequality Constraints on Dependent Variables

Often, the upper and lower limits on dependent variables are specified as

$$x_{\min} \leq x \leq x_{\max}$$

e.g. $|V|_{\min} \leq |V| \leq |V|_{\max}$ on a PQ bus $\tag{7.62}$

Such inequality constraints can be conveniently handled by the *penalty function* method. The objective function is augmented by penalties for inequality constraints violations. This forces the solution to lie sufficiently close to the constraint limits, when these limits are violated. The penalty function method is valid in this case, because these constraints are seldom rigid limits in the strict sense, but are in fact, soft limits (e.g. $|V| \leq 1.0$ on a PQ bus really means V should not exceed 1.0 too much and $|V| = 1.01$ may still be permissible).

The penalty method calls for augmentation of the objective function so that the new objective function becomes

$$C' = C\,(x,\,u) + \sum_j W_j \qquad (7.63)$$

where the penalty W_j is introduced for each violated inequality constraint. A suitable penalty function is defined as

$$W_j = \begin{cases} \gamma_j (x_j - x_{j,\max})^2; & \text{whenever } x_j > x_{j,\max} \\ \gamma_j (x_j - x_{j,\min})^2; & \text{whenever } x_j < x_{j,\min} \end{cases} \qquad (7.64)$$

where γ_j is a real positive number which controls degree of penalty and is called the *penalty factor*.

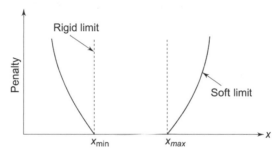

Fig. 7.11 Penalty function

A plot of the proposed penalty function is shown in Fig. 7.11, which clearly indicates how the rigid limits are replaced by soft limits.

The necessary conditions (7.52) and (7.53) would now be modified as given below, while the conditions (7.54), i.e. load flow equations, remain unchanged.

$$\frac{\partial \mathbf{L}}{\partial x} = \frac{\partial C}{\partial x} + \sum_j \frac{\partial W_j}{\partial x} + \left[\frac{\partial f}{\partial x}\right]^T \lambda = 0 \qquad (7.65)$$

$$\frac{\partial \mathbf{L}}{\partial u} = \frac{\partial C}{\partial u} + \sum_j \frac{\partial W_j}{\partial u} + \left[\frac{\partial f}{\partial u}\right]^T \lambda = 0 \qquad (7.66)$$

The vector $\dfrac{\partial W_j}{\partial x}$ obtained from Eq. (7.64) would contain only one non-zero

term corresponding to the dependent variable x_j; while $\dfrac{\partial W_j}{\partial u} = 0$ as the penalty

functions on dependent variables are independent of the control variables.

By choosing a higher value for γ_j, the penalty function can be made steeper so that the solution lies closer to the rigid limits; the convergence, however, will become poorer. A good scheme is to start with a low value of γ_j and to increase it during the optimization process, if the solution exceeds a certain tolerance limit.

This section has shown that the NR method of load flow can be extended to yield the optimal load flow solution that is feasible with respect to all relevant inequality constraints. These solutions are often required for system planning and operation.

7.7 OPTIMAL SCHEDULING OF HYDROTHERMAL SYSTEM

The previous sections have dealt with the problem of optimal scheduling of a power system with thermal plants only. Optimal operating policy in this case can be completely determined at any instant without reference to operation at other times. This, indeed, is the static optimization problem. Operation of a system having both hydro and thermal plants is, however, far more complex as hydro plants have negligible operating cost, but are required to operate under constraints of water available for hydro generation in a given period of time. The problem thus belongs to the realm of dynamic optimization. The problem of minimizing the operating cost of a hydrothermal system can be viewed as one of minimizing the fuel cost of thermal plants under the constraint of water availability (storage and inflow) for hydro generation over a given period of operation.

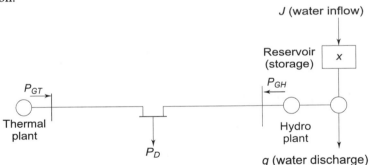

Fig. 7.12 Fundamental hydrothermal system

For the sake of simplicity and understanding, the problem formulation and solution technique are illustrated through a simplified hydrothermal system of Fig. 7.12. This system consists of one hydro and one thermal plant supplying power to a centralized load and is referred to as a *fundamental system*.

Optimization will be carried out with real power generation as control variable, with transmission loss accounted for by the loss formula of Eq. (7.26).

Mathematical Formulation

For a certain period of operation T (one year, one month or one day, depending upon the requirement), it is assumed that (i) storage of hydro reservoir at the beginning and the end of the period are specified, and (ii) water inflow to reservoir (after accounting for irrigation use) and load demand on the system are known as functions of time with complete certainty (deterministic case). The problem is to determine $q(t)$, the water discharge (rate) so as to minimize the cost of thermal generation.

$$C_T = \int_0^T C'(P_{GT}(t))\, dt \tag{7.67}$$

under the following constraints:
(i) Meeting the load demand

$$P_{GT}(t) + P_{GH}(t) - P_L(t) - P_D(t) = 0; \quad t\varepsilon\ [0,T] \tag{7.68}$$

This is called the *power balance equation*.
(ii) Water availability

$$X'(T) - X'(0) - \int_0^T J(t)\, dt + \int_0^T q(t)\, dt = 0 \tag{7.69}$$

where $J(t)$ is the water inflow (rate), $X'(t)$ water storage, and $X'(0)$, $X'(T)$ are specified water storages at the beginning and at the end of the optimization interval.

(iii) The hydro generation $P_{GH}(t)$ is a function of hydro discharge and water storage (or head), i.e.

$$P_{GH}(t) = f\,(X'(t),\, q\,(t)) \tag{7.70}$$

The problem can be handled conveniently by discretization. The optimization interval T is subdivided into M subintervals each of time length ΔT. Over each subinterval it is assumed that all the variables remain fixed in value. The problem is now posed as

$$\min_{q^m\,(m=1,2,...,M)} \Delta T \sum_{m=1}^{M} C'(P_{GT}^m) = \min_{q^m\,(m=1,2,...,M)} \sum_{m=1}^{M} C(P_{GT}^m) \tag{7.71}$$

under the following constraints:
(i) Power balance equation

$$P_{GT}^m + P_{GH}^m - P_L^m - P_D^m = 0 \tag{7.72}$$

where

$\qquad P_{GT}^m$ = thermal generation in the mth interval

$\qquad P_{GH}^m$ = hydro generation in the mth interval

$\qquad P_L^m$ = transmission loss in the mth interval

$$= B_{TT}(P_{GT}^m)^2 + 2B_{TH}P_{GH}^m + B_{HH}(P_{GH}^m)^2$$

P_D^m = load demand in the mth interval

(ii) Water continuity equation

$$X^{\prime m} - X^{\prime (m-1)} - J^m \; \Delta T + q^m \; \Delta T = 0$$

where

$X^{\prime m}$ = water storage at the end of the mth interval

J^m = water inflow (rate) in the mth interval

q^m = water discharge (rate) in the mth interval

The above equation can be written as

$$X^m - X^{m-1} - J^m + q^m = 0; \; m = 1, 2, ..., M \tag{7.73}$$

where $X^m = X^{\prime m}/\Delta T$ = storage in discharge units.

In Eqs. (7.73), X^o and X^M are the specified storages at the beginning and end of the optimization interval.

(iii) Hydro generation in any subinterval can be expressed[*] as

$$P_{GH}^m = h_o\{1 + 0.5e \; (X^m + X^{m-1})\} \; (q^m - \rho) \tag{7.74}$$

where

$$h_o = 9.81 \times 10^{-3} h'_o$$

h_o = basic water head (head corresponding to dead storage)

---

[*] $P_{GH}^m = 9.81 \times 10^{-2} \; h_{av}^m \; (q^m - \rho)$ MW

where

$(q^m - \rho)$ = effective discharge in m^3/s

H_{av}^m = average head in the mth interval

Now

$$h_{av}^m = h'_o + \frac{\Delta T(X^m + X^{m-1})}{2A}$$

where

A = area of cross-section of the reservoir at the given storage

h'_o = basic water head (head corresponding to dead storage,

$h_{av}^m = h'_o\{1 + 0.5e(X^m + X^{m-1})\}$

where

$$e = \frac{\Delta T}{Ah'_o}; \; e \text{ is tabulated for various storage values.}$$

Now

$$P_{GH}^m = h_o \; \{1 + 0.5e(X^m + X^{m-1})\} \; (q^m - P)$$

where

$$h_o = 9.81 \times 10^{-3} h'_o$$

e = water head correction factor to account for head variation with storage

ρ = non-effective discharge (water discharge needed to run hydro generator at no load).

In the above problem formulation, it is convenient to choose water discharges in all subintervals except one as independent variables, while hydro generations, thermal generations and water storages in all subintervals are treated as dependent variables. The fact, that water discharge in one of the subintervals is a dependent variable, is shown below:

Adding Eq. (7.73) for m = 1, 2, ..., M leads to the following equation, known as *water availability equation*

$$X^M - X^0 - \sum_m J^m + \sum_m q^m = 0 \tag{7.75}$$

Because of this equation, only $(M-1)$ qs can be specified independently and the remaining one can then be determined from this equation and is, therefore, a dependent variable. For convenience, q^1 is chosen as a dependent variable, for which we can write

$$q^1 = X^0 - X^M + \sum_m J^m - \sum_{m=2}^M q^m \tag{7.76}$$

Solution Technique

The problem is solved here using non-linear programming technique in conjunction with the first order gradient method. The Lagrangian L is formulated by augmenting the cost function of Eq. (7.71) with equality constraints of Eqs. (7.72)–(7.74) through Lagrange multipliers (dual variables) λ_1^m, λ_2^m and λ_3^m. Thus,

$$\text{L} = \sum_m [C(P_{GT}^m) - \lambda_1^m (P_{GT}^m + P_{GH}^m - P_L^m - P_D^m) + \lambda_2^m (X^m - X^{m-1} - J^m +$$

$$q^m) + \lambda_3^m \{P_{GH}^m - h_o (1 + 0.5e(X^m + X^{m-1})) \times (q^m - p)\}] \tag{7.77}$$

The dual variables are obtained by equating to zero the partial derivatives of the Lagrangian with respect to the dependent variables yielding the following equations

$$\frac{\partial \text{L}}{\partial P_{GT}^m} = \frac{dC(P_{GT}^m)}{dP_{GT}^m} - \lambda_1^m \left(1 - \frac{\partial P_L^m}{\partial P_{GT}^m}\right) = 0 \tag{7.78}$$

[The reader may compare this equation with Eq. (7.23)]

$$\frac{\partial \text{L}}{\partial P_G^m} = \lambda_3^m - \lambda_1^m \left(1 - \frac{\partial P_L^m}{\partial P_{GH}^m}\right) = 0 \tag{7.79}$$

$$\left(\frac{\partial L}{\partial X^m}\right)_{\substack{m\neq M \\ \neq 0}} = \lambda_2^m - \lambda_2^{m+1} - \lambda_3^m \{0.5h_o e(q^m - \rho)\} - \lambda_3^{m+1} \{0.5h_0 e$$

$$(q^{m+1} - \rho)\} = 0 \qquad (7.80)$$

and using Eq. (7.73) in Eq. (7.77), we get

$$\left(\frac{\partial L}{\partial q^1}\right) = \lambda_2^1 - \lambda_3^1 h_o \{1 + 0.5e\ (2X^o + J^1 - 2q^1 + \rho)\} = 0 \qquad (7.81)$$

The dual variables for any subinterval may be obtained as follows:

(i) Obtain λ_1^m from Eq. (7.78).

(ii) Obtain λ_3^m from Eq. (7.79).

(iii) Obtain λ_2^1 from Eq. (7.81) and other values of λ_2^m $(m \neq 1)$ from Eq. (7.80).

The gradient vector is given by the partial derivatives of the Lagrangian with respect to the independent variables. Thus

$$\left(\frac{\partial L}{\partial q^m}\right)_{m\neq 1} = \lambda_2^m - \lambda_3^m h_o \{1 + 0.5e\ (2X^{m-1} + J^m - 2q^m + \rho)\} \qquad (7.82)$$

For optimality the gradient vector should be zero if there are no inequality constraints on the control variables.

Algorithm

1. Assume an initial set of independent variables q^m $(m\neq1)$ for all subintervals except the first.

2. Obtain the values of dependent variables X^m, P_{GH}^m, P_{GT}^m, q^1 using Eqs. (7.73), (7.74), (7.72) and (7.76).

3. Obtain the dual variables λ_1^m, λ_3^m, λ_2^m $(m \neq 1)$ and λ_2^1 using Eqs. (7.78), (7.79), (7.80) and (7.81).

4. Obtain the gradient vector using Eq. (7.82) and check if all its elements are equal to zero within a specified accuracy. If so, optimum is reached. If not, go to step 5.

5. Obtain new values of control variables using the first order gradient method, i.e.

$$q_{new}^m = q_{old}^m - \alpha\left(\frac{\partial L}{\partial q^m}\right); \ m \neq 1 \qquad (7.83)$$

where α is a positive scalar. Repeat from step 2.

In the solution technique presented above, if some of the control variables (water discharges) cross the upper or lower bounds, these are made equal to their respective bounded values. For these control variables, step 4 above is checked in accordance with the Kuhn-Tucker conditions (7.61) given in Sec. 7.6.

The inequality constraints on the dependent variables are treated conveniently by augmenting the cost function with penalty functions as discussed in Sec. 7.6.

The method outlined above is quite general and can be directly extended to a system having multi-hydro and multi-thermal plants. The method, however, has the disadvantage of large memory requirement, since the independent variables, dependent variables and gradients need to be stored simultaneously. A modified technique known as decomposition [24] overcomes this difficulty. In this technique optimization is carried out over each subinterval and the complete cycle of iteration is repeated, if the water availability equation does not check at the end of the cycle.

Example 7.7

Consider the fundamental hydrothermal system shown in Fig. 7.12. The objective is to find the optimal generation schedule for a typical day, wherein load varies in three steps of eight hours each as 7 MW, 10 MW and 5 MW, respectively. There is no water inflow into the reservoir of the hydro plant. The initial water storage in the reservoir is 100 m^3/s and the final water storage should be 60m^3/s, i.e. the total water available for hydro generation during the day is 40 m^3/s.

Basic head is 20 m. Water head correction factor e is given to be 0.005. Assume for simplicity that the reservoir is rectangular so that e does not change with water storage. Let the non-effective water discharge be assumed as 2 m^3/s. Incremental fuel cost of the thermal plant is

$$\frac{dC}{dP_{GT}} = 1.0P_{GT} + 25.0 \text{ Rs/hr}$$

Further, transmission losses may be neglected.

The above problem has been specially constructed (rather oversimplified) to illustrate the optimal hydrothermal scheduling algorithm, which is otherwise computationally involved and the solution has to be worked on the digital computer. Steps of one complete iteration will be given here.

Since there are three subintervals, the control variables are q^2 and q^3. Let us assume their initial values to be

$$q^2 = 15 \text{ m}^3/\text{s}$$

$$q^3 = 15 \text{ m}^2/\text{s}$$

The value of water discharge in the first subinterval can be immediately found out using Eq. (7.76), i.e.

$$q^1 = 100 - 60 - (15 + 15) = 10 \text{ m}^3/\text{s}$$

It is given that $X^0 = 100$ m^3/s and $X^3 = 60$ m^3/s.
From Eq. (7.73)

$$X^1 = X^0 + J^1 - q^1 = 90 \text{ m}^3/\text{s}$$

$$X^2 = X^1 + J^2 - q^2 = 75 \text{ m}^3/\text{s}$$

The values of hydro generations in the subintervals can be obtained using Eq. (7.74) as follows:

$$P_{GH}^1 = 9.81 \times 10^{-3} \times 20 \,\{1 + 0.5 \times 0.005 \,(X^1 + X^0)\} \,\{q^1 - \rho)$$

$$= 0.1962 \,\{1 + 25 \times 10^{-4} \times 190\} \times 8$$

$$= 2.315 \text{ MW}$$

$$P_{GH}^2 = 0.1962 \,\{1 + 25 \times 10^{-4} \times 165\} \times 13$$

$$= 3.602 \text{ MW}$$

$$P_{GH}^3 = 0.1962 \,\{1 + 25 \times 10^{-4} \times 135\} \times 13$$

$$= 3.411 \text{ MW}$$

The thermal generations in the three intervals are then

$$P_{GT}^1 = P_D^1 - P_{GH}^1 = 7 - 2.515 = 4.685 \text{ MW}$$

$$P_{GT}^2 = P_D^2 - P_{GH}^2 = 10 - 3.602 = 6.398 \text{ MW}$$

$$P_{GT}^3 = P_D^3 - P_{GH}^3 = 5 - 3.411 = 1.589 \text{ MW}$$

From Eq. (7.78), we have values of λ_1^m as

$$\frac{dC(P_{GT}^m)}{dP_{GT}^m} = \lambda_1^m$$

or $\qquad\qquad \lambda_1^m = P_{GT}^m + 25$

Calculating λ_1 for all the three subintervals, we have

$$\begin{bmatrix} \lambda_1^1 \\ \lambda_1^2 \\ \lambda_1^3 \end{bmatrix} = \begin{bmatrix} 29.685 \\ 31.398 \\ 26.589 \end{bmatrix}$$

Also from Eq. (7.79), we can write

$$\begin{bmatrix} \lambda_3^1 \\ \lambda_3^2 \\ \lambda_3^3 \end{bmatrix} = \begin{bmatrix} \lambda_1^1 \\ \lambda_1^2 \\ \lambda_1^3 \end{bmatrix} = \begin{bmatrix} 29.685 \\ 31.398 \\ 26.589 \end{bmatrix} \text{ for the lossless case}$$

From Eq. (7.81)

$$\lambda_2^1 = \lambda_3^1 h_o \,\{1 + 0.5e \,(2X^0 + J^1 - 2q^1 + \rho)\}$$

$$= 29.685 \times 0.1962 \,\{1 + 25 \times 10^{-4} \,(200 - 20 + 2)\}$$

$$= 8.474$$

From Eq. (7.80) for $m = 1$ and 2, we have

$$\lambda_2^1 - \lambda_2^2 - \lambda_3^1 \{0.5 h_o e \ (q^1 - \rho)\} - \lambda_3^2 \{0.5 h_o e \ (q^2 - p)\} = 0$$

$$\lambda_2^2 - \lambda_2^3 - \lambda_3^2 \{0.5 \ h_o e \ (q^2 - \rho)\} - \lambda_3^3 \{0.5 h_o e \ (q^3 - \rho)\} = 0$$

Substituting various values, we get

$$\lambda_2^2 = 8.474 - 29.685 \ \{0.5 \times 0.1962 \times 0.005 \times 8\} - 31.398 \ \{0.5 \times$$
$$0.1962 \times 0.005 \times 13\}$$

$$= 8.1574$$

$$\lambda_2^3 = 8.1574 - 31.398 \ (0.5 \times 0.1962 \times 0.005 \times 13)$$
$$- 26.589 \ (0.5 \times 0.1962 \times 0.005 \times 13) = 7.7877$$

Using Eq. (7.82), the gradient vector is

$$\left(\frac{\partial L}{\partial q^2} \right) = \lambda_2^2 - \lambda_3^2 \ h_o \ \{1 + 0.5 \times 0.005 \ (2 \times 90 - 2 \times 15 + 2)\}$$

$$= 8.1574 - 31.398 \times 0.1962 \ \{1 + 25 \times 10^{-4} \times 152\}$$

$$= - \ 0.3437$$

$$\left(\frac{\partial L}{\partial q^3} \right) = \lambda_2^3 - \lambda_3^3 h_o \ \{1 + 0.5e \ (2X^2 + J^3 - 2q^3 + \rho)\}$$

$$= 7.7877 - 26.589 \times 0.1962 \ \{1 + 25 \times 10^{-4} \times 122\}$$

$$= 0.9799$$

If the tolerance for gradient vector is 0.1, then optimal conditions are not yet satisfied, since the gradient vector is not zero, i.e. (≤ 0.1); hence the second iteration will have to be carried out starting with the following new values of the control variables obtained from Eq. (7.83)

$$\begin{bmatrix} q_{new}^2 \\ q_{new}^3 \end{bmatrix} = \begin{bmatrix} q_{old}^2 \\ q_{old}^3 \end{bmatrix} - \alpha \begin{bmatrix} \dfrac{\partial L}{\partial q^2} \\ \dfrac{\partial L}{\partial q^3} \end{bmatrix}$$

Let us take $\alpha = 0.5$, then

$$\begin{bmatrix} q_{new}^2 \\ q_{new}^3 \end{bmatrix} = \begin{bmatrix} 15 \\ 15 \end{bmatrix} - 0.5 \begin{bmatrix} -0.3437 \\ 0.9799 \end{bmatrix} = \begin{bmatrix} 15.172 \\ 14.510 \end{bmatrix}$$

and from Eq. (7.76)

$$q_{new}^1 = 100 - 60 - (15.172 + 14.510) = 10.318 \ m^3/s$$

The above computation brings us to the starting point of the next iteration. Iterations are carried out till the gradient vector becomes zero within specified tolerance.

PROBLEMS

7.1 For Example 7.1 calculate the extra cost incurred in Rs/hr, if a load of 220 MW is scheduled as $P_{G1} = P_{G2} = 110$ MW.

7.2 A constant load of 300 MW is supplied by two 200 MW generators, 1 and 2, for which the respective incremental fuel costs are

$$\frac{dC_1}{dP_{G1}} = 0.10P_{G1} + 20.0$$

$$\frac{dC_2}{dP_{G2}} = 0.12P_{G2} + 15.0$$

with powers P_G in MW and costs C in Rs/hr. Determine (a) the most economical division of load between the generators, and (b) the saving in Rs/day thereby obtained compared to equal load sharing between machines.

7.3 Figure P-7.3 shows the incremental fuel cost curves of generators A and B. How would a load (i) more than $2P_G$, (ii) equal to $2P_G$, and (iii) less than $2P_G$ be shared between A and B if both generators are running.

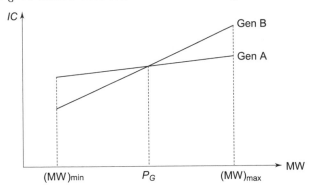

Fig. P-7.3

7.4 Consider the following three IC curves

$$P_{G1} = -100 + 50\ (IC)_1 - 2\ (IC)_1^2$$

$$P_{G2} = -150 + 60\ (IC)_2 - 2.5\ (IC)_2^2$$

$$P_{G3} = -80 + 40\ (IC)_3 - 1.8\ (IC)_3^2$$

where ICs are in Rs/MWh and P_Gs are in MW.

The total load at a certain hour of the day is 400 MW. Neglect transmission loss and develop a computer programme for optimum generation scheduling within and accuracy of $+ 0.05$ MW.

Note: All P_Gs must be real positive.

7.5 For a certain generating unit of a thermal power plant, the fuel input in millions of kilocalories per hour can be expressed as a function of power output P_G in megawatts by the equation

$$0.0001P_G^3 + 0.03P_G^2 + 12.0P_G + 150$$

Find the expression for incremental fuel cost in rupees per megawatt hour as a function of power output in megawatts. Also find a good linear approximation to the incremental fuel cost as a function of P_G.

Given: Fuel cost is Rs 2/million kilocalories.

7.6 For the system of Example 7.4, the system λ is Rs 26/MWh. Assume further the fuel costs at no load to be Rs 250 and Rs 350 per hr, respectively for plants 1 and 2.
    (a) For this value of system λ, what are the values of P_{G1}, P_{G2} and received load for optimum operation.
    (b) For the above value of received load, what are the optimum values of P_{G1} and P_{G2}, if system losses are accounted for but not coordinated.
    (c) Total fuel costs in Rs/hr for parts (a) and (b).

7.7 Figure P-7.7 shows a system having two plants 1 and 2 connected to buses 1 and 2, respectively. There are two loads and a network of three branches. The bus 1 is the reference bus with voltage of $1.0 \angle 0°$ pu. The branch currents and impedances are

$$I_a = 2 - j0.5 \text{ pu}$$

$$I_b = 1.6 - j0.4 \text{ pu}$$

$$I_c = 1.8 - j0.45 \text{ pu}$$

$$Z_a = 0.06 + j0.24 \text{ pu}$$

$$Z_b = 0.03 + j0.12 \text{ pu}$$

$$Z_c = 0.03 + j0.12 \text{ pu}$$

Calculate the loss formula coefficients of the system in per unit and in reciprocal megawatts, if the base is 100 MVA.

Fig. P-7.7   Sample system for Problem P-7.7

7.8 For the power plant of the illustrative example used in Section 7.3, obtain the economically optimum unit commitment for the daily load cycle given in Fig. P-7.8.

Correct the schedule to meet security requirements.

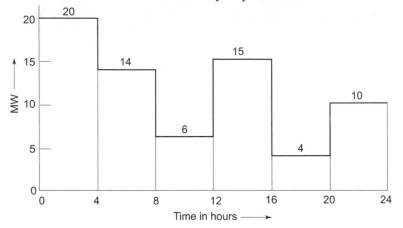

Fig. P-7.8 Daily load curve for Problem P-7.8

7.9 Repeat Example 7.3 with a load of 220 MW from 6 a.m. to 6 p.m. and 40 MW from 6 p.m. to 6 a.m.

7.10 Reformulate the optimal hydrothermal scheduling problem considering the inequality constraints on the thermal generation and water storage employing penalty functions. Find out the necessary equations and gradient vector to solve the problem.

REFERENCES

Books

1. Billinton, R., *Power System Reliability Evaluation*, Gordon and Breach, New York, 1970.
2. Billinton, R., R.J. Ringlee and A.J. Wood. *Power System Reliability Calculations*, The MIT Press, Boston, Mass, 1973.
3. Kusic, G.L., *Computer Aided Power System Analysis*, Prentice-Hall, New Jersey, 1986.
4. Kirchmayer, L.K., *Economic Operation of Power Systems*, John Wiley, New York, 1958.
5. Kirchmayer, L.K., *Economic Control of Interconnected Systems*, Wiley, New York, 1959.
6. Knight, U.G., *Power Systems Engineering and Mathematics*, Pergamon Press, New York, 1972.

7. Neuenswander, J.R., *Modern Power Systems*, International Text Book Co., New York, 1971.
8. Singh, C. and R. Billinton, *System Reliability, Modelling and Evaluation*, Hutchinson of London, 1977.
9. Sullvan, R.L., *Power System Planning*, McGraw-Hill, New York, 1977.
10. Wood, A.J. and B.F. Wollenberg, *Power Generation, Operation and Control*, 2nd edn., Wiley, New York, 1996
11. Mahalanabis, A.K., D.P. Kothari and S.I. Ahson, *Computer Aided Power System Analysis and Control*, Tata McGraw-Hill, New Delhi, 1988.
12. Bergen, A.R., *Power System Analysis*, Prentice-Hall, Inc., New Jersey, 1986.
13. Billinton, R. and R.N. Allan, *Reliability Evaluation of Power System*, Plenum Press, New York, 1984.
14. Sterling, M.J.H., *Power System Control*, I.E.E., England, 1978.
15. Khatib, H., "Economics of Power Systems Reliability", *Technicopy*, 1980.
16. Warwick, K. A.E. Kwue and R. Aggarwal (Eds), *A.I. Techniques in Power Systems*, IEE, UK, 1997.
17. Momoh, J.A., *Electric Power System Applications of Optimization*, Marcel Dekker, Inc., New York., 2001.
18. Yong-Hua Song (Ed.), *Modern Optimization Techniques in Power Systems*, Kluwer Academic Publishers, London, 1999.
19. Debs, A.S., *Modern Power Systems Control and Operation*, KAP, New York, 1988.
20. Berrie, T.W., *Power System Economics*, IEE, London, 1983.
21. Berrie, T.W., *Electricity, Economics and Planning*, IEE, London, 1992.

Papers

22. Meyer, W.S. and V.D. Albertson, "Improved Loss Formula Compuration by Optimally Ordered Elimination Techniques'" *IEEE Transm PAS*, 1971, 90: 716.
23. Hill, E.F. and W.D. Stevenson, J.R., "A New Method of Determining Loss Coefficients", *IEEE Trans. PAS*, July 1968, 87: 1548.
24. Agarwal, S.K. and I.J. Nagrath, "Optimal Scheduling of Hydrothermal Systems", *Proc. IEEE*, 1972, 199: 169.
25. Ayub, A.K. and A.D. Patton, "Optimal Thermal Generating Unit Commitment", *IEEE Trans.*, July-Aug 1971, PAS-90: 1752.
26. Dopazo, J.F. et al., "An Optimization Technique for Real and Reactive Power Allocation", *Proc, IEEE*, Nov 1967, 1877.
27. Happ, H.H., "Optimal Power Dispatch—A Comprehensive Survey", *IEEE Trans.* 1977, PAS-96: 841.
28. Harker, E.C., "A Primer on Loss Formula", *AIEE Trans,* 1958, Pt III, 77: 1434.
29. IEEE Committee Report, "Economy—Security Functions in Power System Operations", *IEEE Special Publication* 75 CHO 969.6 PWR, New York, 1975.
30. Kothari, D.P., "Optimal Hydrothermal Scheduling and Unit Commitment". *Ph. D. Thesis*, B.I.T.S, Pilani, 1975.
31. Kothari, D.P. and I.J. Nagrath, "Security Constrained Economic Thermal Generating Unit Commitment", *J.I.E.* (India), Dec. 1978, 59: 156.
32. Nagrath, I.J. and D.P. Kothari, "Optimal Stochastic Scheduling of Cascaded Hydrothermal Systems", *J.I.E.* (India), June 1976, 56: 264.

33. Peschen, J. et al., "Optimal Control of Reactive Power Flow", *IEEE Trans*, 1968, PAS. 87: 40.

34. Dommel, H.W. and W.F. Trinney, "Optimal Power Flow Solution", *IEEE Trans.*, October 1968, PAS. 87: 1866.

35. Sasson, A.M. and H.M. Merrill, "Some Applications of Optimization Techniques to Power System Problems", *Proc. IEEE*, July 1974, 62: 959.

36. Wu, F. et al., "A Two-Stage Approach to Solving Optimal Power Flows", *Proc. 1979 PICA Conf.* pp. 126-136.

37. Nanda, J., P.R. Bijwe and D.P. Kothari, "Application of Progressive Optimality Algorithm to Optimal Hydrothermal Scheduling Considering Deterministic and Stochastic Data", *International Journal of Electrical Power and Energy Systems*, January 1986, 8: 61.

38. Kothari, D.P. et al., "Some Aspects of Optimal Maintenance Scheduling of Generating Units", *J.I.E.* (India), August 1985, 66: 41.

39. Kothari, D.P. and R.K. Gupta, "Optimal Stochastic Load Flow Studies", *J.I.E.* (India), August 1978, p. 34.

40. "Description and Bibliography of Major Economy—Security Functions—Part I, II, and III", *IEEE Committee Report*, IEEE Trans. Jan 1981. PAS-100, 211-235.

41. Bijwe, P.R., D.P., Kothari, J. Nanda, and K.S. Lingamurthy, "Optimal Voltage Control using Constant Sensitivity Matrix", Electric Power System Research, Vol. II, No. 3, Dec. 1986, pp. 195-203.

42. Nanda, J., D.P. Kothari and K.S. Lingamurthy, "Economic-emission Load Dispatch through Goal Programming Techniques", *IEEE Trans.* on Energy Conversion, Vol. 3, No. 1, March 1988, pp. 26-32.

43. Nanda, J.D.P. Kothari and S.C. Srivastava, "A New Optimal Power Dispatch Algorithm using Fletcher's *QP* Method", *Proc. IEE*, pte, vol. 136. no. 3, May 1989, pp. 153-161.

44. Dhillon, J.S., S.C. Parti and D.P. Kothari, "Stochastic Economic Emission Load Dispatch", *Int. J. of Electric Power System Research,* Vol. 26, No. 3, 1993, pp. 179-183.

45. Dhillon, J.S., S.C. Parti and D.P. Kothari, "Multiobjective Optimal Thermal Power Dispatch", *Int. J. of EPES*, Vol. 16, No. 6, Dec. 1994, pp. 383-389.

46. Kothari, D.P. and Aijaz Ahmad, "An Expert System Approach to the Unit Commitment Problem", *Energy Conversion and Management*", Vol. 36, No. 4, April 1995, pp. 257-261.

47. Sen, Subir, D.P. Kothari and F.A Talukdar, "Environmentally Friendly Thermal Power Dispatch — An Approach", *Int. J. of Energy Sources*, Vol. 19, no. 4, May 1997, pp. 397-408.

48. Kothari, D.P. and A. Ahmad, "Fuzzy Dynamic Programming Based Optimal Generator Maintenance Scheduling Incorporating Load Forecasting", in *Advances in Intelligent Systems*, edited by F.C. Morabito, IOS Press, Ohmsha, 1997, pp. 233-240.

49. Aijaz Ahmad and D.P. Kothari, "A Review of Recent Advances in Generator Maintenance Scheduling", *Electric Machines and Power Systems*, Vol. 26, No. 4, 1998, pp. 373-387.

50. Sen S., and D.P. Kothari, "Evaluation of Benefit of Inter-Area Energy Exchange of Indian Power System Based on Multi-Area Unit Commitment Approach", *Int. J. of EMPS*, Vol. 26, No. 8, Oct. 1998, pp. 801-813.

51. Sen. S., and D.P. Kothari, ''Optimal Thermal Generating Unit Commitment-A Review, "*Int. J. EPES*, Vol. 20, No. 7, Oct. 1998, pp. 443−451.

52. Kulkarni, P.S., A.G. Kothari and D.P. Kothari, "Combined Econonic and Emission Dispatch using Improved BPNN", *Int. J. of EMPS*, Vol. 28, No. 1, Jan 2000, pp. 31−43

53. Arya, L.D., S.C. Chaude and D.P. Kothari, "Economic Despatch Accounting Line Flow Constraints using Functional Link Network," *Int. J. of Electrical Machine & Power Systems*, 28, 1, Jan 2000, pp. 55–68.

54. Ahmad, A. and D.P. Kothari, "A Practical Model for Generator Maintenance Scheduling with Transmission Constraints", *Int. J. of EMPS*, Vol. 28, No. 6, June 2000, pp. 501−514.

55. Dhillon, J.S. and D.P. Kothari, "The Surrogate Worth Trade off Approach for Mutliobjective Thermal Power Dispatch Probelm". *EPSR*, Vol. 56, No. 2, Nov. 2000, pp. 103−110.

56. Son, S. and D.P. Kothari, "Large Scale Thermal Generating Unit Commitment: A New Model", in *The Next Generation of Electric Power Unit Commitment Models*, edited by B.F. Hobbs et. al. KAP, Boston, 2001, pp. 211−225.

57. Dhillon, J.S., S.C. Parti and D.P. Kothari, "Fuzzy Decision Making in Multi Objective Long-term Scheduling of Hydrothermal System", *Int. J. of EPES*, Vol. 23, No. 1, Jan 2001, pp. 19−29.

58. Brar, Y.S., J.S. Dhillon and D.P. Kothari, "Multi-objective Load Dispatch by Fuzzy Logic based Searching Weightage Pattern," Electric Power Systems Research, Vol. 63, 2002, pp. 149–160.

59. Dhillion, J.S., S.C. Parti and D.P. Kothari, "Fuzzy Decision-making in Stochastic Multiobjective Short-term Hydrothermal Scheduling," IEE Proc. GTD, Vol. 149, 2, March 2002, pp 191–200.

60. Kothari, D.P., Application of Neural Networks to Power Systems (Invited Paper), Proc. Int. Conf., ICIT 2000, Jan. 2000, pp. 621–626.

8

Automatic Generation and Voltage Control

8.1 INTRODUCTION

Power system operation considered so far was under conditions of steady load. However, both active and reactive power demands are never steady and they continually change with the rising or falling trend. Steam input to turbo-generators (or water input to hydro-generators) must, therefore, be continuously regulated to match the active power demand, failing which the machine speed will vary with consequent change in frequency which may be highly undesirable[*] (maximum permissible change in power frequency is ± 0.5 Hz). Also the excitation of generators must be continuously regulated to match the reactive power demand with reactive generation, otherwise the voltages at various system buses may go beyond the prescribed limits. In modern large interconnected systems, manual regulation is not feasible and therefore automatic generation and voltage regulation equipment is installed on each generator. Figure 8.1 gives the schematic diagram of load frequency and excitation voltage regulators of a turbo-generator. The controllers are set for a particular operating condition and they take care of small changes in load demand without frequency and voltage exceeding the prescribed limits. With the passage of time, as the change in load demand becomes large, the controllers must be reset either manually or automatically.

It has been shown in previous chapters that for small changes active power is dependent on internal machine angle δ and is independent of bus voltage; while bus voltage is dependent on machine excitation (therefore on reactive

---

[*]Change in frequency causes change in speed of the consumers' plant affecting production processes. Further, it is necessary to maintain network frequency constant so that the power stations run satisfactorily in parallel, the various motors operating on the system run at the desired speed, correct time is obtained from synchronous clocks in the system, and the entertaining devices function properly.

generation Q) and is independent of machine angle δ. Change in angle δ is caused by momentary change in generator speed. Therefore, load frequency and excitation voltage controls are non-interactive for small changes and can be modelled and analysed independently. Furthermore, excitation voltage control is fast acting in which the major time constant encountered is that of the generator field; while the power frequency control is slow acting with major time constant contributed by the turbine and generator moment of inertia—this time constant is much larger than that of the generator field. Thus, the transients in excitation voltage control vanish much faster and do not affect the dynamics of power frequency control.

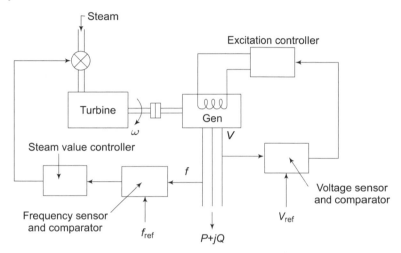

Fig. 8.1 Schematic diagram of load frequency and excitation
voltage regulators of a turbo-generator

Change in load demand can be identified as: (i) slow varying changes in mean demand, and (ii) fast random variations around the mean. The regulators must be designed to be insensitive to fast random changes, otherwise the system will be prone to hunting resulting in excessive wear and tear of rotating machines and control equipment.

8.2 LOAD FREQUENCY CONTROL (SINGLE AREA CASE)

Let us consider the problem of controlling the power output of the generators of a closely knit electric area so as to maintain the scheduled frequency. All the generators in such an area constitute a *coherent* group so that all the generators speed up and slow down together maintaining their relative power angles. Such an area is defined as a *control area*. The boundaries of a control area will generally coincide with that of an individual Electricity Board Company.

To understand the load frequency control problem, let us consider a single turbo-generator system supplying an isolated load.

Turbine Speed Governing System

Figure 8.2 shows schematically the speed governing system of a steam turbine. The system consists of the following components:

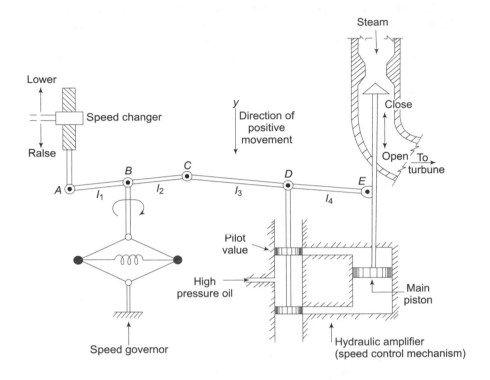

Fig. 8.2 Turbine speed governing system

Reprinted with permission of McGraw-Hill Book Co., New York, from Olle I. Elgerd: *Electric Energy System Theory: An Introduction*, 1971, p. 322.

(i) *Fly ball speed governor:* This is the heart of the system which senses the change in speed (frequency). As the speed increases the fly balls move outwards and the point B on linkage mechanism moves downwards. The reverse happens when the speed decreases.

(ii) *Hydraulic amplifier:* It comprises a pilot valve and main piston arrangement. Low power level pilot valve movement is converted into high power level piston valve movement. This is necessary in order to open or close the steam valve against high pressure steam.

(iii) *Linkage mechanism: ABC* is a rigid link pivoted at B and *CDE* is another rigid link pivoted at D. This link mechanism provides a movement to the control valve in proportion to change in speed. It also provides a feedback from the steam valve movement (link 4).

(iv) *Speed changer:* It provides a steady state power output setting for the turbine. Its downward movement opens the upper pilot valve so that more steam is admitted to the turbine under steady conditions (hence more steady power output). The reverse happens for upward movement of speed changer.

Model of Speed Governing System

Assume that the system is initially operating under steady conditions—the linkage mechanism stationary and pilot valve closed, steam valve opened by a definite magnitude, turbine running at constant speed with turbine power output balancing the generator load. Let the operating conditions be characterized by

f^o = system frequency (speed)

P_G^o = generator output = turbine output (neglecting generator loss)

y_E^o = steam valve setting

We shall obtain a linear incremental model around these operating conditions.

Let the point A on the linkage mechanism be moved downwards by a small amount Δy_A. It is a command which causes the turbine power output to change and can therefore be written as

$$\Delta y_A = k_C \Delta P_C \qquad (8.1)$$

where ΔP_C is the commanded increase in power.

The command signal ΔP_C (i.e. Δy_E) sets into motion a sequence of events—the pilot valve moves upwards, high pressure oil flows on to the top of the main piston moving it downwards; the steam valve opening consequently increases, the turbine generator speed increases, i.e. the frequency goes up. Let us model these events mathematically.

Two factors contribute to the movement of C:

(i) Δy_A contributes $-\left(\dfrac{l_2}{l_1}\right) \Delta y_A$ or $-k_1 \Delta y_A$ (i.e. upwards) of $-k_1 K_C \Delta P_C$

(ii) Increase in frequency Δf causes the fly balls to move outwards so that B moves downwards by a proportional amount $k'_2 \,\Delta f$. The consequent movement of C with A remaining fixed at Δy_A is $+\left(\dfrac{l_1 + l_2}{l_1}\right) k'_2 \Delta f = + k_2 \Delta f$

(i.e. downwards)

The net movement of C is therefore

$$\Delta y_C = -k_1 k_C \Delta P_C + k_2 \Delta f \qquad (8.2)$$

The movement of D, Δy_D, is the amount by which the pilot valve opens. It is contributed by Δy_C and Δy_E and can be written as

$$\Delta y_D = \left(\frac{l_4}{l_3 + l_4}\right) \Delta y_C + \left(\frac{l_3}{l_3 + l_4}\right) \Delta y_E$$

$$= k_3 \Delta y_C + k_4 \Delta y_E \qquad (8.3)$$

The movement Δy_D depending upon its sign opens one of the ports of the pilot valve admitting high pressure oil into the cylinder thereby moving the main piston and opening the steam valve by Δy_E. Certain justifiable simplifying assumptions, which can be made at this stage, are:

(i) Inertial reaction forces of main piston and steam valve are negligible compared to the forces exerted on the piston by high pressure oil.

(ii) Because of (i) above, the rate of oil admitted to the cylinder is proportional to port opening Δy_D.

The volume of oil admitted to the cylinder is thus proportional to the time integral of Δy_D. The movement Δy_E is obtained by dividing the oil volume by the area of the cross-section of the piston. Thus

$$\Delta y_E = k_5 \int_0^t (-\Delta y_D)\, dt \qquad (8.4)$$

It can be verified from the schematic diagram that a positive movement Δy_D, causes negative (upward) movement Δy_E accounting for the negative sign used in Eq. (8.4).

Taking the Laplace transform of Eqs. (8.2), (8.3) and (8.4), we get

$$\Delta Y_C(s) = - k_1 k_C \Delta P_C(s) + k_2 \Delta F(s) \qquad (8.5)$$

$$\Delta Y_D(s) = k_3 \Delta Y_C(s) + k_4 \Delta Y_E(s) \qquad (8.6)$$

$$\Delta y_E(s) = - k_5 \frac{1}{s} \Delta Y_D(s) \qquad (8.7)$$

Eliminating $\Delta Y_C(s)$ and $\Delta Y_D(s)$, we can write

$$\Delta Y_E(s) = \frac{k_1 k_3 k_C \Delta P_C(s) - k_2 k_3 \Delta F(s)}{\left(k_4 + \dfrac{s}{k_5} \right)}$$

$$= \left[\Delta P_C(s) - \frac{1}{R} \Delta F(s) \right] \times \left(\frac{K_{sg}}{1 + T_{sg} s} \right) \qquad (8.8)$$

where

$$R = \frac{k_1 k_C}{k_2} = \text{speed regulation of the governor}$$

$$K_{sg} = \frac{k_1 k_3 k_C}{k_4} = \text{gain of speed governor}$$

$$T_{sg} = \frac{1}{k_4 k_5} = \text{time constant of speed governor}$$

Equation (8.8) is represented in the form of a block diagram in Fig. 8.3.

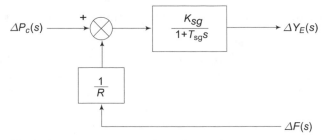

Fig. 8.3 Block diagram representation of speed governor system

The speed governing system of a hydro-turbine is more involved. An additional feedback loop provides temporary droop compensation to prevent instability. This is necessitated by the large inertia of the penstock gate which regulates the rate of water input to the turbine. Modelling of a hydro-turbine regulating system is beyond the scope of this book.

Turbine Model

Let us now relate the dynamic response of a steam turbine in terms of changes in power output to changes in steam valve opening Δy_E. Figure 8.4a shows a two stage steam turbine with a reheat unit. The dynamic response is largely influenced by two factors, (i) entrained steam between the inlet steam valve and first stage of the turbine, (ii) the storage action in the reheater which causes the output of the low pressure stage to lag behind that of the high pressure stage. Thus, the turbine transfer function is characterized by two time constants. For ease of analysis it will be assumed here that the turbine can be modelled to have a single equivalent time constant. Figure 8.4b shows the transfer function model of a steam turbine. Typically the time constant T_t lies in the range 0.2 to 2.5 sec.

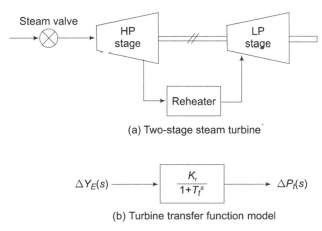

(a) Two-stage steam turbine

(b) Turbine transfer function model

Fig. 8.4

Generator Load Model

The increment in power input to the generator-load system is

$$\Delta P_G - \Delta P_D$$

where $\Delta P_G = \Delta P_t$, incremental turbine power output (assuming generator incremental loss to be negligible) and ΔP_D is the load increment.

This increment in power input to the system is accounted for in two ways:

(i) Rate of increase of stored kinetic energy in the generator rotor. At scheduled frequency (f^o), the stored energy is

$$W^o_{ke} = H \times P_r \text{ kW} = \text{sec (kilojoules)}$$

where P_r is the kW rating of the turbo-generator and H is defined as its inertia constant.

The kinetic energy being proportional to square of speed (frequency), the kinetic energy at a frequency of ($f^o + \Delta f$) is given by

$$W_{ke} = W^o{}_{ke}\left(\frac{f^o + \Delta f}{f^o}\right)^2$$

$$\simeq HP_r\left(1 + \frac{2\Delta f}{f^o}\right) \tag{8.9}$$

Rate of change of kinetic energy is therefore

$$\frac{\mathrm{d}}{\mathrm{d}t}(W_{ke}) = \frac{2HP_r}{f^o}\frac{\mathrm{d}}{\mathrm{d}t}(\Delta f) \tag{8.10}$$

(ii) As the frequency changes, the motor load changes being sensitive to speed, the rate of change of load with respect to frequency, i.e. $\partial P_D/\partial f$ can be regarded as nearly constant for small changes in frequency Δf and can be expressed as

$$(\partial P_D/\partial f)\ \Delta f = B\ \Delta f \tag{8.11}$$

where the constant B can be determined empirically. B is positive for a predominantly motor load.

Writing the power balance equation, we have

$$\Delta P_G - \Delta P_D = \frac{2HP_r}{f^o}\frac{\mathrm{d}}{\mathrm{d}t}(\Delta f) + B\ \Delta f$$

Dividing throughout by P_r and rearranging, we get

$$\Delta P_G(\text{pu}) - \Delta P_D(\text{pu}) = \frac{2H}{f^o}\frac{\mathrm{d}}{\mathrm{d}t}(\Delta f) + B(\text{pu})\ \Delta f \tag{8.12}$$

Taking the Laplace transform, we can write $\Delta F(s)$ as

$$\Delta F(s) = \frac{\Delta P_G(s) - \Delta P_D(s)}{B + \dfrac{2H}{f^o}s}$$

$$= [\varDelta P_G(s) - \varDelta P_D(s)] \times \left(\frac{K_{ps}}{1 + T_{ps}s} \right) \qquad (8.13)$$

where

$$T_{ps} = \frac{2H}{Bf^o} = \text{power system time constant}$$

$$K_{ps} = \frac{1}{B} = \text{power system gain}$$

Equation (8.13) can be represented in block diagram form as in Fig. 8.5.

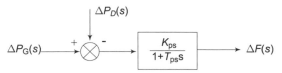

Fig. 8.5 Block diagram representation of generator-load model

Complete Block Diagram Representation of Load Frequency Control of an Isolated Power System

A complete block diagram representation of an isolated power system comprising turbine, generator, governor and load is easily obtained by combining the block diagrams of individual components, i.e. by combining Figs. 8.3, 8.4 and 8.5. The complete block diagram with feedback loop is shown in Fig. 8.6.

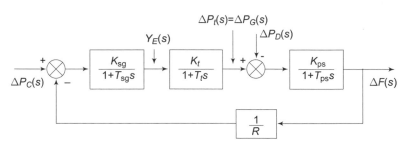

Fig. 8.6 Block diagram model of load frequency control (isolated power system)

Steady States Analysis

The model of Fig. 8.6 shows that there are two important incremental inputs to the load frequency control system – $\varDelta P_C$, the change in speed changer setting; and $\varDelta P_D$, the change in load demand. Let us consider a simple situation in which the speed changer has a fixed setting (i.e. $\Delta P_C = 0$) and the load demand changes. This is known as *free governor operation*. For such an operation the steady change in system frequency for a sudden change in load demand by an amount $\Delta P_D \left(\text{i.e.} \, \Delta P_D(s) = \dfrac{\Delta P_D}{s} \right)$ is obtained as follows:

$$\Delta F(s)\Big|_{\Delta P_C(s)=0} = -\frac{K_{\text{ps}}}{(1+T_{\text{ps}}s)+\dfrac{K_{\text{sg}}K_t K_{\text{ps}} \,/\, R}{(1+T_{\text{sg}}s)(1+T_t s)}} \times \frac{\Delta P_D}{s} \qquad (8.14)$$

$$\Delta f \Big|_{\substack{\text{steady state} \\ \Delta P_C = 0}} = s \, \Delta F(s) \Big|_{\substack{s \to 0 \\ \Delta P_C(s) = 0}}$$

$$= -\left(\frac{K_{\text{ps}}}{1+(K_{\text{sg}}K_t K_{\text{ps}} | R)} \right) \Delta P_D \qquad (8.15)$$

While the gain K_t is fixed for the turbine and K_{ps} is fixed for the power system, K_{sg}, the speed governor gain is easily adjustable by changing lengths of various links. Let it be assumed for simplicity that K_{sg} is so adjusted that

$$K_{\text{sg}} K_t \simeq 1$$

It is also recognized that $K_{\text{ps}} = 1/B$, where $B = \dfrac{\partial P_D}{\partial f} / P_r$ (in pu MW/unit change in frequency). Now

$$\Delta f = -\left(\frac{1}{B+(1/R)} \right) \Delta P_D \qquad (8.16)$$

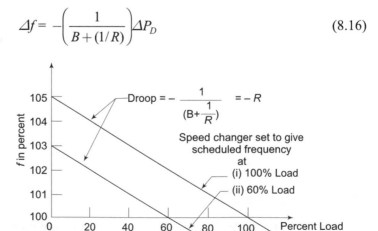

Fig. 8.7 Steady state load-frequency characteristic of a speed governor system

The above equation gives the steady state changes in frequency caused by changes in load demand. Speed regulation R is naturally so adjusted that changes in frequency are small (of the order of 5% from no load to full load). Therefore, the linear incremental relation (8.16) can be applied from no load to full load. With this understanding, Fig. 8.7 shows the linear relationship between frequency and load for free governor operation with speed changer set to give a scheduled frequency of 100% at full load. The 'droop' or slope of this relationship is $-\left(\dfrac{1}{B+(1/R)}\right)$.

Power system parameter B is generally much smaller[*] than $1/R$ (a typical value is $B = 0.01$ pu MW/Hz and $1/R = 1/3$) so that B can be neglected in comparison. Equation (8.16) then simplifies to

$$\Delta f = -R(\Delta P_D) \tag{8.17}$$

The droop of the load frequency curve is thus mainly determined by R, the speed governor regulation.

It is also observed from the above that increase in load demand (ΔP_D) is met under steady conditions partly by increased generation (ΔP_G) due to opening of the steam valve and partly by decreased load demand due to drop in system frequency. From the block diagram of Fig. 8.6 (with $K_{sg}K_t \simeq 1$)

$$\Delta P_G = -\frac{1}{R}\Delta f = \left(\frac{1}{BR+1}\right)\Delta P_D$$

Decrease in system load $= B\Delta f = \left(\dfrac{BR}{BR+1}\right)\Delta P_D$

Of course, the contribution of decrease in system load is much less than the increase in generation. For typical values of B and R quoted earlier

$$\Delta P_G = 0.971\ \Delta P_D$$

Decrease in system load $= 0.029\ \Delta P_D$

Consider now the steady effect of changing speed changer setting $\left(\Delta P_C(s) = \dfrac{\Delta P_C}{s}\right)$ with load demand remaining fixed (i.e. $\Delta P_D = 0$). The steady state change in frequency is obtained as follows.

---

[*]For 250 MW machine with an operating load of 125 MW, let the change in load be 1% for 1% change in frequency (scheduled frequency = 50 Hz). Then

$$\frac{\partial P_D}{\partial f} = \frac{1.25}{0.5} = 2.5 \text{ MW/Hz}$$

$$B = \left(\frac{\partial P_D}{\partial f}\right)\bigg|P_t = \frac{2.5}{250} = 0.01 \text{ pu MW/Hz}$$

$$\Delta F(s)\Big|_{\Delta P_D(s)=0} = \frac{K_{sg}K_tK_{ps}}{(1+T_{sg}s)(1+T_ts)(1+T_{ps}s)+K_{sg}K_tK_{ps}/R} \times \frac{\Delta P_C}{s} \qquad (8.18)$$

$$\Delta f\Big|_{\substack{\text{steady state} \\ \Delta P_D=0}} = \left(\frac{K_{sg}K_tK_{ps}}{1+K_{sg}K_tK_{ps}/R}\right)\Delta P_C \qquad (8.19)$$

If

$$K_{sg}K_t \simeq 1$$

$$\Delta f = \left(\frac{1}{B+1/R}\right)\Delta P_C \qquad (8.20)$$

If the speed changer setting is changed by ΔP_C while the load demand changes by ΔP_D, the steady frequency change is obtained by superposition, i.e.

$$\Delta f = \left(\frac{1}{B+1/R}\right)(\Delta P_C - \Delta P_D) \qquad (8.21)$$

According to Eq. (8.21) the frequency change caused by load demand can be compensated by changing the setting of the speed changer, i.e.

$$\Delta P_C = \Delta P_D, \text{ for } \Delta f = 0$$

Figure 8.7 depicts two load frequency plots—one to give scheduled frequency at 100% rated load and the other to give the same frequency at 60% rated load.

Example 8.1

A 100 MVA synchronous generator operates on full load at at frequency of 50 Hz. The load is suddenly reduced to 50 MW. Due to time lag in governor system, the steam valve begins to close after 0.4 seconds. Determine the change in frequency that occurs in this time.

Given $H = 5$ kW-sec/kVA of generator capacity.

Solution   Kinetic energy stored in rotating parts of generator and turbine

$$= 5 \times 100 \times 1{,}000 = 5 \times 10^5 \text{ kW-sec}$$

Excess power input to generator before the steam valve

begins to close $= 50$ MW

Excess energy input to rotating parts in 0.4 sec

$$= 50 \times 1{,}000 \times 0.4 = 20{,}000 \text{ kW-sec}$$

Stored kinetic energy ∞ (frequency)^2

∴   Frequency at the end of 0.4 sec

$$= 50 \times \left(\frac{500{,}000+20{,}000}{500{,}000}\right)^{1/2} = 51 \text{ Hz}$$

Example 8.2

Two generators rated 200 MW and 400 MW are operating in parallel. The droop characteristics of their governors are 4% and 5%, respectively from no load to full load. Assuming that the generators are operating at 50 Hz at no load, how would a load of 600 MW be shared between them? What will be the system frequency at this load? Assume free governor operation.

Repeat the problem if both governors have a droop of 4%.

Solution   Since the generators are in parallel, they will operate at the same frequency at steady load.

Let load on generator 1 (200 MW) = x MW
and load on generator 2 (400 MW) = $(600 - x)$ MW
Reduction in frequency = Δf

Now

$$\frac{\Delta f}{x} = \frac{0.04 \times 50}{200} \tag{i}$$

$$\frac{\Delta f}{600 - x} = \frac{0.05 \times 50}{400} \tag{ii}$$

Equating Δf in (i) and (ii), we get

$$x = 231 \text{ MW (load on generator 1)}$$

$$600 - x = 369 \text{ MW (load on generator 2)}$$

$$\text{System frequency} = 50 - \frac{0.04 \times 50}{200} \times 231 = 47.69 \text{ Hz}$$

It is observed here that due to difference in droop characteristics of governors, generator 1 gets overloaded while generator 2 is underloaded.

It easily follows from above that if both governors have a droop of 4%, they will share the load as 200 MW and 400 MW respectively, i.e. they are loaded corresponding to their ratings. This indeed is desirable from operational considerations.

Dynamic Response

To obtain the dynamic response giving the change in frequency as function of the time for a step change in load, we must obtain the Laplace inverse of Eq. (8.14). The characteristic equation being of third order, dynamic response can only be obtained for a specific numerical case. However, the characteristic equation can be approximated as first order by examining the relative magnitudes of the time constants involved. Typical values of the time constants of load frequency control system are related as

$$T_{sg} \ll T_t \ll T_{ps}$$

Typically[*] $T_{sg} = 0.4$ sec, $T_t = 0.5$ sec and $T_{ps} = 20$ sec.

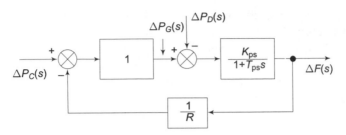

Fig. 8.8   First order approximate block diagram of load frequency control of an isolated area

Letting $T_{sg} = T_t = 0$, (and $K_{sg} K_t \cong 1$), the block diagram of Fig. 8.6 is reduced to that of Fig. 8.8, from which we can write

$$\Delta F(s)|_{\Delta P_C(s)=0} = -\frac{K_{ps}}{(1 + K_{ps}/R) + T_{ps}s} \times \frac{\Delta P_D}{s}$$

$$= -\frac{K_{ps}/T_{ps}}{s\left[s + \dfrac{R + K_{ps}}{RT_{ps}}\right]} \times \Delta P_D$$

$$\Delta f(t) = -\frac{RK_{ps}}{R + K_{ps}}\left\{1 - \exp\left[-t/T_{ps}\left(\frac{R}{R + K_{ps}}\right)\right]\right\} \Delta P_D \qquad (8.22)$$

Taking $R = 3$, $K_{ps} = 1/B = 100$, $T_{ps} = 20$, $\Delta P_D = 0.01$ pu

$$\Delta f(t) = -0.029 \, (1 - e^{-1.717t}) \qquad (8.23a)$$

$$\Delta f|_{\text{steady state}} = -0.029 \text{ Hz} \qquad (8.23b)$$

---

[*]For a 250 MW machine quoted earlier, inertia constant $H = 5\text{kW–sec/kVA}$

$$T_{ps} = \frac{2H}{Bf^o} = \frac{2 \times 5}{0.01 \times 50} = 20 \text{ sec}$$

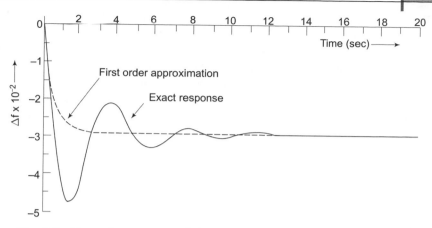

Fig. 8.9 Dynamic response of change in frequency for a step change in load (ΔP_D = 0.01 pu, T_{sg} = 0.4 sec, T_t = 0.5 sec, T_{ps} = 20 sec, K_{ps} = 100, R = 3)

The plot of change in frequency versus time for first order approximation given above and the exact response are shown in Fig. 8.9. First order approximation is obviously a poor approximation.

Control Area Concept

So far we have considered the simplified case of a single turbo-generator supplying an isolated load. Consider now a practical system with a number of generating stations and loads. It is possible to divide an extended power system (say, national grid) into subareas (may be, State Electricity Boards) in which the generators are tightly coupled together so as to form a *coherent* group, i.e. all the generators respond in *unison* to changes in load or speed changer settings. Such a coherent area is called a *control area* in which the frequency is assumed to be the same throughout in static as well as dynamic conditions. For purposes of developing a suitable control strategy, a control area can be reduced to a single speed governor, turbo-generator and load system. All the control strategies discussed so far are, therefore, applicable to an independent control area.

Proportional Plus Integral Control

It is seen from the above discussion that with the speed governing system installed on each machine, the steady load frequency characteristic for a given speed changer setting has considerable droop, e.g. for the system being used for the illustration above, the steady state droop in frequency will be 2.9 Hz [*see* Eq. (8.23b)] from no load to full load (1 pu load). System frequency specifications are rather stringent and, therefore, so much change in frequency cannot be tolerated. In fact, it is expected that the steady change in frequency will be zero. While steady state frequency can be brought back to the scheduled value by adjusting speed changer setting, the system could under go intolerable dynamic frequency changes with changes in load. It leads to the natural suggestion that the speed changer setting be adjusted automatically by monitoring the frequency changes. For this purpose, a signal from Δf is fed through an integrator to the speed changer resulting in the block diagram configuration shown in Fig. 8.10. The system now modifies to a proportional plus integral controller, which, as is well known from control theory, gives zero steady state error, i.e. $\Delta f |_{\text{steady state}} = 0$.

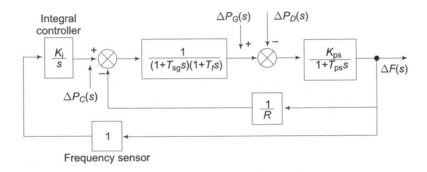

Fig. 8.10 Proportional plus integral load frequency control

The signal $\Delta P_C(s)$ generated by the integral control must be of opposite sign to $\Delta F(s)$ which accounts for negative sign in the block for integral controller. Now

$$\Delta F(s) = - \frac{K_{ps}}{(1 + T_{ps}S) + \left(\frac{1}{R} + \frac{K_i}{s}\right) \times \frac{K_{ps}}{(1 + T_{sg}s)(1 + T_t s)}} \times \frac{\Delta P_D}{s}$$

$$= - \frac{RK_{ps}s(1 + T_{sg}s)(1 + T_t s)}{s(1 + T_{sg}s)(1 + T_t s)(1 + T_{ps}s)R + K_{ps}(K_i R + s)} \times \frac{\Delta P_D}{s} \qquad (8.24)$$

Obviously,

$$\Delta f |_{\text{steady state}} = \lim_{s \to 0} s\, \Delta F(s) = 0 \qquad (8.25)$$

In contrast to Eq. (8.16) we find that the steady state change in frequency has been reduced to zero by the addition of the integral controller. This can be argued out physically as well. Δf reaches steady state (a constant value) only when $\Delta P_C = \Delta P_D = $ constant. Because of the integrating action of the controller, this is only possible if $\Delta f = 0$.

In central load frequency control of a given control area, the change (error) in frequency is known as *Area Control Error* (ACE). The additional signal fed back in the modified control scheme presented above is the integral of ACE.

In the above scheme ACE being zero under steady conditions[*], a logical design criterion is the minimization of $\int ACE\ dt$ for a step disturbance. This integral is indeed the *time error* of a synchronous electric clock run from the power supply. In fact, modern power systems keep track of integrated time error all the time. A corrective action (manual adjustment ΔP_C, the speed changer setting) is taken by a large (preassigned) station in the area as soon as the time error exceeds a prescribed value.

The dynamics of the proportional plus integral controller can be studied numerically only, the system being of fourth order—the order of the system has increased by one with the addition of the integral loop. The dynamic response of the proportional plus integral controller with $K_i = 0.09$ for a step load disturbance of 0.01 pu obtained through digital computer are plotted in Fig. 8.11. For the sake of comparison the dynamic response without integral control action is also plotted on the same figure.

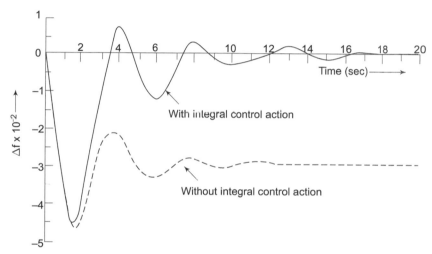

Fig. 8.11 Dynamic response of load frequency controller with and without integral control action ($\Delta P_D = 0.01$ pu, $T_{sg} = 0.4$ sec, $T_t = 0.5$ sec, $T_{ps} = 20$ sec, $K_{ps} = 100$, $R = 3$, $K_i = 0.09$)

8.3 LOAD FREQUENCY CONTROL AND ECONOMIC DESPATCH CONTROL

Load frequency control with integral controller achieves zero steady state frequency error and a fast dynamic response, but it exercises no control over the relative loadings of various generating stations (i.e. *economic despatch*) of the control area. For example, if a sudden small increase in load (say, 1%) occurs

---

[*]Such a control is known as *isochronous control*, but it has its time (integral of frequency) error though steady frequency error is zero.

in the control area, the load frequency control changes the speed changer settings of the governors of all generating units of the area so that, together, these units match the load and the frequency returns to the scheduled value (this action takes place in a few seconds). However, in the process of this change the loadings of various generating units change in a manner independent of economic loading considerations. In fact, some units in the process may even get overloaded. Some control over loading of individual units can be exercised by adjusting the gain factors (K_i) included in the signal representing integral of the area control error as fed to individual units. However, this is not satisfactory.

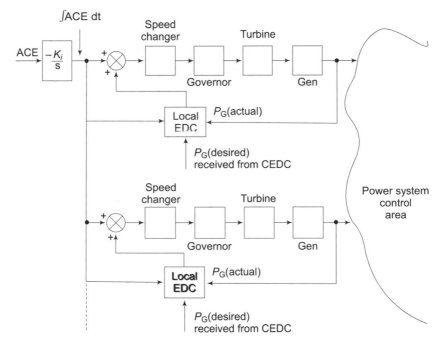

EDC — Economic despatch controller
CEDC — Central economic despatch computer

Fig. 8.12 Control area load frequency and economic despatch control

Reprinted (with modification) with permission of McGraw-Hill Book Company, New York from Olle I. Elgerd: *Electric Energy Systems Theory: An Introduction,* 1971, p. 345.

A satisfactory solution is achieved by using independent controls for load frequency and economic despatch. While the load frequency controller is a fast acting control (a few seconds), and regulates the system around an operating point; the economic despatch controller is a slow acting control, which adjusts the speed changer setting every minute (or half a minute) in accordance with a command signal generated by the central economic despatch computer. Figure 8.12 gives the schematic diagram of both these controls for two typical units of a control area. The signal to change the speed changer setting is constructed in accordance with economic despatch error, $[P_G(\text{desired}) - P_G(\text{actual})]$, suitably modified by the signal representing integral ACE at that instant of time. The signal P_G (desired) is computed by the central economic despatch computer (CEDC) and is transmitted to the local economic despatch controller (EDC) installed at each station. The system thus operates with economic despatch error only for very short periods of time before it is readjusted.

8.4   TWO-AREA LOAD FREQUENCY CONTROL

An extended power system can be divided into a number of load frequency control areas interconnected by means of tie lines. Without loss of generality we shall consider a two-area case connected by a single tie line as illustrated in Fig. 8.13.

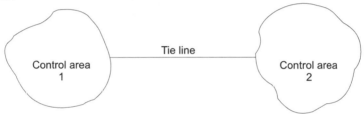

Fig. 8.13   Two interconnected control areas (single tie line)

The control objective now is to regulate the frequency of each area and to simultaneously regulate the tie line power as per inter-area power contracts. As in the case of frequency, proportional plus integral controller will be installed so as to give zero steady state error in tie line power flow as compared to the contracted power.

It is conveniently assumed that each control area can be represented by an equivalent turbine, generator and governor system. Symbols used with suffix 1 refer to area 1 and those with suffix 2 refer to area 2.

In an isolated control area case the incremental power $(\Delta P_G - \Delta P_D)$ was accounted for by the rate of increase of stored kinetic energy and increase in area load caused by increase in frequency. Since a tie line transports power in or out of an area, this fact must be accounted for in the incremental power balance equation of each area.

Power transported out of area 1 is given by

$$P_{\text{tie, }1} = \frac{|V_1||V_2|}{X_{12}} \sin(\delta_1^o - \delta_2^o) \tag{8.26}$$

where

δ_1^o, δ_2^o = power angles of equivalent machines of the two areas.

For incremental changes in δ_1 and δ_2, the incremental tie line power can be expressed as

$$\Delta P_{\text{tie, 1}}(\text{pu}) = T_{12}(\Delta\delta_1 - \Delta\delta_2) \tag{8.27}$$

where

$$T_{12} = \frac{|V_1||V_2|}{P_{r1}X_{12}} \cos(\delta_1^o - \delta_2^o) = \textit{synchronizing coefficient}$$

Since incremental power angles are integrals of incremental frequencies, we can write Eq. (8.27) as

$$\Delta P_{\text{tie, 1}} = 2\pi T_{12}\left(\int \Delta f_1 dt - \int \Delta f_1 dt\right) \tag{8.28}$$

where Δf_1 and Δf_2 are incremental frequency changes of areas 1 and 2, respectively.

Similarly the incremental tie line power out of area 2 is given by

$$\Delta P_{\text{tie, 2}} = 2\pi T_{21}\left(\int \Delta f_2 dt - \int \Delta f_1 dt\right) \tag{8.29}$$

where

$$T_{21} = \frac{|V_2||V_1|}{P_{r2}X_{21}} \cos(\delta_2^o - \delta_1^o) = \left(\frac{P_{r1}}{P_{r2}}\right) T_{12} = a_{12}T_{12} \tag{8.30}$$

With reference to Eq. (8.12), the incremental power balance equation for area 1 can be written as

$$\Delta P_{G1} - \Delta P_{D1} = \frac{2H_1}{f_1^o}\frac{d}{dt}(\Delta f_1) + B_1\Delta f_1 + \Delta P_{\text{tie, 1}} \tag{8.31}$$

It may be noted that all quantities other than frequency are in per unit in Eq. (8.31).

Taking the Laplace transform of Eq. (8.31) and reorganizing, we get

$$\Delta F_1(s) = [\Delta P_{G1}(s) - \Delta P_{D1}(s) - \Delta P_{\text{tie, 1}}(s)] \times \frac{K_{ps1}}{1 + T_{ps1}s} \tag{8.32}$$

where as defined earlier [see Eq. (8.13)]

$$K_{ps1} = 1/B_1$$
$$T_{ps1} = 2H_1/B_1 f^o \tag{8.33}$$

Compared to Eq. (8.13) of the isolated control area case, the only change is the appearance of the signal $\Delta P_{\text{tie,1}}(s)$ as shown in Fig. 8.14.

Taking the Laplace transform of Eq. (8.28), the signal $\Delta P_{\text{tie,1}}(s)$ is obtained as

$$\Delta P_{\text{tie, 1}}(s) = \frac{2\pi T_{12}}{s}[\Delta F_1(s) - \Delta F_2(s)] \tag{8.34}$$

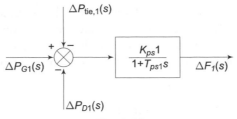

Fig. 8.14

The corresponding block diagram is shown in Fig. 8.15.

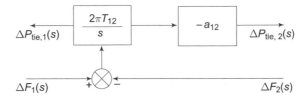

Fig. 8.15

For the control area 2, $\Delta P_{\text{tie, 2}}(s)$ is given by [Eq. (8.29)]

$$\Delta P_{\text{tie, 2}}(s) = \frac{-2\,\pi a_{12} T_{12}}{s}[\Delta F_1(s) - \Delta F_2(s)] \qquad (8.35)$$

which is also indicated by the block diagram of Fig. 8.15.

Let us now turn our attention to ACE (area control error) in the presence of a tie line. In the case of an isolated control area, ACE is the change in area frequency which when used in integral control loop forced the steady state frequency error to zero. In order that the steady state tie line power error in a two-area control be made zero another integral control loop (one for each area) must be introduced to integrate the incremental tie line power signal and feed it back to the speed changer. This is accomplished by a single integrating block by redefining ACE as a linear combination of incremental frequency and tie line power. Thus, for control area 1

$$\text{ACE}_1 = \Delta P_{\text{tie, 1}} + b_1 \Delta f_1 \qquad (8.36)$$

where the constant b_1 is called area *frequency bias*.
Equation (8.36) can be expressed in the Laplace transform as

$$\text{ACE}_1(s) = \Delta P_{\text{tie, 1}}(s) + b_1 \Delta F_1(s) \qquad (8.37)$$

Similarly, for the control are a 2, ACE_2 is expressed as

$$\text{ACE}_2(s) = \Delta P_{\text{tie, 2}}(s) + b_2 \Delta F_2(s) \qquad (8.38)$$

Combining the basic block diagrams of the two control areas corresponding to Fig. 8.6, with $\Delta P_{C1}(s)$ and $\Delta P_{C2}(s)$ generated by integrals of respective ACEs (obtained through signals representing changes in tie line power and local frequency bias) and employing the block diagrams of Figs. 8.14 to 8.15, we easily obtain the composite block diagram of Fig. 8.16.

Let the step changes in loads ΔP_{D1} and ΔP_{D2} be simultaneously applied in control areas 1 and 2, respectively. When steady conditions are reached, the output signals of all integrating blocks will become constant and in order for this to be so, their input signals must become zero. We have, therefore, from Fig. 8.16

$$\Delta P_{\text{tie, 1}} + b_1 \Delta f_1 = 0 \quad \left(\text{input of integrating block} -\frac{K_{i1}}{s}\right) \tag{8.39a}$$

$$\Delta P_{\text{tie, 2}} + b_2 \Delta f_2 = 0 \quad \left(\text{input of integrating block} -\frac{K_{i2}}{s}\right) \tag{8.39b}$$

$$\Delta f_1 - \Delta f_2 = 0 \quad \left(\text{input of integrating block} -\frac{2\pi T_{12}}{s}\right) \tag{8.40}$$

From Eqs. (8.28) and (8.29)

$$\frac{\Delta P_{\text{tie},1}}{\Delta P_{\text{tie, 2}}} = -\frac{T_{12}}{T_{21}} = -\frac{1}{a_{12}} = \text{constant} \tag{8.41}$$

Hence Eqs. (8.39) – (8.41) are simultaneously satisfied only for

$$\Delta P_{\text{tie, 1}} = \Delta P_{\text{tie, 2}} = 0 \tag{8.42}$$

and $\qquad \Delta f_1 = \Delta f_2 = 0$

Thus, under steady condition change in the tie line power and frequency of each area is zero. This has been achieved by integration of ACEs in the feedback loops of each area.

Dynamic response is difficult to obtain by the transfer function approach (as used in the single area case) because of the complexity of blocks and multi-input (ΔP_{D1}, ΔP_{D2}) and multi-output ($\Delta P_{\text{tie, 1}}$, $\Delta P_{\text{tie, 2}}$, Δf_1, Δf_2) situation. A more organized and more conveniently carried out analysis is through the state space approach (a time domain approach). Formulation of the state space model for the two-area system will be illustrated in Sec. 8.5.

The results of the two-area system (ΔP_{tie}, change in tie line power and Δf, change in frequency) obtained through digital computer study are shown in the form of a dotted line in Figs. 8.18 and 8.19. The two areas are assumed to be identical with system parameters given by

$$T_{sg} = 0.4 \text{ sec}, \ T_t = 0.5 \text{ sec}, \ T_{ps} = 20 \text{ sec}$$

$$K_{ps} = 100, \ R = 3, \ b = 0.425, \ K_l = 0.09, \ 2\pi T_{12} = 0.05$$

8.5 OPTIMAL (TWO-AREA) LOAD FREQUENCY CONTROL

Modern control theory is applied in this section to design an optimal load frequency controller for a two-area system. In accordance with modern control terminology ΔP_{C1} and ΔP_{C2} will be referred to as control inputs u_1 and u_2. In the conventional approach u_1 and u_2 were provided by the integral of ACEs. In

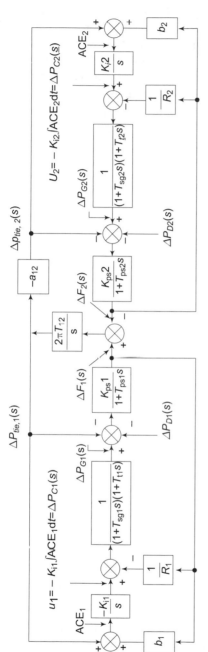

Fig. 8.16 Composite block diagram of two-area load frequency control (feedback loops provided with integral of respective area control errors)

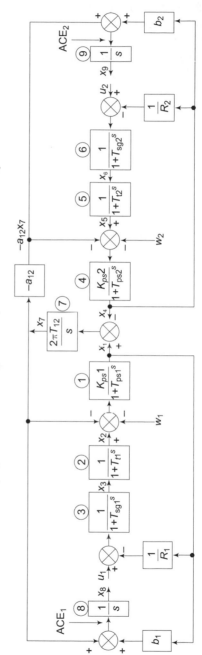

Fig. 8.17 State space model of two-area system

modern control theory approach u_1 and u_2 will be created by a linear combination of all the system states (full state feedback). For formulating the state variable model for this purpose the conventional feedback loops are opened and each time constant is represented by a separate block as shown in Fig. 8.17. State variables are defined as the outputs of all blocks having either an integrator or a time constant. We immediately notice that the system has nine state variables.

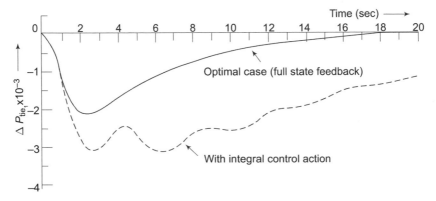

Fig. 8.18   Change in tie line power due to step load (0.01 pu) change in area 1

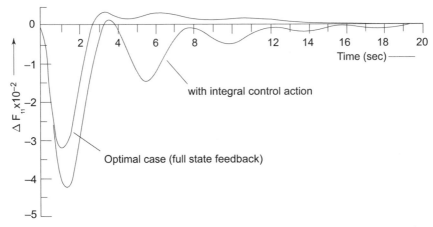

Fig. 8.19   Change in frequency of area 1 due to step load (0.01 pu) change in area 1

Before presenting the optimal design, we must formulate the state model. This is achieved below by writing the differential equations describing each individual block of Fig. 8.17 in terms of state variables (note that differential equations are written by replacing s by $\dfrac{d}{dt}$).

Comparing Figs. 8.16 and 8.17,

$$x_1 = \Delta f_1 \qquad x_4 = \Delta f_2 \qquad x_8 = \int ACE_1 dt$$

$$x_2 = \Delta P_{G1} \qquad x_5 = \Delta P_{G2} \qquad x_9 = \int ACE_2\, dt$$

$$u_1 = \Delta P_{C1} \qquad u_2 = \Delta P_{C2}$$

$$w_1 = \Delta P_{D1} \qquad w_2 = \Delta P_{D2}$$

For block **1**

$$x_1 + T_{ps1}\dot{x}_1 = K_{ps1}(x_2 - x_7 - w_1)$$

or $\qquad \dot{x}_1 = -\dfrac{1}{T_{ps1}}x_1 + \dfrac{K_{ps1}}{T_{ps1}}x_2 - \dfrac{K_{ps1}}{T_{ps1}}x_7 - \dfrac{K_{ps1}}{T_{ps1}}w_1$ (8.43)

For block **2**

$$x_2 + T_{t1}\dot{x}_2 = x_3$$

or $\qquad \dot{x}_2 = -\dfrac{1}{T_{t1}}x_2 + \dfrac{1}{T_{t1}}x_3$ (8.44)

For block **3**

$$x_3 + T_{sg1}\dot{x}_3 = -\dfrac{1}{R_1}x_1 + u_1$$

or $\qquad \dot{x}_3 = -\dfrac{1}{R_1 T_{sg1}}x_1 - \dfrac{1}{T_{sg1}}x_3 + \dfrac{1}{T_{sg1}}u_1$ (8.45)

For block **4**

$$x_4 + T_{ps2}\dot{x}_4 = K_{ps2}(x_5 + a_{12}x_7 - w_2)$$

or $\qquad \dot{x}_4 = -\dfrac{1}{T_{ps2}}x_4 + \dfrac{K_{ps2}}{T_{ps2}}x_5 + \dfrac{a_{12}K_{ps2}}{T_{ps2}}x_7 - \dfrac{K_{ps2}}{T_{ps2}}w_2$ (8.46)

For block **5**

$$x_5 + T_{t2}\dot{x}_5 = x_6$$

or $\qquad \dot{x}_5 = -\dfrac{1}{T_{t2}}x_5 + \dfrac{1}{T_{t2}}x_6$ (8.47)

For block **6**

$$x_6 + T_{sg2}\dot{x}_6 = -\dfrac{1}{R_2}x_4 + u_2$$

or $\qquad \dot{x}_6 = -\dfrac{1}{R_2 T_{sg2}}x_4 - \dfrac{1}{T_{sg2}}x_6 + \dfrac{1}{T_{sg2}}u_2$ (8.48)

For block **7**

$$\dot{x}_7 = 2\pi T_{12}x_1 - 2\pi T_{12}x_4 \qquad (8.49)$$

For block **8**

$$\dot{x}_8 = b_1 x_1 + x_7 \qquad (8.50)$$

For block **9**

$$\dot{x}_9 = b_2 x_4 - a_{12}x_7 \qquad (8.51)$$

The nine equations (8.43) to (8.51) can be organized in the following vector matrix form

$$\dot{x} = Ax + Bu + Fw \qquad (8.52)$$

where

$$x = [x_1 \ x_2 \ ... \ x_9]^T = \text{state vector}$$

$$u = [u_1 \ u_2]^T = \text{control vector}$$

$$w = [w_1 \ w_2]^T = \text{disturbance vector}$$

while the matrices **A**, **B** and **F** are defined below:

$$
A = \begin{array}{c} \\ 1 \\ 2 \\ 3 \\ 4 \\ 5 \\ 6 \\ 7 \\ 8 \\ 9 \end{array}
\begin{bmatrix}
-\dfrac{1}{T_{ps1}} & \dfrac{K_{ps1}}{T_{ps1}} & 0 & 0 & 0 & 0 & -\dfrac{K_{ps1}}{T_{ps1}} & 0 & 0 \\[2ex]
0 & -\dfrac{1}{T_{t1}} & \dfrac{1}{T_{t1}} & 0 & 0 & 0 & 0 & 0 & 0 \\[2ex]
-\dfrac{1}{R_1 T_{sg1}} & 0 & -\dfrac{1}{T_{sg1}} & 0 & 0 & 0 & 0 & 0 & 0 \\[2ex]
0 & 0 & 0 & -\dfrac{1}{T_{ps2}} & \dfrac{K_{ps2}}{T_{ps2}} & 0 & \dfrac{a_{12}K_{ps2}}{T_{ps2}} & 0 & 0 \\[2ex]
0 & 0 & 0 & 0 & -\dfrac{1}{T_{t2}} & \dfrac{1}{T_{t2}} & 0 & 0 & 0 \\[2ex]
0 & 0 & 0 & -\dfrac{1}{R_2 T_{sg2}} & 0 & -\dfrac{1}{T_{sg2}} & 0 & 0 & 0 \\[2ex]
2\pi T_{12} & 0 & 0 & -2\pi T_{12} & 0 & 0 & 0 & 0 & 0 \\[1ex]
b_1 & 0 & 0 & 0 & 0 & 0 & 1 & 0 & 0 \\[1ex]
0 & 0 & 0 & b_2 & 0 & 0 & -a_{12} & 0 & 0
\end{bmatrix}
$$

$$
B^T = \begin{bmatrix}
0 & 0 & \dfrac{1}{T_{sg1}} & 0 & 0 & 0 & 0 & 0 & 0 \\[2ex]
0 & 0 & 0 & 0 & 0 & \dfrac{1}{T_{sg2}} & 0 & 0 & 0
\end{bmatrix}
$$

$$
F^T = \begin{bmatrix}
-\dfrac{K_{ps1}}{T_{ps1}} & 0 & 0 & 0 & 0 & 0 & 0 & 0 & 0 \\[2ex]
0 & 0 & 0 & -\dfrac{K_{ps2}}{T_{ps2}} & 0 & 0 & 0 & 0 & 0
\end{bmatrix}
$$

In the conventional control scheme of Fig. 8.16, the control inputs u_1 and u_2 are constructed as under from the state variables x_8 and x_9 only.

$$u_1 = -K_{i1}x_8 = -K_{i1} \int ACE_1 dt$$

$$u_2 = -K_{i2}x_9 = -K_{i2} \int ACE_2 dt$$

In the optimal control scheme the control inputs u_1 and u_2 are generated by means of feedbacks from all the nine states with feedback constants to be determined in accordance with an optimality criterion.

Examination of Eq. (8.52) reverals that our model is not in the standard form employed in optimal control theory. The standard form is

$$\dot{x} = Ax + Bu$$

which does not contain the disturbance term Fw present in Eq. (8.52). Furthermore, a constant disturbance vector w would drive some of the system states and the control vector u to constant steady values; while the cost function employed in optimal control requires that the system state and control vectors have zero steady state values for the cost function to have a minimum.

For a constant disturbance vector w, the steady state is reached when

$$\dot{x} = 0$$

in Eq. (8.52); which then gives

$$0 = Ax_{ss} + Bu_{ss} + Fw \tag{8.53}$$

Defining x and u as the sum of transient and steady state terms, we can write

$$x = x' + x_{ss} \tag{8.54}$$

$$u = u' + u_{ss} \tag{8.55}$$

Substituting x and u from Eqs. (8.54) and (8.55) in Eq. (8.52), we have

$$\dot{x}' = A(x' + x_{ss}) + B(u' + u_{ss}) + Fw$$

By virtue of relationship (8.53), we get

$$\dot{x}' = Ax' + Bu' \tag{8.56}$$

This represents system model in terms of excursion of state and control vectors from their respective steady state values.

For full state feedback, the control vector u is constructed by a linear combination of all states, i.e.

$$u = -Kx \tag{8.57a}$$

where K is the feedback matrix.
Now

$$u' + u_{ss} = -K(x' + x_{ss})$$

For a stable system both x' and u' go to zero, therefore

$$u_{ss} = -Kx_{ss}$$

Hence

$$u' = -Kx' \tag{8.57b}$$

Examination of Fig. 8.17 easily reveals the steady state values of state and control variables for constant values of disturbance inputs w_1 and w_2. These are

$$x_{1ss} = x_{4ss} = x_{7ss} = 0$$

$$x_{2ss} = x_{3ss} = w_1$$

$$u_{1ss} = w_1$$

$$x_{5ss} = x_{6ss} = w_2 \qquad\qquad (8.58)$$

$$u_{2ss} = w_2$$

$$x_{8ss} = \text{constant}$$

$$x_{9ss} = \text{constant}$$

The values of x_{8ss} and x_{9ss} depend upon the feedback constants and can be determined from the following steady state equations:

$$u_{1ss} = k_{11}x_{1ss} + \ldots + k_{18}x_{8ss} + k_{19}x_{9ss} = w_1$$

$$u_2 ss = k_{21}x_{1ss} + \ldots + k_{28}x_{8ss} + k_{29}x_{9ss} = w_2 \qquad\qquad (8.59)$$

The feedback matrix K in Eq. (8.57b) is to be determined so that a certain performance index (PI) is minimized in transferring the system from an arbitrary initial state $x'(0)$ to origin in infinitie time (i.e. $x'(\infty) = 0$). A convenient PI has the quadratic form

$$\mathbf{PI} = \frac{1}{2}\int_0^\infty (x'^T Q x' + u'^T R u')\, dt \qquad\qquad (8.60)$$

The matrices Q and R are defined for the problem in hand through the following design considerations:

(i) Excursions of ACEs about the steady values $(x'_7 + b_1 x'_1; -a_{12}x'_7 + b_2 x'_4)$ are minimized. The steady values of ACEs are of course zero.

(ii) Excursions of $\int$ACE dt about the steady values (x'_8, x'_9) are minimized. The steady values of $\int$ACE dt are, of course, constants.

(iii) Excursions of the control vector (u'_1, u'_2) about the steady value are minimized. The steady value of the control vector is, of course, a constant. This minimization is intended to indirectly limit the control effort within the physical capability of components. For example, the steam valve cannot be opened more than a certain value without causing the boiler pressure to drop severely.

With the above reasoning, we can write the PI as

$$\text{PI} = \frac{1}{2}\int_0^\infty [(x'_7 + b_1 x'_1)^2 + (-a_{12}x'_7 + b_2 x'_4)^2 + (x'^2_8 + x'^2_9)$$

$$+ k(u'^2_1 + u'^2_2)]\, dt \qquad\qquad (8.61)$$

From the PI of Eq. (8.61), Q and R can be recognized as

$$Q = \begin{bmatrix} b_1^2 & 0 & 0 & 0 & 0 & 0 & b_1 & 0 & 0 \\ 0 & 0 & 0 & 0 & 0 & 0 & 0 & 0 & 0 \\ 0 & 0 & 0 & 0 & 0 & 0 & 0 & 0 & 0 \\ 0 & 0 & 0 & b_2^2 & 0 & 0 & -a_{12}b_2 & 0 & 0 \\ 0 & 0 & 0 & 0 & 0 & 0 & 0 & 0 & 0 \\ 0 & 0 & 0 & 0 & 0 & 0 & 0 & 0 & 0 \\ b_1 & 0 & 0 & -a_{12}b_2 & 0 & 0 & (1+a_{12}^2) & 0 & 0 \\ 0 & 0 & 0 & 0 & 0 & 0 & 0 & 1 & 0 \\ 0 & 0 & 0 & 0 & 0 & 0 & 0 & 0 & 1 \end{bmatrix}$$

= symmetric matrix

$R = kI$ = symmetric matrix

Determination of the feedback matrix K which minimizes the above PI is the standard *optimal regulator problem*. K is obtained from solution of the *reduced matrix Riccati equation** (a set of linear algebraic equations) given below.

$$A^T S + SA - SBR^{-1}B^T S + Q = 0 \tag{8.62}$$

$$K = R^{-1}B^T S \tag{8.63}$$

The acceptable solution of K is that for which the system remains stable. Substituting Eq. (8.57b) in Eq. (8.56), the system dynamics with feedback is defined by

$$\dot{x}' = (A - BK)x' \tag{8.64}$$

For stability all the eigenvalues of the matrix $(A - BK)$ should have negative real parts.

For illustration we consider two identical control areas with the following system parameters:

$T_{sg} = 0.4$ sec; $T_t = 0.5$ sec; $T_{ps} = 20$ sec

$R = 3$; $K_{ps} = 1/B = 100$

$h = 0.425$; $K_i - 0.09$; $a_{12} - 1$; $2\pi T_{12} = 0.05$

Computer solution for the feedback matrix K is presented below, while the system dynamic response is plotted in Figs. 8.18 and 8.19. These figures also give for comparison the dynamic response for the case of integral control action only. The improvement in performance achieved by the optimal controller is evident from these figures.

$$K = \begin{bmatrix} 0.5286 & 1.1419 & 0.6813 & -0.0046 & -0.0211 & -0.0100 & -0.7437 & 0.9999 & 0.0000 \\ -0.0046 & -0.0211 & -0.0100 & 0.5286 & 1.1419 & 0.6813 & 0.7437 & 0.0000 & 0.9999 \end{bmatrix}$$

*Refer Nagrath and Gopal [5].

As the control areas extend over vast geographical regions, there are two ways of obtaining full state information in each area for control purposes.

(i) Transport the state information of the distant area over communication channels. This is, of course, expensive.

(ii) Generate all the states locally (including the distant states) by means of 'observer' or 'Kalman filter' by processing the local output signal. An 'observer' being itself a high order system, renders the overall system highly complex and may, in fact, result in impairing system stability and dynamic response which is meant to be improved through optimal control.

Because of the above mentioned difficulties encountered in implementation of an optimal control load frequency scheme, it is preferably to use a sub-optimal scheme employing only local states of the area. The design of such structurally constrained optimal control schemes is beyond the scope of this book. The conventional control scheme of Fig. 8.16 is, in fact, the local output feedback scheme in which feedback signal derived from changes in local frequency and tie line power is employed in each area.

8.6 AUTOMATIC VOLTAGE CONTROL

Figure 8.20 gives the schematic diagram of an automatic voltage regulator of a generator. It basically consists of a main exciter which excites the alternator field to control the output voltage. The exciter field is automatically controlled through error $e = V_{ref} - V_T$, suitably amplified through voltage and power amplifiers. It is a type-0 system which requires a constant error e for a specified voltage at generator terminals. The block diagram of the system is given in

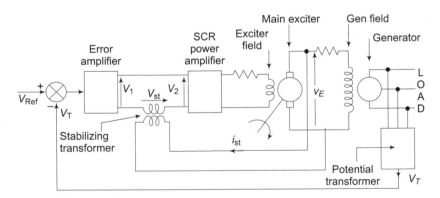

Fig. 8.20 Schematic diagram of alternator voltage regulator scheme

Fig. 8.21. The function of important components and their transfer functions is given below:

Potential transformer: It gives a sample of terminal voltage V_T.

Differencing device: It gives the actuating error

$$e = V_{\text{Ref}} - V_T$$

The error initiates the corrective action of adjusting the alternator excitation. Error wave form is suppressed carrier modulated, the carrier frequency being the system frequency of 50 Hz.

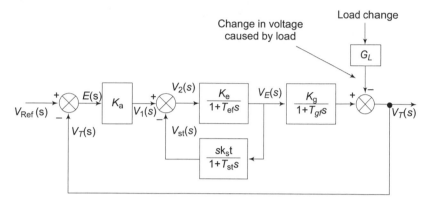

Fig. 8.21 Block diagram of alternator voltage regulator scheme

Error amplifier: It demodulates and amplifies the error signal. Its gain is K_a. *SCR power amplifier and exciter field:* It provides the necessary power amplification to the signal for controlling the exciter field. Assuming the amplifier time constant to be small enough to be neglected, the overall transfer function of these two is

$$\frac{K_e}{1 + T_{ef}s}$$

where T_{ef} is the exciter field time constant.

Alternator: Its field is excited by the main exciter voltage v_E. Under no load it produces a voltage proportional to field current. The no load transfer function is

$$\frac{K_g}{1 + T_{gf}s}$$

where

$$T_{gf} = \text{generator field time constant.}$$

The load causes a voltage drop which is a complex function of direct and quadrature axis currents. The effect is only schematically represented by block G_L. The exact load model of the alternator is beyond the scope of this book.

Stabilizing transformer: T_{ef} and T_{gf} are large enough time constants to impair the system's dynamic response. It is well known that the dynamic response of a control system can be improved by the internal derivative feedback loop. The derivative feedback in this system is provided by means of a stabilizing transformer excited by the exciter output voltage v_E. The output of the stabilizing transformer is fed negatively at the input terminals of the SCR power amplifier. The transfer function of the stabilizing transformer is derived below. Since the secondary is connected at the input terminals of an amplifier, it can be assumed to draw zero current. Now

$$v_E = R_1 i_{st} + L_1 \frac{di_{st}}{dt}$$

$$v_{st} = M \frac{di_{st}}{dt}$$

Taking the Laplace transform, we get

$$\frac{V_{st}(s)}{V_E(s)} = \frac{sM}{R_1 + sL_1} = \frac{sM/R_1}{1 + T_{st}s}$$

$$= \frac{sK_{st}}{1 + T_{st}s}$$

Accurate state variable models of loaded alternator around an operating point are available in literature using which optimal voltage regulation schemes can be devised. This is, of course, beyond the scope of this book.

8.7 LOAD FREQUENCY CONTROL WITH GENERATION RATE CONSTRAINTS (GRCs)

The load frequency control problem discussed so far does not consider the effect of the restrictions on the rate of change of power generation. In power systems having steam plants, power generation can change only at a specified maximum rate. The generation rate (from safety considerations of the equipment) for reheat units is quit low. Most of the reheat units have a generation rate around 3%/min. Some have a generation rate between 5 to 10%/min. If these constraints are not considered, system is likely to chase large momentary disturbances. This results in undue wear and tear of the controller. Several methods have been proposed to consider the effect of GRCs for the design of automatic generation controllers. When GRC is considered, the system dynamic model becomes non-linear and linear control techniques cannot be applied for the optimization of the controller setting.

If the generation rates denoted by P_{Gi} are included in the state vector, the system order will be altered. Instead of augmenting them, while solving the state equations, it may be verified at each step if the GRCs are violated. Another way of considering GRCs for both areas is to add limiters to the governors [15, 17] as shown in Fig. 8.22, i.e., the maximum rate of valve opening or closing speed is restricted by the limiters. Here $T_{sg} g_{max}$ is the power rate limit imposed by valve or gate control. In this model

$$|\Delta \dot{Y}_E| < g_{max} \tag{8.65}$$

The banded values imposed by the limiters are selected to restrict the generation rate by 10% per minute.

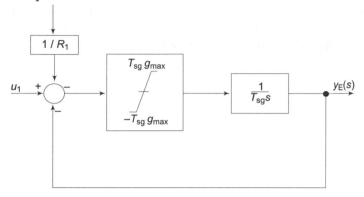

Fig. 8.22 Governor model with GRC

The GRCs result in larger deviations in ACEs as the rate at which generation can change in the area is constrained by the limits imposed. Therefore, the duration for which the power needs to be imported increases considerably as compared to the case where generation rate is not constrained. With GRCs, R should be selected with care so as to give the best dynamic response. In hydro-thermal system, the generation rate in the hydro area normally remains below the safe limit and therefore GRCs for all the hydro plants can be ignored.

8.8 SPEED GOVERNOR DEAD-BAND AND ITS EFFECT ON AGC

The effect of the speed governor dead-band is that for a given position of the governor control valves, an increase/decrease in speed can occur before the position of the valve changes. The governor dead-band can materially affect the system response. In AGC studies, the dead-band effect indeed can be significant, since relatively small signals are under considerations.

The speed governor characteristic, though non-linear, has been approximated by linear characteristics in earlier analysis. Further, there is another non-linearity introduced by the dead-band in the governor operation. Mechanical friction and backlash and also valve overlaps in hydraulic relays cause the governor dead-band. Dur to this, though the input signal increases, the speed governor may not immediately react until the input reaches a particular value. Similar action takes place when the input signal decreases. Thus the governor dead-band is defined as the total magnitude of sustained speed change within which there is no change in valve position. The limiting value of dead-band is specified as 0.06%. It was shown by Concordia *et. al* [18] that one of the effects of governor dead-band is to increase the apparent steady-state speed regulation R.

The effect of the dead-band may be included in the speed governor control loop block diagram as shown in Fig. 8.23. Considering the worst case for the dead-band, (i.e., the system starts responding after the whole dead-band is traversed) and examining the dead-band block in Fig. 8.23, the following set of equations completely define the behaviour of the dead-band [19].

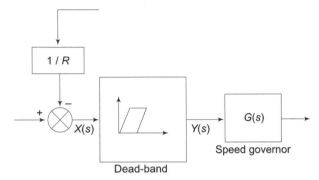

Fig. 8.23 Dead-band in speed-governor control loop

$$y^{(r+1)} = x^{(r)} \text{ if } x^{(r+1)} - x^r \leq \text{dead-band}$$
$$= x^{(r+1)} - \text{dead-band; if } x^{(r+1)} - x^{(r)} > 0 \qquad (8.66)$$
$$= x^{(r+1)}; \text{ if } x^{r+1} - x^r < 0$$

(*r* is the step in the computation)

Reference [20] considers the effect of governor dead-band nonlinearity by using the describing function approach [11] and including the linearised equations in the state space model.

The presence of governor dead-band makes the dynamic response oscillatory. It has been seen [19] that the governor dead-band does not influence the selection of integral controller gain settings in the presence of GRCs. In the presence of GRC and dead band even for small load perturbation, the system becomes highly non-linear and hence the optimization problem becomes rather complex.

8.9 DIGITAL LF CONTROLLERS

In recent years, increasingly more attention is being paid to the question of digital implementation of the automatic generation control algorithms. This is mainly due to the facts that digital control turns out to be more accurate and reliable, compact in size, less sensitive to noise and drift and more flexible. It may also be implemented in a time shared fashion by using the computer systems in load despatch centre, if so desired. The ACE, a signal which is used for AGC is available in the discrete form, i.e., there occurs sampling operation between the system and the controller. Unlike the continuous-time system, the control vector in the discrete mode is constrained to remain constant between the sampling instants. The digital control process is inherently a discontinuous process and the designer has thus to resort to the discrete-time analysis for optimization of the AGC strategies.

Discrete-Time Control Model

The continuous-time dynamic system is described by a set of linear differential equations

$$x = Ax + Bu + \Gamma p \tag{8.67}$$

where x, u, p are state, control and disturbance vectors respectively and A, B and Γ are constant matrices associated with the above vectors.

The discrete-time behaviour of the continuous-time system is modelled by the system of first order linear difference equations:

$$x(k+1) = \phi\, x(k) + \Psi\, u(k) + \gamma\, p(k) \tag{8.68}$$

where $x(k)$, $u(k)$ and $p(k)$ are the state, control and disturbance vectors and are specified at $t = kT$, $k = 0, 1, 2, \dots$ etc. and T is the sampling period. ϕ, ψ and γ are the state, control and disturbance transition matrices and they are evaluated using the following relations.

$$\phi = e^{AT}$$

$$\psi = (e^{AT} - I)\, A^{-1} B$$

$$\gamma = (e^{AT} - I)\, A^{-1} \Gamma$$

where A, B and Γ are the constant matrices associated with x, u, and p vectors in the corresponding continuous-time dynamic system. The matrix e^{AT} can be evaluated using various well-documented approaches like Sylvestor's expansion theorem, series expansion technique etc. The optimal digital load frequency controller design problem is discussed in detail in Ref [7].

8.10 DECENTRALIZED CONTROL

In view of the large size of a modern power system, it is virtually impossible to implement either the classical or the modern LFC algorithm in a centralized manner. In Fig. 8.24, a decentralized control scheme is shown. x_1 is used to find out the vector u_1 while x_2 alone is employed to find out u_2. Thus.

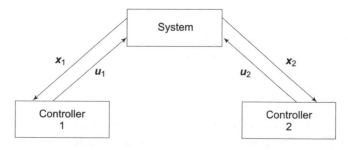

Fig. 8.24 Decentralized control

$$\mathbf{x} = (\mathbf{x}_1 \ \mathbf{x}_2)^T$$

$$\mathbf{u}_1 = -\ \mathbf{k}_1 \ \mathbf{x}_1$$

$$\mathbf{u}_2 = -\ \mathbf{k}_2 \ \mathbf{x}_2$$

A systematic design of the decentralized tie-line bias control solution has been shown possible using the modal control principle. Decentralized or hierarchical implementation of the optimal LFC algorithms seems to have been studied more widely for the stochastic case since the real load disturbances are truely stochastic. A simple approach is discussed in Ref. [7].

It may by noted that other techniques of model simplification are available in the literature on alternative tools to decentralized control. These include the method of "aggregation", "singular perturbation", "moment matching" and other techniques [9] for finding lower order models of a given large scale system.

PROBLEMS

8.1 Two generators rated 200 MW and 400 MW are operating in parallel. The droop characteristics of their governors are 4% and 5% respectively from no load to full load. The speed changers are so set that the generators operate at 50 Hz sharing the full load of 600 MW in the ratio of their ratings. If the load reduces to 400 MW, how will it be shared among the generators and what will the system frequency be? Assume free governor operation.

The speed changers of the governors are reset so that the load of 400 MW is shared among the generators at 50 Hz in the ratio of their ratings. What are the no load frequencies of the generators?

8.2 Consider the block diagram model of load frequency control given in Fig. 8.6. Make the following approximation.

$(1 + T_{sg}s) (1 + T_t s) \simeq 1 + (T_{sg} + T_t)s = 1 + T_{eq}s$

Solve for Δf (t) with parameters given below. Given $\Delta P_D = 0.01$ pu

$$T_{eq} = 0.4 + 0.5 = 0.9 \text{ sec}; \ T_{ps} = 20 \text{ sec}$$

$$K_{sg}K_t = 1; \ K_{ps} = 100; \ R = 3$$

Compare with the exact response given in Fig. 8.9.

8.3 For the load frequency control with proportional plus integral controller as shown in Fig. 8.10, obtain an expression for the steady state error in

cycles, i.e. $\int_0^t \Delta f(t)\,dt$; for a unit step ΔP_D. What is the corresponding time
$\lim t \to \infty$

error in seconds (with respect to 50 Hz)? Comment on the dependence of error in cycles upon the integral controller gain K_i.

$$\left[Hint:\ \text{L} \int_0^t \Delta f(t)\mathrm{d}t = \frac{\Delta F(s)}{s};\quad \int_0^t \underset{\lim t \to \infty}{\Delta f(t)\ \mathrm{d}t} = \lim_{s\to 0} s \times \frac{\Delta F(s)}{s} = \lim_{s\to 0} \Delta F(s) \right]$$

8.4 For the two area load frequency control of Fig. 8.16 assume that integral controller blocks are replaced by gain blocks, i.e. ACE_1 and ACE_2 are fed to the respective speed changers through gains $-K_{i1}$ and $-K_{i2}$. Derive an expression for the steady values of change in frequency and tie line power for simultaneously applied unit step load disturbance inputs in the two areas.

8.5 For the two area load frequency control employing integral of area control error in each area (Fig. 8.16), obtain an expression for $\Delta P_{\text{tie}}(s)$ for unit step disturbance in one of the areas. Assume both areas to be identical. Comment upon the stability of the system for parameter values given below:

$$T_{\text{sg}} = 0.4 \text{ sec};\ T_t = 0.5 \text{ sec};\ T_{\text{ps}} = 20 \text{ sec}$$

$$K_{\text{ps}} = 100;\ R = 3;\ K_i = 1;\ b = 0.425$$

$$a_{12} = 1;\ 2\pi T_{12} = 0.05$$

[*Hint*: Apply Routh's stability criterion to the characteristic equation of the system.]

REFERENCES

Books

1. Elgerd, O.I., *Electric Energy System Theory: An Introduction*, 2nd edn, McGraw-Hill, New York, 1982.
2. Weedy, B.M. and B.J. Cory *Electric Power Systems*, 4th edn, Wiley, New York, 1998.
3. Cohn, N., *Control of Generation and Power Flow on Interconnected Systems*, Wiley, New York, 1971.
4. Wood, A.J., and B.F. Woolenberg, *Power Generation, Operation and Control*, 2nd edn Wiley, New York, 1996.
5. Nagarth, I.J. and M. Gopal, *Control Systems Engineering*, 3rd edn. New Delhi, 2001.
6. Handschin, E. (Ed.), *Real Time Control of Electric Power Systems*, Elsevier, New York 1972.
7. Mahalanabis, A.K., D.P. Kothari and S.I. Ahson, *Computer Aided Power System Analysis and Control*, Tata McGraw-Hill, New Delhi, 1988.
8. Kirchmayer, L.K., *Economic Control of Interconnected Systems*, Wiley, New York, 1959.
9. Jamshidi, M., *Large Scale Systems: Modelling and Control*, North Holland, N.Y., 1983.

10. Singh, M.G. and A. Titli, *Systems Decomposition, Optimization and Control* Pergamon Press, Oxford, 1978.

11. Siljak, D.D., *Non-Linear Systems: The Parameter Analysis and Design*, Wiley, N.Y. 1969.

Papers

12. Elgerd, O.I. and C.E. Fosha, "The Megawatt Frequency Control Problem: A New Approach Via Optimal Control Theory", *IEEE Trans.*, April 1970, No. 4, PAS 89; 556.

13. Bhatti, T.S., C.S Indulkar and D.P. Kothari, "Parameter Optimization of Power Systems for Stochastic Load Demands" *Proc. IFAC*, Bangalore, December 1986.

14. Kothari, M.L., P.S. Satsangi and J. Nanda, "Sampled-Data Automatic Generation Control of Interconnected Reheat Thermal Systems Considering Generation Rate Constraints", *IEEE Trans.*, May 1981, PAS-100; 2334.

15. Nanda, J., M.L. Kothari and P.S. Satsangi, "Automatic Generation Control of an Interconnected Hydro-thermal System in Continuous and Discrete Modes Considering Generation Rate Constraints' *IEE Proc., Pt D*, No. 1, January 1983, 130 : 17.

16. IEEE Committee Report, 'Dynamic Models for Steam and Hydro-turbines in Power System Studies" *IEEE Trans.*, Nov/Dec. 1973, PAS-92, 1904.

17. Hiyama, T., "Optimization of Discrete-type Load Frequency Regulators Considering Generation-Rate Constraints" *Proc. IEE*, Nov. 82, 129, pt C, 285.

18. Concordia, C., L.K. Kirchmayer and E.A. Szyonanski, "Effect of Speed Governor Dead-band on Tie Line Power and Frequency Control Performance" *AIEE Trans.* Aug. 1957, 76, 429.

19. Nanda, J., M.L. Kothari and P.S. Satsangi, "Automatic Control of Reheat Thermal System Considering Generation Rate Constraint and Governor Dead-band", *J.I.E.* (India), June 1983, 63, 245.

20. Tripathy, S.C., G.S. Hope and O.P. Malik, "Optimisation of Load-frequency Control Parameters for Power Systems with Reheat Steam Turbines and Governor Dead-band Nonlinearity", *Proc. IEE*, January 1982, 129, Pt C, No. 1, 10.

21. Kothari, M.L., J. Nanda, D.P. Kothari and D. Das, "Discrete-mode AGC of a `two-area Reheat Thermal System with New Area Control Error", *IEEE Trans. on Power System*, Vol. 4, May 1989, 730

22. Das, D. J. Nanda, M.L. Kothari and D.P. Kothari, "AGC of a Hydro-Thermal System with New ACE considering GRC", *Int. J. EMPS*, 18, No. 5, 1990, 461.

23. Das, D., M.L. Kothari, D.P. Kothari and J. Nanda, "Variable Structure Control Strategy to AGC of an Interconnected Reheat Thermal System", *Proc. IEE*, 138, pt D, 1991, 579.

24. Jalleli, Van Slycik et. al., "Understanding Automatic Generation Control", IEEE Trans. on P.S., Vol 07, 3 Aug. 92, 1106–1122.

25. Kothari, M.L., J. Nanda, D.P. Kothari and D. Das, "Discrete Mode AGC of a two Area Reheat Thermal System with a NACE considering GRC", *J.I.E.* (India), Vol. 72, Feb. 1992, pp 297–303.

26. Bakken, B.H. and Q.S. Grande,"AGC in a Deregulated Power System," IEEE Trans. on Power Systems, 13, 4, Nov. 1998, pp. 1401–1406.

9

Symmetrical Fault Analysis

9.1 INTRODUCTION

So far we have dealt with the steady state behaviour of power system under normal operating conditions and its dynamic behaviour under small scale perturbations. This chapter is devoted to abnormal system behaviour under conditions of symmetrical short circuit (symmetrical three-phase fault*). Such conditions are caused in the system accidentally through insulation failure of equipment or flashover of lines initiated by a lightning stroke or through accidental faulty operation. The system must be protected against flow of heavy short circuit currents (which can cause permanent damage to major equipment) by disconnecting the faulty part of the system by means of circuit breakers operated by protective relaying. For proper choice of circuit breakers and protective relaying, we must estimate the magnitude of currents that would flow under short circuit conditions—this is the scope of fault analysis (study).

The majority of system faults are not three-phase faults but faults involving one line to ground or occasionally two lines to ground. These are unsymmetrical faults requiring special tools like symmetrical components and form the subject of study of the next two chapters. Though the symmetrical faults are rare, the symmetrical fault analysis must be carried out, as this type of fault generally leads to most severe fault current flow against which the system must be protected. Symmetrical fault analysis is, of course, simpler to carry out.

A power network comprises synchronous generators, transformers, lines and loads. Though the operating conditions at the time of fault are important, the loads can be neglected during fault, as voltages dip very low so that currents drawn by loads can be neglected in comparison to fault currents.

---

*Symmetrical fault may be a solid three-phase short circuit or may involve arc impedance.

The synchronous generator during short circuit has a characteristic time-varying behaviour. In the event of a short circuit, the flux per pole undergoes dynamic change with associated transients in damper and field windings. The reactance of the circuit model of the machine changes in the first few cycles from a low subtransient reactance to a higher transient value, finally settling at a still higher synchronous (steady state) value. Depending upon the arc interruption time of circuit breakers, a suitable reactance value is used for the circuit model of synchronous generators for short circuit analysis.

In a modern large interconnected power system, heavy currents flowing during a fault must be interrupted much before the steady state conditions are established. Furthermore, from the considerations of mechanical forces that act on circuit breaker components, the maximum current that a breaker has to carry momentarily must also be determined. For selecting a circuit breaker we must, therefore, determine the initial current that flows on occurrence of a short circuit and also the current in the transient that flows at the time of circuit interruption.

9.2 TRANSIENT ON A TRANSMISSION LINE

Let us consider the short circuit transient on a transmission line. Certain simplifying assumptions are made at this stage.

(i) The line is fed from a constant voltage source (the case when the line is fed from a realistic synchronous machine will be treated in Sec. 9.3).

(ii) Short circuit takes place when the line is unloaded (the case of short circuit on a loaded line will be treated later in this chapter).

(iii) Line capacitance is negligible and the line can be represented by a lumped *RL* series circuit.

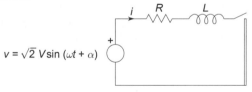

Fig. 9.1

With the above assumptions the line can be represented by the circuit model of Fig. 9.1. The short circuit is assumed to take place at $t = 0$. The parameter α controls the instant on the voltage wave when short circuit occurs. It is known from circuit theory that the current after short circuit is composed of two parts, i.e.

$$i = i_s + i_t$$

where

$$i_s = \text{steady state current}$$

$$= \frac{\sqrt{2}V}{|Z|} \sin (\omega t + \alpha - \theta)$$

$$Z = (R_2 + \omega^2 L^2)^{1/2} \angle \left(\theta = \tan^{-1} \frac{\omega L}{R} \right)$$

i_t = transient current [it is such that $i(0) = i_s(0) + i_t(0) = 0$ being an inductive circuit; it decays corresponding to the time constant L/R].

$$= - i_s(0)e^{-(R/L)t}$$

$$= \frac{\sqrt{2}V}{|Z|} \sin (\theta - \alpha)e^{-(R/L)t}$$

Thus short circuit current is given by

$$i = \underbrace{\frac{\sqrt{2}V}{|Z|} \sin (\omega t + \alpha - \theta)}_{\substack{\text{Symmetrical short} \\ \text{circuit current}}} + \underbrace{\frac{\sqrt{2}V}{|Z|} \sin (\theta - \alpha)e^{-(R/L)t}}_{\text{DC off}-\text{set current}} \qquad (9.1)$$

A plot of i_s, i_t and $i = i_s + i_t$ is shown in Fig. 9.2. In power system terminology, the sinusoidal steady state current is called the *symmetrical short circuit current* and the unidirectional transient component is called the *DC off-set current*, which causes the total short circuit current to be unsymmetrical till the transient decays.

It easily follows from Fig. 9.2 that the *maximum momentary short circuit current* i_{mm} corresponds to the first peak. If the decay of transient current in this short time is neglected,

$$i_{mm} = \frac{\sqrt{2}V}{|Z|} \sin (\theta - \alpha) + \frac{\sqrt{2}V}{|Z|} \qquad (9.2)$$

Since transmission line resistance is small, $\theta \simeq 90°$.

$$\therefore \qquad i_{mm} = \frac{\sqrt{2}V}{|Z|} \cos \alpha + \frac{\sqrt{2}V}{|Z|} \qquad (9.3)$$

This has the maximum possible value for $\alpha = 0$, i.e. short circuit occurring when the voltage wave is going through zero. Thus

$$i_{mm \text{ (mass possible)}} = 2 \frac{\sqrt{2}V}{|Z|} \qquad (9.4)$$

$$= \text{twice the maximum of symmetrical short circuit current}$$

$$(\textit{doubling effect})$$

For the selection of circuit breakers, momentary short circuit current is taken corresponding to its maximum possible value (a safe choice).

The next question is 'what is the current to be interrupted?' As has been pointed out earlier, modern day circuit breakers are designed to interrupt the current in the first few cycles (five cycles or less). With reference to Fig. 9.2 it means that when the current is interrupted, the DC off-set (i_t) has not yet died out and so contributes to the current to be interrupted. Rather than computing the value of the DC off-set at the time of interruption (this would be highly complex in a network of even moderately large size), the symmetrical short circuit current alone is calculated. This figure is then increased by an empirical multiplying factor to account for the DC off-set current. Details are given in Sec. 9.5.

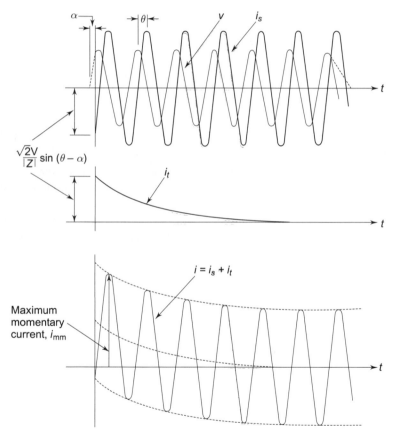

Fig. 9.2 Waveform of a short circuit current on a transmission line

9.3 SHORT CIRCUIT OF A SYNCHRONOUS MACHINE (ON NO LOAD)

Under steady state short circuit conditions, the armature reaction of a synchronous generator produces a demagnetizing flux. In terms of a circuit this effect is modelled as a reactance X_a in series with the induced emf. This reactance when combined with the leakage reactance X_l of the machine is called *synchronous reactance X_d* (direct axis synchronous reactance in the case of salient pole machines). Armature resistance being small can be neglected. The steady state short circuit model of a synchronous machine is shown in Fig. 9.3a on per phase basis.

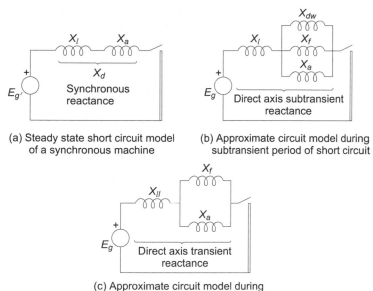

(a) Steady state short circuit model of a synchronous machine

(b) Approximate circuit model during subtransient period of short circuit

(c) Approximate circuit model during transient period of short circuit

Fig. 9.3

Consider now the sudden short circuit (three-phase) of a synchronous generator initially operating under open circuit conditions. The machine undergoes a transient in all the three phase finally ending up in steady state conditions described above. The circuit breaker must, of course, interrupt the current much before steady conditions are reached. Immediately upon short circuit, the DC off-set currents appear in all the three phases, each with a different magnitude since the point on the voltage wave at which short circuit occurs is different for each phase. These DC off-set currents are accounted for separately on an empirical basis and, therefore, for short circuit studies, we need to concentrate our attention on *symmetrical (sinusoidal) short circuit current only*. Immediately in the event of a short circuit, the symmetrical short circuit current is limited only by the leakage reactance of the machine. Since the air gap flux cannot change instantaneously (*theorem of constant flux linkages*), to counter the demagnetization of the armature short circuit current, currents appear in the field winding as well as in the damper winding in a direction to help the main flux. These currents decay in accordance with the winding time constants. The time constant of the damper winding which has low leakage inductance is much less than that of the field winding, which has high leakage inductance. Thus during the initial part of the short circuit, the damper and field windings have transformer currents induced in them so that in the circuit model their reactances—X_f of field winding and X_{dw} of damper winding—appear in parallel* with X_a as shown in Fig. 9.3b. As the damper winding currents are first to die out, X_{dw} effectively becomes open circuited and at a later stage X_f becomes open circuited. The machine reactance thus changes from the parallel combination of X_a, X_f and X_{dw} during the initial period of the short circuit to X_a and X_f in parallel (Fig. 9.3c) in the middle period of the short circuit, and finally to X_a in steady state (Fig. 9.3a). The reactance presented by the machine in the initial period of the short circuit, i.e.

$$X_l + \frac{1}{(1/X_a + 1/X_f + 1/X_{dw})} = X_d'' \qquad (9.5)$$

is called the *subtransient reactance* of the machine. While the reactance effective after the damper winding currents have died out, i.e.

$$X_d' = X_l + (X_a \parallel X_t) \qquad (9.6)$$

is called the transient reactance of the machine. Of course, the reactance under steady conditions is the synchronous reactance of the machine. Obviously $X_d'' < X_d' < X_d$. The machine thus offers a time-varying reactance which changes from X_d'' to X_d' and finally to X_d.

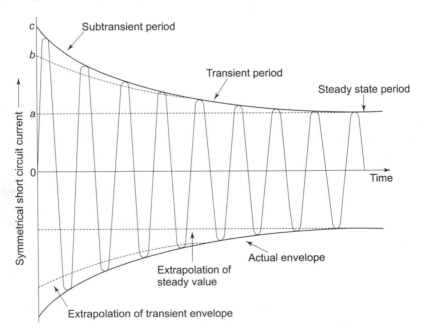

(a) Symmetrical short circuit armature current in synchronous machine

Fig. 9.4 (Contd.)

*Unity turn ratio is assumed here.

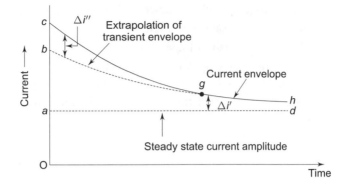

(b) Envelope of synchronous machine symmetrical short circuit current

Fig. 9.4

If we examine the oscillogram of the short circuit current of a synchronous machine after the DC off-set currents have been removed from it, we will find the current wave shape as given in Fig. 9.4a. The envelope of the current wave shape is plotted in Fig. 9.4b. The short circuit current can be divided into three periods—initial subtransient period when the current is large as the machine offers subtransient reactance, the middle transient period where the machine offers transient reactance, and finally the steady state period when the machine offers synchronous reactance.

If the transient envelope is extrapolated backwards in time, the difference between the transient and subtransient envelopes is the current $\Delta i''$ (corresponding to the damper winding current) which decays fast according to the damper winding time constant. Similarly, the difference $\Delta i'$ between the steady state and transient envelopes decays in accordance with the field time constant.

In terms of the oscillogram, the currents and reactances discussed above, we can write

$$|I| = \frac{oa}{\sqrt{2}} = \frac{|E_g|}{X_d} \qquad (9.7a)$$

$$|I'| = \frac{ob}{\sqrt{2}} = \frac{|E_g|}{X_d'} \qquad (9.7b)$$

$$|I''| = \frac{oc}{\sqrt{2}} = \frac{|E_g|}{X_d''} \qquad (9.7c)$$

where

$|I|$ = steady state current (rms)

$|I'|$ = transient current (rms) excluding DC component

$|I''|$ = subtransient current (rms) excluding DC component

X_d = direct axis synchronous reactance

X'_d = direct axis transient reactance

X''_d = direct axis subtransient reactance

$|E_g|$ = per phase no load voltage (rms)

Oa, Ob, Oc = intercepts shown in Figs. 9.4a and b.

The intercept Ob for finding transient reactance can be determined accurately by means of a logarithmic plot. Both $\Delta i''$ and $\Delta i'$ decay exponentially as

$$\Delta i'' = \Delta i''_0 \exp (- t/\tau_{dw})$$

$$\Delta i' = \Delta i'_0 \exp (- t/\tau_f)$$

where τ_{dw} and τ_f are respectively damper, and field winding time constants with $\tau_{dw} \ll \tau_f$. At time $t \gg \tau_{dw}$, $\Delta i''$ practically dies out and we can write

$$\log (\Delta i'' + \Delta t')|_{t \gg \tau_{dw}}, \simeq \log \Delta i' = - \Delta i'_{0\,t}/\tau_f$$

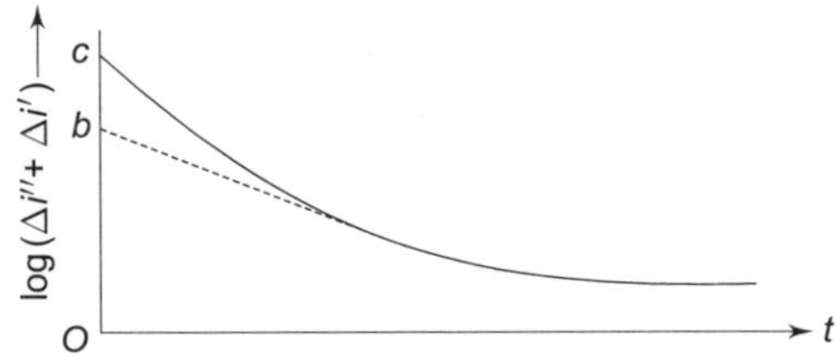

Fig. 9.5

The plot of log $(\Delta i'' + \Delta i')$ versus time for $t \gg \tau_{dw}$ therefore, becomes a straight line with a slope of $(- \Delta i'_0/\tau_f)$ as shown in Fig. 9.5. As the straight line portion of the plot is extrapolated (straight line extrapolation is much more accurate than, the exponential extrapolation of Fig. 9.4), the intercept corresponding to $t = 0$ is

$$\Delta i'|_{t=0} = \Delta i'_0 \exp(-t / \tau_f)|_{t=0} = \Delta i'_0 = ob$$

Table 9.1 Typical values of synchronous machine reactances
(All values expressed in pu of rated MVA)

Type of machine	Turbo-alternator (Turbine generator)	Salient pole (Hydroelectric)	Synchronous compensator (Condenser/ capacitor)	Synchronous motors*
X_s (or X_d)	1.00–2.0	0.6–1.5	1.5–12.5	0.8–1.10
X_q	0.9–1.5	0.4–1.0	0.95–1.5	0.65–0.8
X'_d	0.12–0.35	0.2–0.5	0.3–0.6	0.3–0.35
X''_d	0.1–0.25	0.13–0.35	0.18–0.38	0.18–0.2
X_2	$= X''_d$	$= X''_d$	0.17–0.37	0.19–0.35
X_0	0.04–0.14	0.02–0.2	0.025–0.16	0.05–0.07
r_a	0.003–0.008	0.003–0.015	0.004–0.01	0.003–0.012

r_a = AC resistance of the armature winding per phase.

* High-speed units tend to have low reactance and low speed units high reactance.

Though the machine reactances are dependent upon magnetic saturation (corresponding to excitation), the values of reactances normally lie within certain predictable limits for different types of machines. Table 9.1 gives typical values of machine reactances which can be used in fault calculations and in stability studies.

Normally both generator and motor subtransient reactances are used to determine the momentary current flowing on occurrence of a short circuit. To decide the interrupting capacity of circuit breakers, except those which open instantaneously, subtransient reactance is used for generators and transient reactance for synchronous motors. As we shall see later the transient reactances are used for stability studies.

The machine model to be employed when the short circuit takes place from loaded conditions will be explained in Sec. 9.4.

The method of computing short circuit currents is illustrated through examples given below.

Example 9.1

For the radial network shown in Fig. 9.6, a three-phase fault occurs at F. Determine the fault current and the line voltage at 11 kV bus under fault conditions.

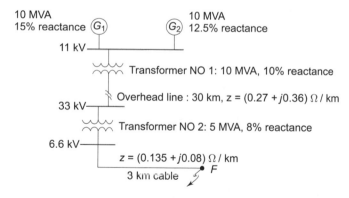

Fig. 9.6   Radial network for Example 9.1

Solution   Select a system base of 100 MVA.
Voltage bases are: 11 kV in generators, 33 kV for overhead line and 6.6 kV for cable.

$$\text{Reactance of } G_1 = j\frac{0.15 \times 100}{10} = j1.5 \text{ pu}$$

$$\text{Reactance of } G_2 = j\frac{0.125 \times 100}{10} = j1.25 \text{ pu}$$

$$\text{Reactance of } T_1 = j\frac{0.1 \times 100}{10} = j1.0 \text{ pu}$$

Reactance of $T_2 = j\dfrac{0.08 \times 100}{5} = j1.6$ pu

Overhead line impedance $= \dfrac{Z \text{ (in ohms)} \times \text{MVA}_{\text{Base}}}{(\text{kV}_{\text{Base}})^2}$

$$= \dfrac{30 \times (0.27 + j0.36) \times 100}{(33)^2}$$

$$= (0.744 + j0.99) \text{ pu}$$

Cable impedance $= \dfrac{3(0.135 + j0.08) \times 100}{(6.6)^2} = (0.93 + j0.55)$ pu

Circuit model of the system for fault calculations is shown in Fig. 9.7. Since the system is on no load prior to occurrence of the fault, the voltages of the two generators are identical (in phase and magnitude) and are equal to 1 pu. The generator circuit can thus be replaced by a single voltage source in series with the parallel combination of generator reactances as shown.

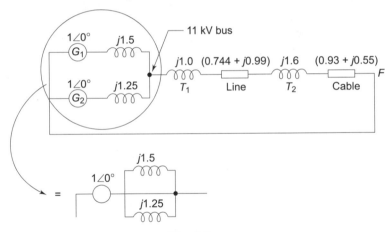

Fig. 9.7

Total impedance $= (j1.5 \parallel j1.25) + (j1.0) + (0.744 + j0.99) + (j1.6) +$
$$(0.93 + j0.55)$$

$$= 1.674 + j4.82 = 5.1 \angle 70.8° \text{ pu}$$

$$I_{SC} = \dfrac{1\angle 0}{5.1\angle 70.8°} = 0.196 \angle -70.8° \text{ pu}$$

$$I_{Base} = \dfrac{100 \times 10^3}{\sqrt{3} \times 6.6} = 8,750 \text{ A}$$

$\therefore$ $\qquad I_{SC} = 0.196 \times 8,750 = 1,715 \text{ A}$

Total impedance between F and 11 kV bus

$$= (0.93 + j055) + (j1.6) + (0.744 + j0.99) + (j1.0)$$

$$= 1.674 + j4.14 = 4.43 \ \angle 76.8° \text{ pu}$$

Voltage at 11 kV bus $= 4.43 \ \angle 67.8° \times 0.196 \ \angle - 70.8°$

$$= 0.88 \ \angle - 3° \text{ pu} = 0.88 \times 11 = 9.68 \text{ kV}$$

Example 9.2

A 25 MVA, 11 kV generator with $X_d'' = 20\%$ is connected through a transformer, line and a transformer to a bus that supplies three identical motors as shown in Fig. 9.8. Each motor has $X_d'' = 25\%$ and $X_d' = 30\%$ on a base of 5 MVA, 6.6 kV. The three-phase rating of the step-up transformer is 25 MVA, 11/66 kV with a leakage reactance of 10% and that of the step-down transformer is 25 MVA, 66/6.6 kV with a leakage reactance of 10%. The bus voltage at the motors is 6.6 kV when a three-phase fault occurs at the point F. For the specified fault, calculate

(a) the subtransient current in the fault,

(b) the subtransient current in the breaker B,

(c) the momentary current in breaker B, and

(d) the current to be interrupted by breaker B in five cycles.

Given: Reactance of the transmission line = 15% on a base of 25 MVA, 66 kV. Assume that the system is operating on no load when the fault occurs.

Fig. 9.8

Solution Choose a system base of 25 MVA.

For a generator voltage base of 11 kV, line voltage base is 66 kV and motor voltage base is 6.6 kV.

(a) For each motor

$$X_{dm}'' = j0.25 \times \frac{25}{5} = j1.25 \text{ pu}$$

Line, transformers and generator reactances are already given on proper base values.

The circuit model of the system for fault calculations is given in Fig. 9.9a. The system being initially on no load, the generator and motor induced emfs are identical. The circuit can therefore be reduced to that of Fig. 9.9b and then to Fig. 9.9c. Now

$$I_{SC} = 3 \times \frac{1}{j1.25} + \frac{1}{j0.55} = -j4.22 \text{ pu}$$

Base current in 6.6 kV circuit $= \dfrac{25 \times 1,000}{\sqrt{3} \times 6.6} = 2,187$ A

$\therefore$ $I_{SC} = 4.22 \times 2,187 = 9,229$ A

(b)   From Fig. 9.9c, current through circuit breaker B is

$$I_{SC}(B) = 2 \times \frac{1}{j1.25} + \frac{1}{j0.55} = -j3.42$$

$$= 3.42 \times 2,187 = 7,479.5 \text{ A}$$

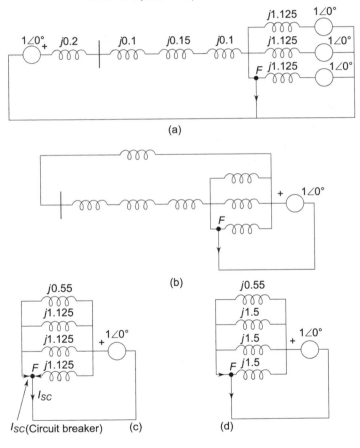

Fig. 9.9

(c)   For finding momentary current through the breaker, we must add the DC off-set current to the symmetrical subtransient current obtained in part (b). Rather than calculating the DC off-set current, allowance is made for it on an empirical basis. As explained in Sec. 9.5, momentary current through breaker $B = 1.6 \times 7,479.5$

$$= 11,967 \text{ A}$$

(d)   To compute the current to be interrupted by the breaker, motor subtransient reactance $(X_d'' = j0.25)$ is now replaced by transient reactance $(X_d' = j0.30)$.

$$X_d' \text{ (motor)} = j0.3 \times \frac{25}{5} = j1.5 \text{ pu}$$

The reactances of the circuit of Fig. 9.9c now modify to that of Fig. 9.9d. Current (symmetrical) to be interrupted by the breaker (as shown by arrow)

$$= 2 \times \frac{1}{j1.5} + \frac{1}{j0.55} = 3.1515 \text{ pu}$$

Allowance is made for the DC off-set value by multiplying with a factor of 1.1 (Sec. 9.5). Therefore, the current to be interrupted is

$$1.1 \times 3.1515 \times 2,187 = 7,581 \text{ A}$$

9.4   SHORT CIRCUIT OF A LOADED SYNCHRONOUS MACHINE

In the previous article on the short circuit of a synchronous machine, it was assumed that the machine was operating at no load prior to the occurrence of short circuit. The analysis of short circuit on a loaded synchronous machine is complicated and is beyond the scope of this book. We shall, however, present here the methods of computing short circuit current when short circuit occurs under loaded conditions.

Figure 9.10 shows the circuit model of a synchronous generator operating under steady conditions supplying a load current I^o to the bus at a terminal voltage of V^o. E_g is the induced emf under loaded condition and X_d is the direct axis synchronous reactance of the machine. When short circuit occurs at the terminals of this machine, the circuit model to be used for computing short circuit current is given in Fig. 9.11a for subtransient current, and in Fig. 9.11b for transient current. The induced emfs to be used in these models are given by

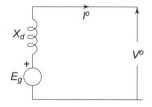

Fig. 9.10 Circuit model of a loaded machine

$$E_g'' = V^o + jI^o X_d'' \tag{9.8}$$

$$E_g' = V^o + jI^o X_d' \tag{9.9}$$

The voltage E_g'' is known as the *voltage behind the subtransient reactance* and the voltage E_g' is known as the *voltage behind the transient reactance*. In fact, if I^o is zero (no load case), $E_g'' = E_g' = E_g$, the no load voltage, in which case the circuit model reduces to that discussed in Sec. 9.3.

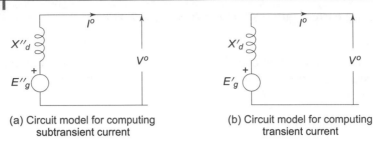

(a) Circuit model for computing   (b) Circuit model for computing
subtransient current   transient current

Fig. 9.11

Synchronous motors have internal emfs and reactances similar to that of a generator except that the current direction is reversed. During short circuit conditions these can be replaced by similar circuit models except that the voltage behind subtransient/transient reactance is given by

$$E''_m = V^\circ - jI^\circ X''_d \tag{9.10}$$

$$E'_m = V^\circ - jI^\circ X'_d \tag{9.11}$$

Whenever we are dealing with short circuit of an interconnected system, the synchronous machines (generators and motors) are replaced by their corresponding circuit models having voltage behind subtransient (transient) reactance in series with subtransient (transient) reactance. The rest of the network being passive remains unchanged.

Example 9.3

A synchronous generator and a synchronous motor each rated 25 MVA, 11 kV having 15% subtransient reactance are connected through transformers and a line as shown in Fig. 9.12a. The transformers are rated 25 MVA, 11/66 kV and 66/11 kV with leakage reactance of 10% each. The line has a reactance of 10% on a base of 25 MVA, 66 kV. The motor is drawing 15 MW at 0.8 power factor leading and a terminal voltage of 10.6 kV when a symmetrical three-phase fault occurs at the motor terminals. Find the subtransient current in the generator, motor and fault.

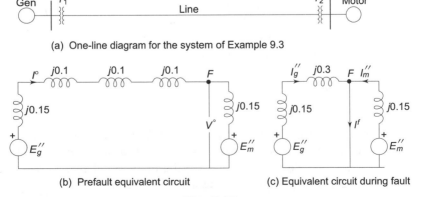

(a) One-line diagram for the system of Example 9.3

(b) Prefault equivalent circuit   (c) Equivalent circuit during fault

Fig. 9.12

Solution All reactances are given on a base of 25 MVA and appropriate voltages.

Prefault voltage $V^o = \dfrac{10.6}{11} = 0.9636 \angle 0°$ pu

Load = 15 MW, 0.8 pf leading

$= \dfrac{15}{25} = 0.6$ pu, 0.8 pf leading

Prefault current $I^o = \dfrac{0.6}{0.9636 \times 0.8} \angle 36.9° = 0.7783 \angle 36.9°$ pu

Voltage behind subtransient reactance (generator)

$E_g'' = 0.9636 \angle 0° + j0.45 \times 0.7783 \angle 36.9°$

$= 0.7536 + j0.28$ pu

Voltage behind subtransient reactance (motor)

$E_m'' = 0.9636 \angle 0° - j0.15 \times 0.7783 \angle 36.9°$

$= 1.0336 - j0.0933$ pu

The prefault equivalent circuit is shown in Fig. 9.12b. Under faulted condition (Fig. 9.12c)

$I_g'' = \dfrac{0.7536 + j0.2800}{j0.45} = 0.6226 - j1.6746$ pu

$I_m'' = \dfrac{1.0336 - j0.0933}{j0.15} = -0.6226 - j6.8906$ pu

Current in fault

$I^f = I_g'' + I_m'' = -j8.5653$ pu

Base current (gen/motor) $= \dfrac{25 \times 10^3}{\sqrt{3} \times 11} = 1{,}312.2$ A

Now

$I_g'' = 1{,}312.0 \,(0.6226 - j1.6746) = (816.4 - j2{,}197.4)$ A

$I_m'' = 1{,}312.2 \,(-0.6226 - j6.8906) = (-816.2 - j9{,}041.8)$ A

$I^f = -j11{,}239$ A

Short Circuit (SC) Current Computation through the Thevenin Theorem

An alternate method of computing short circuit currents is through the application of the Thevenin theorem. This method is faster and easily adopted to systematic computation for large networks. While the method is perfectly general, it is illustrated here through a simple example.

Consider a synchronous generator feeding a synchronous motor over a line. Figure 9.13a shows the circuit model of the system under conditions of steady load. Fault computations are to be made for a fault at F, at the motor terminals. As a first step the circuit model is replaced by the one shown in Fig. 9.13b, wherein the synchronous machines are represented by their transient reactances (or subtransient reactances if subtransient currents are of interest) in series with voltages behind transient reactances. This change does not disturb the prefault current I^o and prefault voltage V^o (at F).

As seen from FG the Thevenin equivalent circuit of Fig. 9.13b is drawn in Fig. 9.13c. It comprises prefault voltage V^o in series with the passive Thevenin impedance network. It is noticed that the prefault current I^o does not appear in the passive Thevenin impedance network. It is therefore to be remembered that this current must be accounted for by superposition after the SC solution is obtained through use of the Thevenin equivalent.

Consider now a fault at F through an impedance Z^f. Figure 9.13d shows the Thevenin equivalent of the system feeding the fault impedance. We can immediately write

$$I^f = \frac{V^o}{jX_{\text{Th}} + Z^t} \qquad (9.12)$$

Current caused by fault in generator circuit

$$\Delta I_g = \frac{X'_{dm}}{(X'_{dg} + X + X'_{dm}} I^f \qquad (9.13)$$

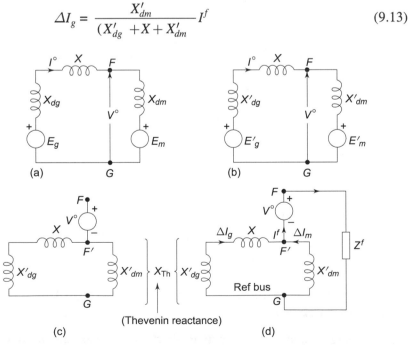

(a)   (b)

(c)   (Thevenin reactance)   (d)

Fig. 9.13   Computation of SC current by the Thevenin equivalent

Current caused by fault in motor circuit

$$\Delta I_m = \frac{X'_{dg} + X}{(X'_{dm} + X + X'_{dg})} I^f \tag{9.14}$$

Postfault currents and voltages are obtained as follows by superposition:

$$I^f_g = I^o + \Delta I_g$$

$$I^f_m = -I^o + \Delta I_m \text{ (in the direction of } \Delta I_m) \tag{9.15}$$

Postfault voltage

$$V^f = V^o + (-jX_{Th}I^f) = V^o + \Delta V \tag{9.16}$$

where $\Delta V = -jX_{Th}I^f$ is the voltage of the fault point F' on the Thevenin passive network (with respect to the reference bus G) caused by the flow of fault current I^f.

An observation can be made here. Since the prefault current flowing out of fault point F is always zero, the postfault current out of F is independent of load for a given prefault voltage at F.

The above approach to SC computation is summarized in the following four steps:

Step 1: Obtain steady state solution of loaded system (load flow study).

Step 2: Replace reactances of synchronous machines by their subtransient/transient values. Short circuit all emf sources. The result is the passive Thevenin network.

Step 3: Excite the passive network of Step 2 at the fault point by negative of prefault voltage (see Fig. 9.13d) in series with the fault impedance. Compute voltages and currents at all points of interest.

Step 4: Postfault currents and voltages are obtained by adding results of Steps 1 and 3.

The following assumptions can be safely made in SC computations leading to considerable computational simplification:

Assumption 1: All prefault voltage magnitudes are 1 pu.

Assumption 2: All prefault currents are zero.

The first assumption is quite close to actual conditions as under normal operation all voltages (pu) are nearly unity.

The changes in current caused by short circuit are quite large, of the order of 10–20 pu and are purely reactive; whereas the prefault load currents are almost purely real. Hence the total postfault current which is the result of the two currents can be taken in magnitude equal to the larger component (caused by the fault). This justifies assumption 2.

Let us illustrate the above method by recalculating the results of Example 9.3.

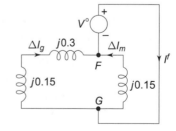

Fig. 9.14 *F* is the fault point on the passive Thevenin network

The circuit model for the system of Example 9.3 for computation of postfault condition is shown in Fig. 9.14.

$$I^f = \frac{V^o}{(j0.15 \| j0.45)} = \frac{0.9636 \times j0.60}{j0.15 \times j0.45} = -j8.565 \text{ pu}$$

Change in generator current due to fault,

$$\Delta I_g = -j8.565 \times \frac{j0.15}{j0.60} = -j2.141 \text{ pu}$$

Change in motor current due to fault,

$$\Delta I_m = -j8.565 \times \frac{j0.45}{j0.60} = -j6.424 \text{ pu}$$

To these changes we add the prefault current to obtain the subtransient current in machines. Thus

$$I_g'' = I^o + \Delta I_g = (0.623 - j1.674) \text{ pu}$$

$$I_m'' = -I^o + \Delta I_m = (-0.623 - j6.891) \text{ pu}$$

which are the same (and should be) as calculated already.

We have thus solved Example 9.3 alternatively through the Thevenin theorem and superposition. This, indeed, is a powerful method for large networks.

9.5 SELECTION OF CIRCUIT BREAKERS

Two of the circuit breaker ratings which require the computation of SC current are: *rated momentary current* and *rated symmetrical interrupting current*. Symmetrical SC current is obtained by using subtransient reactances for synchronous machines. Momentary current (rms) is then calculated by multiplying the symmetrical momentary current by a factor of 1.6 to account for the presence of DC off-set current.

Symmetrical current to be interrupted is computed by using subtransient reactances for synchronous generators and transient reactances for synchronous motors—induction motors are neglected*. The DC off-set value to be added to obtain the current to be interrupted is accounted for by multiplying the symmetrical SC current by a factor as tabulated below:

Circuit Breaker Speed	Multiplying Factor
8 cycles or slower	1.0
5 cycles	1.1
3 cycles	1.2
2 cycles	1.4

*In some recent attempts, currents contributed by induction motors during a short circuit have been accounted for.

If SC MVA (explained below) is more than 500, the above multiplying factors are increased by 0.1 each. The multiplying factor for air breakers rated 600 V or lower is 1.25.

The current that a circuit breaker can interrupt is inversely proportional to the operating voltage over a certain range, i.e.

Amperes at operating voltage

= amperes at rated voltage × rated voltage/operating voltage

Of course, operating voltage cannot exceed the maximum design value. Also, no matter how low the voltage is, the rated interrupting current cannot exceed the *rated maximum interrupting current*. Over this range of voltages, the product of operating voltage and interrupting current is constant. It is therefore logical as well as convenient to express the circuit breaker rating in terms of SC MVA that can be interrupted, defined as

Rated interrupting MVA (three-phase) capacity

$$= \sqrt{3} \; |V(\text{line})|_{\text{rated}} \times |I(\text{line})|_{\text{rated interrupting current}}$$

where V(line) is in kV and I (line) is in kA.

Thus, instead of computing the SC current to be interrupted, we compute three-phase SC MVA to be interrupted, where

SC MVA (3-phase) = $\sqrt{3}$ × prefault line voltage in kV
× SC current in kA.

If voltage and current are in per unit values on a three-phase basis

$$\text{SC MVA (3-phase)} = |V|_{\text{prefault}} \times |I|_{\text{SC}} \times (\text{MVA})_{\text{Base}} \qquad (9.17)$$

Obviously, rated MVA interrupting capacity of a circuit breaker is to be more than (or equal to) the SC MVA required to be interrupted.

For the selection of a circuit breaker for a particular location, we must find the maximum possible SC MVA to be interrupted with respect to type and location of fault and generating capacity (also synchronous motor load) connected to the system. A three-phase fault though rare is generally the one which gives the highest SC MVA and a circuit breaker must be capable of interrupting it. An exception is an LG (line-to-ground) fault close to a synchronous generator*. In a simple system the fault location which gives the highest SC MVA may be obvious but in a large system various possible locations must be tried out to obtain the highest SC MVA requiring repeated SC computations. This is illustrated by the examples that follow.

Example 9.4

Three 6.6 kV generators A, B and C, each of 10% leakage reactance and MVA ratings 40, 50 and 25, respectively are interconnected electrically, as shown in

---

*This will be explained in Chapter 11.

Fig. 9.15, by a tie bar through *current limiting reactors*, each of 12% reactance based upon the rating of the machine to which it is connected. A three-phase feeder is supplied from the bus bar of generator A at a line voltage of 6.6 kV. The feeder has a resistance of 0.06 Ω/phase and an inductive reactance of 0.12 Ω/phase. Estimate the maximum MVA that can be fed into a symmetrical short circuit at the far end of the feeder.

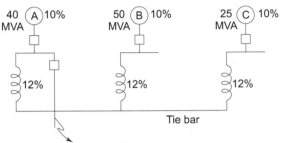

Fig. 9.15

Solution   Choose as base 50 MVA, 6.6 kV.
   Feeder impedance

$$= \frac{(0.06 + j0.12) \times 50}{(6.6)^2} = (0.069 + j0.138) \text{ pu}$$

$$\text{Gen A reactance} = \frac{0.1 \times 50}{40} = 0.125 \text{ pu}$$

$$\text{Gen B reactance} = 0.1 \text{ pu}$$

$$\text{Gen C reactance} = 0.1 \times \frac{50}{25} = 0.2 \text{ pu}$$

$$\text{Reactor A reactance} = \frac{0.12 \times 50}{40} = 0.15 \text{ pu}$$

$$\text{Reactor B reactance} = 0.12 \text{ pu}$$

$$\text{Reactor C reactance} = \frac{0.12 \times 50}{25} = 0.24 \text{ pu}$$

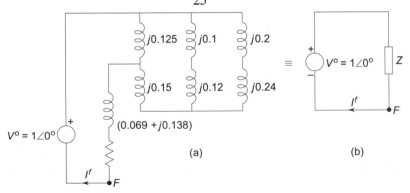

Fig. 9.16

Assume no load prefault conditions, i.e. prefault currents are zero. Postfault currents can then be calculated by the circuit model of Fig. 9.16a corresponding to Fig. 9.13d. The circuit is easily reduced to that of Fig. 9.16b, where

$$Z = (0.069 + j0.138) + j0.125 \parallel (j0.15 + j0.22 \parallel j0.44)$$

$$= 0.069 + j0.226 = 0.236\angle73°$$

$$\text{SC MVA} = V^o I^f = V^o\left(\frac{V^o}{Z}\right) = \frac{1}{Z} \text{ pu (since } V^o = 1 \text{ pu)}$$

$$= \frac{1}{Z} \times (\text{MVA})_{\text{Base}}$$

$$= \frac{50}{0.236} = 212 \text{ MVA}$$

Example 9.5

Consider the 4-bus system of Fig. 9.17. Buses 1 and 2 are generator buses and 3 and 4 are load buses. The generators are rated 11 kV, 100 MVA, with transient reactance of 10% each. Both the transformers are 11/110 kV, 100 MVA with a leakage reactance of 5%. The reactances of the lines to a base of 100 MVA, 110 kV are indicated on the figure. Obtain the short circuit solution for a three-phase solid fault on bus 4 (load bus).

Assume prefault voltages to be 1 pu and prefault currents to be zero.

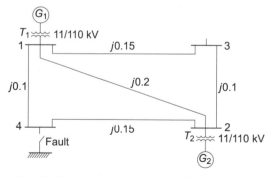

Fig. 9.17 Four-bus system of Example 9.5

Solution Changes in voltages and currents caused by a short circuit can be calculated from the circuit model of Fig. 9.18. Fault current I^f is calculated by systematic network reduction as in Fig. 9.19.

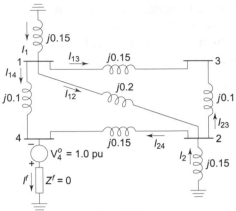

Fig. 9.18

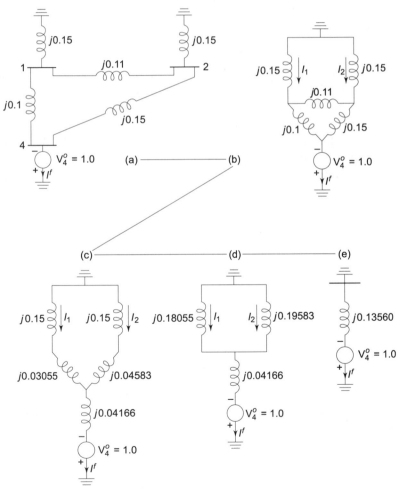

Fig. 9.19 Systematic reduction of the network of Fig. 9.18

From Fig. 9.19e, we get directly the fault current as

$$I^f = \frac{1.0}{j0.13560} = -j7.37463 \text{ pu}$$

From Fig. 9.19d, it is easy to see that

$$I_1 = I_f \times \frac{j0.19583}{j0.37638} = -j3.83701 \text{ pu}$$

$$I_2 = I_f \times \frac{j0.18055}{j0.37638} = -j3.53762 \text{ pu}$$

Let us now compute the voltage changes for buses 1, 2 and 3. From Fig. 9.19b, we give

$$\Delta V_1 = 0 - (j0.15)(-j3.83701) = -0.57555 \text{ pu}$$

$$\Delta V_2 = 0 - (j0.15)(-j3.53762) = -0.53064 \text{ pu}$$

Now

$$V_1^f = 1 + \Delta V_1 = 0.42445 \text{ pu}$$

$$V_2^f = 1 + \Delta V_2 = 0.46936 \text{ pu}$$

∴

$$I_{13} = \frac{V_1^f - V_2^f}{j0.15 + j0.1} = j0.17964 \text{ pu}$$

Now

$$\Delta V_3 = 0 - [(j0.15)(-j3.83701) + (j0.15)(j0.17964)]$$

$$= -0.54860 \text{ pu}$$

∴

$$V_3^f = 1 - 0.54860 = 0.4514 \text{ pu}$$

$$V_4^f = 0$$

The determination of currents in the remaining lines is left as an exercise to the reader.

Short circuit study is complete with the computation of SC MVA at bus 4.

$$(\text{SC MVA})_4 = 7.37463 \times 100 = 737.463 \text{ MVA}$$

It is obvious that the heuristic network reduction procedure adopted above is not practical for a real power network of even moderate size. It is, therefore, essential to adopt a suitable algorithm for carrying out short circuit study on a digital computer. This is discussed in Sec. 9.6.

9.6 ALGORITHM FOR SHORT CIRCUIT STUDIES

So far we have carried out short circuit calculations for simple systems whose passive networks can be easily reduced. In this section we extend our study to large systems. In order to apply the four steps of short circuit computation developed earlier to large systems, it is necessary to evolve a systematic general algorithm so that a digital computer can be used.

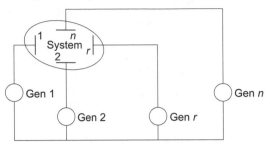

Fig. 9.20 *n*-bus system under steady load

Consider an *n*-bus system shown schematically in Fig. 9.20 operating at steady load. The first step towards short circuit computation is to obtain prefault voltages at all buses and currents in all lines through a load flow study. Let us indicate the prefault bus voltage vector as

$$V^0_{\text{BUS}} = \begin{bmatrix} V^0_1 \\ V^0_2 \\ \vdots \\ V^0_n \end{bmatrix} \qquad (9.18)$$

Let us assume that the *r*th bus is faulted through a fault impedance Z^f. The postfault bus voltage vector will be given by

$$V^f_{\text{BUS}} = V^0_{\text{BUS}} + \Delta V \qquad (9.19)$$

where ΔV is the vector of changes in bus voltages caused by the fault.

As step 2, we drawn the passive Thevenin network of the system with generators replaced by transient/subtransient reactances with their emfs shorted (Fig. 9.21).

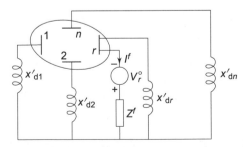

Fig. 9.21 Network of the system of Fig. 9.20 for computing changes in bus voltages caused by the fault

As per step 3 we now excite the passive Thevenin network with $-V_r^o$ in series with Z^f as in Fig. 9.21. The vector ΔV comprises the bus voltages of this network.

Now

$$\Delta V = Z_{BUS} J^f \tag{9.20}$$

where

$$Z_{BUS} = \begin{bmatrix} Z_{11} & \cdots & Z_{1n} \\ \vdots & & \vdots \\ Z_{n1} & \cdots & Z_{nn} \end{bmatrix} = \text{bus impedance matrix of the} \atop \text{passive Thevenin network} \tag{9.21}$$

$J^f = $ bus current injection vector

Since the network is injected with current $-I^f$ only at the rth bus, we have

$$J^f = \begin{bmatrix} 0 \\ 0 \\ \vdots \\ I_r^f = -I^f \\ \vdots \\ 0 \end{bmatrix} \tag{9.22}$$

Substituting Eq. (9.22) in Eq. (9.20), we have for the rth bus

$$\Delta V_r = -Z_{rr}I^f$$

By step 4, the voltage at the rth bus under fault is

$$V_r^f = V_r^0 + \Delta V_r^0 = V_r^0 - Z_{rr}I^f \tag{9.23}$$

However, this voltage must equal

$$V_r^f = Z^f I^f \tag{9.24}$$

We have from Eqs. (9.23) and (9.24)

$$Z^f I^f = V_r^0 - Z_{rr}I^f$$

or

$$I^f = \frac{V_r^0}{Z_{rr} + Z^f} \tag{9.25}$$

At the ith bus (from Eqs (9.20) and (9.22))

$$\Delta V_t = -Z_{ir}I^f$$

∴

$$V_i^f = V_i^0 - Z_{ir}I^f, \quad i = 1, 2, ..., n \tag{9.26}$$

substituting for I^f from Eq. (9.25), we have

$$V_i^f = V_i^0 - \frac{Z_{ir}}{Z_{rr} + Z^f} V_r^0 \tag{9.27}$$

For $i = r$ in Eq. (9.27)

$$V_r^f = \frac{Z^f}{Z_{rr} + Z^f} V_r^0 \tag{9.28}$$

In the above relationship V_i^0's, the prefault bus voltages are assumed to be known from a load flow study. Z_{BUS} matrix of the short-circuit study network of Fig. 9.21 can be obtained by the inversion of its Y_{BUS} matrix as in Example 9.6 or the Z_{BUS} building algorithm presented in Section 9.7. It should be observed here that the SC study network of Fig. 9.21 is different from the corresponding load flow study network by the fact that the shunt branches corresponding to the generator reactances do not appear in the load flow study network. Further, in formulating the SC study network, the load impedances are ignored, these being very much larger than the impedances of lines and generators. Of course synchronous motors must be included in Z_{BUS} formulation for the SC study.

Postfault currents in lines are given by

$$I_{ij}^f = Y_{ij} (V_i^f - V_j^f) \tag{9.29}$$

For calculation of postfault generator current, examine Figs. 9.22(a) and (b). From the load flow study (Fig. 9.22(a))

Prefault generator output $= P_{Gi} + jQ_{Gi}$

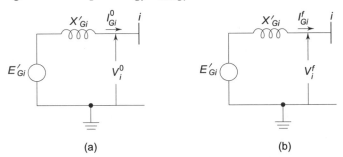

(a)   (b)

Fig. 9.22

$\therefore$

$$I_{Gi}^0 = \frac{P_{Gi} - jQ_{Gi}}{V_i^0}; \quad \text{(prefault generator output} = P_{Gi} + jQ_{Gi}) \tag{9.30}$$

$$E_{Gi}' = V_i + jX_{Gi}'I_{Gi}^0 \tag{9.31}$$

From the SC study, V_i^f is obtained. It then follows from Fig. 9.22(b) that

$$I_{Gi}^f = \frac{E_{Gi}' - V_i^f}{jX_{Gi}'} \tag{9.32}$$

Example 9.6

To illustrate the algorithm discussed above, we shall recompute the short circuit solution for Example 9.5 which was solved earlier using the network reduction technique.

First of all the bus admittance matrix for the network of Fig. 9.18 is formed as follows:

$$Y_{11} = \frac{1}{j0.15} + \frac{1}{j0.15} + \frac{1}{j0.1} + \frac{1}{j0.2} = -j28.333$$

$$Y_{12} = Y_{21} = \frac{-1}{j0.2} = j5.000$$

$$Y_{13} = Y_{31} = \frac{-1}{j0.15} = j6.667$$

$$Y_{14} = Y_{41} = \frac{-1}{j0.1} = j10.000$$

$$Y_{22} = \frac{1}{j0.15} + \frac{1}{j0.15} + \frac{1}{j0.1} + \frac{1}{j0.2} = -j28.333$$

$$Y_{23} = Y_{32} = \frac{-1}{j0.1} = j10.000$$

$$Y_{24} = Y_{42} = \frac{-1}{j0.15} = j6.667$$

$$Y_{33} = \frac{1}{j0.15} + \frac{1}{j0.1} = -j16.667$$

$$Y_{34} = Y_{43} = 0.000$$

$$Y_{44} = \frac{1}{j0.1} + \frac{1}{j0.15} = -j16.667$$

$$Y_{\text{BUS}} = \begin{bmatrix} -j28.333 & j5.000 & j6.667 & j10.000 \\ j5.000 & -j28.333 & j10.000 & j6.667 \\ j6.667 & j10.000 & -j16.667 & j0.000 \\ j10.000 & j6.667 & j0.000 & -j16.667 \end{bmatrix}$$

By inversion we get Z_{BUS} as

$$Z_{\text{BUS}} = \begin{bmatrix} j0.0903 & j0.0597 & j0.0719 & j0.0780 \\ j0.0597 & j0.0903 & j0.0780 & j0.0719 \\ j0.0719 & j0.0780 & j0.1356 & j0.0743 \\ j0.0780 & j0.0719 & j0.0743 & j0.1356 \end{bmatrix}$$

Now, the postfault bus voltages can be obtained using Eq. (9.27) as

$$V_1^f = V_1^0 - \frac{Z_{14}}{Z_{44}} V_4^0$$

The prefault condition being no load, $V_1^0 = V_2^0 = V_3^0 = V_4^0 = 1$ pu

$$V_1^f = 1.0 - \frac{j0.0780}{j0.1356} \times 1.0 = 0.4248 \text{ pu}$$

$$V_2^f = V_2^0 - \frac{Z_{24}}{Z_{44}} V_4^0$$

$$= 1.0 - \frac{j0.0719}{j0.1356} \times 1.0 = 0.4698 \text{ pu}$$

$$V_3^f = V_3^0 - \frac{Z_{34}}{Z_{44}} V_4^0$$

$$= 1.0 - \frac{j0.0743}{j0.1356} 1.0 = 0.4521 \text{ pu}$$

$$V_4^f = 0.0$$

Using Eq. (9.25) we can obtain the fault current as

$$I^f = \frac{1.000}{j0.1356} = -j7.37463 \text{ pu}$$

These values agree with those obtained earlier in Example 9.5. Let us also calculate the short circuit current in lines 1-3, 1-2, 1-4, 2-4 and 2-3.

$$I_{12}^f = \frac{V_1^f - V_3^f}{z_{13}} = \frac{0.4248 - 0.4521}{j0.15} = j0.182 \text{ pu}$$

$$I_{12}^f = \frac{V_1^f - V_2^f}{z_{12}} = \frac{0.4248 - 0.4698}{j0.2} = j0.225 \text{ pu}$$

$$I_{14}^f = \frac{V_1^f - V_4^f}{z_{14}} = \frac{0.4248 - 0}{j0.1} = -j4.248 \text{ pu}$$

$$I_{24}^f = \frac{V_2^f - V_4^f}{z_{24}} = \frac{0.4698 - 0}{j0.15} = -j3.132 \text{ pu}$$

$$I_{23}^f = \frac{V_2^f - V_3}{z_{23}} = \frac{0.4698 - 0.4521}{j.01} = -j0.177 \text{ pu}$$

For the example on hand this method may appear more involved compared to the heuristic network reduction method employed in Example 9.5. This, however, is a systematic method and can be easily adopted on the digital computer for practical networks of large size. Further, another important feature of the method is that having computed Z_{BUS}, we can at once obtain all the required short circuit data for a fault on any bus. For example, in this particular system, the fault current for a fault on bus 1 (or bus 2) will be

$$I^f = \frac{1.000}{Z_{11}(\text{or } Z_{22})} = \frac{1.00}{j0.0903} = -j11.074197 \text{ pu}$$

9.7 Z_{BUS} FORMULATION

By Inventing Y_{BUS}

$$J_{BUS} = Y_{BUS} \, V_{BUS}$$

or $\qquad V_{BUS} = [Y_{BUS}]^{-1} \, J_{BUS} = Z_{BUS} \, J_{BUS}$ $\qquad$ (9.33)

or $\qquad Z_{BUS} = [Y_{BUS}]^{-1}$

The sparsity of Y_{BUS} may be retained by using an efficient inversion technique [1] and nodal impedance matrix can then be calculated directly from the factorized admittance matrix. This is beyond the scope of this book.

Current Injection Technique

Equation (9.33) can be written in the expanded form

$$V_1 = Z_{11}I_1 + Z_{12}I_2 + \ldots + Z_{1n}I_n \qquad (9.34)$$

$$V_2 = Z_{21}I_1 + Z_{22}I_2 + \ldots + Z_{2n}I_n$$

$$\text{-------------------------}$$

$$V_n = Z_{n1}I_1 + Z_{n2}I_1 + \ldots + Z_{nn}I_n$$

It immediately follows from Eq. (9.34) that

$$Z_{ij} = \frac{V_i}{I_j}\bigg|_{\substack{I_1 = I_2 = \cdots = I_n = 0 \\ I_j \neq 0}} \qquad (9.35)$$

Also $Z_{ij} = Z_{ji}$; (Z_{BUS} is a symmetrical matrix).

As per Eq. (9.35) if a unit current is injected at bus (node) j, while the other buses are kept open circuited, the bus voltages yield the values of the jth column of Z_{BUS}. However, no organized computerizable techniques are possible for finding the bus voltages. The technique had utility in AC Network Analyzers where the bus voltages could be read by a voltmeter.

Example 9.7

Consider the network of Fig. 9.23(a) with three buses one of which is a reference. Evaluate Z_{BUS}.

Solution Inject a unit current at bus 1 keeping bus 2 open circuit, i.e., $I_1 = 1$ and $I_2 = 0$ as in Fig. 9.22(b). Calculating voltages at buses 1 and 2, we have

$$Z_{11} = V_1 = 7$$

$$Z_{21} = V_2 = 4$$

Now let $I_1 = 0$ and $I_2 = 1$. It similarly follows that

$$Z_{12} = V_1 = 4 = Z_{12}$$
$$Z_{22} = V_2 = 6$$

Collecting the above values

$$Z_{BUS} = \begin{bmatrix} 7 & 4 \\ 4 & 6 \end{bmatrix}$$

Because of the above computational procedure, the Z_{BUS} matrix is referred to as the 'open-circuit impedance matrix'.

Z_{BUS} Building Algorithm

It is a step-by-step programmable technique which proceeds branch by branch. It has the advantage that any modification of the network does not require complete rebuilding of Z_{BUS}.

Consider that Z_{BUS} has been formulated upto a certain stage and another branch is now added. Then

$$Z_{BUS} \text{ (old)} \xrightarrow{\quad Z_b = \text{branch impedance} \quad} Z_{BUS} \text{ (new)}$$

Upon adding a new branch, one of the following situations is presented.

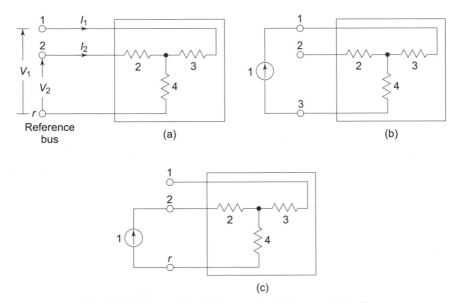

Fig. 9.23   Current injection method of computing Z_{BUS}

1. Z_b is added from a new bus to the reference bus (i.e. a new branch is added and the dimension of Z_{BUS} goes up by one). This is *type-1 modification*.

2. Z_b is added from a new bus to an old bus (i.e., a new branch is added and the dimension of Z_{BUS} goes up by one). This is *type-2 modification*.

3. Z_b connects an old bus to the reference branch (i.e., a new loop is formed but the dimension of Z_{BUS} does not change). This is *type-3 modification*.

4. Z_b connects two old buses (i.e., new loop is formed but the dimension of Z_{BUS} does not change). This is *type-4 modification*.

5. Z_b connects two new buses (Z_{BUS} remains unaffected in this case). This situation can be avoided by suitable numbering of buses and from now onwards will be ignored.

Notation: i, j—old buses; *r*—reference bus; *k*—new bus.

Type-1 Modification

Figure 9.24 shows a passive (linear) n-bus network in which branch with impedance Z_b is added to the new bus k and the reference bus r. Now

$$V_k = Z_b I_k$$

$$Z_{ki} = Z_{ik} = 0; \; i = 1, 2, ..., n$$

$\therefore \qquad Z_{kk} = Z_b$

Hence

$$Z_{BUS} \text{ (new)} = \begin{bmatrix} Z_{Bus}(old) & \begin{matrix} 0 \\ \vdots \\ 0 \end{matrix} \\ \hline 0 \qquad\quad ...0 & Z_b \end{bmatrix} \qquad (9.36)$$

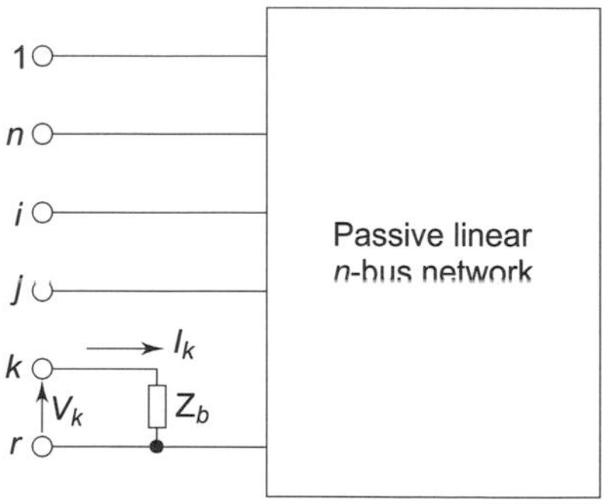

Fig. 9.24 Type-1 modification

Type-2 Modification

Z_b is added from new bus k to the old bus j as in Fig. 9.25. It follows from this figure that

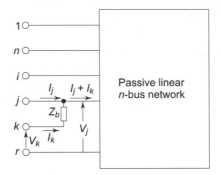

Fig. 9.25 Type-2 modification

$$V_k = Z_b I_k + V_j$$
$$= Z_b I_k + Z_{j1}I_1 + Z_{j2}I_2 + \dots + Z_{jj}(I_j + I_k) + \dots + Z_{jn}I_n$$

Rearranging,

$$V_k = Z_{j1}I_1 + Z_{j2}I_2 + \dots + Z_{jj}I_j + \dots + Z_{jn}I_n + (Z_{jj} + Z_b)I_k$$

Consequently

$$Z_{\text{BUS}} \text{ (new)} = \left[\begin{array}{c|c} Z_{\text{BUS}}\text{(old)} & \begin{matrix} Z_{1j} \\ Z_{2j} \\ \vdots \\ Z_{nj} \end{matrix} \\ \hline Z_{ji}Z_{j2}\dots Z_{jn} & Z_{jj} + Z_b \end{array} \right] \qquad (9.37)$$

Type-3 Modification

Z_b connects an old bus (j) to the reference bus (r) as in Fig. 9.26. This case follows from Fig. 9.25 by connecting bus k to the reference bus r, i.e. by setting $V_k = 0$.

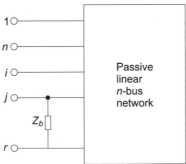

Fig. 9.26 Type-3 modification

Thus

$$
\begin{bmatrix} V_1 \\ V_2 \\ \vdots \\ V_n \\ \hline 0 \end{bmatrix} = \left[\begin{array}{c|c} Z_{\text{BUS}}(\text{old}) & \begin{matrix} Z_{1j} \\ Z_{2}j \\ \vdots \\ Z_{nj} \end{matrix} \\ \hline Z_{j1}Z_{j2}...Z_{jn} & Z_{jj}+Z_b \end{array} \right] \begin{bmatrix} I_1 \\ I_2 \\ \vdots \\ I_n \\ \hline I_k \end{bmatrix} \tag{9.38}
$$

Eliminate I_k in the set of equations contained in the matrix operation (9.38),
$$
0 = Z_{j1}I_1 + Z_{j2}I_2 + ... + Z_{jn}I_n + (Z_{jj} + Z_b)I_k
$$
or
$$
I_k = -\frac{1}{Z_{jj} - Z_b}(Z_{j1}I_1 + Z_{j2}I_2 + ... + Z_{jn}I_n) \tag{9.39}
$$

Now
$$
V_i = Z_{i1}I_1 + Z_{i2}I_2 + ... + Z_{in}I_n + Z_{ij}I_k \tag{9.40}
$$
Substituting Eq. (9.40) in Eq. (9.39)

$$
V_i = \left[Z_{i1} - \frac{1}{Z_{jj} + Z_b}(Z_{ij}Z_{j1}) \right]I_1 + \left[Z_{i2} - \frac{1}{Z_{jj} + Z_b}(Z_{ij}Z_{j2}) \right]I_2
$$

$$
+ ... + \left[Z_{in} - \frac{1}{Z_{jj} + Z_b}(Z_{ij} Z_{jn}) \right]I_n \tag{9.41}
$$

Equation (9.37) can be written in matrix form as

$$
Z_{\text{BUS}}(\text{new}) = Z_{\text{BUS}}\left(\text{old} - \frac{1}{Z_{jj} + Z_b} \begin{bmatrix} Z_{1j} \\ \vdots \\ Z_{nj} \end{bmatrix}[Z_{j1}...Z_{jn}] \right) \tag{9.42}
$$

Type-4 Modification

Z_b connects two old buses as in Fig. 9.27. Equations can be written as follows for all the network buses.

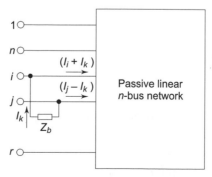

Fig. 9.27 Type-4 modification

$$
V_i = Z_{i1}I_1 + Z_{i2}I_2 + ... + Z_{1i}(I_i + I_k) + Z_{ij}(I_j - I_k) + ...+ Z_{in}I_n \tag{9.43}
$$
Similar equations follow for other buses.

The voltages of the buses i and j are, however, constrained by the equation (Fig. 9.27)

$$V_j = Z_b I_k + V_i \qquad (9.44)$$

or $\quad Z_{j1}I_1 + Z_{j2}I_2 + \ldots + Z_{ji}(I_i + I_k) + Z_{jj}(I_j - I_k) + \ldots + Z_{jn}I_n$
$$= Z_b I_k + Z_{i1}I_1 + Z_{i2}I_2 + \ldots + Z_{ii}(I_i + I_k) + Z_{ij}(I_j - I_k) + \ldots + Z_{in}I_n$$

Rearranging

$$0 = (Z_{i1} - Z_{j1})I_1 + \ldots + (Z_{ii} - Z_{ji})I_i + (Z_{ij} - Z_{jj})I_j$$
$$+ \ldots + (Z_{in} - Z_{jn})I_n + (Z_b + Z_{ii} + Z_{jj} - Z_{ij} - Z_{ji})I_k \qquad (9.45)$$

Collecting equations similar to Eq. (9.43) and Eq. (9.45) we can write

$$
\begin{bmatrix} V_1 \\ V_2 \\ \vdots \\ V_n \\ - \\ 0 \end{bmatrix}
=
\left[
\begin{array}{c|c}
Z_{\text{BUS}} & \begin{matrix} (Z_{1i} - Z_{1j}) \\ | \\ (Z_{ni} - Z_{nj}) \end{matrix} \\
\hline
(Z_{i1} - Z_{j1})\ldots(Z_{in} - Z_{jn}) & Z_b + Z_{ii} + Z_{jj} - 2Z_{ij}
\end{array}
\right]
\begin{bmatrix} I_1 \\ I_2 \\ \vdots \\ I_n \\ I_j \end{bmatrix}
\qquad (9.46)
$$

Eliminating I_k in Eq. (9.46) on lines similar to what was done in Type-2 modification, it follows that

$$Z_{\text{BUS}}(\text{new}) = Z_{\text{BUS}}(\text{old}) - \frac{1}{Z_b + Z_{ii} + Z_{jj} - 2Z_{ij}}
\begin{bmatrix} Z_{1i} & - & Z_{1j} \\ & \vdots & \\ Z_{ni} & - & Z_{nj} \end{bmatrix}$$

$$[Z_{i1} - Z_{j1}] \ldots (Z_{in} - Z_{jn})] \qquad (9.47)$$

With the use of four relationships Eqs (9.36), (9.37), (9.42) and (9.47) bus impedance matrix can be built by a step-by-step procedure (bringing in one branch at a time) as illustrated in Example 9.8. This procedure being a mechanical one can be easily computerized.

When the network undergoes changes, the modification procedures can be employed to revise the bus impedance matrix of the network. The opening of a line (Z_{ij}) is equivalent to adding a branch in parallel to it with impedance $-Z_{ij}$ (see Example 9.8).

Example 9.8

For the 3-bus network shown in Fig. 9.28 build Z_{BUS}.

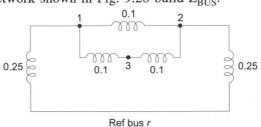

Ref bus r

Fig. 9.28

Solution

Step 1: Add branch $z_{1r} = 0.25$ (from bus 1 (new) to bus r)

$$Z_{BUS} = [0.25] \tag{i}$$

Step 2: Add branch $z_{21} = 0.1$ (from bus 2 (new) to bus 1 (old)); type-2 modification

$$Z_{BUS} = \begin{matrix} 1 \\ 2 \end{matrix} \begin{bmatrix} 0.25 & 0.25 \\ 0.25 & 0.35 \end{bmatrix} \tag{ii}$$

Step 3: Add branch $z_{13} = 0.1$ (from bus 3 (new) to bus 1 (old)); type-2 modification

$$Z_{BUS} = \begin{bmatrix} 0.25 & 0.25 & 0.25 \\ 0.25 & 0.35 & 0.25 \\ 0.25 & 0.25 & 0.35 \end{bmatrix} \tag{iii}$$

Step 4: Add branch z_{2r} (from bus 2 (old) to bus r); type-3 modification

$$Z_{BUS} = \begin{bmatrix} 0.25 & 0.25 & 0.25 \\ 0.25 & 0.35 & 0.25 \\ 0.25 & 0.25 & 0.35 \end{bmatrix} - \frac{1}{0.35+0.25} \begin{bmatrix} 0.25 \\ 0.35 \\ 0.25 \end{bmatrix} [0.25 \ 0.35 \ 0.25]$$

$$= \begin{bmatrix} 0.1458 & 0.1042 & 0.1458 \\ 0.1042 & 0.1458 & 0.1042 \\ 0.1458 & 0.1042 & 0.2458 \end{bmatrix}$$

Step 5: Add branch $z_{23} = 0.1$ (from bus 2 (old) to bus 3 (old)); type-4 modification

$$Z_{BUS} = \begin{bmatrix} 0.1458 & 0.1042 & 0.1458 \\ 0.1042 & 0.1458 & 0.1042 \\ 0.1458 & 0.1042 & 0.2458 \end{bmatrix} - \frac{1}{0.1+0.1458+0.2458-2\times0.1042}$$

$$= \begin{bmatrix} -0.1042 \\ 0.0417 \\ -0.0417 \end{bmatrix} [-0.1042 \quad 0.0417 \quad -0.0417]$$

$$= \begin{bmatrix} 0.1397 & 0.1103 & 0.1250 \\ 0.1103 & 0.1397 & 0.1250 \\ 0.1250 & 0.1250 & 0.1750 \end{bmatrix}$$

Opening a line (line 3-2): This is equivalent to connecting an impedance $- 0.1$ between bus 3 (old) and bus 2 (old) i.e. type-4 modification.

$$Z_{BUS} = Z_{BUS} \text{ (old)} - \frac{1}{(-0.1)+0.175+0.1397-2\times0.125}$$

$$\begin{bmatrix} 0.0147 \\ -0.0147 \\ 0.0500 \end{bmatrix} [0.0147 - 0.0147 \ 0.0500]$$

$$= \begin{bmatrix} 0.1458 & 0.1042 & 0.1458 \\ 0.1042 & 0.1458 & 0.1042 \\ 0.1458 & 0.1042 & 0.2458 \end{bmatrix}; \ \text{(same as in step 4)}$$

Example 9.9

For the power system shown in Fig. 9.29 the pu reactances are shown therein. For a solid 3-phase fault on bus 3, calculate the following

(a) Fault current

(b) V_1^f and V_2^f

(c) I_{12}^f, I_{13}^f and I_{23}^f

(d) I_{G1}^f, and I_{G2}^f

Assume prefault voltage to be 1 pu.

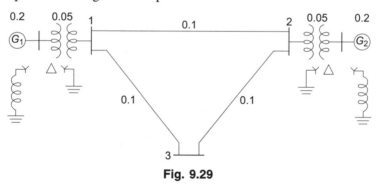

Fig. 9.29

Solution   The Thevenin passive network for this system is drawn in Fig. 9.28 with its Z_{BUS} given in Eq. (iv) of Example 9.8.

(a) As per Eq. (9.25)

$$I^f = \frac{V_r^0}{Z_{rr} + Z^f}$$

or

$$I^f = \frac{V_3^0}{Z_{33}} = \frac{1}{j0.175} = -j5.71$$

(b) As per Eq. (9.26)

$$V_i^f = V_i^0 - \frac{Z_{ir}}{Z_{rr} + Z^f} V_r^0$$

Now
$$V_1^f = \left(1 - \frac{Z_{13}}{Z_{33}}\right) = 1 - \frac{0.125}{0.175}$$

$$= 0.286$$

and
$$V_2^f = \left(1 - \frac{Z_{23}}{Z_{33}}\right) = 0.286$$

These two voltages are equal because of the symmetry of the given power network

(c) From Eq. (9.29)

$$I_{ij}^f = Y_{ij} \, (V_i^f - V_j^f)$$

$$I_{12}^f = \frac{1}{j0.1}(0.286 - 0.286) = 0$$

and
$$I_{13}^f = I_{31}^f = \frac{1}{j0.1}(0.286 - 0)$$

$$= -j2.86$$

(d) As per Eq. (9.32)

$$I_{G1}^f = \frac{E'_{G1} - V_1^f}{jX'_{iG} + jX_T}$$

But
$$E'_{G1} = 1 \text{ pu (prefault no load)}$$

$$\therefore \quad I_{G1}^f = \frac{1 - 0.286}{j0.2 + j0.05} = -j2.86$$

Similarly

$$I_{G2}^f = j2.86$$

PROBLEMS

9.1 A transmission line of inductance 0.1 H and resistance 5 ohms is suddenly short circuited at $t = 0$ at the bar end as shown in Fig. P-9.1. Write the expression for short circuit current $i(t)$. Find approximately the value of the first current maximum (maximum momentary current).

[**Hint:** Assume that the first current maximum occurs at the same time as the first current maximum of the symmetrical short circuit current.)

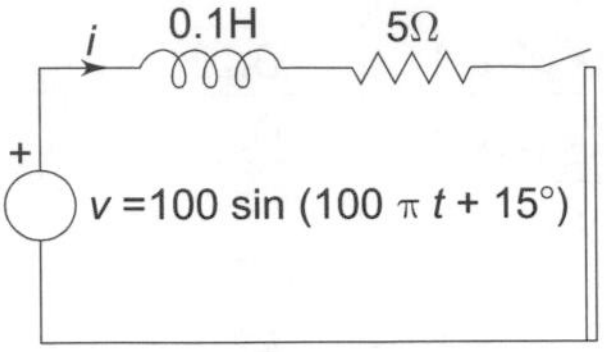

$$i \quad \text{0.1H} \quad 5\Omega$$

$$v = 100 \sin (100 \pi t + 15°)$$

Fig. P-9.1

9.2 (a) What should the instant of short circuit be in Fig. P-9.1 so that the DC off-set current is zero?

   (b) What should the instant of short circuit be in Fig. P-9.1 so that the DC off-set current is maximum?

9.3 For the system of Fig. 9.8 (Example 9.2) find the symmetrical currents to be interrupted by circuit breakers A and B for a fault at (i) P and (ii) Q.

9.4 For the system in Fig. P-9.4 the ratings of the various components are:

   Generator: 25 MVA, 12.4 kV, 10% subtransient reactance

   Motor: 20 MVA, 3.8 kV, 15% subtransient reactance

   Transformer T_1:   25 MVA, 11/33 kV, 8% reactance

   Transformer T_2:   20 MVA, 33/3.3 kV, 10% reactance

   Line: 20 ohms reactance

The system is loaded so that the motor is drawing 15 MW at 0.9 loading power factor, the motor terminal voltage being 3.1 kV. Find the subtransient current in generator and motor for a fault at generator bus.

[**Hint:** Assume a suitable voltage base for the generator. The voltage base for transformers, line and motor would then be given by the transformation ratios. For example, if we choose generator voltage base as 11 kV, the line voltage base is 33 kV and motor voltage base is 3.3 kV. Per unit reactances are calculated accordingly.]

Fig. P-9.4

9.5 Two synchronous motors are connected to the bus of a large system through a short transmission line as shown in Fig. P-9.5. The ratings of various components are:

   Motors (each):   1 MVA, 440 V, 0.1 pu transient reactance

   Line: 0.05 ohm reactance

   Large system: Short circuit MVA at its bus at 440 V is 8.

When the motors are operating at 440 V, calculate the short circuit current (symmetrical) fed into a three-phase fault at motor bus.

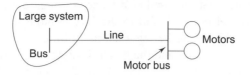

Fig. P-9.5

9.6 A synchronous generator rated 500 kVA, 440 V, 0.1 pu subtransient reactance is supplying a passive load of 400 kW at 0.8 lagging power factor. Calculate the initial symmetrical rms current for a three-phase fault at generator terminals.

9.7 A generator-transformer unit is connected to a line through a circuit breaker. The unit ratings are:

Generator: 10 MVA, 6.6 kV; $X''_d = 0.1$ pu, $X'_d = 0.20$ pu and $X_d = 0.80$ pu

Transformer: 10 MVA, 6.9/33 kV, reactance 0.08 pu

The system is operating no load at a line voltage of 30 kV, when a three-phase fault occurs on the line just beyond the circuit breaker. Find

(a) the initial symmetrical rms current in the breaker,
(b) the maximum possible DC off-set current in the breaker,
(c) the momentary current rating of the breaker,
(d) the current to be interrupted by the breaker and the interrupting kVA, and
(e) the sustained short circuit current in the breaker.

9.8 The system shown in Fig. P-9.8 is delivering 50 MVA at 11 kV, 0.8 lagging power factor into a bus which may be regarded as infinite. Particulars of various system components are:

Generator: 60 MVA, 12 kV, $X'_d = 0.35$ pu

Transformers (each): 80 MVA, 12/66 kV, $X = 0.08$ pu

Line: Reactance 12 ohms, resistance negligible.

Calculate the symmetrical current that the circuit breakers A and B will be called upon to interrupt in the event of a three-phase fault occurring at F near the circuit breaker B.

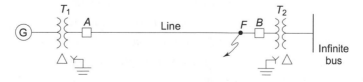

Fig. P-9.8

9.9 A two generator station supplies a feeder through a bus as shown in Fig. P-9.9. Additional power is fed to the bus through a transformer from a large system which may be regarded as infinite. A reactor X is included between the transformer and the bus to limit the SC rupturing capacity of the feeder circuit breaker B to 333 MVA (fault close to breaker). Find the inductive reactance of the reactor required. System data are:

Generator G_1:	25 MVA, 15% reactance
Generator G_2:	50 MVA, 20% reactance
Transformer T_1:	100 MVA; 8% reactance
Transformer T_2:	40 MVA; 10% reactance.

Assume that all reactances are given on appropriate voltage bases. Choose a base of 100 MVA.

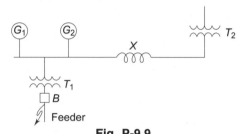

Fig. P-9.9

9.10 For the three-phase power network shown in Fig. P-9.10, the ratings of the various components are:

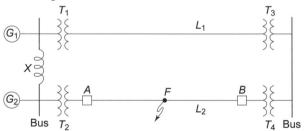

Fig. P-9.10

Generators G_1: 100 MVA, 0.30 pu reactance

G_2: 60 MVA, 0.18 pu reactance

Transformers (each): 50 MVA, 0.10 pu reactance

Inductive reactor X: 0.20 pu on a base of 100 MVA

Lines (each): 80 ohms (reactive); neglect resistance.

With the network initially unloaded and a line voltage of 110 kV, a symmetrical short circuit occurs at mid point F of line L_2.

Calculate the short circuit MVA to be interrupted by the circuit breakers A and B at the ends of the line. What would these values be, if the reactor X were eliminated? Comment.

9.11 A synchronous generator feeds bus 1 of a system. A power network feeds bus 2 of the system. Buses 1 and 2 are connected through a transformer and a transmission line. Per unit reactances of the various components are:

Generator (connected to bus bar 1)   0.25

Transformer   0.12

Transmission line   0.28

The power network can be represented by a generator with a reactance (unknown) in series.

With the generator on no load and with 1.0 pu voltage at each bus under operating condition, a three-phase short circuit occurring on bus 1 causes a current of 5.0 pu to flow into the fault. Determine the equivalent reactance of the power network.

9.12 Consider the 3-bus system of Fig. P-9.12. The generators are 100 MVA, with transient reactance 10% each. Both the transformers are 100 MVA with a leakage reactance of 5%. The reactance of each of the lines to a base of 100 MVA, 110 KV is 10%. Obtain the short circuit solution for a three-phase solid short circuit on bus 3.

Assume prefault voltages to be 1 pu and prefault currents to be zero.

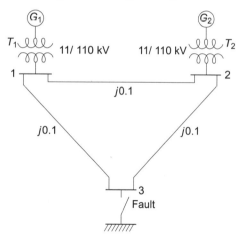

Fig. P-9.12

9.13 In the system configuration of Fig. P-9.12, the system impedance data are given below:

Transient reactance of each generator = 0.15 pu

Leakage reactance of each transformer = 0.05 pu

$z_{12} = j0.1$, $z_{13} = j0.12$, $z_{23} = j0.08$ pu

For a solid 3-phase fault on bus 3, find all bus voltages and sc currents in each component.

9.14 For the fault (solid) location shown in Fig. P-9.14, find the sc currents in lines 1.2 and 1.3. Prefault system is on no-load with 1 pu voltage and prefault currents are zero. Use Z_{BUS} method and compute its elements by the current injection technique.

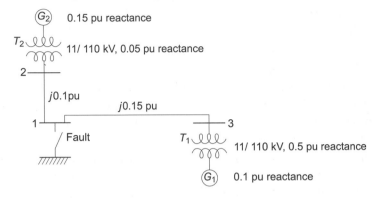

Fig. P-9.14

REFERENCES

Books

1. Brown, H.E., *Solution of Large Network by Matrix Methods*, Wiley, New York, 1975.
2. Neuenswander, J.R., *Modern Power Systems*, International Textbook Company, New York, 1971.
3. Stagg, G.W. and A.H. El-Abiad, *Computer Methods in Power Systems Analysis*, McGraw-Hill Book Co., New York, 1968.
4. Anderson, P.M., *Analysis of Faulted Power Systems*, Iowa State Press, Ames, Iowa, 1973.
5. Clarke, E., *Circuit Analysis of Alternating Current Power Systems*, Vol. 1, Wiley, New York, 1943.
6. Stevenson, W.D. Jr., *Elements of Power Systems Analysis*, 4th edn, McGraw-Hill, New York, 1982.

Paper

7. Brown, H.E. et al. "Digital Calculation of Three-Phase Short Circuits by Matrix Methods", *AIEE Trans.*, 1960, 79 : 1277.

10

Symmetrical Components

10.1 INTRODUCTION

In our work so far, we have considered both normal and abnormal (short circuit) operations of power system under completely balanced (symmetrical) conditions. Under such operation the system impedances in each phase are identical and the three-phase voltages and currents throughout the system are completely balanced, i.e. they have equal magnitudes in each phase and are progressively displaced in time phase by 120° (phase a leads/lags phase b by 120° and phase b leads/lags phase c by 120°). In a balanced system, analysis can proceed on a single-phase basis. The knowledge of voltage and current in one phase is sufficient to completely determine voltages and currents in the other two phases. Real and reactive powers are simply three times the corresponding per phase values.

Unbalanced system operation can result in an otherwise balanced system due to unsymmetrical fault, e.g. line-to-ground fault or line-to-line fault. These faults are, in fact, of more common occurrence* than the symmetrical (three phase) fault. System operation may also become unbalanced when loads are unbalanced as in the presence of large single-phase loads. Analysis under unbalanced conditions has to be carried out on a three-phase basis. Alternatively, a more convenient method of analyzing unbalanced operation is through symmetrical components where the three-phase voltages (and currents) which may be unbalanced are transformed into three sets of balanced voltages (and

---

* Typical relative frequencies of occurrence of different kinds of faults in a power system (in order of decreasing severity) are:

Three-phase (3L) faults	5%
Double line-to-ground (LLG) faults	10%
Double line (LL) faults	15%
Single line-to-ground (LG) faults	70%

currents) called symmetrical components. Fortunately, in such a transformation the impedances presented by various power system elements (synchronous generators, transformers, lines) to symmetrical components are decoupled from each other resulting in independent system networks for each component (balanced set). This is the basic reason for the simplicity of the symmetrical component method of analysis.

10.2 SYMMETRICAL COMPONENT TRANSFORMATION

A set of three balanced voltages (phasors) V_a, V_b, V_c is characterized by equal magnitudes and interphase differences of 120°. The set is said to have a phase sequence *abc* (*positive sequence*) if V_b lags V_a by 120° and V_c lags V_b by 120°. The three phasors can then be expressed in terms of the reference phasor V_a as

$$V_a = V_a, \ V_b = \alpha^2 V_a, \ V_c = \alpha V_a$$

where the complex number operator α is defined as

$$\alpha = e^{j120°}$$

It has the following properties

$$\left.\begin{array}{l} \alpha^2 = e^{j240°} = e^{-j120°} = \alpha^* \\ \left(\alpha^2\right)^* = \alpha \\ \alpha^3 = 1 \\ 1 + \alpha + \alpha^2 = 0 \end{array}\right] \tag{10.1}$$

If the phase sequence is *acb* (*negative sequence*), then

$$V_a = V_a, \ V_b = \alpha V_a, \ V_c = \alpha^2 V_a$$

Thus a set of balanced phasors is fully characterized by its reference phasor (say V_a) and its phase sequence (positive or negative).

Suffix 1 is commonly used to indicate positive sequence. A set of (balanced) positive sequence phasors is written as

$$V_{a1}, \ V_{b1} = \alpha^2 V_{a1}, \ V_{c1} = \alpha V_{a1} \tag{10.2}$$

Similarly, suffix 2 is used to indicate negative sequence. A set of (balanced) negative sequence phasors is written as

$$V_{a2}, \ V_{b2} = \alpha V_{a2}, \ V_{c2} = \alpha^2 V_{a2} \tag{10.3}$$

A set of three voltages (phasors) equal in magnitude and having the same phase is said to have zero sequence. Thus a set of *zero sequence* phasors is written as

$$V_{a0}, \ V_{b0} = V_{a0}, \ V_{c0} = V_{a0} \tag{10.4}$$

Consider now a set of three voltages (phasors) V_a, V_b, V_c which in general may be unbalanced. According to Fortesque's theorem* the three phasors can be

---

* The theorem is a general one and applies to the case of *n* phasors [6].

expressed as the sum of positive, negative and zero sequence phasors defined above. Thus

$$V_a = V_{a1} + V_{a2} + V_{a0} \tag{10.5}$$

$$V_b = V_{b1} + V_{b2} + V_{b0} \tag{10.6}$$

$$V_c = V_{c1} + V_{c2} + V_{c0} \tag{10.7}$$

The three phasor sequences (positive, negative and zero) are called the *symmetrical components* of the original phasor set V_a, V_b, V_c. The addition of symmetrical components as per Eqs. (10.5) to (10.7) to generate V_a, V_b, V_c is indicated by the phasor diagram of Fig. 10.1.

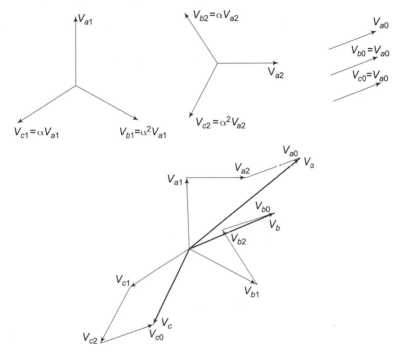

Fig. 10.1 Graphical addition of the symmetrical components to obtain the set of phasors V_a, V_b, V_c (unbalanced in general)

Let us now express Eqs. (10.5) to (10.7) in terms of reference phasors V_{a1}, V_{a2} and V_{a0}. Thus

$$V_a = V_{a1} + V_{a2} + V_{a0} \tag{10.8}$$

$$V_b = \alpha^2 V_{a1} + \alpha V_{a2} + V_{a0} \tag{10.9}$$

$$V_c = \alpha V_{a1} + \alpha^2 V_{a2} + V_{a0} \tag{10.10}$$

These equations can be expressed in the matrix form

$$\begin{bmatrix} V_a \\ V_b \\ V_c \end{bmatrix} = \begin{bmatrix} 1 & 1 & 1 \\ \alpha^2 & \alpha & 1 \\ \alpha & \alpha^2 & 1 \end{bmatrix} \begin{bmatrix} V_{a1} \\ V_{a2} \\ V_{a0} \end{bmatrix}$$

(10.11)

or

$$\mathbf{V}_p = \mathbf{A}\mathbf{V}_s$$

(10.12)

where

$$V_p = \begin{bmatrix} V_a \\ V_b \\ V_c \end{bmatrix} = \text{vector of original phasors}$$

$$V_s = \begin{bmatrix} V_{a1} \\ V_{a2} \\ V_{a0} \end{bmatrix} = \text{vector of symmetrical components}$$

$$A = \begin{bmatrix} 1 & 1 & 1 \\ \alpha^2 & \alpha & 1 \\ \alpha & \alpha^2 & 1 \end{bmatrix}$$

(10.13)

We can write Eq. (10.12) as

$$\mathbf{V}_s = A^{-1}V_p$$

(10.14)

Computing A^{-1} and utilizing relations (10.1), we get

$$A^{-1} = \frac{1}{3} \begin{bmatrix} 1 & \alpha & \alpha^2 \\ 1 & \alpha^2 & \alpha \\ 1 & 1 & 1 \end{bmatrix}$$

(10.15)

In expanded form we can write Eq. (10.14) as

$$V_{a1} = \frac{1}{3} (V_a + \alpha V_b + \alpha^2 V_c)$$

(10.16)

$$V_{a2} = \frac{1}{3} (V_a + \alpha^2 V_b + \alpha V_c)$$

(10.17)

$$V_{a0} = \frac{1}{3} (V_a + V_b + V_c)$$

(10.18)

Equations (10.16) to (10.18) give the necessary relationships for obtaining symmetrical components of the original phasors, while Eqs. (10.5) to (10.7) give the relationships for obtaining original phasors from the symmetrical components.

The symmetrical component transformations though given above in terms of voltages hold for any set of phasors and therefore automatically apply for a set of currents. Thus

$$I_p = AI_s \tag{10.19}$$

and

$$I_s = A^{-1}I_p \tag{10.20}$$

where

$$I_p = \begin{bmatrix} I_a \\ I_b \\ I_c \end{bmatrix} ; \text{ and } I_s = \begin{bmatrix} I_{a1} \\ I_{a2} \\ I_{a0} \end{bmatrix}$$

Of course A and A^{-1} are the same as given earlier.

In expanded form the relations (10.19) and (10.20) can be expressed as follows:

(i) Construction of current phasors from their symmetrical components:

$$I_a = I_{a1} + I_{a2} + I_{a0} \tag{10.21}$$
$$I_b = \alpha^2 I_{a1} + \alpha I_{a2} + I_{a0} \tag{10.22}$$
$$I_c = \alpha I_{a1} + \alpha^2 I_{a2} + I_{a0} \tag{10.23}$$

(ii) Obtaining symmetrical components of current phasors:

$$I_{a1} = \frac{1}{3} \ (I_a + \alpha I_b + \alpha^2 I_c) \tag{10.24}$$

$$I_{a2} = \frac{1}{3} \ (I_a + \alpha^2 I_b + \alpha I_c) \tag{10.25}$$

$$I_{a0} = \frac{1}{3} \ (I_a + I_b + I_c) \tag{10.26}$$

Certain observations can now be made regarding a three-phase system with neutral return as shown in Fig. 10.2.

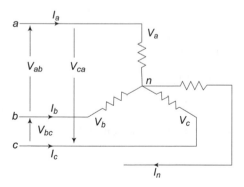

Fig. 10.2 Three-phase system with neutral return

The sum of the three line voltages will always be zero. Therefore, the *zero sequence component of line voltages is always zero*, i.e.

$$V_{ab0} = \frac{1}{3} (V_{ab} + V_{bc} + V_{ca}) = 0 \tag{10.27}$$

On the other hand, the sum of phase voltages (line to neutral) may not be zero so that their zero sequence component V_{a0} may exist.

Since the sum of the three line currents equals the current in the neutral wire, we have

$$I_{a0} = \frac{1}{3}(I_a + I_b + I_c) = \frac{1}{3}I_n \tag{10.28}$$

i.e. the current in the neutral is three times the zero sequence line current. If the neutral connection is severed,

$$I_{a0} = \frac{1}{3}I_n = 0 \tag{10.29}$$

i.e. *in the absence of a neutral connection the zero sequence line current is always zero.*

Power Invariance

We shall now show that the symmetrical component transformation is power invariant, which means that the sum of powers of the three symmetrical components equals the three-phase power.

Total complex power in a three-phase circuit is given by

$$S = V_p^T I_p^* = V_a I_a^* + V_b I_b^* + V_c I_c^* \tag{10.30}$$

or

$$S = [AV_s]^T[AI_s]^*$$
$$= V_s^T A^T A^* I_s^* \tag{10.31}$$

Now

$$A^T A^* = \begin{bmatrix} 1 & \alpha^2 & \alpha \\ 1 & \alpha & \alpha^2 \\ 1 & 1 & 1 \end{bmatrix} \begin{bmatrix} 1 & 1 & 1 \\ \alpha & \alpha^2 & 1 \\ \alpha^2 & \alpha & 1 \end{bmatrix} = 3\begin{bmatrix} 1 & 0 & 0 \\ 0 & 1 & 0 \\ 0 & 0 & 1 \end{bmatrix} = 3U \tag{10.32}$$

$$\therefore \quad S = 3V_s^T U I_s^* = 3V_s^T I_s^*$$
$$= 3V_{a1}I_{a1}^* + 3V_{a2}I_{a2}^* + 3V_{a0}I_{a0}^* \tag{10.33}$$

= sum of symmetrical component powers

Example 10.1

A delta connected balanced resistive load is connected across an unbalanced three-phase supply as shown in Fig. 10.3. With currents in lines A and B specified, find the symmetrical components of line currents. Also find the symmetrical components of delta currents. Do you notice any relationship between symmetrical components of line and delta currents ? Comment.

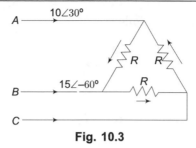

Fig. 10.3

Solution $I_A + I_B + I_C = 0$

or

$$10\angle 30° + 15\angle - 60° + I_C = 0$$

∴ $$I_C = -16.2 + j8.0 = 18 \angle 154° \text{ A}$$

From Eqs. (10.24) to (10.26)

$$I_{A1} = \frac{1}{3}(10\angle 30° + 15\angle(-60° + 120°) + 18\angle(154° + 240°))$$

$$= 10.35 + j9.3 = 14\angle 42° \text{ A} \tag{i}$$

$$I_{A2} = \frac{1}{3}(10\angle 30° + 15\angle(-60° + 240°) + 18\angle(154° + 120°))$$

$$= -1.7 - j4.3 = 4.65\angle 248° \text{ A} \tag{ii}$$

$$I_{A0} = \frac{1}{3}(I_A + I_B + I_C) = 0 \tag{iii}$$

From Eq. (10.2)

$$I_{B1} = 14\angle 282° \text{ A} \qquad\qquad I_{C1} = 14\angle 162° \text{ A}$$
$$I_{B2} = 4.65\angle 8° \text{ A} \qquad\qquad I_{C2} = 4.65\angle 128° \text{ A}$$
$$I_{B0} = 0 \text{ A} \qquad\qquad\qquad I_{C0} = 0 \text{ A}$$

Check:

$$I_A = I_{A1} + I_{A2} + I_{A0} = 8.65 + j5 = 10\angle 30°$$

Converting delta load into equivalent star, we can redraw Fig. 10.3 as in Fig. 10.4.

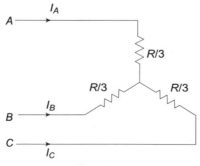

Fig. 10.4

Delta currents are obtained as follows

$$V_{AB} = \frac{1}{3} R \ (I_A - I_B)$$

Now

$$I_{AB} = V_{AB}/R = \frac{1}{3}(I_A - I_B)$$

Similarly,

$$I_{BC} = \frac{1}{3}(I_B - I_C)$$

$$I_{CA} = \frac{1}{3}(I_C - I_A)$$

Substituting the values of I_A, I_B and I_C, we have

$$I_{AB} = \frac{1}{3}(10\angle 30° - 15\angle{-60°}) = 6\angle 86° \text{ A}$$

$$I_{BC} = \frac{1}{3}(15\angle{-60°} - 18\angle 154°) = 10.5\angle{-41.5°} \text{ A}$$

$$I_{CA} = \frac{1}{3}(18\angle 154° - 10\angle 30°) = 8.3\angle 173° \text{ A}$$

The symmetrical components of delta currents are

$$I_{AB1} = \frac{1}{3}(6\angle 86° + 10.5\angle(-41.5° + 120°) + 8.3\angle(173° + 240°)) \text{ (iv)}$$

$$= 8\angle 72° \text{ A}$$

$$I_{AB2} = \frac{1}{3}(6\angle 86° + 10.5\angle(-41.5° + 240°) + 8.3\angle(173° + 120°)) \text{ (v)}$$

$$= 2.7\angle 218° \text{ A}$$

$$I_{AB0} = 0 \tag{vi}$$

I_{BC1}, I_{BC2}, I_{BC0}, I_{CA1}, I_{CA2} and I_{CA0} can be found by using Eq. (10.2).

Comparing Eqs. (i) and (iv), and (ii) and (v), the following relationship between symmetrical components of line and delta currents are immediately observed:

$$I_{AB1} = \frac{I_{A1}}{\sqrt{3}} \angle 30° \tag{vii}$$

$$I_{AB2} = \frac{I_{A2}}{\sqrt{3}} \angle{-30°} \tag{viii}$$

The reader should verify these by calculating I_{AB1} and I_{AB2} from Eqs. (vii) and (viii) and comparing the results with Eqs. (iv) and (v).

10.3 PHASE SHIFT IN STAR-DELTA TRANSFORMERS

Positive and negative sequence voltages and currents undergo a phase shift in passing through a star-delta transformer which depends upon the labelling of terminals. Before considering this phase shift, we need to discuss the standard polarity marking of a single-phase transformer as shown in Fig. 10.5. The transformer ends marked with a dot have the same polarity. Therefore, voltage $V_{HH'}$ is in phase with voltage $V_{LL'}$. Assuming that the small amount of magnetizing current can be neglected, the primary current I_1, entering the dotted end cancels the demagnetizing ampere-turns of the secondary current I_2 so that I_1 and I_2 with directions of flow as indicated in the diagram are in phase. If the direction of I_2 is reversed, I_1 and I_2 will be in phase opposition.

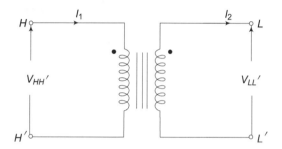

Fig. 10.5 Polarity marking of a single-phase transformer

Consider now a star/delta transformer with terminal labelling as indicated in Fig. 10.6 (a). Windings shown parallel to each other are magnetically coupled. Assume that the transformer is excited with positive sequence voltages and carries positive sequence currents. With the polarity marks shown, we can immediately draw the phasor diagram of Fig. 10.7. The following interrelationship between the voltages on the two sides of the transformer is immediately observed from the phasor diagram

$$V_{AB1} = x \ V_{ab1} \ \angle 30°, \ x = \text{phase transformation ratio} \quad (10.34)$$

As per Eq. (10.34), the positive sequence line voltages on star side lead the corresponding voltages on the delta side by 30° (The same result would apply to line-to-neutral voltages on the two sides). The same also applies for line currents.

If the delta side is connected as in Fig. 10.6(b) the phase shift reverses (the reader should draw the phasor diagram); the delta side quantities lead the star side quantities by 30°.

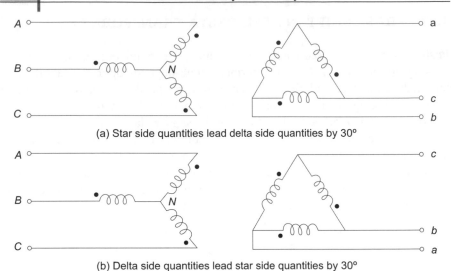

(a) Star side quantities lead delta side quantities by 30°

(b) Delta side quantities lead star side quantities by 30°

Fig. 10.6 Labelling of star/delta transformer

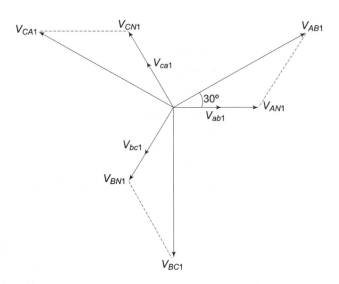

Fig. 10.7 Positive sequence voltages on a star/delta transformer

Instead, if the transformer of Fig. 10.6(a) is now excited by negative sequence voltages and currents, the voltage phasor diagram will be as in Fig. 10.8. The phase shift in comparison to the positive sequence case now reverses, i.e., the star side quantities lag the delta side quantities by 30°. The result for Fig. 10.6(b) also correspondingly reverses.

It shall from now onwards be assumed that *a star/delta transformer is so labelled that the positive sequence quantities on the HV side lead their corresponding positive sequence quantities on the LV side by 30°. The reverse is the case for negative sequence quantities wherein HV quantities lag the corresponding LV quantities by 30°.*

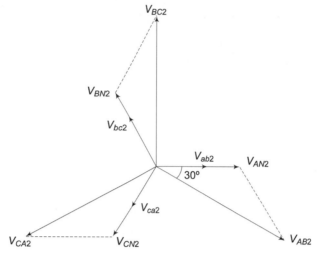

Fig. 10.8 Negative sequence voltages on a star/delta transformer

10.4 SEQUENCE IMPEDANCES OF TRANSMISSION LINES

Figure 10.9 shows the circuit of a fully transposed line carrying unbalanced currents. The return path for I_n is sufficiently away for the mutual effect to be ignored. Let

X_s = self reactance of each line

X_m = mutual reactance of any line pair

The following KVL equations can be written down from Fig. 10.9.

$$V_a - V'_a = jX_sI_a + jX_mI_b + jX_mI_c$$

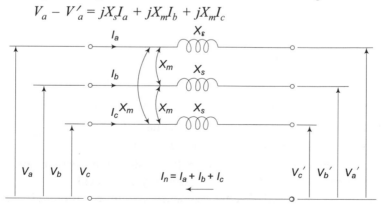

Fig. 10.9

$$V_b - V_b' = jX_m I_a + jX_s I_b + jX_m I_c \qquad (10.35)$$

$$V_c - V_c' = jX_m I_a + jX_m I_b + jX_s I_c$$

or in matrix form

$$
\begin{bmatrix} V_a \\ V_b \\ V_c \end{bmatrix} - \begin{bmatrix} V_a' \\ V_b' \\ V_c' \end{bmatrix} = j \begin{bmatrix} X_s & X_m & X_m \\ X_m & X_s & X_m \\ X_m & X_m & X_s \end{bmatrix} \begin{bmatrix} I_a \\ I_b \\ I_c \end{bmatrix} \qquad (10.36)
$$

or $\qquad V_p - V_p' = ZI_p \qquad (10.37)$

or $\qquad A\,(V_s - V_s') = ZAI_s \qquad (10.38)$

or $\qquad V_s V_s' = A^{-1} ZAI_s \qquad (10.39)$

Now

$$
A^{-1}ZA = \frac{1}{3} \begin{bmatrix} 1 & \alpha & \alpha^2 \\ 1 & \alpha^2 & \alpha \\ 1 & 1 & 1 \end{bmatrix} \begin{bmatrix} jX_s & jX_m & jX_m \\ JX_m & jX_s & jX_m \\ jX_m & jX_m & jX_s \end{bmatrix} \begin{bmatrix} 1 & 1 & 1 \\ \alpha^2 & \alpha & 1 \\ \alpha & \alpha^2 & 1 \end{bmatrix} (10.40)
$$

$$
= j \begin{bmatrix} X_s - X_m & 0 & 0 \\ 0 & X_s - X_m & 0 \\ 0 & 0 & X_s + 2X_m \end{bmatrix}
$$

Thus Eq. (10.37) can be written as

$$
\begin{bmatrix} V_1 \\ V_2 \\ V_0 \end{bmatrix} - \begin{bmatrix} V_1' \\ V_2' \\ V_0' \end{bmatrix} = j \begin{bmatrix} X_s - X_m & 0 & 0 \\ 0 & X_s - X_m & 0 \\ 0 & 0 & X_s + 2X_m \end{bmatrix} \begin{bmatrix} I_1 \\ I_2 \\ I_0 \end{bmatrix} \qquad (10.41)
$$

$$
= \begin{bmatrix} Z_1 & 0 & 0 \\ 0 & Z_2 & 0 \\ 0 & 0 & Z_0 \end{bmatrix} \begin{bmatrix} I_1 \\ I_2 \\ I_0 \end{bmatrix} \qquad (10.42)
$$

wherein

$$Z_1 = j(X_s - X_m) = \textit{positive sequence impedance} \qquad (10.43)$$

$$Z_2 = j(X_s - X_m) = \textit{negative sequence impedance} \qquad (10.44)$$

$$Z_0 = j(X_s + 2X_m) = \textit{zero sequence impedance} \qquad (10.45)$$

We conclude that a fully transposed transmission has:

(i) equal positive and negative sequence impedances.

(ii) zero sequence impedance much larger than the positive (or negative) sequence impedance (it is approximately 2.5 times).

It is further observed that the sequence circuit equations (10.42) are in *decoupled* form, i.e. there are no mutual sequence inductances. Equation (10.42) can be represented in network form as in Fig. 10.10.

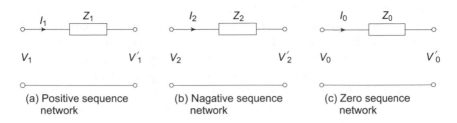

(a) Positive sequence network (b) Nagative sequence network (c) Zero sequence network

Fig. 10.10

The decoupling between sequence networks of a fully transposed transmission holds also in 3-phase synchronous machines and 3-phase transformers. This fact leads to considerable simplications in the use of symmetrical components method in unsymmetrical fault analysis.

In case of three static unbalanced impedances, coupling appears between sequence networks and the method is no more helpful than a straight forward 3-phase analysis.

10.5 SEQUENCE IMPEDANCES AND SEQUENCE NETWORK OF POWER SYSTEM

Power system elements—transmission lines, transformers and synchronous machines—have a three-phase symmetry because of which when currents of a particular sequence are passed through these elements, voltage drops of the same sequence appear, i.e. the elements possess only self impedances to sequence currents. Each element can therefore be represented by three decoupled *sequence networks* (on single-phase basis) pertaining to positive, negative and zero sequences, respectively. EMFs are involved only in a positive sequence network of synchronous machines. For finding a particular sequence impedance, the element in question is subjected to currents and voltages of that sequence only. With the element operating under these conditions, the sequence impedance can be determined analytically or through experimental test results.

With the knowledge of sequence networks of elements, complete positive, negative and zero sequence networks of any power system can be assembled. As will be explained in the next chapter, these networks are suitably interconnected to simulate different unsymmetrical faults. The sequence currents and voltages during the fault are then calculated from which actual fault currents and voltages can be found.

10.6 SEQUENCE IMPEDANCES AND NETWORKS OF SYNCHRONOUS MACHINE

Figure 10.11 depicts an unloaded synchronous machine (generator or motor) grounded through a reactor (impedance Z_n). E_a, E_b and E_c are the induced emfs of the three phases. When a fault (not shown in the figure) takes place at machine terminals, currents I_a, I_b and I_c flow in the lines. Whenever the fault involves ground, current $I_n = I_a + I_b + I_c$ flows to neutral from ground via Z_n. Unbalanced line currents can be resolved into their symmetrical components I_{a1}, I_{a2} and I_{a0}. Before we can proceed with fault analysis (Chapter 11), we must know the equivalent circuits presented by the machine to the flow of positive, negative and zero sequence currents, respectively. Because of winding symmetry currents of a particular sequence produce voltage drops of that sequence only. Therefore, there is a no coupling between the equivalent circuits of various sequences*.

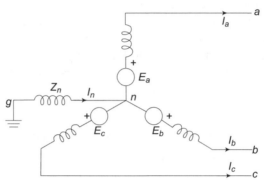

Fig. 10.11   Three-phase synchronous generator with grounded neutral

Positive Sequence Impedance and Network

Since a synchronous machine is designed with symmetrical windings, it induces emfs of positive sequence only, i.e. no negative or zero sequence voltages are induced in it. When the machine carries positive sequence currents only, this mode of operation is the balanced mode discussed at length in Chapter 9. The armature reaction field caused by positive sequence currents rotates at synchronous speed in the same direction as the rotor, i.e., it is stationary with respect to field excitation. The machine equivalently offers a direct axis reactance whose value reduces from subtransient reactance (X''_d) to transient reactance (X'_d) and finally to steady state (synchronous) reactance (X_d), as the short circuit transient progresses in time. If armature resistance is assumed negligible, the positive sequence impedance of the machine is

$$Z_1 = jX''_d \text{(if 1 cycle transient is of interest)} \tag{10.46}$$

$$= jX'_d \text{ (if 3-4 cycle transient is of interest)} \tag{10.47}$$

$$= jX_d \text{ (if steady state value is of interest)} \tag{10.48}$$

If the machine short circuit takes place from unloaded conditions, the terminal voltage constitutes the positive sequence voltage; on the other hand, if

---

*This can be shown to be so by synchronous machine theory [5].

the short circuit occurs from loaded conditions, the voltage behind appropriate reactance (subtransient, transient or synchronous) constitutes the positive sequence voltage.

Figure 10.12a shows the three-phase positive sequence network model of a synchronous machine. Z_n does not appear in the model as $I_n = 0$ for positive sequence currents. Since it is a balanced network it can be represented by the single-phase network model of Fig. 10.12b for purposes of analysis. The reference bus for a positive sequence network is at neutral potential. Further, since no current flows from ground to neutral, the neutral is at ground potential.

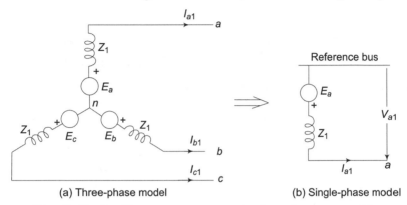

(a) Three-phase model   (b) Single-phase model

Fig. 10.12   Positive sequence network of synchronous machine

With reference to Fig. 10.12b, the positive sequence voltage of terminal a with respect to the reference bus is given by

$$V_{a1} = E_a - Z_1 I_{a1} \tag{10.49}$$

Negative Sequence Impedance and Network

It has already been said that a synchronous machine has zero negative sequence induced voltages. With the flow of negative sequence currents in the stator a rotating field is created which rotates in the opposite direction to that of the positive sequence field and, therefore, at double synchronous speed with respect to rotor. Currents at double the stator frequency are therefore induced in rotor field and damper winding. In sweeping over the rotor surface, the negative sequence mmf is alternately presented with reluctances of direct and quadrature axes. The negative sequence impedance presented by the machine with consideration given to the damper windings, is often defined as

$$Z_2 = j \frac{X_q'' + X_d''}{2}; \ |Z_2| < |Z_1| \tag{10.50}$$

Negative sequence network models of a synchronous machine, on a three-phase and single-phase basis are shown in Figs. 10.13a and b, respectively. The reference bus is of course at neutral potential which is the same as ground potential.

From Fig. 10.13b the negative sequence voltage of terminal a with respect to reference bus is

$$V_{a2} = - Z_2 I_{a2} \qquad\qquad (10.51)$$

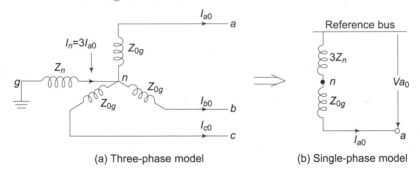

(a) Three-phase model   (b) Single-phase model

Fig. 10.13   Negative sequence network of a synchronous machine

Zero Sequence Impedance and Network

We state once again that no zero sequence voltages are induced in a synchronous machine. The flow of zero sequence currents creates three mmfs which are in time phase but are distributed in space phase by 120°. The resultant air gap field caused by zero sequence currents is therefore zero. Hence, the rotor windings present leakage reactance only to the flow of zero sequence currents ($Z_{0g} < Z_2 < Z_1$).

(a) Three-phase model   (b) Single-phase model

10.14   Zero sequence network of a synchronous machine

Zero sequence network models on a three- and single-phase basis are shown in Figs. 10.14a and b. In Fig. 10.14a, the current flowing in the impedance Z_n between neutral and ground is $I_n = 3I_{a0}$. The zero sequence voltage of terminal a with respect to ground, the reference bus, is therefore

$$V_{a0} = - 3Z_n I_{a0} - Z_{0g} I_{a0} = - (3Z_n + Z_{0g}) I_{a0} \qquad (10.52)$$

where Z_{0g} is the zero sequence impedance per phase of the machine.

Since the single-phase zero sequence network of Fig. 10.14b carries only per phase zero sequence current, its total zero sequence impedance must be

$$Z_0 = 3Z_n + Z_{0g} \qquad (10.53)$$

in order for it to have the same voltage from a to reference bus. The reference bus here is, of course, at ground potential.

From Fig. 10.14b zero sequence voltage of point a with respect to the reference bus is

$$V_{a0} = - Z_0 I_{a0} \qquad (10.54).$$

Order of Values of Sequence Impedances of a Synchronous Generator

Typical values of sequence impedances of a turbo-generator rated 5 MVA, 6.6 kV, 3,000 rpm are:

$$Z_1 = 12\% \text{ (subtransient)}$$

$$Z_1 = 20\% \text{ (transient)}$$

$$Z_1 = 110\% \text{ (synchronous)}$$

$$Z_2 = 12\%$$

$$Z_0 = 5\%$$

For typical values of positive, negative and zero sequence reactances of a synchronous machine refer to Table 9.1.

10.7 SEQUENCE IMPEDANCES OF TRANSMISSION LINES

A fully transposed three-phase line is completely symmetrical and therefore the per phase impedance offered by it is independent of the phase sequence of a balanced set of currents. In other words, the impedances offered by it to positive and negative sequence currents are identical. The expression for its per phase inductive reactance accounting for both self and mutual linkages has been derived in Chapter 2.

When only zero sequence currents flow in a transmission line, the currents in each phase are identical in both magnitude and phase angle. Part of these currents return via the ground, while the rest return through the overhead ground wires. The ground wires being grounded at several towers, the return currents in the ground wires are not necessarily uniform along the entire length. The flow of zero sequence currents through the transmission lines, ground wires and ground creates a magnetic field pattern which is very different from that caused by the flow of positive or negative sequence currents where the currents have a phase difference of 120° and the return current is zero. The zero sequence impedance of a transmission line also accounts for the ground impedance ($Z_0 = Z_{l0} + 3Z_{g0}$). Since the ground impedance heavily depends on soil conditions, it is essential to make some simplifying assumptions to obtain analytical results. The zero sequence impedance of transmission lines usually ranges from 2 to 3.5 times the positive sequence impedance*. This ratio is on the higher side for double circuit lines without ground wires.

10.8 SEQUENCE IMPEDANCES AND NETWORKS OF TRANSFORMERS

It is well known that almost all present day installations have three-phase transformers since they entail lower initial cost, have smaller space requirements and higher efficiency.

The positive sequence series impedance of a transformer equals its leakage impedance. Since a transformer is a static device, the leakage impedance does not change with alteration of phase sequence of balanced applied voltages. The transformer negative sequence impedance is also therefore equal to its leakage reactance. Thus, for a transformer

$$Z_1 = Z_2 = Z_{\text{leakage}} \tag{10.55}$$

Assuming such transformer connections that zero sequence currents can flow on both sides, a transformer offers a zero sequence impedance which may differ slightly from the corresponding positive and negative sequence values. It is, however, normal practice to assume that the series impedances of all sequences are equal regardless of the type of transformer.

The zero sequence magnetizing current is somewhat higher in a core type than in a shell type transformer. This difference does not matter as the magnetizing current of a transformer is always neglected in short circuit analysis.

Above a certain rating (1,000 kVA) the reactance and impedance of a transformer are almost equal and are therefore not distinguished.

---

*We can easily compare the forward path positive and zero sequence impedances of a transmission line with ground return path infinitely away. Assume that each line has a self inductance L and mutual inductance M between any two lines (completely symmetrical case). The voltage drop in line a caused by positive sequence currents is

$$V_{Aa1} = \omega L I_{a1} + \omega M I_{b1} + \omega M I_{c1}$$
$$= [\omega L + (\alpha^2 + \alpha)\,\omega M] I_{a1} = \omega(L-M) I_{a1}$$

∴ Positive sequence reactance = $\omega(L - M)$
The voltage drop in line a caused by zero sequence currents is

$$V_{Aa0} = \omega L I_{a0} + \omega M I_{b0} + \omega M I_{c0}$$
$$= \omega(L + 2M) I_{a0}$$

∴ Zero sequence reactance = $\omega(L + 2M)$
Obviously, zero sequence reactance is much more than positive sequence reactance. This result has already been derived in Eq. (10.45).

Zero Sequence Networks of Transformers

Before considering the zero sequence networks of various types of transformer connections, three important observations are made:

(i) When magnetizing current is neglected, transformer primary would carry current only if there is current flow on the secondary side.

(ii) Zero sequence currents can flow in the legs of a star connection only if the star point is grounded which provides the necessary return path for zero sequence currents. This fact is illustrated by Figs. 10.15a and b.

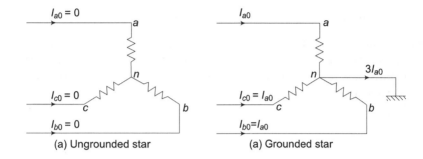

(a) Ungrounded star   (a) Grounded star

Fig. 10.15 Flow of zero sequence currents in a star connection

(iii) No zero sequence currents can flow in the lines connected to a delta connection as no return path is available for these currents. Zero sequence currents can, however, flow in the legs of a delta—such currents are caused by the presence of zero sequence voltages in the delta connection. This fact is illustrated by Fig. 10.16.

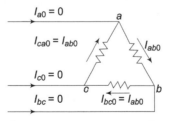

Fig. 10.16 Flow of zero sequence currents in a delta connection

Let us now consider various types of transformer connections.

Case 1: Y-Y transformer bank with any one neutral grounded.

If any one of the two neutrals of a *Y-Y* transformer is ungrounded, zero sequence currents cannot flow in the ungrounded star and consequently, these cannot flow in the grounded star. Hence, an open circuit exists in the zero sequence network between *H* and *L*, i.e. between the two parts of the system connected by the transformer as shown in Fig. 10.17.

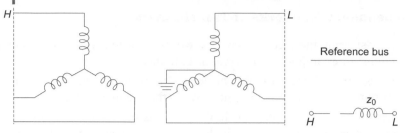

Fig. 10.17   *Y-Y* transformer bank with one neutral grounded and its zero sequence network

Case 2: Y-Y transformer bank both neutrals grounded

When both the neutrals of a *Y–Y* transformer are grounded, a path through the transformer exists for zero sequence currents in both windings via the two grounded neutrals. Hence, in the zero sequence network *H* and *L* are connected by the zero sequence impedance of the transformer as shown in Fig. 10.18.

Case 3: Y-Δ transformer bank with grounded Y neutral

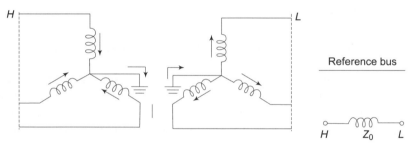

Fig. 10.18   *Y-Y* transformer bank with neutrals grounded and its zero sequence network

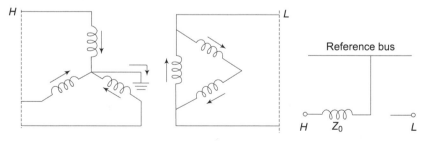

Fig. 10.19   *Y-Δ* transformer bank with grounded *Y* neutral and its zero sequence network

If the neutral of star side is grounded, zero sequence currents can flow in star because a path is available to ground and the balancing zero sequence currents can flow in delta. Of course no zero sequence currents can flow in the line on the delta side. The zero sequence network must therefore have a path from the line *H* on the star side through the zero sequence impedance of the transformer to the reference bus, while an open circuit must exist on the line L side of delta (*see* Fig. 10.19). If the star neutral is grounded through Z_n, an impedance $3Z_n$ appears in series with Z_0 in the sequence network.

Case 4: Y-Δ transformer bank with ungrounded star

This is the special case of Case 3 where the neutral is grounded through $Z_n = \infty$. Therefore no zero sequence current can flow in the transformer windings. The zero sequence network then modifies to that shown in Fig. 10.20.

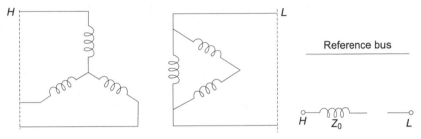

Fig. 10.20   Y-Δ transformer bank with ungrounded star and its zero sequence network

Case 5: Δ-Δ transformer bank

Since a delta circuit provides no return path, the zero sequence currents cannot flow in or out of Δ-Δ transformer; however, it can circulate in the delta windings*. Therefore, there is an open circuit between H and L and Z_0 is connected to the reference bus on both ends to account for any circulating zero sequence current in the two deltas (*see* Fig. 10.21).

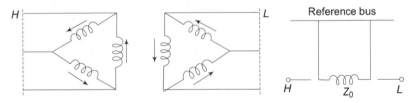

Fig. 10.21 Δ-Δ transformer bank and its zero sequence network

10.9   CONSTRUCTION OF SEQUENCE NETWORKS OF A POWER SYSTEM

In the previous sections the sequence networks for various power system elements—synchronous machines, transformers and lines—have been given. Using these, complete sequence networks of a power system can be easily constructed. To start with, the positive sequence network is constructed by

---

*Such circulating currents would exist only if zero sequence voltages are somehow induced in either delta winding.

examination of the one-line diagram of the system. It is to be noted that positive sequence voltages are present in synchronous machines (generators and motors) only. The transition from positive sequence network to negative sequence network is straightforward. Since the positive and negative sequence imped-ances are identical for static elements (lines and transformers), the only change necessary in positive sequence network to obtain negative sequence network is in respect of synchronous machines. Each machine is represented by its negative sequence impedance, the negative sequence voltage being zero.

The reference bus for positive and negative sequence networks is the system neutral. Any impedance connected between a neutral and ground is not included in these sequence networks as neither of these sequence currents can flow in such an impedance.

Zero sequence subnetworks for various parts of a system can be easily combined to form complete zero sequence network. No voltage sources are present in the zero sequence network. Any impedance included in generator or transformer neutral becomes three times its value in a zero sequence network. Special care needs to be taken of transformers in respect of zero sequence network. Zero sequence networks of all possible transformer connections have been dealt with in the preceding section.

The procedure for drawing sequence networks is illustrated through the following examples.

Example 10.2

A 25 MVA, 11 kV, three-phase generator has a subtransient reactance of 20%. The generator supplies two motors over a transmission line with transformers at both ends as shown in the one-line diagram of Fig. 10.22. The motors have rated inputs of 15 and 7.5 MVA, both 10 kV with 25% subtransient reactance. The three-phase transformers are both rated 30 MVA, 10.8/121 kV, connection Δ–Y with leakage reactance of 10% each. The series reactance of the line is 100 ohms. Draw the positive and negative sequence networks of the system with reactances marked in per unit.

Fig. 10.22

Assume that the negative sequence reactance of each machine is equal to its subtransient reactance. Omit resistances. Select generator rating as base in the generator circuit.

Solution A base of 25 MVA, 11 kV in the generator circuit requires a 25 MVA base in all other circuits and the following voltage bases.

$$\text{Transmission line voltage base} = 11 \times \frac{121}{10.8} = 123.2 \text{ kV}$$

$$\text{Motor voltage base} = 123.2 \times \frac{10.8}{121} = 11 \text{ kV}$$

The reactances of transformers, line and motors are converted to pu values on appropriate bases as follows:

$$\text{Transformer reactance} = 0.1 \times \frac{25}{30} \times \left(\frac{10.8}{11}\right)^2 = 0.0805 \text{ pu}$$

$$\text{Line reactance} = \frac{100 \times 25}{(123.2)^2} = 0.164 \text{ pu}$$

$$\text{Reactance of motor 1} = 0.25 \times \frac{25}{15} \times \left(\frac{10}{11}\right)^2 = 0.345 \text{ pu}$$

$$\text{Reactance of motor 2} = 0.25 \times \frac{25}{7.5} \times \left(\frac{10}{11}\right)^2 = 0.69 \text{ pu}$$

The required positive sequence network is presented in Fig. 10.23.

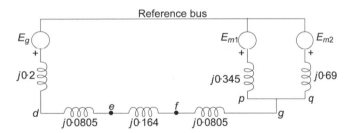

Fig. 10.23 Positive sequence network for Example 10.3

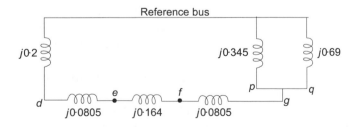

Fig. 10.24 Negative sequence network for Example 10.3

Since all the negative sequence reactances of the system are equal to the positive sequence reactances, the negative sequence network is identical to the positive sequence network but for the omission of voltage sources. The negative sequence network is drawn in Fig. 10.24.

Example 10.3

For the power system whose one-line diagram is shown in Fig. 10.25, sketch the zero sequence network.

Fig. 10.25

Solution   The zero sequence network is drawn in Fig. 10.26.

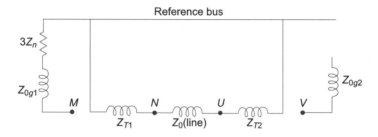

Fig. 10.26   Zero sequence network of the system presented in Fig. 10.25

Example 10.4

Draw the zero sequence network for the system described in Example 10.3. Assume zero sequence reactances for the generator and motors of 0.06 per unit. Current limiting reactors of 2.5 ohms each are connected in the neutral of the generator and motor No. 2. The zero sequence reactance of the transmission line is 300 ohms.

Solution   The zero sequence reactance of the transformer is equal to its positive sequence reactance. Hence

     Transformer zero sequence reactance = 0.0805 pu

     Generator zero sequence reactances = 0.06 pu

$$\text{Zero sequence reactance of motor 1} = 0.06 \times \frac{25}{15} \times \left(\frac{10}{11}\right)^2$$

$$= 0.082 \text{ pu}$$

$$\text{Zero sequence reactance of motor 2} = 0.06 \times \frac{25}{7.5} \times \left(\frac{10}{11}\right)^2$$

$$= 0.164 \text{ pu}$$

$$\text{Reactance of current limiting reactors} = \frac{2.5 \times 25}{(11)^2} = 0.516 \text{ pu}$$

Reactance of current limiting reactor included in zero sequence network
$$= 3 \times 0.516 = 1.548 \text{ pu}$$

$$\text{Zero sequence reactance of transmission line} = \frac{300 \times 25}{(123.2)^2}$$

$$= 0.494 \text{ pu}$$

The zero sequence network is shown in Fig. 10.27.

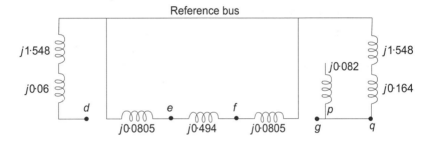

Fig. 10.27   Zero sequence network of Example 10.5

PROBLEMS

10.1 Compute the following in polar form

(i) $\alpha^2 - 1$ (ii) $1 - \alpha - \alpha^2$ (iii) $3\alpha^2 + 4\alpha + 2$ (iv) $j\alpha$

10.2 Three identical resistors are star connected and rated 2,500 V, 750 kVA. This three-phase unit of resistors is connected to the Y side of a Δ-Y transformer. The following are the voltages at the resistor load

$$|V_{ab}| = 2,000 \text{ V}; \ |V_{bc}| = 2,900 \text{ V}; \ |V_{ca}| = 2,500 \text{ V}$$

Choose base as 2,500 V, 750 kVA and determine the line voltages and currents in per unit on the delta side of the transformer. It may be assumed that the load neutral is not connected to the neutral of the transformer secondary.

10.3 Determine the symmetrical components of three voltages

$$V_a = 200\angle 0°, \ V_b = 200\angle 245° \text{ and } V_c = 200\angle 105° \ V$$

10.4 A single-phase resistive load of 100 kVA is connected across lines *bc* of a balanced supply of 3 kV. Compute the symmetrical components of the line currents.

10.5 A delta connected resistive load is connected across a balanced three-phase supply

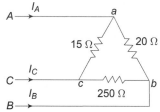

Fig. P-10.5   Phase sequence *ABC*

of 400 V as shown in Fig. P-10.5. Find the symmetrical components of line currents and delta currents.

10.6 Three resistances of 10, 15 and 20 ohms are connected in star across a three-phase supply of 200 V per phase as shown in Fig. P-10.6. The supply neutral is earthed while the load neutral is isolated. Find the currents in each load branch and the voltage of load neutral above earth. Use the method of symmetrical components.

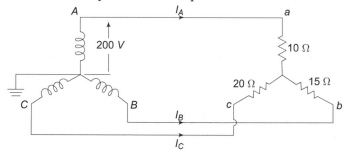

Fig. P-10.6

10.7 The voltages at the terminals of a balanced load consisting of three 20 ohm *Y*-connected resistors are 200∠0°, 100 ∠255.5° and 200 ∠151° V. Find the line currents from the symmetrical components of the line voltages if the neutral of the load is isolated. What relation exists between the symmetrical components of the line and phase voltages? Find the power expanded in three 20 ohm resistors from the symmetrical components of currents and voltages.

10.8. Draw the positive, negative and zero sequence impedance networks for the power system of Fig. P-10.8.

Choose a base of 50 MVA, 220 kV in the 50 Ω transmission lines, and mark all reactances in pu. The ratings of the generators and transformers are:

Generator 1:   25 MVA, 11 kV, $X'' = 20\%$

Generator 2:   25 MVA, 11 kV, $X'' = 20\%$

Three-phase transformer (each): 20 MVA, 11 Y/220 Y kV, $X = 15\%$

The negative sequence reactance of each synchronous machine is equal to its subtransient reactance. The zero sequence reactance of each machine is 8%. Assume that the zero sequence reactances of lines are 250% of their positive sequence reactances.

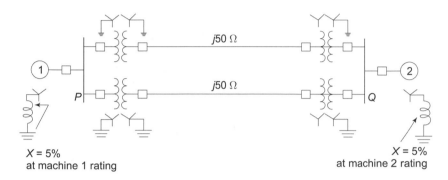

Fig. P-10.8

10.9 For the power system of Fig. P-10.9 draw the positive, negative and zero sequence networks. The generators and transformers are rated as follows:

Generator 1: 25 MVA, 11 kV, $X'' = 0.2$, $X_2 = 0.15$, $X_0 = 0.03$ pu

Generator 2: 15 MVA, 11 kV, $X'' = 0.2$, $X_2 = 0.15$, $X_0 = 0.05$ pu

Synchronous Motor 3: 25 MVA, 11 kV, $X'' = 0.2$, $X_2 = 0.2$, $X_0 = 0.1$ pu

Transformer 1: 25 MVA, 11 Δ/120 Y kV, $X = 10\%$

　　　　　　 2: 12.5 MVA, 11 Δ/120 Y kV, $X = 10\%$

　　　　　　 3: 10 MVA, 120 Y/11 Y kV, $X = 10\%$

Choose a base of 50 MVA, 11 kV in the circuit of generator 1.

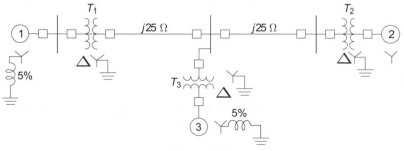

Fig. P-10.9

Note: Zero sequence reactance of each line is 250% of its positive sequence reactance.

10.10 Consider the circuit shown in Fig. P-10.10. Suppose

$$V_{an} = 100 \angle 0 \qquad\qquad X_s = 12 \ \Omega$$
$$V_{bn} = 60 \angle 60° \qquad\qquad X_{ab} = X_{bc} = X_{ca} = 5 \ \Omega$$
$$V_{cn} = 60 \angle 120°$$

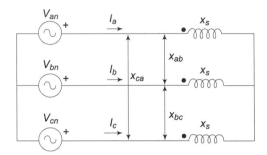

Fig. P-10.10

(a) Calculate I_a, I_b, and I_c without using symmetrical component.
(b) Calculate I_a, I_b, and I_c using symmetrical component.

REFERENCES

Books

1. Wagner, C.F. and R.D. Evans, *Symmetrical Components*, McGraw-Hill, New York 1933.
2. Clarke, E., *Circuit Analysis of Alternating Current Power Systems*, Vol. 1, Wiley, New York, 1943.
3. Austin Stigant, S., *Master Equations and Tables for Symmetrical Component Fault Studies*, Macdonald, London, 1964.
4. Stevenson, W.D., *Elements of Power System Analysis,* 4th edn, McGraw-Hill, New York, 1982.
5. Nagrath, I.J. and D.P. Kothari, *Electric Machines*, 2nd edn., Tata McGraw-Hill, New Delhi, 1997.

Paper

6. Fortescue, C.L., "Method of Symmetrical Coordinates Applies to the Solution of Polyphase Networks', *AIEE*, 1918, 37: 1027.

11

Unsymmetrical Fault Analysis

11.1 INTRODUCTION

Chapter 9 was devoted to the treatment of symmetrical (three-phase) faults in a power system. Since the system remains balanced during such faults, analysis could conveniently proceed on a single-phase basis. In this chapter, we shall deal with unsymmetrical faults. Various types of unsymmetrical faults that occur in power systems are:

Shunt Type Faults

  (i) Single line-to-ground (LG) fault
 (ii) Line-to-line (LL) fault
(iii) Double line-to-ground (LLG) fault

Series Type Faults

  (i) Open conductor (one or two conductors open) fault.

It was stated in Chapter 9, that a three-phase (3L) fault being the most severe must be used to calculate the rupturing capacity of circuit breakers, even though this type of fault has a low frequency of occurrence, when compared to the unsymmetrical faults listed above. There are, however, situations when an LG fault can cause greater fault current than a three-phase fault (this may be so when the fault location is close to large generating units). Apart from this, unsymmetrical fault analysis is important for relay setting, single-phase switching and system stability studies (Chapter 12).

The probability of two or more simultaneous faults (cross-country faults) on a power system is remote and is therefore ignored in system design for abnormal conditions.

The method of symmetrical components presented in Chapter 10, is a powerful tool for study of unsymmetrical faults and will be fully exploited in this chapter.

11.2 SYMMETRICAL COMPONENT ANALYSIS OF UNSYMMETRICAL FAULTS

Consider a general power network shown in Fig. 11.1. It is assumed that a shund type fault occurs at point F in the system, as a result of which currents I_a, I_b, I_c flow out of the system, and V_a, V_b, V_c are voltages of lines a, b, c with respect to ground.

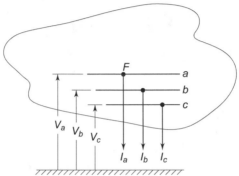

Fig. 11.1 A general power network

Let us also assume that the system is operating at no load before the occurrence of a fault. Therefore, the positive sequence voltages of all synchronous machines will be identical and will equal the prefault voltage at F. Let this voltage be labelled as E_a.

As seen from F, the power system will present positive, negative and zero sequence networks, which are schematically represented by Figs. 11.2a, b and c. The reference bus is indicated by a thick line and the point F is identified on each sequence network. Sequence voltages at F and sequence currents flowing out of the networks at F are also shown on the sequence networks. Figures 11.3a, b, and c respectively, give the Thevenin equivalents of the three sequence networks.

Recognizing that voltage E_a is present only in the positive sequence network and that there is no coupling between sequence networks, the sequence voltages at F can be expressed in terms of sequence currents and Thevenin sequence impedances as

$$
\begin{bmatrix} V_{a1} \\ V_{a2} \\ V_{a0} \end{bmatrix} = \begin{bmatrix} E_a \\ 0 \\ 0 \end{bmatrix} - \begin{bmatrix} Z_1 & 0 & 0 \\ 0 & Z_2 & 0 \\ 0 & 0 & Z_0 \end{bmatrix} \begin{bmatrix} I_{a1} \\ I_{a2} \\ I_{a0} \end{bmatrix}
\tag{11.1}
$$

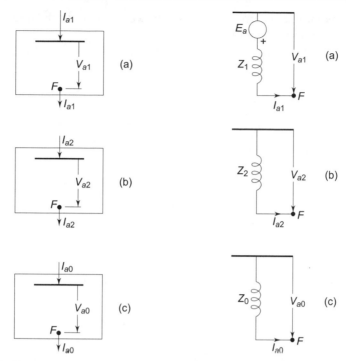

Fig. 11.2 Sequence networks as seen from the fault point F

Fig. 11.3 Thevenin equivalents of the sequence networks as seen from the fault point F

Depending upon the type of fault, the sequence currents and voltages are constrained, leading to a particular connection of sequence networks. The sequence currents and voltages and fault currents and voltages can then be easily computed. We shall now consider the various types of faults enumerated earlier.

11.3   SINGLE LINE-TO-GROUND (LG) FAULT

Figure 11.4 shows a line-to-ground fault at F in a power system through a fault impedance Z^f. The phases are so labelled that the fault occurs on phase a.

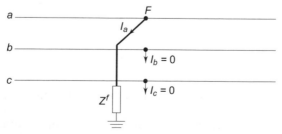

Fig. 11.4 Single line-to-ground (LG) fault at F

At the fault point F, the currents out of the power system and the line to ground voltages are constrained as follows:

$$I_b = 0 \tag{11.2}$$

$$I_c = 0 \tag{11.3}$$

$$V_a = Z^f I_a \tag{11.4}$$

The symmetrical components of the fault currents are

$$\begin{bmatrix} I_{a1} \\ I_{a2} \\ I_{a0} \end{bmatrix} = \frac{1}{3} \begin{bmatrix} 1 & \alpha & \alpha^2 \\ 1 & \alpha^2 & \alpha \\ 1 & 1 & 1 \end{bmatrix} \begin{bmatrix} I_a \\ 0 \\ 0 \end{bmatrix}$$

from which it is easy to see that

$$I_{a1} = I_{a2} = I_{a0} = \frac{1}{3} I_a \tag{11.5}$$

Expressing Eq. (11.4) in terms of symmetrical components, we have

$$V_{a1} + V_{a2} + V_{a0} = Z^f I_a = 3Z^f I_{a1} \tag{11.6}$$

As per Eqs. (11.5) and (11.6) all sequence currents are equal and the sum of sequence voltages equals $3Z^f I_{a1}$. Therefore, these equations suggest a series connection of sequence networks through an impedance $3Z^f$ as shown in Figs. 11.5a and b.

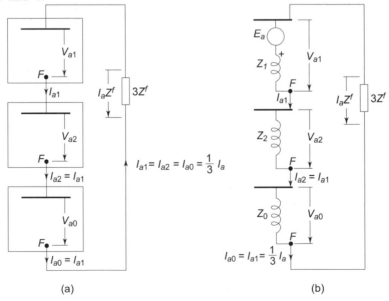

(a)   (b)

Fig. 11.5 Connection of sequence network for a single line-to-ground (LG) fault

In terms of the Thevenin equivalent of sequence networks, we can write from Fig. 11.5b.

$$I_{a1} = \frac{E_a}{(Z_1 + Z_2 + Z_0) + 3Z^f} \tag{11.7}$$

Fault current I_a is then given by

$$I_a = 3I_{a1} = \frac{3E_a}{(Z_1 + Z_2 + Z_0) + 3Z^f} \tag{11.8}$$

The above results can also be obtained directly from Eqs. (11.5) and (11.6) by using V_{a1}, V_{a2} and V_{a0} from Eq. (11.1). Thus

$$(E_a - I_{a1}Z_1) + (-I_{a2}Z_2) + (-I_{a0}Z_0) = 3Z^f I_{a1}$$

or

$$[(Z_1 + Z_2 + Z_0) + 3Z^f]I_{a1} = E_a$$

or

$$I_{a1} = \frac{E_a}{(Z_1 + Z_2 + Z_0) + 3Z^f}$$

The voltage of line b to ground under fault condition is

$$V_b = \alpha^2 V_{a1} + \alpha V_{a2} + V_{a0}$$

$$= \alpha^2 \left(E_a - Z_1 \frac{I_a}{3} \right) + \alpha \left(-Z_2 \frac{I_a}{3} \right) + \left(-Z_0 \frac{I_a}{3} \right)$$

Substituting for I_a from Eq. (11.8) and reorganizing, we get

$$V_b = E_a \frac{3\alpha^2 Z^f + Z_2(\alpha^2 - \alpha) + Z_0(\alpha^2 - 1)}{(Z_1 + Z_2 + Z_0) + 3Z^f} \tag{11.9}$$

The expression for V_c can be similarly obtained.

Fault Occurring Under Loaded Conditions

When a fault occurs under balanced load conditions, positive sequence currents alone flow in power system before the occurrence of the fault. Therefore, negative and zero sequence networks are the same as without load. The positive sequence network must of course carry the load current. To account for load current, the synchronous machines in the positive sequence network are replaced by subtransient, transient or synchronous reactances (depending upon the time after the occurrence of fault, when currents are to be determined) and voltages behind appropriate reactances. This change does not disturb the flow of prefault positive sequence currents (*see* Chapter 9). This positive sequence network would then be used in the sequence network connection of Fig. 11.5a for computing sequence currents under fault.

In case the positive sequence network is replaced by its Thevenin equivalent as in Fig. 11.5b, the Thevenin voltage equals the prefault voltage V_f^o at the fault point F (under loaded conditions). The Thevenin impedance is the impedance between F and the reference bus of the passive positive sequence network (with voltage generators short circuited).

This is illustrated by a two machine system in Fig. 11.6. It is seen from this figure that while the prefault currents flow in the actual positive sequence network of Fig. 11.6a, the same do not exist in its Thevenin equivalent network of Fig. 11.6b. Therefore, when the Thevenin equivalent of positive sequence network is used for calculating fault currents, the positive sequence currents within the network are those due to fault alone and we must superimpose on these the prefault currents. Of course, the positive sequence current into the fault is directly the correct answer, the prefault current into the fault being zero.

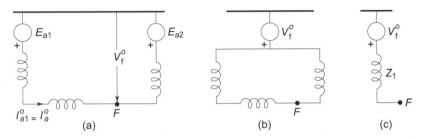

Fig. 11.6 Positive sequence network and its Thevenin equivalent before occurrence of a fault

The above remarks are valid for the positive sequence network, independent of the type of fault.

11.4 LINE-TO-LINE (LL) FAULT

Figure 11.7 shows a line-to-line fault at F in a power system on phases b and c through a fault impedance Z^f. The phases can always be relabelled, such that the fault is on phases b and c.

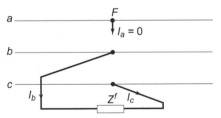

Fig. 11.7 Line-to-line (*LL*) fault through impedance Z^f

The currents and voltages at the fault can be expressed as

$$I_p = \begin{bmatrix} I_a = 0 \\ I_b \\ I_c = -I_b \end{bmatrix}; \quad V_b - V_c = I_b Z^f \qquad (11.10)$$

The symmetrical components of the fault currents are

$$\begin{bmatrix} I_{a1} \\ I_{a2} \\ I_{a0} \end{bmatrix} = \frac{1}{3} \begin{bmatrix} 1 & \alpha & \alpha^2 \\ 1 & \alpha^2 & \alpha \\ 1 & 1 & 1 \end{bmatrix} \begin{bmatrix} 0 \\ I_b \\ -I_b \end{bmatrix}$$

from which we get

$$I_{a2} = -I_{a1} \tag{11.11}$$

$$I_{a0} = 0 \tag{11.12}$$

The symmetrical components of voltages at F under fault are

$$\begin{bmatrix} V_{a1} \\ V_{a2} \\ V_{a0} \end{bmatrix} = \frac{1}{3} \begin{bmatrix} 1 & \alpha & \alpha^2 \\ 1 & \alpha^2 & \alpha \\ 1 & 1 & 1 \end{bmatrix} \begin{bmatrix} V_a \\ V_b \\ V_b - Z^f I_b \end{bmatrix} \tag{11.13}$$

Writing the first two equations, we have

$$3V_{a1} = V_a + (\alpha + \alpha^2) V_b - \alpha^2 Z^f I_b$$

$$3V_{a2} = V_a + (\alpha + \alpha^2) V_b - \alpha Z^f I_b$$

from which we get

$$3(V_{a1} - V_{a2}) = (\alpha - \alpha^2) Z^f I_b = j\sqrt{3}\, Z^f I_b \tag{11.14}$$

Now

$$I_b = (\alpha^2 - \alpha) I_{a1} \quad (\because I_{a2} = -I_{a1};\ I_{a0} = 0)$$

$$= -j\sqrt{3} I_{a1} \tag{11.15}$$

Substituting I_b from Eq. (11.15) in Eq. (11.14), we get

$$V_{a1} - V_{a2} = Z^f I_{a1} \tag{11.16}$$

Equations (11.11) and (11.16) suggest parallel connection of positive and negative sequence networks through a series impedance Z^f as shown in Figs. 11.8a and b. Since $I_{a0} = 0$ as per Eq. (11.12), the zero sequence network is unconnected.

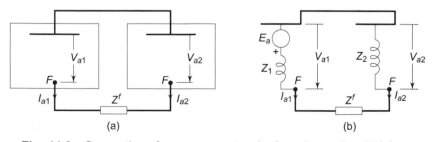

(a)   (b)

Fig. 11.8 Connection of sequence networks for a line-to-line (*LL*) fault

In terms of the Thvenin equivalents, we get from Fig. 11.8b

$$I_{a1} = \frac{E_a}{Z_1 + Z_2 + Z^f} \tag{11.17}$$

From Eq. (11.15), we get

$$I_b = - I_c = \frac{-j\sqrt{3}E_a}{Z_1 + Z_2 + Z^f} \tag{11.18}$$

Knowing I_{a1}, we can calculate V_{a1} and V_{a2} from which voltages at the fault can be found.

If the fault occurs from loaded conditions, the positive sequence network can be modified on the lines of the later portion of Sec. 11.3.

11.5 DOUBLE LINE-TO-GROUND (LLG) FAULT

Figure 11.9 shows a double line-to-ground fault at F in a power system. The fault may in general have an impedance Z^f as shown.

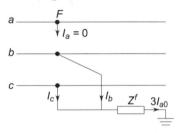

Fig. 11.9 Double line-to-ground (LLG) fault through impedance Z^f

The current and voltage (to ground) conditions at the fault are expressed as

$$\left.\begin{array}{c} I_a = 0 \\ \text{or} \quad I_{a1} + I_{a2} + I_{a0} = 0 \end{array}\right\} \tag{11.19}$$

$$V_b = V_c = Z^f (I_b + I_c) = 3Z^f I_{a0} \tag{11.20}$$

The symmetrical components of voltages are given by

$$\begin{bmatrix} V_{a1} \\ V_{a2} \\ V_{a0} \end{bmatrix} = \frac{1}{3} \begin{bmatrix} 1 & \alpha & \alpha^2 \\ 1 & \alpha^2 & \alpha \\ 1 & 1 & 1 \end{bmatrix} \begin{bmatrix} V_a \\ V_b \\ V_b \end{bmatrix} \tag{11.21}$$

from which it follows that

$$V_{a1} = V_{a2} = \frac{1}{3}[V_a + (\alpha + \alpha^2)V_b] \tag{11.22a}$$

$$V_{a0} = \frac{1}{3}(V_a + 2V_b) \tag{11.22b}$$

From Eqs. (11.22a) and (11.22b)

$$V_{a0} - V_{a1} = \frac{1}{3}(2 - \alpha - \alpha^2) \, V_b = V_b = 3Z^f I_{a0}$$

or

$$V_{a0} = V_{a1} + 3Z^f I_{a0} \qquad (11.23)$$

From Eqs. (11.19), (11.22a) and (11.23), we can draw the connection of sequence networks as shown in Figs. 11.10a and b. The reader may verify this by writing mesh and nodal equations for these figures.

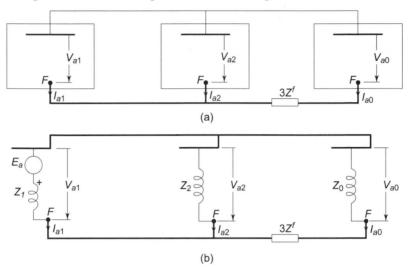

(a)

(b)

Fig. 11.10 Connection of sequence networks for a double line-to-ground (LLG) fault

In terms of the Thevenin equivalents, we can write from Fig. 11.10b

$$I_{a1} = \frac{E_a}{Z_1 + Z_2 \| (Z_0 + 3Z^f)}$$

$$= \frac{E_a}{Z_1 + Z_2(Z_0 + 3Z^f)/(Z_2 + Z_0 + 3Z^f)} \qquad (11.24)$$

The above result can be obtained analytically as follows:

Substituting for V_{a1}, V_{a2} and V_{a0} in terms of E_a in Eq. (11.1) and premultiplying both sides by Z^{-1} (inverse of sequence impedance matrix), we get

$$\begin{bmatrix} Z_1^{-1} & 0 & 0 \\ 0 & Z_2^{-1} & 0 \\ 0 & 0 & Z_0^{-1} \end{bmatrix} \begin{bmatrix} E_a - Z_1 I_{a1} \\ E_a - Z_1 I_{a1} \\ E_a - Z_1 I_{a1} + 3Z^f I_{a0} \end{bmatrix}$$

$$= \begin{bmatrix} Z_1^{-1} & 0 & 0 \\ 0 & Z_2^{-1} & 0 \\ 0 & 0 & Z_0^{-1} \end{bmatrix} \begin{bmatrix} E_a \\ 0 \\ 0 \end{bmatrix} - \begin{bmatrix} I_{a1} \\ I_{a2} \\ I_{a0} \end{bmatrix} \qquad (11.25)$$

Premultiplying both sides by row matrix [1 1 1] and using Eqs. (11.19) and (11.20), we get

$$-\frac{3Z^f}{Z_0}I_{a0}+\left(1+\frac{Z_1}{Z_0}+\frac{Z_1}{Z_2}\right)I_{a1}=\left(\frac{1}{Z_2}+\frac{1}{Z_0}\right)E_a \qquad (11.26)$$

From Eq. (11.22a), we have

$$E_a - Z_1I_{a1} = -Z_2I_{a2}$$

Substituting $\qquad I_{a2} = -(I_{a1} + I_{a0})$ [see Eq. (11.19)]

$$E_a - Z_1I_{a1} = Z_2(I_{a1} + I_{a0})$$

or

$$I_{a0} = \frac{E_a}{Z_2} - \left(\frac{Z_1+Z_2}{Z_2}\right)I_{a1}$$

Substituting this value of I_{a0} in Eq. (11.26) and simplifying, we finally get

$$I_{a1} = \frac{E_a}{Z_1 + Z_2(Z_0 + 3Z^f)/(Z_2 + Z_0 + 3Z^f)}$$

If the fault takes place from loaded conditions, the positive sequence network will be modified as discussed in Sec. 11.3.

Example 11.1

Figure 11.11 shows a synchronous generator whose neutral is grounded through a reactance X_n. The generator has balanced emfs and sequence reactances X_1, X_2 and X_0 such that $X_1 = X_2 \gg X_0$.

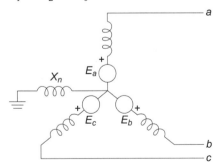

Fig. 11.11 Synchronous generator grounded through neutral reactance

(a) Draw the sequence networks of the generator as seen from the terminals.
(b) Derive expression for fault current for a solid line-to-ground fault on phase a.

(c) Show that, if the neutral is grounded solidly, the LG fault current would be more than the three-phase fault current.

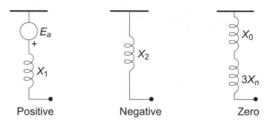

Fig. 11.12 Sequence networks of synchronous generator grounded through neutral impedance

(d) Write expression for neutral grounding reactance, such that the LG fault current is less than the three-phase fault current.

Solution (a) Figure 11.12 gives the sequence networks of the generator. As stated earlier voltage source is included in the positive sequence network only.

(b) Connection of sequence networks for a solid LG fault ($Z^f = 0$) is shown in Fig. 11.13, from which we can write the fault current as

$$|I_a|_{LG} = \frac{3|E_a|}{2X_1 + X_0 + 3X_n} \qquad (i)$$

(c) If the neutral is solidly grounded

$$|I_a|_{LG} = \frac{3|E_a|}{2X_1 + X_0} \qquad (ii)$$

For a solid three-phase fault (*see* Fig. 11.14)

$$|I_a|_{3L} = \frac{|E_a|}{X_1} = \frac{3|E_a|}{3X_1} \qquad (iii)$$

Comparing (ii) and (iii), it is easy to see that

$$|I_a|_{LG} > |I_a|_{3L}$$

An important observation is made here that, when the generator neutral is solidly grounded, LG fault is more severe than a 3L fault. It is so because, $X_0 \ll X_1 = X_2$ in generator. However, for a line $X_0 \gg X_1 = X_2$, so that for a fault on a line sufficiently away from generator, 3L fault will be more severe than an LG fault.

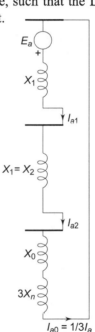

Fig. 11.13 LG fault

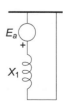

Fig. 11.14 Three-phase fault

(d) With generator neutral grounded through reactance, comparing Eqs. (i) and (iii), we have for LG fault current to be less than 3L fault

$$\frac{3|E_a|}{2X_1 + X_0 + 3X_n} < \frac{3|E_a|}{3X_1}$$

or

$$2X_1 + X_0 + 3X_n > 3X_1$$

or

$$X_n > \frac{1}{3}(X_1 - X_0) \tag{iv}$$

Example 11.2

Two 11 kV, 20 MVA, three-phase, star connected generators operate in parallel as shown in Fig. 11.15; the positive, negative and zero sequence reactances of each being, respectively, $j0.18$, $j0.15$, $j0.10$ pu. The star point of one of the generators is isolated and that of the other is earthed through a 2.0 ohm resistor. A single line-to-ground fault occurs at the terminals of one of the generators. Estimate (i) the fault current, (ii) current in grounding resistor, and (iii) the voltage across grounding resistor.

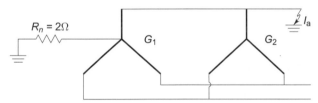

Fig. 11.15

Solution (*Note*: All values are given in per unit.)
Since the two identical generators operate in parallel,

$$X_{1eq} = \frac{j0.18}{2} = j0.09, \ X_{2eq} = \frac{j0.15}{2} = j0.075$$

Since the star point of the second generator is isolated, its zero sequence reactance does not come into picture. Therefore,

$$Z_{0eq} = j0.10 + 3R_n = j0.10 + 3 \times \frac{2 \times 20}{(11)^2} = 0.99 + j0.1$$

For an LG fault, using Eq. (11.18), we get

$$I_f \text{ (fault current for LG fault)} = I_a = 3I_{a1} = \frac{3E_a}{X_{1eq} + X_{2eq} + Z_{0eq}}$$

(a) $I_f = \dfrac{3 \times 1}{j0.09 + j0.075 + j0.1 + 0.99} = \dfrac{3}{0.99 + j0.265}$

$\qquad = 2.827 - j0.756$

(b) Current in the grounding resistor $= I_f = 2.827 - j0.756$

$$|I_f| = 2.926 \times \dfrac{20}{\sqrt{3} \times 11} = 3.07 \text{ kA}$$

(c) Voltage across grounding resistor $= \dfrac{40}{121}(2.827 - j0.756)$

$$= 0.932 - j0.249$$

$$= 0.965 \times \dfrac{11}{\sqrt{3}} = 6.13 \text{ kV}$$

Example 11.3

For the system of Example 10.3 the one-line diagram is redrawn in Fig. 11.16. On a base of 25 MVA and 11 kV in generator circuit, the positive, negative and zero sequence networks of the system have been drawn already in Figs. 10.23, 10.24 and 10.27. Before the occurrence of a solid LG at bus g, the motors are loaded to draw 15 and 7.5 MW at 10 kV, 0.8 leading power factor. If prefault current is neglected, calculate the fault current and subtransient current in all parts of the system.

What voltage behind subtransient reactances must be used in a positive sequence network if prefault current is to be accounted for?

Fig. 11.16 One-line diagram of the system of Example 11.3

Solution The sequence networks given in Figs. 10.23, 10.24 and 10.27 are connected in Fig. 11.17 to simulate a solid LG fault at bus g (*see* Fig. 11.16). If prefault currents are neglected

$$E_g'' = E_{m1}'' = E_{m2}'' = V_f^o \text{ (prefault voltage at } g)$$

$$= \dfrac{10}{11} = 0.909 \text{ pu}$$

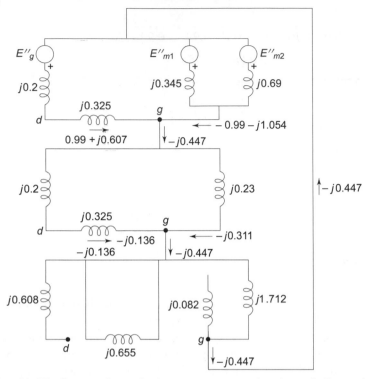

Fig. 11.17 Connection of the sequence networks of Example 11.3. Subtransient currents are shown on the diagram in pu for a solid line-to-ground fault at *g*

The positive sequence network can now be easily replaced by its Thevenin equivalent as shown in Fig. 11.18.

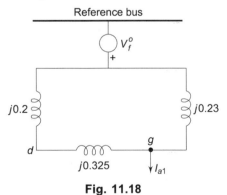

Fig. 11.18

Now

$$Z_1 = \frac{j0.525 \times j0.23}{j0.755} = j0.16 \text{ pu}$$

$$Z_2 = Z_1 = j0.16 \text{ pu}$$

From the sequence network connection

$$I_{a1} = \frac{V_f^o}{Z_1 + Z_2 + Z_0}$$

$$= \frac{0.909}{j2.032} = -j0.447 \text{ pu}$$

$$I_{a2} = I_{a0} = I_{a1} = -j0.447 \text{ pu}$$

Fault current $= 3I_{a0} = 3 \times (-j0.447) = -j1.341$ pu

The component of I_{a1} flowing towards g from the generator side is

$$-j0.447 \times \frac{j0.23}{j0.755} = -j0.136 \text{ pu}$$

and its component flowing towards g from the motors side is

$$-j0.447 \times \frac{j0.525}{j0.755} = -j0.311 \text{ pu}$$

Similarly, the component of I_{a2} from the generator side is $-j0.136$ pu and its component from the motors side is $j0.311$. All of I_{a0} flows towards g from motor 2.

Fault currents from the generator towards g are

$$\begin{bmatrix} I_a \\ I_b \\ I_c \end{bmatrix} = \begin{bmatrix} 1 & 1 & 1 \\ \alpha^2 & \alpha & 1 \\ \alpha & \alpha^2 & 1 \end{bmatrix} \begin{bmatrix} -j0.136 \\ -j0.136 \\ 0 \end{bmatrix} = \begin{bmatrix} -j0.272 \\ j0.136 \\ j0.136 \end{bmatrix} \text{pu}$$

and to g from motors are

$$\begin{bmatrix} I_a \\ I_b \\ I_c \end{bmatrix} = \begin{bmatrix} 1 & 1 & 1 \\ \alpha^2 & \alpha & 1 \\ \alpha & \alpha^2 & 1 \end{bmatrix} \begin{bmatrix} -j0.311 \\ -j0.311 \\ -j0.447 \end{bmatrix} = \begin{bmatrix} -j1.069 \\ -j0.136 \\ -j0.136 \end{bmatrix} \text{pu}$$

The positive and negative sequence components of the transmission line currents are shifted $-90°$ and $+90°$ respectively, from the corresponding components on the generator side of T_2, i.e.

Positive sequence current $= -j(-j0.136) = -0.136$ pu
Negative sequence current $= j(-j0.136) = 0.136$ pu
Zero sequence current $= 0$ ($\because$ there are no zero sequence currents on the transmission line, *see* Fig. 11.17)

$\therefore$ Line a current on the transmission line

$$= -0.136 + 0.136 + 0 = 0$$

I_b and I_c can be similarly calculated.

Let us now calculate the voltages behind subtransient reactances to be used if the load currents are accounted for. The per unit motor currents are:

Motor 1: $\dfrac{15}{25 \times 0.909 \times 0.8} \angle 36.86° = 0.825 \angle 36.86° = 0.66 + j0.495$ pu

Motor 2: $\dfrac{7.5}{25 \times 0.909 \times 0.8} \angle 36.86° = 0.4125 \angle 36.86° = 0.33 + j0.248$ pu

Total current drawn by both motors $= 0.99 + j0.743$ pu

The voltages behind subtransient reactances are calculated below:

Motor 1: $E''_{m1} = 0.909 - j0.345 \times 0.825 \angle 36.86°$

$\qquad = 1.08 - j0.228 = 1.104 \angle -11.92°$ pu

Motor 2: $E''_{m2} = 0.909 - j0.69 \times 0.4125 \angle 36.86°$

$\qquad = 1.08 - j0.228 = 1.104 \angle -11.92°$ pu

Generator: $E''_g = 0.909 + j0.525 \times 1.2375 \angle 36.86°$

$\qquad = 0.52 + j0.52 = 0.735 \angle 45°$ pu

It may be noted that with these voltages behind subtransient reactances, the Thevenin equivalent circuit will still be the same as that of Fig. 11.18. Therefore, in calculating fault currents taking into account prefault loading condition, we need not calculate E''_{m1}, E''_{m2} and E''_g. Using the Thevenin equivalent approach, we can first calculate currents caused by fault to which the load currents can then be added.

Thus, the actual value of positive sequence current from the generator towards the fault is

$\qquad 0.99 + -j0.743 - j0.136 = 0.99 + j0.607$

and the actual value of positive sequence current from the motors to the fault is

$\qquad -0.99 - j0.743 - j0.311 = -0.99 - j1.054$

In this problem, because of large zero sequence reactance, load current is comparable with (in fact, more than) the fault current. In a large practical system, however, the reverse will be the case, so that it is normal practice to neglect load current without causing an appreciable error.

Example 11.4

For Example 11.2, assume that the grounded generator is solidly grounded. Find the fault current and voltage of the healthy phase for a line-to-line fault on terminals of the generators. Assume solid fault ($Z^f = 0$).

Solution For the LL fault, using Eq. (11.17) and substituting the values of X_{1eq} and X_{2eq} from Example 11.2, we get

$$I_{a1} = \frac{E_a}{X_{1eq} + X_{2eq}} = \frac{1}{j0.09 + j0.075} = -j6.06$$

Using Eq. (11.15), we have

$$I_f \text{ (fault current)} = I_b = -j\sqrt{3}I_{a1} = (-j\sqrt{3})(-j6.06) = -10.496$$

Now

$$V_{a1} = V_{a2} = E_a - I_{a1}X_{1eq} = 1.0 - (-j6.06)(j0.09)$$

$$= 0.455$$

$$V_{a0} = -I_{a0}Z_0 = 0 \qquad\qquad (\because I_{a0} = 0)$$

Voltage of the healthy phase,

$$V_a = V_{a1} + V_{a2} + V_{a0} = 0.91$$

For Example 11.2, assume that the grounded generator is solidly grounded. Find the fault current in each phase and voltage of the healthy phase for a double line-to-ground fault on terminals of the generator. Assume solid fault ($Z^f = 0$).

Solution Using Eq. (11.24) and substituting the values of Z_{1eq}, Z_{2eq} and Z_{0eq} from Example 11.2, we get (note $Z^f = 0$, $Z_{0eq} = j0.1$)

$$I_{a1} = \frac{1 + j0}{j0.09 + \dfrac{j0.075 \times j0.10}{j0.075 + j0.10}} = -j7.53$$

$$V_{a1} = V_{a2} = V_{a0} = E_a - I_{a1}Z_{1eq} = 1 - (-j7.53)(j0.09)$$

$$= 0.323$$

$$I_{a2} = -\frac{V_{a2}}{Z_{2eq}} = -\frac{0.323}{j0.075} = j4.306$$

$$I_{a0} = -\frac{V_{a0}}{Z_{0eq}} = -\frac{0.323}{j0.10} = j3.23$$

Now

$$I_b = \alpha^2 I_{a1} + \alpha I_{a2} + I_{a0}$$

$$= (-0.5 - j0.866)(-j7.53) + (-0.5 + j0.866)(j4.306) + j3.23$$

$$= -10.248 + j4.842 = 11.334 \angle 154.74°$$

$$I_c = \alpha I_{a1} + \alpha^2 I_{a2} + I_{a0}$$

$$= (-0.5 + j0.866)(-j7.53) + (-0.5 - j0.866)(j4.306) + j3.23$$

$$= 10.248 + j4.842 = 11.334\angle25.28°$$

Voltage of the healthy phase,

$$V_a = 3V_{a1} = 3 \times 0.323 = 0.969$$

11.6 OPEN CONDUCTOR FAULTS

An open conductor fault is in series with the line. Line currents and series voltages between broken ends of the conductors are required to be determined.

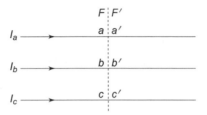

Fig. 11.19 Currents and voltages in open conductor fault

Figure 11.19 shows currents and voltages in an open conductor fault. The ends of the system on the sides of the fault are identified as F, F', while the conductor ends are identified as aa', bb' and cc'. The set of series currents and voltages at the fault are

$$I_p = \begin{bmatrix} I_a \\ I_b \\ I_c \end{bmatrix}; \ V_p = \begin{bmatrix} V_{aa'} \\ V_{bb'} \\ V_{cc'} \end{bmatrix}$$

The symmetrical components of currents and voltages are

$$I_s = \begin{bmatrix} I_{a1} \\ I_{a2} \\ I_{a0} \end{bmatrix}; \ V_s = \begin{bmatrix} V_{aa'1} \\ V_{aa'2} \\ V_{aa'0} \end{bmatrix}$$

The sequence networks can be drawn for the power system as seen from FF' and are schematically shown in Fig. 11.20. These are to be suitably connected depending on the type of fault (one or two conductors open).

Two Conductors Open

Figure 11.21 represents the fault at FF' with conductors b and c open. The currents and voltages due to this fault are expressed as

$$V_{aa'} = 0 \tag{11.27}$$

$$I_b = I_c = 0 \tag{11.28}$$

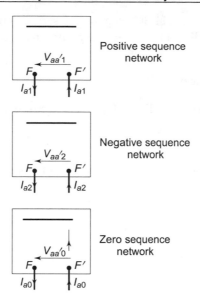

Fig. 11.20 Sequence networks for open conductor fault at *FF'*

In terms of symmetrical components, we can write

$$V_{aa'1} + V_{aa'2} + V_{aa'0} = 0 \qquad (11.29)$$

$$I_{a1} = I_{a2} = I_{a0} = \frac{1}{3} I_a \qquad (11.30)$$

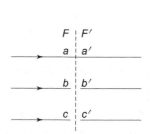

Fig 11.21 Two conductors open

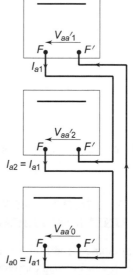

Fig. 11.22 Connection of sequence networks for two conductors open

Equations (11.29) and (11.30) suggest a series connection of sequence networks as shown in Fig. 11.22. Sequence currents and voltages can now be computed.

One Conductor Open

For one conductor open as in Fig. 11.23, the circuit conditions require

$$V_{bb'} = V_{cc'} = 0 \tag{11.31}$$
$$I_a = 0 \tag{11.32}$$

In terms of symmetrical components these conditions can be expressed as

$$V_{aa'1} = V_{aa'2} = V_{aa'0} = \frac{1}{3} V_{aa'} \tag{11.33}$$

$$I_{a1} + I_{a2} + I_{a0} = 0 \tag{11.34}$$

Equations (11.33) and (11.34) suggest a parallel connection of sequence networks as shown in Fig. 11.24.

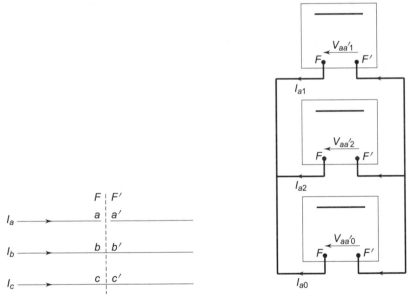

Fig. 11.23 One conductor open

Fig. 11.24 Connection of sequence networks for one conductor open

11.7 BUS IMPEDANCE MATRIX METHOD FOR ANALYSIS OF UNSYMMETRICAL SHUNT FAULTS

Bus impedance method of fault analysis, given for symmetrical faults in Chapter 9, can be easily extended to the case of unsymmetrical faults. Consider for example an LG fault on the rth bus of a n-bus system. The connection of sequence networks to simulate the fault is shown in Fig. 11.25. The positive sequence network has been replaced here by its Thevenin equivalent, i.e.

prefault voltage V_{1-r}^o of bus r in series with the passive positive sequence network (all voltage sources short circuited). Since negative and zero sequence prefault voltages are zero, both these are passive networks only.

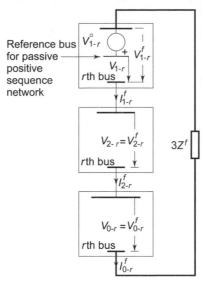

Fig. 11.25 Connection of sequence networks for LG fault on the r th bus (positive sequence network represented by its Thevenin equivalent)

It may be noted that subscript a has been dropped in sequence currents and voltages, while integer subscript is introduced for bus identification. Superscripts o and f respectively, indicate prefault and postfault values.

For the passive positive sequence network

$$\mathbf{V}_{1-BUS} = \mathbf{Z}_{1-BUS}\,\mathbf{J}_{1-BUS} \qquad (11.35)$$

where

$$\mathbf{V}_{1-BUS} = \begin{bmatrix} V_{1-1} \\ V_{1-2} \\ \vdots \\ V_{1-n} \end{bmatrix} - \text{positive sequence bus voltage vector} \quad (11.36)$$

$$\mathbf{Z}_{1-BUS} = \begin{bmatrix} Z_{1-11} & \cdots & Z_{1-1n} \\ \vdots & & \vdots \\ Z_{1-n1} & \cdots & Z_{1-nn} \end{bmatrix} = \text{positive sequence bus impedance matrix}$$

$$(11.37)$$

and

$$\mathbf{J}_{1-BUS} = \begin{bmatrix} J_{1-1} \\ J_{1-2} \\ \vdots \\ J_{1-n} \end{bmatrix} = \text{positive sequence bus current injection vector} \quad (11.38)$$

As per the sequence network connection, current $-I^f_{1-r}$ is injected only at the faulted rth bus of the positive sequence network, we have therefore

$$
\mathbf{J}_{1-\text{BUS}} =
\begin{bmatrix}
0 \\
0 \\
\vdots \\
-I^f_{1-r} \\
\vdots \\
0
\end{bmatrix}
\tag{11.39}
$$

Substituting Eq. (11.39) in Eq. (11.35), we can write the positive sequence voltage at the rth bus of the passive positive sequence network as

$$
\mathbf{V}_{1-r} = - \mathbf{Z}_{1-rr}\mathbf{I}^f_{1-r}
\tag{11.40}
$$

Thus the passive positive sequence network presents an impedance $\mathbf{Z}_{1-rr}$ to the positive sequence current $\mathbf{I}^f_{1-r}$.

For the negative sequence network

$$
\mathbf{V}_{2-\text{BUS}} = \mathbf{Z}_{2-\text{BUS}}\, \mathbf{J}_{2-\text{BUS}}
\tag{11.41}
$$

The negative sequence network is injected with current $\mathbf{I}^f_{2-r}$ at the rth bus only. Therefore,

$$
\mathbf{J}_{2-\text{BUS}} =
\begin{bmatrix}
0 \\
0 \\
\vdots \\
-I^f_{2-r} \\
\vdots \\
0
\end{bmatrix}
\tag{11.42}
$$

The negative sequence voltage at the rth bus is then given by

$$
\mathbf{V}_{2-r} = - \mathbf{Z}_{2-rr}\mathbf{I}^f_{2-r}
\tag{11.43}
$$

Thus, the negative sequence network offers an impedance $\mathbf{Z}_{2-rr}$ to the negative sequence current $\mathbf{I}^f_{2-r}$

Similarly, for the zero sequence network

$$
\mathbf{V}_{0-\text{BUS}} = \mathbf{Z}_{0-\text{BUS}}\, \mathbf{J}_{0-\text{BUS}}
\tag{11.44}
$$

$$
\mathbf{J}_{0-\text{BUS}} =
\begin{bmatrix}
0 \\
0 \\
\vdots \\
-I^f_{0-r} \\
\vdots \\
0
\end{bmatrix}
\tag{11.45}
$$

and $\qquad\qquad \mathbf{V}_{0-r} = - \mathbf{Z}_{0-rr}\mathbf{I}^f_{0-rr}$ $\qquad\qquad$ (11.46)

That is, the zero sequence network offers an impedance $\mathbf{Z}_{0-rr}$ to the zero sequence current $\mathbf{I}^f_{0-r}$.

From the sequence network connection of Fig. 11.25, we can now write

$$I^f_{1-r} = I^f_{2-r} = I^f_{0-r} = \frac{V^0_{1-r}}{Z_{1-rr} + Z_{2-rr} + Z_{0-rr} + 3Z^f} \qquad (11.47)$$

Sequence currents for other types of faults can be similarly computed using Z_{1-rr}, Z_{2-rr} and Z_{0-rr} in place of Z_1, Z_2 and Z_0 in Eqs. (11.7), (11.17) and (11.24) with $E_a = V^0_{1-r}$.

Postfault sequence voltages at any bus can now be computed by superposing on prefault bus voltage, the voltage developed owing to the injection of appropriate sequence current at bus r.

For passive positive sequence network, the voltage developed at bus i owing to the injection of $-I^f_{1-r}$ at bus r is

$$V_{1-i} = -Z_{1-ir}I^f_{1-r} \qquad (11.48)$$

Hence postfault positive sequence voltage at bus i is given by

$$V^f_{1-i} = V^0_{1-i} - Z_{1-ir}I^f_{1-r} ; \ i = 1, 2, ..., n \qquad (11.49)$$

where

$$V^0_{1-i} = \text{prefault positive sequence voltage at bus } i$$

$$Z_{1-ir} = ir\text{th component of } Z_{1-\text{BUS}}$$

Since the prefault negative sequence bus voltages are zero, the postfault negative sequence bus voltages are given by

$$V^f_{2-i} = 0 + V_{2-i}$$
$$= -Z_{2-ir}I^f_{2-r} \qquad (11.50)$$

where

$$Z_{2-ir} = ir\text{th component of } Z_{2-\text{BUS}}$$

Similarly, the postfault zero sequence bus voltages are given by

$$V^f_{0-i} = -Z_{0-ir}I^f_{0-r}; \ i = 1, 2, ..., n \qquad (11.51)$$

where

$$Z_{0-ir} = ir\text{th component of } Z_{0-\text{BUS}}$$

With postfault sequence voltages known at the buses, sequence currents in lines can be computed as:

For line uv, having sequence admittances y_{1-uv}, y_{2-uv} and y_{0-uv}

$$I^f_{1-uv} = y_{1-uv} \ (V^f_{1-u} - V^f_{1-v})$$
$$I^f_{2-uv} = y_{2-uv} \ (V^f_{2-u} - V^f_{2-v}) \qquad (11.52)$$
$$I^f_{0-uv} = y_{0-uv} \ (V^f_{0-u} - V^f_{0-v})$$

Knowing sequence voltages and currents, phase voltages and currents can be easily computed by the use of the symmetrical component transformation

$$\mathbf{V}_p = \mathbf{A}\mathbf{V}_s$$
$$\mathbf{I}_p = \mathbf{A}\mathbf{I}_s$$

It appears at first, as if this method is more laborious than computing fault currents from Thevenin impedances of the sequence networks, as it requires computation of bus impedance matrices of all the three sequence networks. It must, however, be pointed out here that once the bus impedance matrices have been assembled, fault analysis can be conveniently carried out for all the buses, which, in fact, is the aim of a fault study. Moreover, bus impedance matrices can be easily modified to account for changes in power network configuration.

Example 11.6

For Example 10.3, positive, negative and zero sequence networks have been drawn in Figs. 10.23, 10.24 and 10.27. Using the bus impedance method of fault analysis, find fault currents for a solid LG fault at (i) bus e and (ii) bus f. Also find bus voltages and line currents in case (i). Assume the prefault currents to be zero and the prefault voltages to be 1 pu.

Solution   Figure 11.26 shows the connection of the sequence networks of Figs. 10.23, 10.24 and 10.27 for a solid LG fault at bus e.

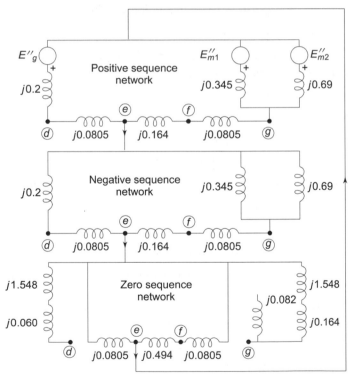

Fig. 11.26   Connection of the sequence networks of Example 11.6 for an LG fault at bus e

Refer to Fig. 11.26 to find the elements of the bus admittance matrices of the three sequence networks, as follows:

$$Y_{1-dd} = \frac{1}{j0.2} + \frac{1}{j0.0805} = -j17.422$$

$$Y_{1-fg} = Y_{1-de} = \frac{-1}{j0.0805} = j12.422$$

$$Y_{1-ff} = Y_{1-ee} = \frac{1}{j0.0805} + \frac{1}{j0.164} = -j18.519$$

$$Y_{1-ef} = \frac{-1}{j0.164} = j6.097$$

$$Y_{1-gg} = \frac{1}{j0.085} + \frac{1}{j0.345} + \frac{1}{j0.69} = -j16.769$$

$$Y_{1-\text{BUS}} = Y_{2-\text{BUS}} = j\begin{array}{c} d \\ e \\ f \\ g \end{array}\begin{array}{cccc} d & e & f & g \\ \begin{bmatrix} -17.422 & 12.422 & 0 & 0 \\ 12.422 & -18.519 & 6.097 & 0 \\ 0 & 6.097 & -18.519 & 12.422 \\ 0 & 0 & 12.422 & -16.769 \end{bmatrix} \end{array}$$

$$Y_{0-dd} = \frac{1}{j1.608} = -j0.621$$

$$Y_{0-ee} = Y_{0\ ff} = \frac{1}{j0.0805} + \frac{1}{j0.494} = -j14.446$$

$$Y_{0-gg} = \frac{1}{j1.712} = -j0.584$$

$$Y_{0-de} = 0.0$$

$$Y_{0-ef} = \frac{-1}{j0.494} = j2.024$$

$$Y_{0-fg} = 0.0$$

$$Y_{0-\text{BUS}} = j\begin{array}{c} d \\ e \\ f \\ g \end{array}\begin{array}{cccc} d & e & f & g \\ \begin{bmatrix} -0.621 & 0 & 0 & 0 \\ 0 & -14.446 & 2.024 & 0 \\ 0 & 2.024 & -14.446 & 0 \\ 0 & 0 & 0 & -0.584 \end{bmatrix} \end{array}$$

Inverting the three matrices above renders the following three bus impedance matrices

$$Z_{1\text{-BUS}} = Z_{2\text{-BUS}} = j \begin{bmatrix} 0.14706 & 0.12575 & 0.08233 & 0.06102 \\ 0.12575 & 0.17636 & 0.11547 & 0.08558 \\ 0.08233 & 0.11547 & 0.18299 & 0.13563 \\ 0.06102 & 0.08558 & 0.13563 & 0.16019 \end{bmatrix}$$

$$Z_{0\text{-BUS}} = j \begin{bmatrix} 1.61031 & 0 & 0 & 0 \\ 0 & 0.07061 & 0.00989 & 0 \\ 0 & 0.00989 & 0.07061 & 0 \\ 0 & 0 & 0 & 1.71233 \end{bmatrix}$$

The fault current with LG fault on bus e is

$$I_e^f = \frac{3 \times 1}{j0.17636 + j0.17636 + j0.07061} = -j7.086 \text{ pu} \qquad \text{(i)}$$

The fault current with LG fault on bus f is

$$I_f^f = \frac{3 \times 1}{j0.18299 + j0.18299 + j0.07061} = \frac{3}{j0.43659}$$

$$= -j6.871 \text{ pu} \qquad \text{(ii)}$$

Bus voltages and line currents in case (i) can easily be computed using Eqs. (11.49)–(11.52). Given below is a sample calculation for computing voltage at bus f and current in line ef.

From Eq. (11.49)

$$V_{1-d}^f = V_{1-d}^o - Z_{1-de} - I_{1-e}^f$$

$$= 1.0 - j0.12575\left(-j\frac{7.086}{3}\right) = 0.703 \text{ pu}$$

$$V_{1-f}^f = V_{1-f}^o - Z_{1-fe} - I_{1-e}^f$$

$$= 1.0 - j0.11547\left(-j\frac{7.086}{3}\right) = 0.728 \text{ pu}$$

$$V_{1-e}^f = V_{1-e}^o - Z_{1-ee} - I_{1-e}^f$$

$$= 1.0 - j0.17638\,(-j2.363) = 0.584 \text{ pu}$$

$$V_{1-g}^f = V_{1-g}^o - Z_{1-ge} - I_{1-e}^f$$

$$= 1.0 - j0.08558\,(-j2.363) = 0.798 \text{ pu}$$

$$V_{2-f}^f = - Z_{2-fe} I_{2-e}^f$$

$$= - j0.11547 \times (-j2.362) = -0.272 \text{ pu}$$

$$V_{0-f}^f = - Z_{0-fe} I_{0-e}^f = -j0.00989 \times (-j2.362)$$

$$= -0.023 \text{ pu}$$

$$V^f_{2-e} = - Z_{2-ee}I^f_{2-e} = - j0.17636 \times (- j2.362)$$
$$= - 0.417 \text{ pu}$$

$$V^f_{0-e} = - Z_{0-ee}I^f_{0-e} = - j0.0706 \times (- j2.362)$$
$$= - 0.167 \text{ pu}$$

$$V^f_{2-g} = - Z_{2-ge}I^f_{2-e} = - j0.08558 \times (- j2.362)$$
$$= - 0.202 \text{ pu}$$

$$V^f_{0-g} = - Z_{0-ge}I^f_{0-e} = 0$$

Using Eq. (11.52), the currents in various parts of Fig. 11.26 can be computed as follows:

$$I^f_{1-fe} = Y_{1-fe} (V^f_{1-f} - V^f_{1-fe})$$
$$= - j6.097 (0.728 - 0.584)$$
$$= - j0.88$$

$$I^f_{1-de} = Y_{1-de} (V^f_{1-d} - V^f_{1-e})$$
$$= - j12.422 (0.703 - 0.584) = - j1.482$$

$$\therefore \quad I_{a1} = I^f_{1-fe} + I^f_{1 \, dc} = - j0.88 + (- j1.482)$$
$$= - j2.362$$

which is the same as obtained earlier [see Eq. (i)] where $I^f_e = 3I_{a1}$.

$$I^f_{1-gf} = Y_{1-gf} (V^f_{1-g} - V^f_{1-f})$$
$$= j12.422 (-0.798 - 0.728) = - j0.88$$

Notice that as per Fig. 11.26, it was required to be the same as I^f_{1-fe}.

$$I_{2-fe} = Y_{2-fe} (V^f_{2-f} - V^f_{2-e})$$
$$= - j6.097 (- 0.272 + 0.417) = - j0.884$$

$$I^f_{0-fe} = \gamma_{0-fe} (V^f_{0-f} - V^f_{0-e})$$
$$= - j2.024 (- 0.023 + 0.167) = - j0.291 \text{ pu}$$

$$\therefore \quad I^f_{fe} \ (a) = I^f_{1-fe} + I^f_{2-fe} + I^f_{0-fe}$$
$$= - j0.88 + (- j0.88) + (- j0.291)$$
$$= - j2.05$$

Similarly, other currents can be computed.

Example 11.7

A single line to ground fault (on phase a) occurs on the bus 1 of the system of Fig. 11.27. Find

Fig. 11.27

(a)   Current in the fault.

(b)   SC current on the transmission line in all the three phases.

(c)   SC current in phase a of the generator.

(d)   Voltage of the healthy phases of the bus 1.

Given: Rating of each machine 1200 kVA, 600 V with $X' = X_2 = 10\%$, $X_0 = 5\%$. Each three-phase transformer is rated 1200 kVA, 600 V – Δ/3300 V–Y with leakage reactance of 5%. The reactances of the transmission line are $X_1 = X_2 = 20\%$ and $X_0 = 40\%$ on a base of 1200 kVA, 3300 V. The reactances of the neutral grounding reactors are 5% on the kVA and voltage base of the machine.

Note: Use Z_{BUS} method.

Solution   Figure 11.28 shows the passive positive sequence network of the system of Fig.11.27. This also represents the negative sequence network for the system. Bus impedance matrices are computed below:

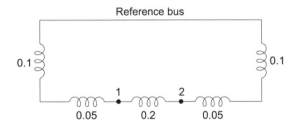

Fig. 11.28

Bus 1 to reference bus

$$Z_{1-BUS} = j[0.15]$$

Bus 2 to Bus 1

$$Z_{1-BUS} = j\begin{bmatrix} 0.15 & 0.15 \\ 0.15 & 0.35 \end{bmatrix}$$

Bus 2 to reference bus

$$Z_{1-BUS} = j\begin{bmatrix} 0.15 & 0.15 \\ 0.15 & 0.35 \end{bmatrix} - \frac{j}{0.35 + 0.15}\begin{bmatrix} 0.15 \\ 0.35 \end{bmatrix}[0.15 \quad 0.35]$$

or $\quad\quad Z_{1-\text{BUS}} = j \begin{bmatrix} 0.105 & 0.045 \\ 0.045 & 0.105 \end{bmatrix} = Z_{2-\text{BUS}}$ $\quad\quad$ (i)

Zero sequence network of the system is drawn in Fig. 11.29 and its bus impedance matrix is computed below.

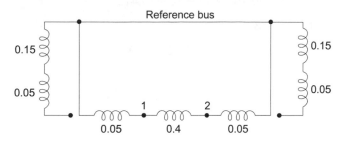

Fig. 11.29

Bus 1 to reference bus

$$Z_{0-\text{BUS}} = j\,[0.05]$$

Bus 2 to bus 1

$$Z_{0-\text{BUS}} = j \begin{bmatrix} 0.05 & 0.05 \\ 0.05 & 0.45 \end{bmatrix}$$

Bus 2 to reference bus

$$Z_{0-\text{BUS}} = j \begin{bmatrix} 0.05 & 0.05 \\ 0.05 & 0.45 \end{bmatrix} - \frac{j}{0.45+0.05} \begin{bmatrix} 0.05 \\ 0.45 \end{bmatrix} [0.05 \quad 0.45]$$

or

$$Z_{0-\text{BUS}} = j \begin{matrix} 1 \\ 2 \end{matrix} \begin{bmatrix} 0.045 & 0.005 \\ 0.005 & 0.045 \end{bmatrix}$$ $\quad\quad$ (ii)

As per Eq. (11.47)

$$I^f_{1-1} = \frac{V^0_1}{Z_{1-11} + Z_{2-11} + Z_{0-11} + 3Z^f}$$

But $V^o_1 = 1$ pu (system unloaded before fault)
Then

$$I^f_{1-1} = \frac{-j1.0}{0.105 + 0.105 + 0.045} = -j3.92 \text{ pu}$$

$$I^f_{1-1} = I^f_{2-1} = I^f_{a-1} = -j3.92 \text{ pu}$$

(a) Fault current, $I^f_1 = 3I^f_{1-1} = -j11.7$ pu

(b) $V^f_{1-1} = V^o_{1-1} = Z_{1-11}\,I^f_{1-1}$

$$= 1.0 - j0.105 \times -j3.92 = 0.588; \; V^o_{1-1} = 1 \text{ pu}$$

$$V^f_{1-2} = V^o_{1-2} - Z_{1-21}I^f_{2-1}; \quad V^o_{1-2} = 1.0 \text{ (system unloaded before fault)}$$

$$= 1.0 - j0.045 \times -j3.92 = 0.824$$

$$V^f_{2-1} = -Z_{2-11}I^f_{2-1}$$

$$= -j0.105 \times -j3.92 = 0.412$$

$$V^f_{2-2} = -Z_{2-21} I^f_{2-1}$$

$$= -j0.045 \times -j3.92 = -0.176$$

$$V^f_{0-1} = -Z_{0-11}I^f_{0-1}$$

$$= -j0.045 \times -j3.92 = -0.176$$

$$V^f_{0-2} = -Z_{0-21} I^f_{0-1}$$

$$= -j0.005 \times -j3.92 = -0.02$$

$$I^f_{1-12} = y_{1-12} (V^f_{1-1} - V^f_{1-3})$$

$$= \frac{1}{j0.2} (0.588 - 0.824) = j1.18$$

$$I^f_{2-12} = y_{2-12} (V^f_{2-1} - V^f_{2-2})$$

$$= \frac{1}{j0.2}(-0.412 + 0.176) = j1.18$$

$$I^f_{0-12} = y_{0-12} (V^f_{0-1} - V^f_{0-2})$$

$$= \frac{1}{j0.4} (-0.176 + 0.020) = j0.39$$

$$\begin{bmatrix} I^f_{a-12} \\ I^f_{b-12} \\ I^f_{c-12} \end{bmatrix} = \begin{bmatrix} 1 & 1 & 1 \\ \alpha^2 & \alpha & 1 \\ \alpha & \alpha^2 & 1 \end{bmatrix} \begin{bmatrix} I^f_{1-12} \\ I^f_{2-12} \\ I^f_{0-12} \end{bmatrix}$$

$$= \begin{bmatrix} 1 & 1 & 1 \\ \alpha^2 & \alpha & 1 \\ \alpha & \alpha^2 & 1 \end{bmatrix} \begin{bmatrix} j1.18 \\ j1.18 \\ j0.39 \end{bmatrix}$$

$$I^f_{a-12} = j1.18 + j1.18 + j0.39 = j2.75$$

$$I^f_{b-12} = j1.18 \angle 240° + j1.18 \angle 120° + j0.39$$

$$= -j079$$

$$I^f_{c-12} = j1.18 \angle 120° + j1.18 \angle 240° + j0.39$$

$$= j0.79$$

(c) $I^f_{1-G} = \dfrac{1}{j0.15} (1 - 0.588) \angle{-33°}$

$$= - 1.37 - j2.38$$

$$I^f_{0-G} = \dfrac{1}{j0.15} [0 - (- 0.412)] \angle{30°}$$

$$= 1.37 - j2.38$$

$$I^f_{0-G} = 0 \ (\textit{see} \text{ Fig. } 11.29)$$

$\therefore$ $\qquad I^f_{a-G} = (-1.37 - j2.38) + (1.37 - j2.38)$

$$= - j4.76$$

Current in phases b and c of the generator can be similarly calculated.

(d) $V^f_{b-1} = 2V^f_{1-1} + V^f_{2-1} + V^f_{0-1}$

$$= 0.588 \ \angle{240°} - 0.412 \ \angle{120°} - 0.176$$

$$= - 0.264 - j0.866 = 0.905 \ \angle{- 107°}$$

$V^f_{c-1} = V^f_{1-1} + V^f_{2-1} + V^f_{0-1}$

$$- 0.588 \ \angle{120°} - 0.412 \ \angle{240°} - 0.176$$

$$= - 0.264 + j0.866 = 0.905 \ \angle{107°}$$

PROBLEMS

11.1 A 25 MVA, 11 kV generator has a $X''_d = 0.2$ pu. Its negative and zero sequence reactances are respectively 0.3 and 0.1 pu. The neutral of the generator is solidly grounded. Determine the subtransient current in the generator and the line-to-line voltages for subtransient conditions when an LG fault occurs at the generator terminals. Assume that before the occurrence of the fault, the generator is operating at no load at rated voltage. Ignore resistances.

11.2 Repeat Problem 11.1 for (a) an LL fault; and (b) an LLG fault.

11.3 A synchronous generator is rated 25 MVA, 11 kV. It is star-connected with the neutral point solidly grounded. The generator is operating at no load at rated voltage. Its reactances are $X'' = X_2 = 0.20$ and $X_0 = 0.08$ pu. Calculate the symmetrical subtransient line currents for (i) single line-to-ground fault; (ii) double line fault; (iii) double line-to-ground fault; and (iv) symmetrical three-phase fault. Compare these currents and comment.

11.4 For the generator of Problem 11.3, calculate the value of reactance to be included in the generator neutral and ground, so that line-to-ground fault current equals the three-phase fault current. What will be the value of the grounding resistance to achieve the same condition?

With the reactance value (as calculated above) included between neutral and ground, calculate the double line fault current and also double line-to-ground fault current.

11.5 Two 25 MVA, 11 kV synchronous generators are connected to a common bus bar which supplies a feeder. The star point of one of the generators is grounded through a resistance of 1.0 ohm, while that of the other generator is isolated. A line-to-ground fault occurs at the far end of the feeder. Determine: (a) the fault current; (b) the voltage to ground of the sound phases of the feeder at the fault point; and (c) voltage of the star point of the grounded generator with respect to ground.

The impedances to sequence currents of each generator and feeder are given below:

	Generator (per unit)	Feeder (ohms/phase)
Positive sequence	$j0.2$	$j0.4$
Negative sequence	$j0.15$	$j0.4$
Zero sequence	$j0.08$	$j0.8$

11.6 Determine the fault currents in each phase following a double line-to-ground short circuit at the terminals of a star-connected synchronous generator operating initially on an open circuit voltage of 1.0 pu. The positive, negative and zero sequence reactance of the generator are, respectively, $j0.35$, $j0.25$ and $j0.20$, and its star point is isolated from ground.

11.7 A three-phase synchronous generator has positive, negative and zero sequence reactances per phase respectively, of 1.0, 0.8 and 0.4 ohm. The winding resistances are negligible. The phase sequence of the generator is RYB with a no load voltage of 11 kV between lines. A short circuit occurs between lines Y and B and earth at the generator terminals.

Calculate sequence currents in phase R and current in the earth return circuit, (a) if the generator neutral is solidly earthed; and (b) if the generator neutral is isolated.

Use R phase voltage as reference.

11.8 A generator supplies a group of identical motors as shown in Fig. P-11.8. The motors are rated 600 V, 90% efficiency at full load unity power factor with sum of their output ratings being 5 MW. The motors are sharing equally a load of 4 MW at rated voltage, 0.8 power factor lagging and 90% efficiency when an LG fault occurs on the low voltage side of the transformer.

Specify completely the sequence networks to simulate the fault so as to include the effect of prefault current. The group of motors can be treated as a single equivalent motor.

Find the subtransient line currents in all parts of the system with prefault current ignored.

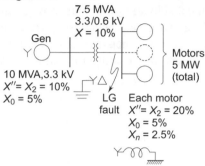

Fig. P-11.8

11.9 A double line-to-ground fault occurs on lines b and c at point F in the system of Fig. P-11.9. Find the subtransient current in phase c of machine 1, assuming prefault currents to be zero. Both machines are rated 1,200 kVA, 600 V with reactances of $X'' = X_2 = 10\%$ and $X_0 = 5\%$. Each three-phase transformer is rated 1,200 kVA, 600 V–Δ/3,300 V–Y with leakage reactance of 5%. The reactances of the transmission line are $X_1 = X_2 = 20\%$ and $X_0 = 40\%$ on a base of 1,200 kVA, 3,300 V. The reactances of the neutral grounding reactors are 5% on the kVA base of the machines.

Fig. P-11.9

11.10 A synchronous machine 1 generating 1 pu voltage is connected through a Y/Y transformer of reactance 0.1 pu to two transmission lines in parallel. The other ends of the lines are connected through a Y/Y transformer of reactance 0.1 pu to a machine 2 generating 1 pu voltage. For both transformers $X_1 = X_2 = X_0$.

Calculate the current fed into a double line-to-ground fault on the line side terminals of the transformer fed from machine 2. The star point of machine 1 and of the two transformers are solidly grounded. The reactances of the machines and lines referred to a common base are

	X_1	X_2	X_0
Machine 1	0.35	0.25	0.05
Machine 2	0.30	0.20	0.04
Line (each)	0.40	0.40	0.80

11.11 Figure P-11.11 shows a power network with two generators connected in parallel to a transformer feeding a transmission line. The far end of the line is connected to an infinite bus through another transformer. Star point of each transformer, generator 1 and infinite bus are solidly grounded. The positive, negative and zero sequence reactances of various components in per unit on a common base are:

	Positive	Negative	Zero
Generator 1	0.15	0.15	0.08
Generator 2	0.25	0.25	∞ (i.e. neutral isolated)
Each transformer	0.15	0.15	0.15
Infinite bus	0.15	0.15	0.05
Line	0.20	0.20	0.40

(a) Draw the sequence networks of the power system.
(b) With both generators and infinite bus operating at 1.0 pu voltage on no load, a line-to-ground fault occurs at one of the terminals of the star-connected winding of the transformer A. Calculate the currents flowing (i) in the fault; and (ii) through the transformer A.

Fig. P-11.11

11.12 A star connected synchronous generator feeds bus bar 1 of a power system. Bus bar 1 is connected to bus bar 2 through a star/delta transformer in series with a transmission line. The power network connected to bus bar 2 can be equivalently represented by a star-connected generator with equal positive and negative sequences reactances. All star points are solidly connected to ground. The per unit sequence reactances of various components are given below:

	Positive	Negative	Zero
Generator	0.20	0.15	0.05
Transformer	0.12	0.12	0.12
Transmission Line	0.30	0.30	0.50
Power Network	X	X	0.10

Under no load condition with 1.0 pu voltage at each bus bar, a current of 4.0 pu is fed to a three-phase short circuit on bus bar 2. Determine the positive sequence reactance X of the equivalent generator of the power network.

For the same initial conditions, find the fault current for single line-to-ground fault on bus bar 1.

11.13 The reactance data for the three-phase system of Fig. P-11.13 is:

Generator: $X_1 = X_2 = 0.1$ pu; $X_0 = 0.05$ pu

X_g (grounding reactance) = 0.02 pu

Transformer: $X_1 = X_2 = X_0 = 0.1$ pu

X_g (grounding reactance) = 0.04 pu

Form the positive, negative and zero sequence bus impedance matrices. For a solid LG fault at bus 1, calculate the fault current and its contributions from the generator and transformer.

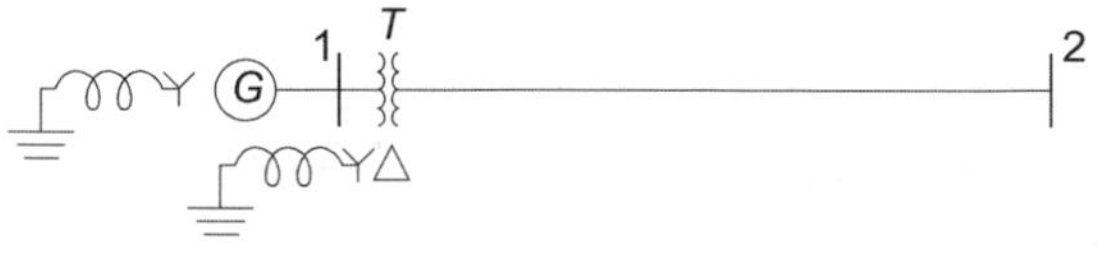

Fig. P-11.13

Hint: Notice that the line reactances are not given. Therefore it is convenient to obtain $Z_{1, \text{BUS}}$ directly rather than by inverting $Y_{1, \text{BUS}}$. Also $Y_{0, \text{BUS}}$ is singular and $Z_{0, \text{BUS}}$ cannot be obtained from it. In such situations the method of unit current injection outlined below can be used.

For a two-bus case

$$\begin{bmatrix} V_1 \\ V_2 \end{bmatrix} = \begin{bmatrix} Z_{11} & Z_{12} \\ Z_{21} & Z_{22} \end{bmatrix} \begin{bmatrix} I_1 \\ I_2 \end{bmatrix}$$

Injecting unit current at bus 1 (i.e. $I_1 = 1$, $I_2 = 0$), we get

$$Z_{11} = V_1$$

$$Z_{21} = V_2$$

Similarly injecting unit current at bus 2 (i.e. $I_1 = 0$, $I_2 = 1$), we get

$$Z_{12} = V_1$$

$$Z_{22} = V_2$$

Z_{BUS} could thus be directly obtained by this technique.

11.14 Consider the 2-bus system of Example 11.3. Assume that a solid LL fault occurs on bus *f*. Determine the fault current and voltage (to ground) of the healthy phase.

11.15 Write a computer programme to be employed for studying a solid LG fault on bus 2 of the system shown in Fig. 9.17. Our aim is to find the fault current and all bus voltages and the line currents following the fault. Use the impedance data given in Example 9.5. Assume all transformers to be Y/Δ type with their neutrals (on HV side) solidly grounded.

Assume that the positive and negative sequence reactances of the generators are equal, while their zero sequence reactance is one-fourth of their positive sequence reactance. The zero sequence reactances of the lines are to be taken as 2.5 times their positive sequence reactances. Set all prefault voltages = 1 pu.

REFERENCES

Books

1. Stevenson, W.D., *Elements of Power System Analysis*, 4th edn., McGraw-Hill, New York, 1982.
2. Elgerd, O.I., *Electric Energy Systems Theory: An Introduction*, 2nd edn., McGraw-Hill, New York, 1982.
3. Gross, C.A., *Power System Analysis*, Wiley, New York, 1979.
4. Neuenswander, J.R., *Modern Power Systems*, International Textbook Co., New York, 1971.
5. Bergan, A.R. and V. Vittal, *Power System Analysis*, 2nd edn., Pearson Education Asia, Delhi, 2000.
6. Soman, S.A, S.A. Khaparde and Shubha Pandit, *Computational Methods for Large Sparse Power Systems Analysis*, KAP, Boston, 2002.

Papers

7. Brown, H.E. and C.E. Person, "Short Circuit Studies of Large Systems by the Impedance Matrix Method", Proc. PICA, 1967, p. 335.
8. Smith, D.R., "Digital Simulation of Simultaneous Unbalances Involving Open and Faulted Conductors", IEEE Trans. PAS, 1970, 1826.

12

Power System Stability

12.1 INTRODUCTION

The stability of an interconnected power system is its ability to return to normal or stable operation after having been subjected to some form of disturbance. Conversely, instability means a condition denoting loss of synchronism or falling out of step. Stability considerations have been recognized as an essential part of power system planning for a long time. With interconnected systems continually growing in size and extending over vast geographical regions, it is becoming increasingly more difficult to maintain synchronism between various parts of a power system.

The dynamics of a power system are characterised by its basic features given below:

1. Synchronous tie exhibits the typical behaviour that as power transfer is gradually increased a maximum limit is reached beyond which the system cannot stay in synchronism, i.e., it falls out of step.

2. The system is basically a spring-inertia oscillatory system with inertia on the mechanical side and spring action provided by the synchronous tie wherein power transfer is proportional to sin δ or δ (for small δ; δ being the relative internal angle of machines).

3. Because of power transfer being proportional to sin δ, the equation determining system dynamics is nonlinear for disturbances causing large variations in angle δ. Stability phenomenon peculiar to non-linear systems as distinguished from linear systems is therefore exhibited by power systems (stable up to a certain magnitude of disturbance and unstable for larger disturbances).

Accordingly power system stability problems are classified into three basic types*—steady state, dynamic and transient.

---

*There are no universally accepted precise definitions of this terminology. For a definition of some important terms related to power system stability, refer to IEEE Standard Dictionary of Electrical and Electronic Terms, IEEE, New York, 1972.

The study of steady state stability is basically concerned with the determination of the upper limit of machine loadings before losing synchronism, provided the loading is increased gradually.

Dynamic instability is more probable than steady state instability. Small disturbances are continually occurring in a power system (variations in loadings, changes in turbine speeds, etc.) which are small enough not to cause the system to lose synchronism but do excite the system into the state of natural oscillations. The system is said to be dynamically stable if the oscillations do not acquire more than certain amplitude and die out quickly (i.e., the system is well-damped). In a dynamically unstable system, the oscillation amplitude is large and these persist for a long time (i.e., the system is underdamped). This kind of instability behaviour constitutes a serious threat to system security and creates very difficult operating conditions. Dynamic stability can be significantly improved through the use of power system stabilizers. Dynamic system study has to be carried out for 5–10 s and sometimes up to 30 s. Computer simulation is the only effective means of studying dynamic stability problems. The same simulation programmes are, of course, applicable to transient stability studies.

Following a sudden disturbance on a power system rotor speeds, rotor angular differences and power transfer undergo fast changes whose magnitudes are dependent upon the severity of disturbance. For a large disturbance, changes in angular differences may be so large as to cause the machines to fall out of step. This type of instability is known as transient instability and is a fast phenomenon usually occurring within 1 s for a generator close to the cause of disturbance. There is a large range of disturbances which may occur on a power system, but a fault on a heavily loaded line which requires opening the line to clear the fault is usually of greatest concern. The tripping of a loaded generator or the abrupt dropping of a large load may also cause instability.

The effect of short circuits (faults), the most severe type of disturbance to which a power system is subjected, must be determined in nearly all stability studies. During a fault, electrical power from nearby generators is reduced drastically, while power from remote generators is scarcely affected. In some cases, the system may be stable even with a sustained fault, whereas other systems will be stable only if the fault is cleared with sufficient rapidity. Whether the system is stable on occurrence of a fault depends not only on the system itself, but also on the type of fault, location of fault, rapidity of clearing and method of clearing, i.e., whether cleared by the sequential opening of two or more breakers or by simultaneous opening and whether or not the faulted line is reclosed. The transient stability limit is almost always lower than the steady state limit, but unlike the latter, it may exhibit different values depending on the nature, location and magnitude of disturbance.

Modern power systems have many interconnected generating stations, each with several generators and many loads. The machines located at any one point in a system normally act in unison. It is, therefore, common practice in stability studies to consider all the machines at one point as one large machine. Also machines which are not separated by lines of high reactance are lumped together and considered as one equivalent machine. Thus a multimachine system can often be reduced to an equivalent few machine system. If synchronism is lost, the machines of each group stay together although they go out of step with other groups. Qualitative behaviour of machines in an actual system is usually that of a two machine system. Because of its simplicity, the two machine system is extremely useful in describing the general concepts of power system stability and the influence of various factors on stability. It will be seen in this chapter that a two machine system can be regarded as a single machine system connected to infinite system.

Stability study of a multimachine system must necessarily be carried out on a digital computer.

12.2 DYNAMICS OF A SYNCHRONOUS MACHINE

The kinetic energy of the rotor at synchronous machine is

$$\text{KE} = \frac{1}{2}\, J\omega_{sm}^2 \times 10^{-6} \text{ MJ}$$

where $\qquad J$ = rotor moment of inertia in kg-m^2

$\qquad\qquad \omega_{sm}$ = synchronous speed in rad (mech)/s

But $\qquad\qquad \omega_s = \left(\dfrac{P}{2}\right)\omega_{sm}$ = rotor speed in rad (elect)/s

where $\qquad\qquad P$ = number of machine poles

$\therefore \qquad\qquad \text{KE} = \frac{1}{2}\left(J\left(\frac{2}{P}\right)^2 \omega_s \times 10^{-6}\right)\omega_s$

$$= \frac{1}{2}\, M\,\omega_s$$

where $\qquad\qquad M = J\left(\dfrac{2}{P}\right)^2 \omega_s \times 10^{-6}$

$\qquad\qquad\qquad$ = moment of inertia in MJ-s/elect rad

We shall define the inertia constant H such that

$$GH = \text{KE} = \frac{1}{2}\, M\,\omega_s \text{ MJ}$$

where $\qquad\qquad G$ = machine rating (base) in MVA (3-phase)

$\qquad\qquad\qquad$ H = *inertia constant* in MJ/MVA or MW-s/MVA

It immediately follows that

$$M = \frac{2GH}{\omega_s} = \frac{GH}{\pi f} \text{ MJ-s/elect rad} \tag{12.1}$$

$$= \frac{GH}{180 f} \text{MJ-s/elect degree}$$

M is also called the inertia constant.

Taking G as base, the inertia constant in pu is

$$M(\text{pu}) = \frac{H}{\pi f} \text{ s}^2/\text{elect rad} \tag{12.2}$$

$$= \frac{H}{180\text{f}} \text{ s}^2/\text{elect degree}$$

The inertia constant H has a characteristic value or a range of values for each class of machines. Table 12.1 lists some typical inertia constants.

Table 12.1 Typical inertia constants of synchronous machines*

Type of Machine		Intertia Constant H Stored Energy in MW Sec per MVA**
Turbine Generator		
Condensing	1,800 rpm	9-6
	3,000 rpm	7-4
Non-Condensing	3,000 rpm	4-3
Water wheel Generator		
Slow-speed (< 200 rpm)		2-3
High-speed (> 200 rpm)		2-4
Synchronous Condenser***		
Large		1.25
Small		1.00
Synchronous Motor with load varying from		
1.0 to 5.0 and higher for heavy flywheels		2.00

It is observed from Table 12.1 that the value of H is considerably higher for steam turbogenerator than for water wheel generator. Thirty to sixty per cent of the total inertia of a steam turbogenerator unit is that of the prime mover, whereas only 4 –15% of the inertia of a hydroelectric generating unit is that of the waterwheel, including water.

---

  * Reprinted with permission of the Westinghous Electric Corporation from Electrical Transmission and Distribution Reference Book.

  ** Where range is given, the first figure applies to the smaller MVA sizes.

 *** Hydrogen-Cooled, 25 per cent less.

The Swing Equation

Figure 12.1 shows the torque, speed and flow of mechanical and electrical powers in a synchronous machine. It is assumed that the windage, friction and iron-loss torque is negligible. The differential equation governing the rotor dynamics can then be written as

$$J\frac{d^2\theta_m}{dt^2} = T_m - T_e \text{ Nm} \tag{12.3}$$

where

θ_m = angle in rad (mech)

T_m = turbine torque in Nm; it acquires a negative value for a motoring machine

T_e = electromagnetic torque developed in Nm; it acquires negative value for a motoring machine

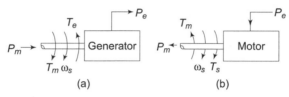

Fig. 12.1 Flow of mechanical and electrical powers in a synchronous machine

While the rotor undergoes dynamics as per Eq. (12.3), the rotor speed changes by insignificant magnitude for the time period of interest (1s) [Sec. 12.1]. Equation (12.3) can therefore be converted into its more convenient power form by assuming the rotor speed to remain constant at the synchronous speed (ω_{sm}). Multiplying both sides of Eq. (12.3) by ω_{sm}' we can write

$$J\omega_{sm}\frac{d^2\theta_m}{dt^2} \times 10^{-6} = P_m - P_e \text{ MW} \tag{12.4}$$

where

P_m = mechanical power input in MW

P_e = electrical power output in MW; stator copper loss is assumed negligible.

Rewriting Eq. (12.4)

$$\left(J\left(\frac{2}{P}\right)^2 \omega_s \times 10^{-6}\right)\frac{d^2\theta_e}{dt^2} = P_m - P_e \text{ MW}$$

where $\qquad \theta_e$ = angle in rad (elect)

or $\qquad M\dfrac{d^2\theta_e}{dt^2} = P_m - P_e \tag{12.5}$

It is more convenient to measure the angular position of the rotor with respect to a synchronously rotating frame of reference. Let

$$\delta = \theta_e - \omega_s t; \text{ rotor angular displacement from synchronously rotating reference frame}$$
(called *torque angle/power angle*) (12.6)

From Eq. (12.6)

$$\frac{d^2\theta_e}{dt^2} = \frac{d^2\delta}{dt^2} \tag{12.7}$$

Hence Eq. (12.5) can be written in terms of δ as

$$M\frac{d^2\delta}{dt^2} = P_m - P_e \text{ MW} \tag{12.8}$$

With M as defined in Eq. (12.1), we can write

$$\frac{GH}{\pi f}\frac{d^2\delta}{dt^2} = P_m - P_e \text{ MW} \tag{12.9}$$

Dividing throughout by G, the MVA rating of the machine,

$$M(pu)\frac{d^2\delta}{dt^2} = P_m - P_e;$$

$$\text{in pu of machine rating as base} \tag{12.10}$$

where

$$M(pu) = \frac{H}{\pi f}$$

or $\qquad \dfrac{H}{\pi f}\dfrac{d^2\delta}{dt^2} = P_m - P_e \text{ pu}$ (12.11)

This equation (Eq. (12.10)/Eq. (12.11)), is called the *swing equation* and it describes the rotor dynamics for a synchronous machine (generating/motoring). It is a second-order differential equation where the damping term (proportional to $d\delta/dt$) is absent because of the assumption of a lossless machine and the fact that the torque of *damper winding* has been ignored. This assumption leads to pessimistic results in transient stability analysis—damping helps to stabilize the system. Damping must of course be considered in a dynamic stability study. Since the electrical power P_e depends upon the sine of angle δ (see Eq. (12.29)), the swing equation is a non-linear second-order differential equation.

Multimachine System

In a multimachine system a common system base must be chosen.
Let

$$G_{mach} = \text{machine rating (base)}$$
$$G_{system} = \text{system base}$$

Equation (12.11) can then be written as

$$\frac{G_{mach}}{G_{system}} \left(\frac{H_{mach}}{f} \frac{d^2\delta}{dt^2} \right) = (P_m - P_e) \frac{G_{mach}}{G_{system}}$$

or

$$\frac{H_{system}}{\pi f} \frac{d^2\delta}{dt^2} = P_m - P_e \text{ pu in system base} \tag{12.12}$$

where

$$H_{system} = H_{mach} \left(\frac{G_{mach}}{G_{system}} \right) \tag{12.13}$$

$$= \text{machine inertia constant in system base}$$

Machines Swinging Coherently

Consider the swing equations of two machines on a *common system base*.

$$\frac{H_1}{\pi f} \frac{d^2\delta_1}{dt^2} = P_{m1} - P_{e1} \text{ pu} \tag{12.14}$$

$$\frac{H_2}{\pi f} \frac{d^2\delta_2}{dt^2} = P_{m2} - P_{e2} \text{ pu} \tag{12.15}$$

Since the machine rotors swing together (coherently or in unison)

$$\delta_1 = \delta_2 = \delta$$

Adding Eqs (12.14) and (12.15)

$$\frac{H_{eq}}{\pi f} \frac{d^2\delta}{dt^2} = P_m - P_e \tag{12.16}$$

where

$$P_m = P_{m1} + P_{m2}$$
$$P_e = P_{e1} + P_{e2} \tag{12.17}$$
$$H_{eq} = H_1 + H_2$$

The two machines swinging coherently are thus reduced to a single machine as in Eq. (12.16). The equivalent inertia in Eq. (12.17) can be written as

$$H_{eq} = H_{1 \text{ mach}} \, G_{1 \text{ mach}}/G_{system} + H_{2 \text{ mach}} \, G_{2 \text{ mach}}/G_{system} \tag{12.18}$$

The above results are easily extendable to any number of machines swinging coherently.

Example 12.1

A 50 Hz, four pole turbogenerator rated 100 MVA, 11 kV has an inertia constant of 8.0 MJ/MVA.

(a) Find the stored energy in the rotor at synchronous speed.

(b) If the mechanical input is suddenly raised to 80 MW for an electrical load of 50 MW, find rotor acceleration, neglecting mechanical and electrical losses.

(c) If the acceleration calculated in part (b) is maintained for 10 cycles, find the change in torque angle and rotor speed in revolutions per minute at the end of this period.

Solution

(a) Stored energy = GH = 100×8 = 800 MJ

(b) $P_a = 80 - 50 = 30$ MW = $M \dfrac{d^2\delta}{dt^2}$

$$M = \frac{GH}{180f} = \frac{800}{180 \times 50} = \frac{4}{45} \text{ MJ-s/elect deg}$$

$\therefore$ $\quad \dfrac{4}{45} \dfrac{d^2\delta}{dt^2} = 30$

or

$$\alpha = \frac{d^2\delta}{dt^2} = 337.5 \text{ elect deg/s}^2$$

(c) 10 cycles = 0.2 s

Change in $\delta = \frac{1}{2}(337.5) \times (0.2)^2 = 6.75$ elect degrees

$$= 60 \times \frac{337.5}{2 \times 360°} = 28.125 \text{ rpm/s}$$

$\therefore$ Rotor speed at the end of 10 cycles

$$= \frac{120 \times 50}{4} + 28.125 \times 0.2$$

$$= 1505.625 \text{ rpm}$$

12.3 POWER ANGLE EQUATION

In solving the swing equation (Eq. (12.10)), certain simplifying assumptions are usually made. These are:

1. Mechanical power input to the machine (P_m) remains constant during the period of electromechanical transient of interest. In other words, it means that the effect of the turbine governing loop is ignored being much slower than the speed of the transient. This assumption leads to pessimistic result—governing loop helps to stabilize the system.

2. Rotor speed changes are insignificant—these have already been ignored in formulating the swing equation.

3. Effect of voltage regulating loop during the transient is ignored, as a consequence the generated machine emf remains constant. This assumption also leads to pessimistic results—voltage regulator helps to stabilize the system.

Before the swing equation can be solved, it is necessary to determine the dependence of the electrical power output (P_e) upon the rotor angle.

Simplified Machine Model

For a nonsalient pole machine, the per phase induced emf-terminal voltage equation under steady conditions is

$$E = V + jX_dI_d + jX_qI_q; \quad X_d > X_q \tag{12.19}$$

where

$$I = I_d + I_g \tag{12.20}$$

and usual symbols are used.

Under transient condition

$$X_d \rightarrow X_d' < X_d$$

but

$X_q' = X_q$ since the main field is on the d-axis

$X_d' < X_q$; but the difference is less than in Eq. (12.19)

Equation (12.19) during the transient modifies to

$$E' = V + jX_d'I_d + jX_qI_q \tag{12.21}$$

$$= V + jX_q(I - I_d) + jX_d'I_d$$

$$= (V + jX_qI) + j(X_d' - X_q)I_d \tag{12.22}$$

The phasor diagram corresponding to Eqs. (12.21) and (12.22) is drawn in Fig. 12.2.

Since under transient condition, $X_d' < X_d$ but X_q remains almost unaffected, it is fairly valid to assume that

$$X_d' \approx X_q \tag{12.23}$$

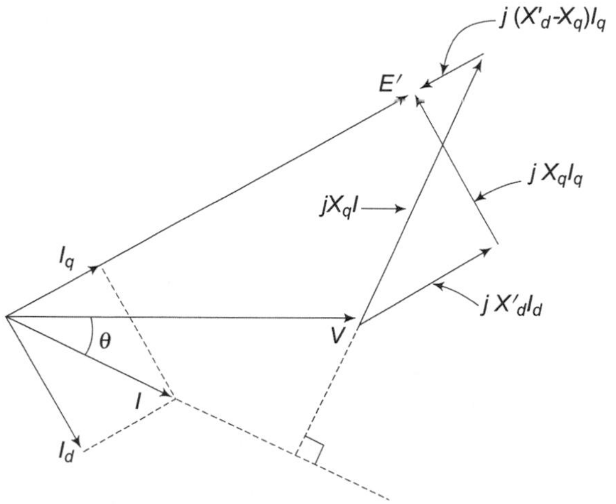

Fig. 12.2 Phasor diagram—salient pole machine

Equation (12.22) now becomes

$$E' = V + jX_qI$$

$$= V + jX_d'I \qquad (12.24)$$

The machine model corresponding to Eq. (12.24) is drawn in Fig. 12.3 which also applies to a cylindrical rotor machine where $X_d' = X_q' = X_s'$(transient synchronous reactance)

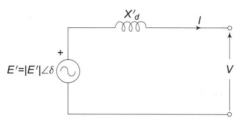

Fig. 12.3 Simplified machine model

The simplified machine of Fig. 12.3 will be used in all stability studies.

Power Angle Curve

For the purposes of stability studies $|E'|$, transient emf of generator motor, remains constant or is the independent variable determined by the voltage regulating loop but V, the generator determined terminal voltage is a dependent variable. Therefore, the nodes (buses) of the stability study network pertain to the emf terminal in the machine model as shown in Fig. 12.4, while the machine reactance (X_d') is absorbed in the system network as different from a load flow study. Further, the loads (other than large synchronous motors) will be replaced by equivalent static admittances (connected in shunt between transmission network buses and the reference bus). This is so because load voltages vary during a stability study (in a load flow study, these remain constant within a narrow band).

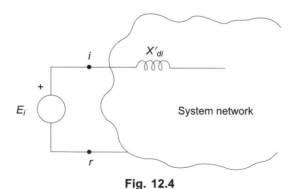

Fig. 12.4

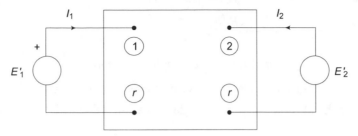

Fig. 12.5　Two-bus stability study network

For the 2-bus system of Fig. 12.5

$$\mathbf{Y}_{\text{BUS}} = \begin{bmatrix} Y_{11} & Y_{12} \\ Y_{21} & Y_{22} \end{bmatrix}; \ Y_{12} = Y_{21} \tag{12.25}$$

Complex power into bus is given by

$$P_i + jQ_i = E_i I_i^*$$

At bus 1

$$P_1 + jQ_1 = E_1' \, (Y_{11}E_1')^* + E_1 \, (Y_{12}E_2')^* \tag{12.26}$$

But

$$E_1' = |E_1'| \, \angle \, \delta_1; \ E_2' = |E_2'| \, \angle \delta_2$$

$$Y_{11} = G_{11} + jB_{11}; \ Y_{12} = |Y_{12}| \, \angle \theta_{12}$$

Since in solution of the swing equation only real power is involved, we have from Eq. (12.26)

$$P_1 = |E_1'|^2 \, G_{11} + |E_1'| \, |E_2'| \, |Y_{12}| \, \cos \, (\delta_1 - \delta_2 - \theta_{12}) \tag{12.27}$$

A similar equation will hold at bus 2.

Let

$$|E_1'|^2 G_{11} = P_c$$

$$|E_1'| \, |E_2'| \, |Y_{12}| = P_{\text{max}}$$

$$\delta_1 - \delta_2 = \delta$$

and

$$\phi_{12} = \pi/2 - \gamma$$

Then Eq. (12.27) can be written as

$$P_1 = P_c + P_{\text{max}} \, \sin \, (\delta - \gamma); \ \textit{Power Angle Equation} \tag{12.28}$$

For a purely reactive network

$$G_{11} = 0 \ (\therefore P_c = 0); \ \text{lossless network}$$

$$\theta_{12} = \pi/2, \ \therefore \gamma = 0$$

Hence

$$P_e = P_{\text{max}} \, \sin \, \delta \tag{12.29a}$$

where $$P_{max} = \frac{|E_1'|\,|E_2'|}{X};$$

simplified *power angle equation* (12.29b)

where X = transfer reactance between nodes (i.e., between E_1' and E_2')

The graphical plot of power angle equation (Eq.(12.29)) is shown in Fig. 12.6.

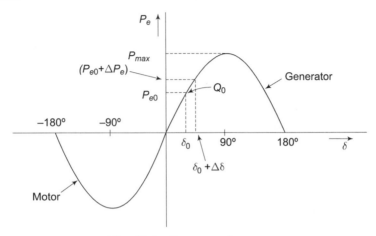

Fig. 12.6 Power angle curve

The swing equation (Eq. (12.10)) can now be written as

$$\frac{H}{\pi f}\frac{d^2\delta}{dt^2} = P_m - P_{max}\sin\delta\;\text{pu} \qquad (12.30)$$

which, as already stated, is a non-linear second-order differential equation with no damping.

12.4 NODE ELIMINATION TECHNIQUE

In stability studies, it has been indicated that the buses to be considered are those which are excited by the internal machine voltages (transient emf's) and not the load buses which are excited by the terminal voltages of the generators. Therefore, in Y_{BUS} formulation for the stability study, the load buses must be eliminated. Three methods are available for bus elimination. These are illustrated by the simple system of Fig. 12.7(a) whose reactance diagram is drawn in Fig. 12.7(b). In this simple situation, bus 3 gets easily eliminated by parallel combination of the lines. Thus

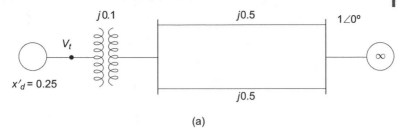

(a)

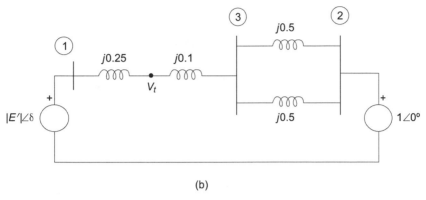

(b)

Fig. 12.7 A simple system with its reactance diagram

$$X_{12} = 0.25 + 0.1 + \frac{0.5}{2}$$

$$= 0.6$$

Consider now a more complicated case wherein a 3-phase fault occurs at the midpoint of one of the lines in which case the reactance diagram becomes that of Fig. 12.8 (a).

Star-Delta Conversion

Converting the star at the bus 3 to delta, the network transforms to that of Fig. 12.8(b) wherein

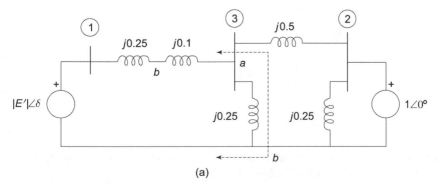

(a)

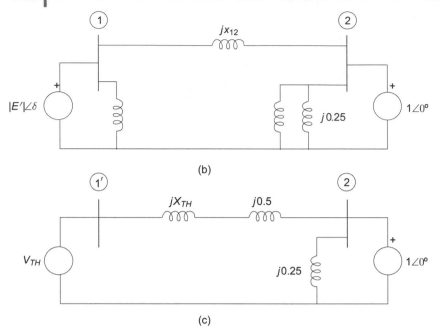

(b)

(c)

Fig. 12.8

$$X_{12} = \frac{0.25 \times 0.35 + 0.35 \times 0.5 + 0.5 \times 0.25}{0.25}$$

$$= 1.55$$

This method for a complex network, however, cannot be mechanized for preparing a computer programme.

Thevenin's Equivalent

With reference to Fig. 12.8(a), the Thevenin's equivalent for the network portion to the left of terminals a b as drawn in Fig. 12.8(c) wherein bus 1 has been modified to $1'$.

$$V_{\text{Th}} = \frac{0.25}{0.25 + 0.35} \, |E'| \, \angle \delta$$

$$= 0.417 \, |E'| \, \angle \delta$$

$$X_{\text{Th}} = \frac{0.35 \times 0.25}{0.35 + 0.25} = 0.146$$

Now

$$X_{12} = 0.146 + 0.5 = 0.646*$$

---

*This value is different from that obtained by star delta transformation as V_{Th} is no longer $|E'| \, \angle \delta$; in fact it is $0.417 \, |E'| \, \angle \delta$.

This method obviously is cumbersome to apply for a network of even small complexity and cannot be computerized.

Node Elimination Technique

Formulate the bus admittances for the 3-bus system of Fig. 12.8(a). This network is redrawn in Fig. 12.9 wherein instead of reactance branch, admittances are shown. For this network,

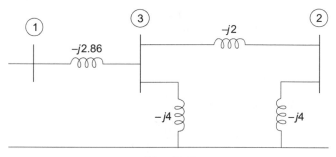

Fig. 12.9

$$Y_{\text{BUS}} = j \begin{array}{c} 1 \\ 2 \\ 3 \end{array} \begin{bmatrix} -2.86 & 0 & 2.86 \\ 0 & -6 & 2 \\ 2.86 & 2 & -8.86 \end{bmatrix}$$

The bus 3 is to be eliminated.

In general for a 3-bus system

$$\begin{bmatrix} I_1 \\ I_2 \\ I_3 \end{bmatrix} = \begin{bmatrix} Y_{11} & Y_{12} & Y_{13} \\ Y_{21} & Y_{22} & Y_{23} \\ Y_{31} & Y_{32} & Y_{33} \end{bmatrix} \begin{bmatrix} V_1 \\ V_2 \\ V_3 \end{bmatrix} \tag{12.31}$$

Since no source is connected at the bus 3

$$I_3 = 0$$

or

$$Y_{31}V_1 + Y_{32}V_2 + Y_{33}V_3 = 0$$

or

$$V_3 = -\frac{Y_{31}}{Y_{33}} V_1 - \frac{Y_{32}}{Y_{33}} V_2 \tag{12.32}$$

Substituting this value of V_3 in the remaining two equations of Eq. (12.31), thereby eliminating V_3,

$$I_1 = Y_{11}V_1 + Y_{12}V_2 + Y_{13}V_3$$

$$= \left(Y_{11} - \frac{Y_{13}Y_{31}}{Y_{33}} \right) V_1 + \left(Y_{12} - \frac{Y_{13}Y_{32}}{Y_{33}} \right) V_2$$

In compact form

$$\mathbf{Y}_{BUS}\text{ (reduced)} = \begin{bmatrix} Y'_{11} & Y'_{12} \\ Y'_{21} & Y'_{22} \end{bmatrix} \tag{12.33}$$

where

$$Y'_{11} = Y_{11} - \frac{Y_{13}Y_{31}}{Y_{33}} \tag{12.34a}$$

$$Y'_{12} = Y'_{21} = Y_{12} - \frac{Y_{13}Y_{32}}{Y_{33}} \tag{12.34b}$$

$$Y'_{22} = Y_{22} - \frac{Y_{23}Y_{32}}{Y_{33}} \tag{12.34c}$$

In general, in eliminating node n

$$Y_{kj}\text{ (new)} = Y_{kj}\text{ (old)} - \frac{Y_{kn}(\text{old})Y_{nj}(\text{old})}{Y_{nn}(\text{old})} \tag{12.35}$$

Applying Eq. (12.34) to the example in hand

$$Y_{BUS}\text{ (reduced)} = j \begin{bmatrix} -1.937 & 0.646 \\ 0.646 & -5.549 \end{bmatrix}$$

It then follows that

$$X_{12} = \frac{1}{0.646} = 1.548\ (\approx 1.55)$$

Example 12.2

In the system shown in Fig. 12.10, a three-phase static capacitive reactor of reactance 1 pu per phase is connected through a switch at motor bus bar. Calculate the limit of steady state power with and without reactor switch closed. Recalculate the power limit with capacitive reactor replaced by an inductive reactor of the same value.

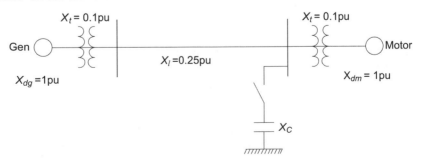

Fig. 12.10

Assume the internal voltage of the generator to be 1.2 pu and that of the motor to be 1.0 pu.

Solution

(1) Steady state power limit without reactor

$$= \frac{|E_g||E_m|}{X(\text{total})} = \frac{1.2 \times 1}{1 + 0.1 + 0.25 + 0.1 + 1} = 0.49 \text{ pu}$$

(2) Equivalent circuit with capacitive reactor is shown in Fig. 12.11 (a).

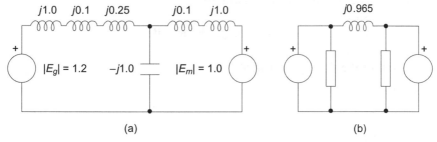

(a)   (b)

Fig. 12.11

Converting star to delta, the network of Fig. 12.11(a) is reduced to that of Fig. 12.11(b) where

$$jX(\text{transfer}) = \frac{j1.35 \times j1.1 + j1.1 \times (-j1.0) + (-j1.0) \times j1.35}{-j1.0}$$

$$= j0.965$$

Steady state power limit $= \dfrac{1.2 \times 1}{0.965} = 1.244$ pu

(3) With capacitive reactance replaced by inductive reactance, we get the equivalent circuit of Fig. 12.12. Converting star to delta, we have the transfer reactance of

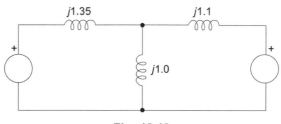

Fig. 12.12

$$jX(\text{transfer}) = \frac{j1.35 \times j1.1 + j1.1 \times j1.0 + j1.0 \times j1.35}{j1.0}$$

$$= j3.935$$

Steady state power limit $= \dfrac{1.2 \times 1}{3.935} = 0.304$ pu

Example 12.3

The generator of Fig. 12.7(a) is delivering 1.0 pu power to the infinite bus ($|V|$ = 1.0 pu), with the generator terminal voltage of $|V_t|$ = 1.0 pu. Calculate the generator emf behind transient reactance. Find the maximum power that can be transferred under the following conditions:

(a) System healthy
(b) One line shorted (3-phase) in the middle
(c) One line open.

Plot all the three power angle curves.

Solution

Let $\qquad\qquad V_t = |V_t| \angle \alpha = 1 \angle \alpha$

From power angle equation

$$\frac{|V_t||V|}{X} \sin \alpha = P_e$$

or $\qquad \left(\dfrac{1 \times 1}{0.25 + 0.1} \right) \sin \alpha = 1$

or $\qquad\qquad \alpha = 20.5°$

Current into infinite bus,

$$I = \frac{|V_t| \angle \alpha - |V| \angle 0°}{jX}$$

$$= \frac{1 \angle 20.5° - 1 \angle 0°}{j0.35}$$

$$= 1 + j0.18 = 1.016 \angle 10.3°$$

Voltage behind transient reactance,

$$E' = 1 \angle 0° + j0.6 \times (1 + j0.18)$$

$$= 0.892 + j0.6 = 1.075 \angle 33.9°$$

(a) System healthy

$$P_{\max} = \frac{|V||E'|}{X_{12}} = \frac{1 \times 1.075}{0.6} = 1.79 \text{ pu}$$

$\therefore \qquad\qquad P_e = 1.79 \sin \delta$ \hfill (i)

(b) One line shorted in the middle:

As already calculated in this section,

$$X_{12} = 1.55$$

$$\therefore \qquad P_{max} = \frac{1 \times 1.075}{1.55} = 0.694 \text{ pu}$$

or $\qquad P_e = 0.694 \sin \delta$ \hfill (ii)

(c) One line open:

It easily follows from Fig. 12.7(b) that

$$X_{12} = 0.25 + 0.1 + 0.5 = 0.85$$

$$\therefore \qquad P_{max} = \frac{1 \times 1.075}{0.85} = 1.265$$

or $\qquad P_e = 1.265 \sin \delta$ \hfill (iii)

The plot of the three power angle curves (Eqs. (i), (ii) and (iii)) is drawn in Fig. 12.13. Under healthy condition, the system is operated with $P_m = P_e = 1.0$ pu and $\delta_0 = 33.9°$, i.e., at the point P on the power angle curve 1.79 sin δ. As one line is shorted in the middle, P_m remains fixed at 1.0 pu (governing system act instantaneously) and is further assumed to remain fixed throughout the transient (governing action is slow), while the operating point instantly shifts to Q on the curve 0.694 sin δ at $\delta = 33.9°$. Notice that because of machine inertia, the rotor angle can not change suddenly.

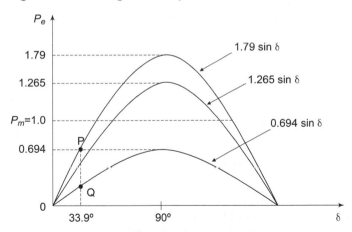

Fig. 12.13 Power angle curves

12.5 SIMPLE SYSTEMS

Machine Connected to Infinite Bus

Figure 12.14 is the circuit model of a single machine connected to infinite bus through a line of reactance X_e. In this simple case

$$X_{\text{transfer}} = X_d' + X_e$$

From Eq. (12.29b)

$$P_e = \frac{|E'||V|}{X_{\text{transfer}}} \sin \delta = P_{\max} \sin \delta \qquad (12.36)$$

The dynamics of this system are described in Eq. (12.11) as

$$\frac{H}{\pi f} \frac{d^2 \delta}{dt^2} = P_m - P_e \text{ pu} \qquad (12.37)$$

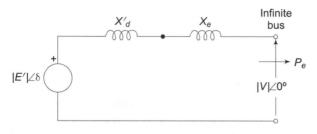

Fig. 12.14   Machine connected to infinite bus

Two Machine System

The case of two finite machines connected through a line (X_e) is illustrated in Fig. 12.15 where one of the machines must be generating and the other must be motoring. Under steady condition, before the system goes into dynamics and

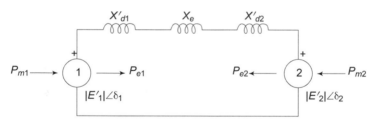

Fig. 12.15   Two-machine system

$$P_{m1} = -P_{m2} = P_m \qquad (12.38a)$$

the mechanical input/output of the two machines is assumed to remain constant at these values throughout the dynamics (governor action assumed slow). During steady state or in dynamic condition, the electrical power output of the generator must be absorbed by the motor (network being lossless). Thus at all time

$$P_{e1} = -P_{e2} = P_e \qquad (12.38b)$$

The swing equations for the two machines can now be written as

$$\frac{d^2 \delta_1}{dt^2} = \pi f \left(\frac{P_{m1} - P_{e1}}{H_1} \right) = \pi f \left(\frac{P_m - P_e}{H_1} \right) \qquad (12.39a)$$

and
$$\frac{d^2\delta_2}{dt^2} = \pi f \left(\frac{P_{m2} - P_{e2}}{H_2} \right) = \pi f \left(\frac{P_e - P_m}{H_2} \right) \tag{12.39b}$$

Subtracting Eq. (12.39b) from Eq. (12.39a)

$$\frac{d^2(\delta_1 - \delta_2)}{dt^2} = \pi f \left(\frac{H_1 + H_2}{H_1 H_2} \right) (P_m - P_e) \tag{12.40}$$

or
$$\frac{H_{eq}}{\pi f} \frac{d^2\delta}{dt^2} = P_m - P_e \tag{12.41}$$

where
$$\delta = \delta_1 - \delta_2 \tag{12.42}$$

$$H_{eq} = \frac{H_1 H_2}{H_1 + H_2} \tag{12.43}$$

The electrical power interchange is given by expression

$$P_e = \frac{|E_1'||E_2'|}{X_{d1}' + X_e + X_{d2}'} \sin \delta \tag{12.44}$$

The swing equation Eq. (12.41) and the power angle equation Eq. (12.44) have the same form as for a single machine connected to infinite bus. Thus a two-machine system is equivalent to a single machine connected to infinite bus. Because of this, the single-machine (connected to infinite bus) system would be studied extensively in this chapter.

Example 12.4

In the system of Example 12.3, the generator has an inertia constant of 4 MJ/MVA, write the swing equation upon occurrence of the fault. What is the initial angular acceleration? If this acceleration can be assumed to remain constant for $\Delta t = 0.05$s, find the rotor angle at the end of this time interval and the new acceleration.

Solution

Swing equation upon occurrence of fault

$$\frac{H}{180 f} \frac{d^2\delta}{dt^2} = P_m - P_e$$

$$\frac{4}{180 \times 50} \frac{d^2\delta}{dt^2} = 1 - 0.694 \sin \delta$$

or
$$\frac{d^2\delta}{dt^2} = 2250 (1 - 0.694 \sin \delta).$$

Initial rotor angle $\delta_0 = 33.9°$ (calculated in Example 12.3)

$$\left.\frac{d^2\delta}{dt^2}\right|_{t=0^+} = 2250 \ (1 - 0.694 \sin 33.9°)$$

$$= 1379 \text{ elect deg/s}^2$$

$$\left.\frac{d\delta}{dt}\right|_{t=0^+} = 0; \text{ rotor speed cannot change suddenly}$$

$$\Delta\delta \ (\text{in } \Delta t = 0.05s) = \frac{1}{2} \times 1379 \times (0.05)^2$$

$$= 1.7°$$

$$\delta_1 = \delta_0 + \Delta\delta = 33.9 + 1.7° = 35.6°$$

$$\left.\frac{d^2\delta}{dt^2}\right|_{t=0.05s} = 2250 \ (1 - 0.694 \sin 35.6°)$$

$$= 1341 \text{ elect deg/s}^2$$

Observe that as the rotor angle increases, the electrical power output of the generator increases and so the acceleration of the rotor reduces.

12.6 STEADY STATE STABILITY

The steady state stability limit of a particular circuit of a power system is defined as the maximum power that can be transmitted to the receiving end without loss of synchronism.

Consider the simple system of Fig. 12.14 whose dynamics is described by equations

$$M\frac{d^2\delta}{dt^2} = P_m - P_e \text{ MW; Eq. (12.8)}$$

$$M = \frac{H}{\pi f} \text{ in pu system} \tag{12.45}$$

and
$$P_e = \frac{|E||V|}{X_d} \sin \delta = P_{max} \sin \delta \tag{12.46}$$

For determination of steady state stability, the direct axis reactance (X_d) and voltage behind X_d are used in the above equations.

The plot of Eq. (12.46) is given in Fig. 12.6. Let the system be operating with steady power transfer of $P_{e0} = P_m$ with torque angle δ_0 as indicated in the figure. Assume a small increment ΔP in the electric power with the input from the prime mover remaining fixed at P_m (governor response is slow compared to the speed of energy dynamics), causing the torque angle to change to $(\delta_0 + \Delta\delta)$. Linearizing about the operating point Q_0 (P_{e0}, δ_0) we can write

$$\Delta P_e = \left(\frac{\partial \mathbf{P}_e}{\partial \delta}\right)_0 \Delta\delta$$

The excursions of $\Delta\delta$ are then described by

$$M \frac{d^2\Delta\delta}{dt^2} = P_m - (P_{e0} + \Delta P_e) = -\Delta P_e$$

or

$$M \frac{d^2\Delta\delta}{dt^2} + \left[\frac{\partial \mathbf{P}_e}{\partial \delta}\right]_0 \Delta\delta = 0 \tag{12.47}$$

or

$$\left[Mp^2 + \left(\frac{\partial P_e}{\partial \delta}\right)_0\right]\Delta\delta = 0$$

where

$$P = \frac{d}{dt}$$

The system stability to small changes is determined from the characteristic equation

$$Mp^2 + \left[\frac{\partial P_e}{\partial \delta}\right]_0 = 0$$

whose two roots are

$$P = \pm \left[\frac{-(\partial P_e/\partial\delta)_0}{M}\right]^{\frac{1}{2}}$$

As long as $(\partial \mathbf{P}_e/\partial \delta)_0$ is positive, the roots are purely imaginary and conjugate and the system behaviour is oscillatory about δ_0. Line resistance and damper windings of machine, which have been ignored in the above modelling, cause the system oscillations to decay. The system is therefore stable for a small increment in power so long as

$$(\partial P_e/\partial \delta)_0 > 0 \tag{12.48}$$

When $(\partial P_e/\partial \delta)_0$ is negative, the roots are real, one positive and the other negative but of equal magnitude. The torque angle therefore increases without bound upon occurrence of a small power increment (disturbance) and the synchronism is soon lost. The system is therefore unstable for

$$(\partial P_e/\partial \delta)_0 < 0$$

$(\partial P_e/\partial \delta)_0$ is known as *synchronizing coefficient*. This is also called *stiffness* (electrical) of synchronous machine.

Assuming $|E|$ and $|V|$ to remain constant, the system is unstable, if

$$\frac{|E||V|}{X} \cos \delta_0 < 0$$

or

$$\delta_0 > 90° \tag{12.49}$$

The maximum power that can be transmitted without loss of stability (steady state) occurs for

$$\delta_0 = 90° \tag{12.50}$$

and is given by

$$P_{max} = \frac{|E||V|}{X} \tag{12.51}$$

If the system is operating below the limit of steady stability condition (Eq. 12.48), it may continue to oscillate for a long time if the damping is low. Persistent oscillations are a threat to system security. The study of system damping is the study of dynamical stability.

The above procedure is also applicable for complex systems wherein governor action and excitation control are also accounted for. The describing differential equation is linearized about the operating point. Condition for steady state stability is then determined from the corresponding characteristic equation (which now is of order higher than two).

It was assumed in the above account that the internal machine voltage $|E|$ remains constant (i.e., excitation is held constant). The result is that as loading increases, the terminal voltage $|V_t|$ dips heavily which cannot be tolerated in practice. Therefore, we must consider the steady state stability limit by assuming that excitation is adjusted for every load increase to keep $|V_t|$ constant. This is how the system will be operated practically. It may be understood that we are still not considering the effect of automatic excitation control.

Steady state stability limit with $|V_t|$ and $|V|$ constant is considered in Example 12.6.

Example 12.5

A synchronous generator of reactance 1.20 pu is connected to an infinite bus bar ($|V| = 1.0$ pu) through transformers and a line of total reactance of 0.60 pu. The generator no load voltage is 1.20 pu and its inertia constant is H = 4 MW-s/MVA. The resistance and machine damping may be assumed negligible. The system frequency is 50 Hz.

Calculate the frequency of natural oscillations if the generator is loaded to (i) 50% and (ii) 80% of its maximum power limit.

Solution

(i) For 50% loading

$$\sin \delta_0 = \frac{P_e}{P_{max}} = 0.5 \text{ or } \delta_0 = 30°$$

$$\left[\frac{\partial P_e}{\partial \delta}\right]_{30°} = \frac{1.2 \times 1}{1.8} \cos 30°$$

$$= 0.577 \text{ MW (pu)/elect rad}$$

$$M(\text{pu}) = \frac{H}{\pi \times 50} = \frac{4}{\pi \times 50} \text{ s}^2/\text{elect rad}$$

From characteristic equation

$$p = \pm j\left[\left(\frac{\partial P_e}{\partial \delta}\right)_{30°} \Big/ M\right]^{\frac{1}{2}}$$

$$= \pm j \left(\frac{0.577 \times 50\,\pi}{4}\right)^{\frac{1}{2}} = \pm j4.76$$

Frequency of oscillations = 4.76 rad/sec

$$= \frac{4.76}{2\pi} = 0.758 \text{ Hz}$$

(ii) For 80% loading

$$\sin \delta_0 = \frac{P_e}{P_{max}} = 0.8 \text{ or } \delta_0 = 53.1°$$

$$\left(\frac{\partial P_e}{\partial \delta}\right)_{53.1°} = \frac{1.2 \times 1}{1.8} \cos 53.1°$$

$$= 0.4 \text{ MW (pu)/elect rad}$$

$$p = \pm j \left(\frac{0.4 \times 50\,\pi}{4}\right)^{\frac{1}{2}} = \pm j3.96$$

Frequency of oscillations = 3.96 rad/sec

$$= \frac{3.96}{2\pi} = 0.63 \text{ Hz}$$

Example 12.6

Find the steady state power limit of a system consisting of a generator equivalent reactance 0.50 pu connected to an infinite bus through a series reactance of 1.0 pu. The terminal voltage of the generator is held at 1.20 pu and the voltage of the infinite bus is 1.0 pu.

Solution

The system is shown in Fig. 12.16. Let the voltage of the infinite bus be taken as reference.

Then $\qquad V = 1.0 \angle 0°, \; V_t = 1.2 \angle \theta$

Now $\qquad I = \dfrac{V_t - V}{jX} = \dfrac{1.2 \angle \theta - 1.0}{j1}$

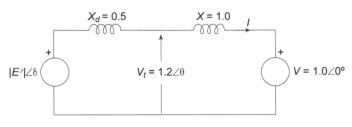

Fig. 12.16

$$E = V_t + jX_d I = 1.2 \angle \theta + j0.5 \left[\dfrac{1.2 \angle \theta - 1.0}{j1} \right]$$

or

$$E = 1.8 \angle \theta - 0.5 = (1.8 \cos \theta - 0.5) + j1.8 \sin \theta$$

Steady state power limit is reached when E has an angle of $\delta = 90°$, i.e., its real part is zero. Thus,

$$1.8 \cos \theta - 0.5 = 0$$

or $\qquad \theta = 73.87°$

Now $\qquad V_t = 1.2 \angle 73.87° = 0.332 + j1.152$

$$I = \dfrac{0.332 + j1.152 - 1}{j1} = 1.152 + j0.668$$

$$E = 0.332 + j1.152 + j0.5 \,(1.152 + j0.668)$$

$$= -0.002 + j1.728 \simeq 1.728 \angle 90°$$

Steady state power limit is given by

$$P_{max} = \dfrac{|E|\,|V|}{X_d + X} = \dfrac{1.728 \times 1}{1.5} = 1.152 \text{ pu}$$

If instead, the generator emf is held fixed at a value of 1.2 pu, the steady state power limit would be

$$P_{max} = \dfrac{1.2 \times 1}{1.5} = 0.8 \text{ pu}$$

It is observed that regulating the generator emf to hold the terminal generator voltage at 1.2 pu raises the power limit from 0.8 pu to 1.152 pu; this is how the voltage regulating loop helps in power system stability.

Some Comments on Steady State Stability

A knowledge of steady state stability limit is important for various reasons. A system can be operated above its transient stability limit but not above its steady state limit. Now, with increased fault clearing speeds, it is possible to make the transient limit closely approach the steady state limit.

As is clear from Eq. (12.51), the methods of improving steady state stability limit of a system are to reduce X and increase either or both $|E|$ and $|V|$. If the transmission lines are of sufficiently high reactance, the stability limit can be raised by using two parallel lines which incidently also increases the reliability of the system. Series capacitors are sometimes employed in lines to get better voltage regulation and to raise the stability limit by decreasing the line reactance. Higher excitation voltages and quick excitation system are also employed to improve the stability limit.

12.7 TRANSIENT STABILITY

It has been shown in Sec. 12.4 that the dynamics of a single synchronous machine connected to infinite bus bars is governed by the nonlinear differential equation

$$M \frac{d^2\delta}{dt^2} = P_m - P_e$$

where

$$P_e = P_{max} \sin \delta$$

or

$$M \frac{d^2\delta}{dt^2} = P_m - P_{max} \sin \delta \qquad (12.52)$$

As said earlier, this equation is known as the *swing equation*. No closed form solution exists for swing equation except for the simple case $P_m = 0$ (not a practical case) which involves elliptical integrals. For small disturbance (say, gradual loading), the equation can be linearized (see Sec. 12.6) leading to the concept of steady state stability where a unique criterion of stability $(\partial P_e/\partial \delta > 0)$ could be established. No generalized criteria are available* for determining system stability with large disturbances (called transient stability). The practical approach to the transient stability problem is therefore to list all important severe disturbances along with their possible locations to which the system is likely to be subjected according to the experience and judgement of the power system analyst. Numerical solution of the swing equation (or equations for a multimachine case) is then obtained in the presence of such disturbances giving a plot of δ vs. t called the *swing curve*. If δ starts to decrease after reaching a maximum value, it is normally assumed that the system is stable and the oscillation of δ around the equilibrium point will decay

---

*Recent literature gives methods of determining transient stability through Liapunov and Popov's stability criteria, but these have not been of practical use so far.

and finally die out. As already pointed out in the introduction, important severe disturbances are a short circuit or a sudden loss of load.

For ease of analysis certain assumptions and simplifications are always made (some of these have already been made in arriving at the swing equation (Eq. (12.52)). All the assumptions are listed, below along with their justification and consequences upon accuracy of results.

1. Transmission line as well as synchronous machine resistance are ignored. This leads to pessimistic result as resistance introduces damping term in the swing equation which helps stability. In Example 12.11, line resistance has been taken into account.

2. Damping term contributed by synchronous machine damper windings is ignored. This also leads to pessimistic results for the transient stability limit.

3. Rotor speed is assumed to be synchronous. In fact it varies insignificantly during the course of the stability transient.

4. Mechanical input to machine is assumed to remain constant during the transient, i.e., regulating action of the generator loop is ignored. This leads to pessimistic results.

5. Voltage behind transient reactance is assumed to remain constant, i.e., action of voltage regulating loop is ignored. It also leads to pessimistic results.

6. Shunt capacitances are not difficult to account for in a stability study. Where ignored, no greatly significant error is caused.

7. Loads are modelled as constant admittances. This is a reasonably accurate representation.

Note: Since rotor speed and hence frequency vary insignificantly, the network parameters remain fixed during a stability study.

A digital computer programme to compute the transient following sudden disturbance can be suitably modified to include the effect of governor action and excitation control.

Present day power systems are so large that even after lumping of machines (Eq. (12.17)), the system remains a multimachine one. Even then, a simple two-machine system greatly aids the understanding of the transient stability problem. It has been shown in Section 12.4 that an equivalent single-machine infinite bus system can be found for a two-machine system (Eqs. (12.41) to (12.43)).

Upon occurrence of a severe disturbance, say a short circuit, the power transfer between machines is greatly reduced, causing the machine torque angles to swing relatively. The circuit breakers near the fault disconnect the unhealthy part of the system so that power transfer can be partially restored, improving the chances of the system remaining stable. The shorter the time to breaker operating, called *clearing time*, the higher is the probability of the system being stable. Most of the line faults are transient in nature and get cleared on opening the line. Therefore, it is common practice now to employ *autoreclose breakers* which automatically close rapidly after each of the two sequential openings. If the fault still persists, the circuit breakers open and lock permanently till cleared manually. Since in the majority of faults the first reclosure will be successful, the chances of system stability are greatly enhanced by using autoreclose breakers.

Fig. 12.17

The procedure of determining the stability of a system upon occurrence of a disturbance followed by various switching off and switching on actions is called a *stability study*. Steps to be followed in a stability study are outlined below for a single-machine infinite bus bar system shown in Fig. 12.17. The fault is assumed to be a transient one which is cleared by the time of first reclosure. In the case of a permanent fault, this system completely falls apart. This will not be the case in a multimachine system. The steps listed, in fact, apply to a system of any size.

1. From prefault loading, determine the voltage behind transient reactance and the torque angle δ_0 of the machine with reference to the infinite bus.
2. For the specified fault, determine the power transfer equation $P_e(\delta)$ during fault. In this system $P_e = 0$ for a three-phase fault.
3. From the swing equation starting with δ_0 as obtained in step 1, calculate δ as a function of time using a numerical technique of solving the non-linear differential equation.
4. After clearance of the fault, once again determine $P_e (\delta)$ and solve further for $\delta (t)$. In this case, $P_e (\delta) = 0$ as when the fault is cleared, the system gets disconnected.
5. After the transmission line is switched on, again find $P_e (\delta)$ and continue to calculate $\delta (t)$.
6. If $\delta (t)$ goes through a maximum value and starts to reduce, the system is regarded as stable. It is unstable if $\delta (t)$ continues to increase. Calculation is ceased after a suitable length of time.

An important numerical method of calculating $\delta (t)$ from the swing equation will be given in Section 12.9. For the single machine infinite bus bar system, stability can be conveniently determined by the equal area criterion presented in the following section.

12.8 EQUAL AREA CRITERION

In a system where one machine is swinging with respect to an infinite bus, it is possible to study transient stability by means of a simple criterion, without resorting to the numerical solution of a swing equation.

Consider the swing equation

$$\frac{d^2\delta}{dt^2} = \frac{1}{M}(P_m - P_e) = \frac{P_a}{M}; \; P_a = \text{accelerating power}$$

$$M = \frac{H}{\pi f} \text{ in pu system} \qquad (12.53)$$

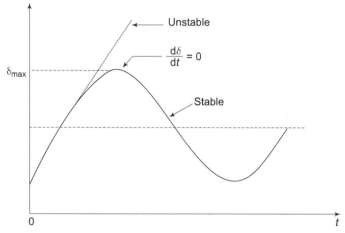

Fig. 12.18 Plot of δ vs t for stable and unstable systems

If the system is unstable δ continues to increase indefinitely with time and the machine loses synchronism. On the other hand, if the system is stable, $\delta(t)$ performs oscillations (nonsinusoidal) whose amplitude decreases in actual practice because of damping terms (not included in the swing equation). These two situations are shown in Fig. 12.18. Since the system is non-linear, the nature of its response [$\delta(t)$] is not unique and it may exhibit instability in a fashion different from that indicated in Fig. 12.18, depending upon the nature and severity of disturbance. However, experience indicates that the response $\delta(t)$ in a power system generally falls in the two broad categories as shown in the figure. It can easily be visualized now (this has also been stated earlier) that for a stable system, indication of stability will be given by observation of the first swing where δ will go to a maximum and will start to reduce. This fact can be stated as a stability criterion, that the system is stable if at some time

$$\frac{d\delta}{dt} = 0 \qquad (12.54)$$

and is unstable, if

$$\frac{d\delta}{dt} > 0 \qquad (12.55)$$

for a sufficiently long time (more than 1 s will generally do).

The stability criterion for power systems stated above can be converted into a simple and easily applicable form for a single machine infinite bus system.

Multiplying both sides of the swing equation by $\left(2\dfrac{d\delta}{dt}\right)$, we get

$$2\,\frac{d\delta}{dt} \cdot \frac{d^2\delta}{dt^2} = \frac{2P_a}{M}\frac{d\delta}{dt}$$

Integrating, we have

$$\left(\frac{d\delta}{dt}\right)^2 = \frac{2}{M}\int_{\delta_0}^{\delta} P_a\, d\delta$$

or

$$\frac{d\delta}{dt} = \left(\frac{2}{M}\int_{\delta_0}^{\delta} P_a\, d\delta\right)^{\frac{1}{2}} \tag{12.56}$$

where δ_0 is the initial rotor angle before it begins to swing due to disturbance.

From Eqs. (12.55) and (12.56), the condition for stability can be written as

$$\left(\frac{2}{M}\int_{\delta_0}^{\delta} P_a d\,\delta\right)^{\frac{1}{2}} = 0$$

or

$$\int_{\delta_0}^{\delta} P_a\, d\delta = 0 \tag{12.57}$$

The condition of stability can therefore be stated as: the system is stable if the area under P_a (accelerating power) – δ curve reduces to zero at some value of δ. In other words, the positive (accelerating) area under $P_a - \delta$ curve must equal the negative (decelerating) area and hence the name 'equal area' criterion of stability.

To illustrate the equal area criterion of stability, we now consider several types of disturbances that may occur in a single machine infinite bus bar system.

Sudden Change in Mechanical Input

Figure 12.19 shows the transient model of a single machine tied to infinite bus bar. The electrical power transmitted is given by

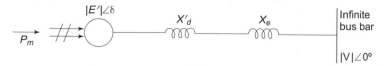

Fig. 12.19

$$P_e = \frac{|E'|\,|V|}{X'_d + X_e} \sin \delta = P_{\max} \sin \delta$$

Under steady operating condition

$$P_{m0} = P_{e0} = P_{\max} \sin \delta_0$$

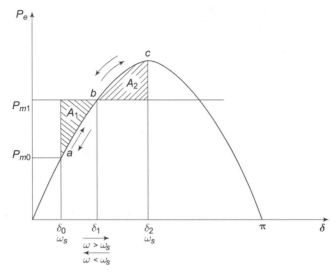

Fig. 12.20 $P_e - \delta$ diagram for sudden increase in mechanical input to generator of Fig. 12.19

This is indicated by the point a in the $P_e - \delta$ diagram of Fig. 12.20.

Let the mechanical input to the rotor be suddenly increased to P_{m1} (by opening the steam valve). The accelerating power $P_a = P_{m1} - P_e$ causes the rotor speed to increase ($\omega > \omega_s$) and so does the rotor angle. At angle δ_1, $P_a = P_{m1} - P_e\,(= P_{\max} \sin \delta_1) = 0$ (state point at b) but the rotor angle continues to increase as $\omega > \omega_s$. P_a now becomes negative (decelerating), the rotor speed begins to reduce but the angle continues to increase till at angle δ_2, $\omega = \omega_s$ once again (state point at c. At c), the decelerating area A_2 equals the accelerating area A_1 (areas are shaded), i.e., $\displaystyle\int_{\delta_0}^{\delta_2} P_a\, d\delta = 0$. Since the rotor is decelerating, the speed reduces below ω_s and the rotor angle begins to reduce. The state point now traverses the $P_e - \delta$ curve in the opposite direction as indicated by arrows in Fig. 12.20. It is easily seen that the system oscillates about the new steady state point b ($\delta = \delta_1$) with angle excursion up to δ_0 and δ_2 on the two sides. These oscillations are similar to the simple harmonic motion of an inertia-spring system except that these are not sinusoidal.

As the oscillations decay out because of inherent system damping (not modelled), the system settles to the new steady state where

$$P_{m1} = P_e = P_{\max} \sin \delta_1$$

From Fig. 12.20, areas A_1 and A_2 are given by

$$A_1 = \int_{\delta_0}^{\delta_1} (P_{m1} - P_e)\,d\delta$$

$$A_2 = \int_{\delta_1}^{\delta_2} (P_e - P_{m1})\,d\delta$$

For the system to be stable, it should be possible to find angle δ_2 such that $A_1 = A_2$. As P_{m1} is increased, a limiting condition is finally reached when A_1 equals the area above the P_{m1} line as shown in Fig. 12.21. Under this condition, δ_2 acquires the maximum value such that

$$\delta_2 = \delta_{\max} = \pi - \delta_1 = \pi - \sin^{-1} \frac{P_{m1}}{P_{\max}} \qquad (12.58)$$

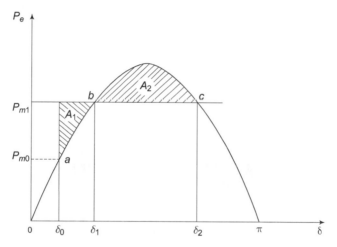

Fig. 12.21   Limiting case of transient stability with mechanical input suddenly increased

Any further increase in P_{m1} means that the area available for A_2 is less than A_1, so that the excess kinetic energy causes δ to increase beyond point c and the decelerating power changes over to accelerating power, with the system consequently becoming unstable. It has thus been shown by use of the equal area criterion that there is an upper limit to sudden increase in mechanical input $(P_{m1} - P_{m0})$, for the system in question to remain stable.

It may also be noted from Fig. 12.21 that the system will remain stable even though the rotor may oscillate beyond $\delta = 90°$, so long as the equal area criterion is met. The condition of $\delta = 90°$ is meant for use in steady state stability only and does not apply to the transient stability case.

Effect of Clearing Time on Stability

Let the system of Fig. 12.22 be operating with mechanical input P_m at a steady angle of δ_0 $(P_m = P_e)$ as shown by the point a on the $P_e - \delta$ diagram of Fig. 12.23. If a 3-phase fault occurs at the point P of the outgoing radial line, the electrical output of the generator instantly reduces to zero, i.e., $P_e = 0$ and the state point drops to b. The acceleration area A_1 begins to increase and so does the rotor angle while the state point moves along bc. At time t_c corresponding to angle δ_c the faulted line is cleared by the opening of the line circuit breaker. The values of t_c and δ_c are respectively known as *clearing time* and *clearing angle*. The system once again becomes healthy and transmits $P_e = P_{max} \sin \delta$ i.e. the state point shifts to d on the original $P_e - \delta$ curve. The rotor now decelerates and the decelerating area A_2 begins while the state point moves along de.

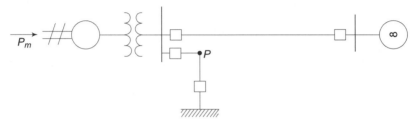

Fig. 12.22

If an angle δ_1 can be found such that $A_2 = A_1$, the system is found to be stable. The system finally settles down to the steady operating point a in an oscillatory manner because of inherent damping.

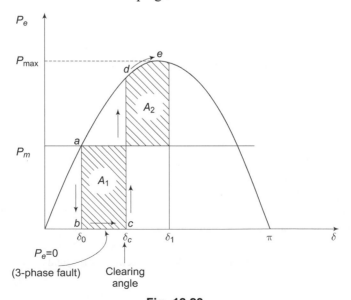

Fig. 12.23

The value of clearing time corresponding to a clearing angle can be established only by numerical integration except in this simple case. The equal area criterion therefore gives only qualitative answer to system stability as the time when the breaker should be opened is hard to establish.

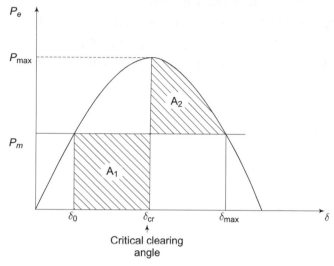

Fig. 12.24 Critical clearing angle

As the clearing of the faulty line is delayed, A_1 increases and so does δ_1 to find $A_2 = A_1$ till $\delta_1 = \delta_{max}$ as shown in Fig. 12.24. For a clearing time (or angle) larger than this value, the system would be unstable as $A_2 < A_1$. The maximum allowable value of the clearing time and angle for the system to remain stable are known respectively as *critical clearing time* and *angle*.

For this simple case ($P_e = 0$ during fault), explicit relationships for δ_c (critical) and t_c (critical) are established below. All angles are in *radians*.

It is easily seen from Fig. 12.24 that

$$\delta_{max} = \pi - \delta_0 \tag{12.59}$$

and

$$P_m = P_{max} \sin \delta_0 \tag{12.60}$$

Now

$$A_1 = \int_{\delta_0}^{\delta_{cr}} (P_m - 0)\, d\delta = P_m\, (\delta_{cr} - \delta_0)$$

and

$$A_2 = \int_{\delta_{cr}}^{\delta_{max}} (P_{max} \sin \delta - P_m)\, d\delta$$

$$= P_{max}\, (\cos \delta_{cr} - \cos \delta_{max}) - P_m\, (\delta_{max} - \delta_{cr})$$

For the system to be stable, $A_2 = A_1$, which yields

$$\cos \delta_{cr} = \frac{P_m}{P_{max}} (\delta_{max} - \delta_0) + \cos \delta_{max} \qquad (12.61)$$

where δ_{cr} = critical clearing angle

Substituting Eqs. (12.59) and (12.60) in Eq. (12.61), we get

$$\delta_{cr} = \cos^{-1} [(\pi - 2\delta_0) \sin \delta_0 - \cos \delta_0] \qquad (12.62)$$

During the period the fault is persisting, the swing equation is

$$\frac{d^2\delta}{dt^2} = \frac{\pi f}{H} P_m; \; P_e = 0 \qquad (12.63)$$

Integrating twice

$$\delta = \frac{\pi f}{2H} P_m t^2 + \delta_0$$

or

$$\delta_{cr} = \frac{\pi f}{2H} P_m t^2_{cr} + \delta_0 \qquad (12.64)$$

where

t_{cr} = critical clearing time

δ_{cr} = critical clearing angle

From Eq. (12.64)

$$t_{cr} = \sqrt{\frac{2H (\delta_{cr} - \delta_0)}{\pi f P_m}} \qquad (12.65)$$

where δ_{cr} is given by the expression of Eq. (12.62)

An explicit relationship for determining t_{cr} is possible in this case as during the faulted condition $P_e = 0$ and so the swing equation can be integrated in closed form. This will not be the case in most other situations.

Sudden Loss of One of Parallel Lines

Consider now a single machine tied to infinite bus through two parallel lines as in Fig. 12.25a. Circuit model of the system is given in Fig. 12.25b.

Let us study the transient stability of the system when one of the lines is suddenly switched off with the system operating at a steady load. Before switching off, power angle curve is given by

$$P_{eI} = \frac{|E'||V|}{X_d + X_1 || X_2} \sin \delta = P_{maxI} \sin \delta$$

Immediately on switching off line 2, power angle curve is given by

$$P_{eII} = \frac{|E'||V|}{X'_d + X_1} \sin \delta = P_{maxII} \sin \delta$$

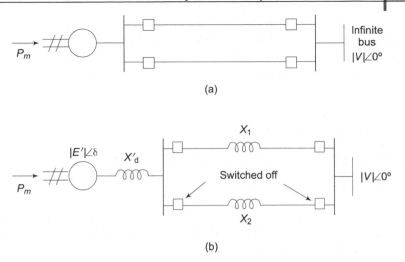

(a)

(b)

Fig. 12.25 Single machine tied to infinite bus through two parallel lines

Both these curves are plotted in Fig. 12.26, wherein $P_{\text{maxII}} < P_{\text{maxI}}$ as $(X'_d + X_1)$ $> (X'_d + X_1 \parallel X_2)$. The system is operating initially with a steady power transfer $P_e = P_m$ at a torque angle δ_0 on curve I.

Immediately on switching off line 2, the electrical operating point shifts to curve II (point b). Accelerating energy corresponding to area A_1 is put into rotor followed by decelerating energy for $\delta > \delta_1$. Assuming that an area A_2 corresponding to decelerating energy (energy out of rotor) can be found such that $A_1 = A_2$, the system will be stable and will finally operate at c corresponding to a new rotor angle $\delta_1 > \delta_0$. This is so because a single line offers larger reactance and larger rotor angle is needed to transfer the same steady power.

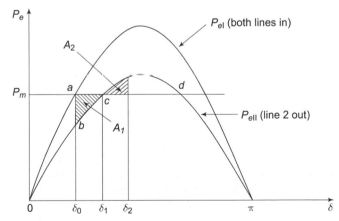

Fig. 12.26 Equal area criterion applied to the opening of one of the two lines in parallel

It is also easy to see that if the steady load is increased (line P_m is shifted upwards in Fig. 12.26), a limit is finally reached beyond which decelerating area equal to A_1 cannot be found and therefore, the system behaves as an unstable one. For the limiting case of stability, δ_1 has a maximum value given by

$$\delta_1 = \delta_{max} = \pi - \delta_c$$

which is the same condition as in the previous example.

Sudden Short Circuit on One of Parallel Lines

Case a: Short circuit at one end of line

Let us now assume the disturbance to be a short circuit at the generator end of line 2 of a double circuit line as shown in Fig. 12.27a. We shall assume the fault to be a three-phase one.

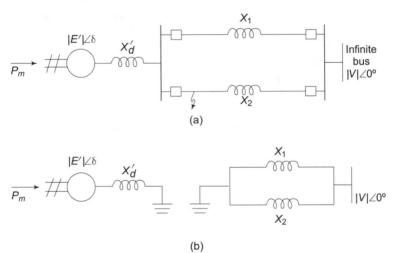

Fig. 12.27 Short circuit at one end of the line

Before the occurrence of a fault, the power angle curve is given by

$$P_{el} = \frac{|E'||V|}{X_d' + X_1||X_2} \sin \delta = P_{maxI} \sin \delta$$

which is plotted in Fig. 12.25.

Upon occurrence of a three-phase fault at the generator end of line 2 (see Fig. 12.24a), the generator gets isolated from the power system for purposes of power flow as shown by Fig. 12.27b. Thus during the period the fault lasts,

$$P_{eII} = 0$$

The rotor therefore accelerates and angles δ increases. Synchronism will be lost unless the fault is cleared in time.

The circuit breakers at the two ends of the faulted line open at time t_c (corresponding to angle δ_c), the clearing time, disconnecting the faulted line.

The power flow is now restored via the healthy line (through higher line reactance X_2 in place of $X_1 \parallel X_2$), with power angle curve

$$P_{e\text{III}} = \frac{|E'||V|}{X_d' + X_1} \sin \delta = P_{\max \text{ II}} \sin \delta$$

Obviously, $P_{\max\text{II}} < P_{\max\text{I}}$. The rotor now starts to decelerate as shown in Fig. 12.28. The system will be stable if a decelerating area A_2 can be found equal to accelerating area A_1 before δ reaches the maximum allowable value $\delta_{\max}$. As area A_1 depends upon clearing time t_c (corresponding to clearing angle δ_c), clearing time must be less than a certain value (critical clearing time) for the system to be stable. It is to be observed that the equal area criterion helps to determine critical clearing angle and not critical clearing time. Critical clearing time can be obtained by numerical solution of the swing equation (discussed in Section 12.8).

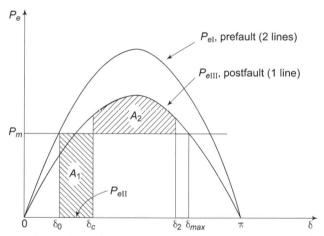

Fig. 12.28   Equal area criterion applied to the system of Fig. 12.24a, I system normal, II fault applied, III faulted line isolated.

It also easily follows that larger initial loading (P_m) increases A_1 for a given clearing angle (and time) and therefore quicker fault clearing would be needed to maintain stable operation.

Case b: Short circuit away from line ends

When the fault occurs away from line ends (say in the middle of a line), there is some power flow during the fault though considerably reduced, as different from case a where $P_{e\text{II}} = 0$. Circuit model of the system during fault is now shown in Fig. 12.29a. This circuit reduces to that of Fig. 12.29c through one delta-star and one star-delta conversion. Instead, node elimination technique of Section 12.3 could be employed profitably. The power angle curve during fault is therefore given by

$$P_{e\text{II}} = \frac{|E'||V|}{X_{\text{II}}} \sin \delta = P_{\max\text{II}} \sin \delta$$

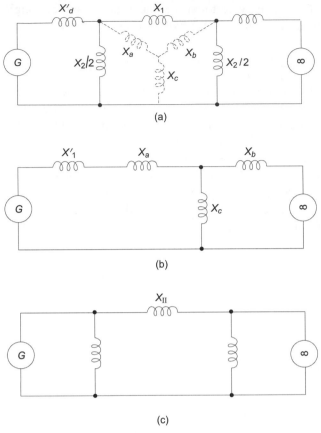

Fig. 12.29

P_{eI} and P_{eIII} as in Fig. 12.28 and P_{eII} as obtained above are all plotted in Fig. 12.30. Accelerating area A_1 corresponding to a given clearing angle δ_c is less

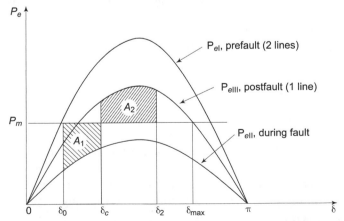

Fig. 12.30   Fault on middle of one line of the system of Fig. 12.24a with $\delta_c < \delta_{cr}$

in this case, than in case *a*, giving a better chance for stable operation. Stable system operation is shown in Fig. 12.30, wherein it is possible to find an area A_2 equal to A_1 for $\delta_2 < \delta_{max}$. As the clearing angle δ_c is increased, area A_1 increases and to find $A_2 = A_1$, δ_2 increases till it has a value δ_{max}, the maximum allowable for stability. This case of critical clearing angle is shown in Fig. 12.31.

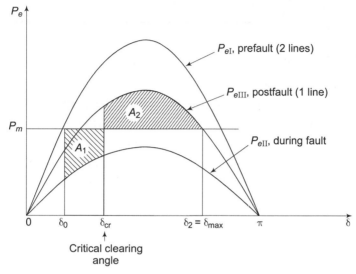

Fig. 12.31 Fault on middle of one line of the system of Fig. 12.24a, case of critical clearing angle

Applying equal area criterion to the case of critical clearing angle of Fig. 12.31, we can write

$$\int_{\delta_0}^{\delta_{cr}} (P_m - P_{maxII} \sin \delta)\, d\delta = \int_{\delta_{cr}}^{\delta_{max}} (P_{maxIII} \sin \delta - P_m)\, d\delta$$

where

$$\delta_{max} = \pi - \sin^{-1} \left(\frac{P_m}{P_{max\,III}} \right) \tag{12.66}$$

Integrating, we get

$$(P_{m\delta} + P_{maxII} \cos \delta) \bigg|_{\delta_o}^{\delta_{cr}} + (P_{maxIII} \cos \delta + P_m \delta) \bigg|_{\delta_{cr}}^{\delta_{max}} = 0$$

or

$$P_m (\delta_{cr} - \delta_0) + P_{maxII} (\cos \delta_{cr} - \cos \delta_0)$$
$$+ P_m (\delta_{max} - \delta_{cr}) + P_{maxIII} (\cos \delta_{max} - \cos \delta_{cr}) = 0$$

or

$$\cos \delta_{cr} = \frac{P_m(\delta_{max} - \delta_0) - P_{max\,II}\cos\delta_0 + P_{max\,III}\cos\delta_{max}}{P_{max\,III} - P_{max\,II}} \qquad (12.67)$$

Critical clearing angle can be calculated from Eq. (12.67) above. The angles in this equation are in radians. The equation modifies as below if the angles are in degrees.

$$\cos \delta_{cr} = \frac{\dfrac{\pi}{180} P_m(\delta_{max} - \delta_0) - P_{max\,II}\cos\delta_0 + P_{max\,III}\cos\delta_{max}}{P_{max\,III} - P_{max\,II}}$$

Case c: Reclosure

If the circuit breakers of line 2 are reclosed successfully (i.e., the fault was a transient one and therefore vanished on clearing the faulty line), the power transfer once again becomes

$$P_{eIV} = P_{eI} = P_{maxI}\sin\delta$$

Since reclosure restores power transfer, the chances of stable operation improve. A case of stable operation is indicated by Fig. 12.32.

For critical clearing angle

$$\delta_1 = \delta_{max} = \pi - \sin^{-1}(P_m/P_{maxI})$$

$$\int_{\delta_0}^{\delta_{cr}} (P_m - P_{maxII}\sin\delta)\,d\delta = \int_{\delta_{cr}}^{\delta_{rc}} (P_{maxIII}\sin\delta - P_m)\,d\delta$$

$$+ \int_{\delta_{rc}}^{\delta_{max}} (P_{maxI}\sin\delta - P_m)\,d\delta$$

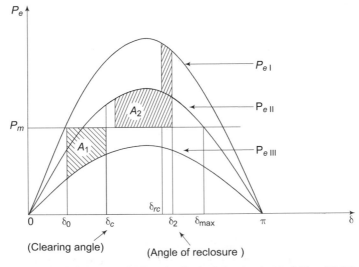

Fig. 12.32 Fault in middle of a line of the system of Fig. 12.27a

where $t_{rc} = t_{cr} + \tau$; τ = time between clearing and reclosure.

Example 12.7

Give the system of Fig. 12.33 where a three-phase fault is applied at the point P as shown.

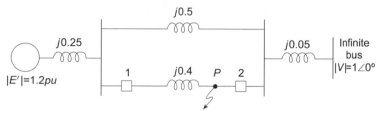

Fig. 12.33

Find the critical clearing angle for clearing the fault with simultaneous opening of the breakers 1 and 2. The reactance values of various components are indicated on the diagram. The generator is delivering 1.0 pu power at the instant preceding the fault.

Solution

With reference to Fig. 12.31, three separate power angle curves are involved.

I. Normal operation (prefault)

$$X_I = 0.25 + \frac{0.5 \times 0.4}{0.5 + 0.4} + 0.05$$

$$= 0.522 \text{ pu}$$

$$P_{eI} = \frac{|E'| |V|}{X_I} \sin \delta = \frac{1.2 \times 1}{0.522} \sin \delta$$

$$= 2.3 \sin \delta \qquad \qquad \text{(i)}$$

Prefault operating power angle is given by

$$1.0 = 2.3 \sin \delta_0$$

or $\qquad \delta_0 = 25.8° = 0.45 \text{ radians}$

II. During fault

It is clear from Fig. 12.31 that no power is transferred during fault, i.e.,

$$P_{eII} = 0$$

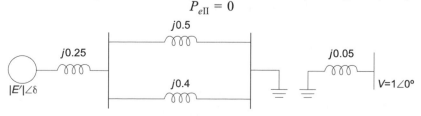

Fig. 12.34

III. Post fault operation (fault cleared by opening the faulted line)

$$X_{III} = 0.25 + 0.5 + 0.05 = 0.8$$

$$P_{eIII} = \frac{1.2 \times 1.0}{0.8} \sin \delta = 1.5 \sin \delta \qquad \text{(iii)}$$

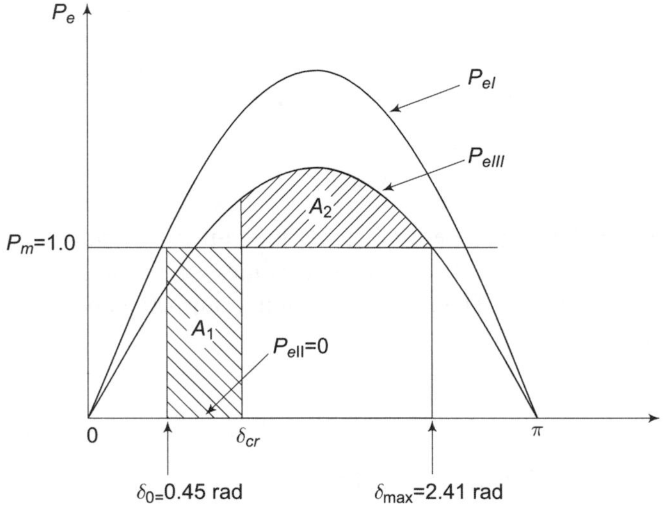

Fig. 12.35

The maximum permissible angle δ_{max} for area $A_1 = A_2$ (see Fig. 12.35) is given by

$$\delta_{max} = \pi - \sin^{-1} \frac{1}{1.5} = 2.41 \text{ radians}$$

Applying equal area criterion for critical clearing angle δ_c

$$A_1 = P_m (\delta_{cr} - \delta_0)$$

$$= 1.0 (\delta_{cr} - 0.45) = \delta_{cr} - 0.45$$

$$A_2 = \int_{\delta_{cr}}^{\delta_{max}} (P_{eIII} - P_m) \, d\delta$$

$$= \int_{\delta_{cr}}^{2.41} (1.5 \sin \delta - 1) \, d\delta$$

$$= -1.5 \cos \delta - \delta \Big|_{\delta_{cr}}^{2.41}$$

$$= - 1.5 \ (\cos 2.41 - \cos \delta_{cr}) - (2.41 - \delta_{cr})$$

$$= 1.5 \cos \delta_{cr} + \delta_{cr} - 1.293$$

Setting $A_1 = A_2$ and solving

$$\delta_{cr} - 0.45 = 1.5 \cos \delta_{cr} + \delta_{cr} - 1.293$$

or $\qquad \cos \delta_{cr} = 0.843/1.5 = 0.562$

or $\qquad \delta_{cr} = 55.8°$

The corresponding power angle diagrams are shown in Fig. 12.35.

Example 12.8

Find the critical clearing angle for the system shown in Fig. 12.36 for a three-phase fault at the point P. The generator is delivering 1.0 pu power under prefault conditions.

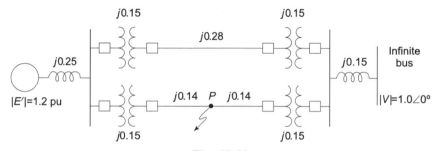

Fig. 12.36

Solution

I. Prefault operation Transfer reactance between generator and infinite bus is

$$X_I = 0.25 + 0.17 + \frac{0.15 + 0.28 + 0.15}{2} = 0.71$$

$\therefore \qquad P_{eI} = \dfrac{1.2 \times 1}{0.71} \sin \delta = 1.69 \sin \delta \qquad\qquad\qquad\qquad \text{(i)}$

The operating power angle is given by

$$1.0 = 1.69 \sin \delta_0$$

or $\qquad \delta_0 = 0.633$ rad

II. During fault The positive sequence reactance diagram during fault is presented in Fig. 12.37a.

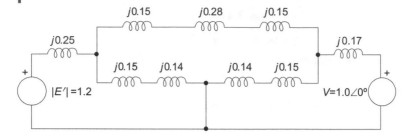

(a) Positive sequence reactance diagram during fault

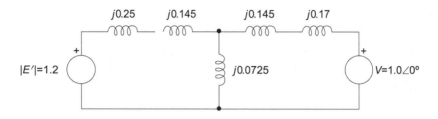

(b) Network after delta-star conversion

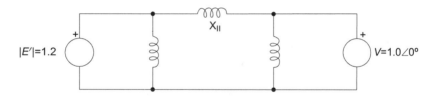

(c) Network after star-delta conversion

Fig. 12.37

Converting delta to star*, the reactance network is changed to that of Fig. 12.37(b). Further, upon converting star to delta, we obtain the reactance network of Fig. 12.37(c). The transfer reactance is given by

$$X_{II} = \frac{(0.25+0.145)\,0.0725+(0.145+0.17)\,0.0725+(0.25+0.145)\dfrac{(0.145+0.17)}{0.075}}{0.075}$$

$$= 2.424$$

$$\therefore \qquad P_{eII} = \frac{1.2 \times 1}{2.424}\ \sin\,\delta = 0.495\ \sin\,\delta \qquad \qquad \text{(ii)}$$

III. Postfault operation (faulty line switched off)

$$X_{III} = 0.25 + 0.15 + 0.28 + 0.15 + 0.17 = 1.0$$

*Node elimination technique would be used for complex network.

$$P_{eIII} = \frac{1.2 \times 1}{1} \sin \delta = 1.2 \sin \delta \qquad \text{(iii)}$$

With reference to Fig. 12.30 and Eq. (12.66), we have

$$\delta_{max} = \pi - \sin^{-1}\frac{1}{1.2} = 2.155 \text{ rad}$$

To find the critical clearing angle, areas A_1 and A_2 are to be equated.

$$A_1 = 1.0 \ (\delta_{cr} - 0.633) - \int_{\delta_0}^{\delta_{cr}} 0.495 \sin \delta \ d\delta$$

and

$$A_2 = \int_{\delta_{cr}}^{\delta_{max}} 1.2 \sin \delta \ d\delta - 1.0 \ (2.155 - \delta_c)$$

Now

$$A_1 = A_2$$

or

$$\delta_{cr} = 0.633 - \int_{0.633}^{\delta_{cr}} 0.495 \sin \delta \ d\delta$$

$$= \int_{\delta_{cr}}^{2.155} 1.2 \sin \delta \ d\delta - 2.155 + \delta_{cr}$$

or $- 0.633 + 0.495 \cos \delta \Big|_{0.633}^{\delta_{cr}} = - 1.2 \cos \delta \Big|_{\delta_{cr}}^{2.155} - 2.155$

or $- 0.633 + 0.495 \cos \delta_{cr} - 0.399 = 0.661 + 1.2 \cos \delta_{cr} - 2.155$

or $\cos \delta_{cr} = 0.655$

or $\delta_{cr} = 49.1°$

Example 12.9

A generator operating at 50 Hz delivers 1 pu power to an infinite bus through a transmission circuit in which resistance is ignored. A fault takes place reducing the maximum power transferable to 0.5 pu whereas before the fault, this power was 2.0 pu and after the clearance of the fault, it is 1.5 pu. By the use of equal area criterion, determine the critical clearing angle.

Solution

All the three power angle curves are shown in Fig. 12.30.

Here $P_{\text{maxI}} = 2.0$ pu, $P_{\text{maxII}} = 0.5$ pu and $P_{\text{maxIII}} = 1.5$ pu

Initial loading $P_m = 1.0$ pu

$$\delta_0 = \sin^{-1}\left(\frac{P_m}{P_{\text{max} I}}\right) = \sin^{-1}\frac{1}{2} = 0.523 \text{ rad}$$

$$\delta_{\text{max}} = \pi \; \sin^{-1}\left(\frac{P_m}{P_{\text{max III}}}\right)$$

$$= \pi - \sin^{-1}\frac{1}{1.5} = 2.41 \text{ rad}$$

Applying Eq. (12.67)

$$\cos \delta_{cr} = \frac{1.0(2.41 - 0.523) - 0.5 \cos 0.523 + 1.5 \cos 2.41}{1.5 - 0.5} = 0.337$$

or $\delta_{cr} = 70.3°$

12.9 NUMERICAL SOLUTION OF SWING EQUATION

In most practical systems, after machine lumping has been done, there are still more than two machines to be considered from the point of view of system stability. Therefore, there is no choice but to solve the swing equation of each machine by a numerical technique on the digital computer. Even in the case of a single machine tied to infinite bus bar, the critical clearing time cannot be obtained from equal area criterion and we have to make this calculation numerically through swing equation. There are several sophisticated methods now available for the solution of the swing equation including the powerful Runge-Kutta method. Here we shall treat the point-by-point method of solution which is a conventional, approximate method like all numerical methods but a well tried and proven one. We shall illustrate the point-by-point method for one machine tied to infinite bus bar. The procedure is, however, general and can be applied to every machine of a multimachine system.

Consider the swing equation

$$\frac{d^2\delta}{dt^2} = \frac{1}{M}(P_m - P_{\text{max}} \sin \delta) = P_a/M;$$

$$\left(M = \frac{GH}{\pi} \text{ or in pu system } M = \frac{H}{\pi f}\right)$$

The solution $\delta(t)$ is obtained at discrete intervals of time with interval spread of Δt uniform throughout. Accelerating power and change in speed which are continuous functions of time are discretized as below:

1. The accelerating power P_a computed at the beginning of an interval is assumed to remain constant from the middle of the preceding interval to the middle of the interval being considered as shown in Fig. 12.38.

2. The angular rotor velocity $\omega = d\delta/dt$ (over and above synchronous velocity ω_s) is assumed constant throughout any interval, at the value computed for the middle of the interval as shown in Fig. 12.38.

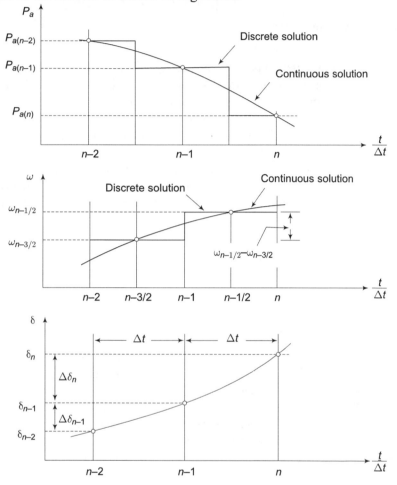

Fig. 12.38 Point-by-point solution of swing equation

In Fig. 12.38, the numbering on $t/\Delta t$ axis pertains to the end of intervals. At the end of the $(n-1)$th interval, the acceleration power is

$$P_{a\,(n-1)} = P_m - P_{\max} \sin \delta_{n-1} \tag{12.68}$$

where δ_{n-1} has been previously calculated. The change in velocity ($\omega = d\delta/dt$), caused by the $P_{a(n-1)}$, assumed constant over Δt from $(n-3/2)$ to $(n-1/2)$ is

$$w_{n\,-1/2} - w_{n-3/2} = (\Delta t/M)\, P_{a(n-1)} \tag{12.69}$$

The change in δ during the $(n-1)$th interval is

$$\Delta\delta_{n-1} = \delta_{n-1} - \delta_{n-2} = \Delta t \omega_{n-3/2} \tag{12.70a}$$

and during the nth interval

$$\Delta\delta_n = \delta_n - \delta_{n-1} = \Delta t \omega_{n-1/2} \tag{12.70b}$$

Subtracting Eq. (12.70a) from Eq. (12.70b) and using Eq. (12.69), we get

$$\Delta \delta_n = \Delta \delta_{n-1} + \frac{(\Delta t)^2}{M} P_{a(n-1)} \qquad (12.71)$$

Using this, we can write

$$\delta_n = \delta_{n-1} + \Delta \delta_n \qquad (12.72)$$

The process of computation is now repeated to obtain $P_{a(n)}$, $\Delta \delta_{n+1}$ and δ_{n+1}. The time solution in discrete form is thus carried out over the desired length of time, normally 0.5 s. Continuous form of solution is obtained by drawing a smooth curve through discrete values as shown in Fig. 12.38. Greater accuracy of solution can be achieved by reducing the time duration of intervals.

The occurrence or removal of a fault or initiation of any switching event causes a discontinuity in accelerating power P_a. If such a discontinuity occurs at the beginning of an interval, then the average of the values of P_a before and after the discontinuity must be used. Thus, in computing the increment of angle occurring during the first interval after a fault is applied at $t = 0$, Eq. (12.71) becomes

$$\Delta \delta_1 = \frac{(\Delta t)^2}{M} + \frac{P_{a0+}}{2}$$

where P_{a0+} is the accelerating power immediately after occurrence of fault. Immediately before the fault the system is in steady state, so that $P_{a0-} = 0$ and δ_0 is a known value. If the fault is cleared at the beginning of the nth interval, in calculation for this interval one should use for $P_{a(n-1)}$ the value $\frac{1}{2}[P_{a(n-1)-} + P_{a(n-1)+}]$, where $P_{a(n-1)-}$ is the accelerating power immediately before clearing and $P_{a(n-1)+}$ is that immediately after clearing the fault. If the discontinuity occurs at the middle of an interval, no special procedure is needed. The increment of angle during such an interval is calculated, as usual, from the value of P_a at the beginning of the interval.

The procedure of calculating solution of swing equation is illustrated in the following example.

Example 12.10

A 20 MVA, 50 Hz generator delivers 18 MW over a double circuit line to an infinite bus. The generator has kinetic energy of 2.52 MJ/MVA at rated speed. The generator transient reactance is $X'_d = 0.35$ pu. Each transmission circuit has $R = 0$ and a reactance of 0.2 pu on a 20 MVA base. $|E'| = 1.1$ pu and infinite bus voltage $V = 1.0 \angle 0°$. A three-phase short circuit occurs at the mid point of one of the transmission lines. Plot swing curves with fault cleared by simultaneous opening of breakers at both ends of the line at 2.5 cycles and 6.25 cycles after the occurrence of fault. Also plot the swing curve over the period of 0.5 s if the fault is sustained.

Solution Before we can apply the step-by-step method, we need to calculate the inertia constant M and the power angle equations under prefault and postfault conditions.

Base MVA = 20

Inertia constant, $M(\text{pu}) = \dfrac{H}{180\,f} = \dfrac{1.0 \times 2.52}{180 \times 50}$

$$= 2.8 \times 10^{-4}\ s^2/\text{elect degree}$$

I *Prefault*

$$X_{\mathrm{I}} = 0.35 + \frac{0.2}{2} = 0.45$$

∴ $\qquad P_{e\mathrm{I}} = P_{\max \mathrm{I}}\ \sin\ \delta$

$$= \frac{1.1 \times 1}{0.45}\sin\ \delta = 2.44\ \sin\ \delta \tag{i}$$

Prefault power transfer $= \dfrac{18}{20} = 0.9$ pu

Initial power angle is given by

$$2.44\ \sin\ \delta_0 = 0.9$$

or $\qquad\qquad \delta_0 = 21.64°$

II *During fault* A positive sequence reactance diagram is shown in Fig. 12.39a. Converting star to delta, we obtain the network of Fig. 12.39b, in which

$$X_{\mathrm{II}} = \frac{0.35 \times 0.1 + 0.2 \times 0.1 + 0.35 \times 0.2}{0.1} = 1.25\ \text{pu}$$

∴ $\qquad P_{e\mathrm{II}} = P_{\max \mathrm{II}}\ \sin\ \delta$

$$= \frac{1.1 \times 1}{1.25}\ \sin\ \delta = 0.88\ \sin\ \delta \tag{ii}$$

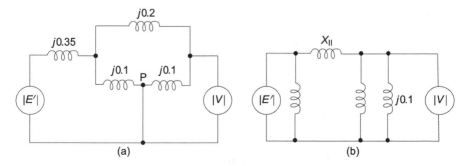

Fig. 12.39

III *Postfault*   With the faulted line switched off,

$$X_{III} = 0.35 + 0.2 = 0.55$$

$\therefore$ $$P_{eIII} = P_{maxIII} \sin \delta$$

$$= \frac{1.1 \times 1}{0.55} \sin \delta = 2.0 \sin \delta \qquad \text{(iii)}$$

Let us choose $\Delta t = 0.05$ s

The recursive relationships for step-by-step swing curve calculation are reproduced below.

$$P_{a(n-1)} = P_m - P_{max} \sin \delta_{n-1} \qquad \text{(iv)}$$

$$\Delta \delta_n = \Delta \delta_{n-1} + \frac{(\Delta t)^2}{M} P_{a(n-1)} \qquad \text{(v)}$$

$$\delta_n = \delta_{n-1} + \Delta \delta_n \qquad \text{(vi)}$$

Since there is a discontinuity in P_e and hence in P_a, the average value of P_a must be used for the first interval.

$$P_a(0_-) = 0 \text{ pu and } P_a(0_+) = 0.9 - 0.88 \sin 21.64° = 0.576 \text{ pu}$$

$$P_a(0_{average}) = \frac{0 + 0.576}{2} = 0.288 \text{ pu}$$

Sustained Fault

Calculations are carried out in Table 12.2 in accordance with the recursive relationship (iv), (v) and (vi) above. The second column of the table shows P_{max} the maximum power that can be transferred at time t given in the first column. P_{max} in the case of a sustained fault undergoes a sudden change at $t = 0_+$ and remains constant thereafter. The procedure of calculations is illustrated below by calculating the row corresponding to $t = 0.15$ s.

$$(0.1 \text{ sec}) = 31.59°$$

$$P_{max} = 0.88$$

$$\sin \delta \, (0.1 \text{ s}) = 0.524$$

$$P_e \, (0.1 \text{ s}) = P_{max} \sin \delta \, (0.1 \text{ s}) = 0.88 \times 0.524 = 0.461$$

$$P_a \, (0.1 \text{ s}) = 0.9 - 0.461 = 0.439$$

$$\frac{(\Delta t)^2}{M} P_a \, (0.1 \text{ s}) = 8.929 \times 0.439 = 3.92°$$

$$\delta \, (0.15 \text{ s}) = \Delta \delta \, (0.1 \text{ s}) + \frac{(\Delta t)^2}{M} P_a \, (0.1 \text{ s})$$

$$= 7.38° + 3.92° = 11.33°$$

$$\delta \, (0.15 \text{ s}) = \delta \, (0.1 \text{ s}) + \Delta \delta \, (0.15 \text{ s})$$

$$= 31.59° + 11.30° = 42.89°$$

$\delta(t)$ for sustained fault as calculated in Table 12.2 is plotted in Fig. 12.40 from which it is obvious that the system is unstable.

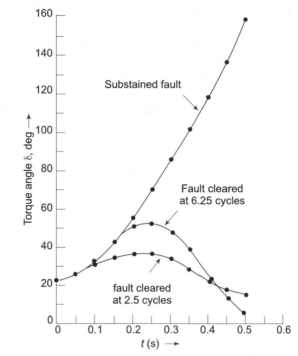

Fig. 12.40 Swing curves for Example 12.10 for a sustained fault and for clearing in 2.5 and 6.25 cycles

Table 12.2 Point-by-point computations of swing curve for sustained fault, $\Delta t = 0.05$ s

t	P_{max}	$\sin \delta$	$P_e = P_{max}\sin\delta$	$P_a = 0.9 - P_e$	$\dfrac{(\Delta t)^2}{M}P_a$	$\Delta\delta$	δ
sec	pu		pu	pu	$= 8.929\,P_a$ deg	deg	deg
0_-	2.44	0.368	0.9	0.0	—	—	21.64
0_+	0.88	0.368	0.324	0.576	—	—	21.64
0_{avg}	—	0.368	—	0.288	2.57	2.57	21.64
0.05	0.88	0.41	0.361	0.539	4.81	7.38	24.21
0.10	0.88	0.524	0.461	0.439	3.92	11.30	31.59
0.15	0.88	0.680	0.598	0.301	2.68	13.98	42.89
0.20	0.88	0.837	0.736	0.163	1.45	15.43	56.87
0.25	0.88	0.953	0.838	0.06	0.55	15.98	72.30
0.30	0.88	0.999	0.879	0.021	0.18	16.16	88.28
0.35	0.88	0.968	0.852	0.048	0.426	16.58	104.44
0.40	0.88	0.856	0.754	0.145	1.30	17.88	121.02
0.45	0.88	0.657	0.578	0.321	2.87	20.75	138.90
0.50	0.88	—	—	—	—	—	159.65

Fault Cleared in 2.5 Cycles

Time to clear fault = $\dfrac{2.5}{50}$ = 0.05 s

P_{max} suddenly changes from 0.88 at $t = 0.05$ to 2.0 at $t = 0.05_+$. Since the discontinuity occurs at the beginning of an interval, the average value of P_a will be assumed to remain constant from 0.025 s to 0.075 s. The rest of the procedure is the same and complete calculations are shown in Table 12.3. The swing curve is plotted in Fig. 12.40 from which we find that the generator undergoes a maximum swing of 37.5° but is stable as δ finally begins to decrease.

Table 12.3 Computations of swing curves for fault cleared at 2.5 cycles (0.05 s), $\Delta t = 0.05$ s

t	P_{max}	$\sin\delta$	$P_e = P_{max}\sin\delta$	$P_a = 0.9 - P_e$	$\dfrac{(\Delta t)^2}{M}P_a$ = 8.929 P_a	$\Delta\delta$	δ
sec	pu		pu	pu	deg	deg	deg
0_	2.44	0.368	0.9	0.0	—	—	21.64
0_+	0.88	0.368	0.324	0.576	—	—	21.64
0_avg	—	0.368	—	0.288	2.57	2.57	21.64
0.05_	0.88	0.41	0.36	0.54	—	—	24.21
0.05_+	2.00	0.41	0.82	0.08	—	—	24.21
0.05_avg				0.31	2.767	5.33	24.21
0.10	2.00	0.493	0.986	− 0.086	− 0.767	4.56	29.54
0.15	2.00	0.56	1.12	− 0.22	− 1.96	2.60	34.10
0.20	2.00	0.597	1.19	− 0.29	− 2.58	0.02	36.70
0.25	2.00	0.597	1.19	− 0.29	− 2.58	− 2.56	37.72
0.30	2.00	0.561	1.12	− 0.22	− 1.96	− 4.52	34.16
0.35	2.00	0.494	0.989	− 0.089	− 0.79	− 5.31	29.64
0.40	2.00	0.41	0.82	0.08	0.71	− 4.60	24.33
0.45	2.00	0.337	0.675	0.225	2.0	− 2.6	19.73
0.50							17.13

Fault Cleared in 6.25 Cycles

Time to clear fault = $\dfrac{6.25}{50}$ = 0.125 s

Since the discontinuity now lies in the middle of an interval, no special procedure is necessary, as in deriving Eqs. (iv) – (vi) discontinuity is assumed to occur in the middle of the time interval. The swing curve as calculated in Table 12.4 is also plotted in Fig. 12.40. It is observed that the system is stable with a maximum swing of 52.5° which is much larger than that in the case of 2.5 cycle clearing time.

To find the critical clearing time, swing curves can be obtained, similarly, for progressively greater clearing time till the torque angle δ increases without bound. In this example, however, we can first find the critical clearing angle using Eq. (12.67) and then read the critical clearing time from the swing curve corresponding to the sustained fault case. The values obtained are:

Critical clearing angle = 118.62°

Critical clearing time = 0.38 s

Table 12.4 Computations of swing curve for fault cleared at
6.25 cycles (0.125s), $\Delta t = 0.05$ s

t	P_{max}	$sin\ \delta$	$P_e = P_{max}sin\delta$	$P_a = 0.9 - P_e$	$\dfrac{(\Delta t)^2}{M}P_a$ $= 8.929\ P_a$	$\Delta\delta$	δ
sec	pu		pu	pu	deg	deg	deg
0_-	2.44	0.368	0.9	0.0	–	–	21.64
0_+	0.88	0.368	0.324	0.576	–	–	21.64
0_{avg}	–	0.368	–	0.288	2.57	2.57	21.64
0.05	0.88	0.41	0.361	0.539	4.81	7.38	24.21
0.10	0.88	0.524	0.461	0.439	3.92	11.30	31.59
0.15	2.00	0.680	1.36	– 4.46	– 4.10	7.20	42.89
0.20	2.00	0.767	1.53	– 0.63	– 5.66	1.54	50.09
0.25	2.00	0.78	1.56	– 0.66	– 5.89	– 4.35	51.63
0.30	2.00	0.734	1.46	– 0.56	– 5.08	– 9.43	47.28
0.35	2.00	0.613	1.22	– 0.327	– 2.92	– 12.35	37.85
0.40	2.00	0.430	0.86	0.04	0.35	– 12.00	25.50
0.45	2.00	0.233	0.466	0.434	3.87	– 8.13	13.50
0.50	2.00						5.37

12.10 MULTIMACHINE STABILITY

From what has been discussed so far, the following steps easily follow for determining multimachine stability.

1. From the prefault load flow data determine E'_k voltage behind transient reactance for all generators. This establishes generator emf magnitudes $|E_k|$ which remain constant during the study and initial rotor angle $\delta^o_k = \angle E_k$. Also record prime mover inputs to generators, $P_{mk} = P^o_{Gk}$.

2. Augment the load flow network by the generator transient reactances. Shift network buses behind the transient reactances.

3. Find Y_{BUS} for various network conditions—during fault, post fault (faulted line cleared), after line reclosure.

4. For faulted mode, find generator outputs from power angle equations (generalized forms of Eq. (12.27)) and solve swing equations step by step (point-by-point method).

5. Keep repeating the above step for post fault mode and after line reclosure mode.
6. Examine $\delta(t)$ plots of all generators and establish the answer to the stability question.

The above steps are illustrated in the following example.

Example 12.11

A 50 Hz, 220 kV transmission line has two generators and an infinite bus as shown in Fig. 12.41. The transformer and line data are given in Table 12.5. A three-phase fault occurs as shown. The prefault load flow solution is presented in Table 12.6. Find the swing equation for each generator during the fault period.

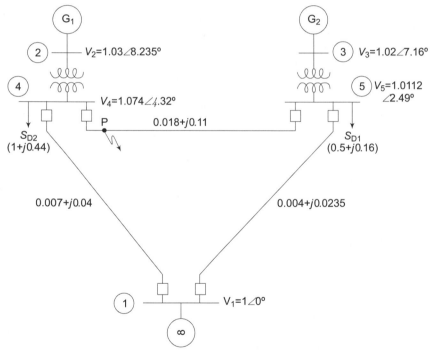

Fig. 12.41

Data are given below for the two generators on a 100 MVA base.
     Gen 1 500 MVA, 25 kV, $X_d' = 0.067$ pu, H = 12 MJ/MVA
     Gen 2 300 MVA, 20 kV, $X_d' = 0.10$ pu, H = 9 MJ/MVA
     Plot the swing curves for the machines at buses 2 and 3 for the above fault which is cleared by simultaneous opening of the circuit breakers at the ends of the faulted line at (i) 0.275 s and (ii) 0.08 s.

Table 12.5 Line and transformer data for Ex. 12.11. All values are in pu on 220 kV, 100 MVA base

Bus to bus	Series Z		Half line charging
	R	X	
Line 4-5	0.018	0.11	0.113
Line 5-1	0.004	0.0235	0.098
Line 4-1	0.007	0.04	0.041
Trans: 2-4	—	0.022	—
Trans: 3-5	—	0.04	—

Table 12.6 Bus data and prefault load-flow values in pu on 220 kV, 100 MVA base

S.No. and Bus No.	Voltage Polar Form	Bus type	Voltage		Generation		Load	
			Real e	Imaginary f	P	Q	P	Q
1	$1.0\angle0°$	Slack	1.00	0.0	− 3.8083	−0.2799	0	0
2	$1.03\angle8.35°$	PV	1.0194	0.1475	3.25	0.6986	0	0
3	$1.02\angle7.16°$	PV	1.0121	0.1271	2.10	0.3110	0	0
4	$1.0174\angle4.32°$	PQ	1.0146	0.767	0	1.0	1.0	0.44
5	$1.0112\angle2.69°$	PQ	1.0102	0.0439	0	0	0.5	0.16

Solution Before determining swing equations, we have to find transient internal voltages.

The current into the network at bus 2 based on the data in Table 12.6 is

$$I_2 = \frac{P_2 - jQ_2}{V_2^*} = \frac{3.25 - j0.6986}{1.03\angle - 8.23519°}$$

$$E_2' = (1.0194 + j0.1475) + \frac{3.25 - j0.6986}{1.03 \angle - 8.23519°} \times 0.067\angle90°$$

$$- 1.0340929 + j0.3632368$$

$$= 1.0960333 \angle19.354398° = 1.0960 \angle0.3377 \text{ rad}$$

$$E_1' = 1.0 \angle0° \text{ (slack bus)}$$

$$E_3' = (1.0121 + j0.1271) + \frac{2.1 - j0.311}{1.02\angle - 7.15811°} \times 0.1\angle90°$$

$$= 1.0166979 + j0.335177 = 1.0705 \angle18.2459°$$

$$= 1.071 \angle0.31845 \text{ rad}$$

The loads at buses 4 and 5 are represented by the admittances calculated as follows:

$$Y_{L4} = \frac{1.0 - j0.44}{(1.0174)^2}(0.9661 - j0.4251)$$

$$Y_{L5} = \frac{0.5 - j0.16}{(1.0112)^2} \quad (0.4889 - j0.15647)$$

Prefault Bus Matrix

Load admittances, along with the transient reactances, are used with the line and transformer admittances to form the prefault augmented bus admittance matrix which contains the transient reactances of the machines. We will, therefore, now designate as buses 2 and 3, the fictitious internal nodes between the internal voltages and the transient reactances of the machines. Thus we get

$$Y_{22} = \frac{1}{(j0.067 + j0.022)} = -j11.236$$

$$Y_{24} = j11.236 = Y_{42}$$

$$Y_{33} = \frac{1}{j0.04 + j0.1} = -j7.143$$

$$Y_{35} = j7.143 = Y_{53}$$

$$Y_{44} = Y_{L4} + Y_{41} + Y_{45} + \frac{B_{41}}{2} + \frac{B_{45}}{2} + Y_{24}$$

$$= 0.9660877 - j0.4250785 + 4.245 - j24.2571 + 1.4488 - j8.8538 + j0.041 + j0.113 - j11.2359$$

$$= 6.6598977 - j44.6179$$

$$Y_{55} = Y_{L5} + Y_{54} + Y_{51} + \frac{B_{54}}{2} + \frac{B_{51}}{2} + Y_{35}$$

$$= 0.4889 - j0.1565 + 1.4488 - j8.8538 + 7.0391 - j41.335 + j0.113 + j0.098 - j7.1428$$

$$= 8.976955 - j57.297202$$

The complete augmented prefault Y_{BUS} matrix is shown in Table 12.7.

Table 12.7 The augmented prefault bus admittance matrix for Ex. 12.11, admittances in pu

Bus	1	2	3	4	5
1	11.284–j65.473	0	0	−4.245 + j24.257	−7.039 + j41.355
2	0	−j11.2359	0	j11.2359	0
3	0	0	− j7.1428	0	j7.1428
4	−4.245 + j24.257	j11.2359	0	6.6598–j44.617	−1.4488 +j8.8538
5	−7.039 + j41.355	0	0 + j7.1428	−1.4488 + j8.8538	8.9769 + j57.2972

During Fault Bus Matrix

Since the fault is near bus 4, it must be short circuited to ground. The Y_{BUS} during the fault conditions would, therefore, be obtained by deleting 4th row and 4th column from the above augmented prefault Y_{BUS} matrix. Reduced fault matrix (to the generator internal nodes) is obtained by eliminating the new 4th row and column (node 5) using the relationship

$$Y_{kj(new)} = Y_{kj(old)} - Y_{kn(old)}\, Y_{nj(Old)}/Y_{nn(old)}$$

The reduced faulted matrix (Y_{BUS} during fault) (3 × 3) is given in Table 12.8, which clearly depicts that bus 2 decouples from the other buses during the fault and that bus 3 is directly connected to bus 1, showing that the fault at bus 4 reduces to zero the power pumped into the system from the generator at bus 2 and renders the second generator at bus 3 to give its power radially to bus 1.

Table 12.8 Elements of Y_{BUS} (during fault) and Y_{BUS} (post fault) for Ex. 12.11, admittances in pu.

	Reduced during fault Y_{BUS}		
Bus	1	2	3
1	5.7986–j35.6301	0	– 0.0681 + j5.1661
2	0	– j11.236	0
3	– 0.0681+ j5.1661	0	0.1362 – j6.2737
	Reduced post fault Y_{BUS}		
1	1.3932 – j13.8731	– 0.2214 + j7.6289	– 0.0901 + j6.0975
2	– 0.2214 + j7.6289	0.5 – j7.7898	0
3	– 0.0901 + j6.0975	0	0.1591 – j6.1168

Post Fault Bus Matrix

Once the fault is cleared by removing the line, simultaneously opening the circuit breakers at the either ends of the line between buses 4 and 5, the prefault Y_{BUS} has to be modified again. This is done by substituting $Y_{45} = Y_{54} = 0$ and subtracting the series admittance of line 4–5 and the capacitive susceptance of half the line from elements Y_{44} and Y_{55}.

$$Y_{44(post\ fault)} = Y_{44(prefault)} - Y_{45} - B_{45}/2$$

$$= 6.65989 - j44.6179 - 1.448 + j8.853 - j0.113$$

$$= 5.2111 - j35.8771$$

Similarly, $Y_{55(post\ fault)} = 7.5281 - j48.5563$

The reduced post fault Y_{BUS} is shown in the lower half to Table 12.8. It may be noted that 0 element appears in 2nd and 3rd rows. This shows that, physically, the generators 1 and 2 are not interconnected when line 4–5 is removed.

During Fault Power Angle Equation

$$P_{e2} = 0$$

$$P_{e3} = Re\ [Y_{33}E_3'E_3'^* + E_3'^*\ Y_{31}E_1'];\ \text{since}\ Y_{32} = 0$$

$$= |E_3'|^2\ G_{33} + |E_1'|\ |E_3'|\ |Y_{31}|\ \cos\ (\delta_{31} - \theta_{31})$$

$$= (1.071)^2\ (0.1362) + 1 \times 1.071 \times 5.1665\ \cos\ (\delta_3 - 90.755°)$$

$$P_{e3} = 0.1561 + 5.531\ \sin\ (\delta_3 - 0.755°)$$

Postfault Power Angle Equations

$$P_{e2} = |E_2'|^2\ G_{22} + |E_1'|\ |E_2'|\ |Y_{21}|\ \cos\ (\delta_{21} - \theta_{21})$$

$$= 1.096^2 \times 0.5005 + 1 \times 1.096 \times 7.6321\ \cos\ (\delta_2 - 91.662°)$$

$$= 0.6012 + 8.365\ \sin\ (\delta_2 - 1.662°)$$

$$P_{e3} = |E_3'|^2 G_{33} + |E_1'|\ |E_3'|\ |Y_{31}|\ \cos\ (\delta_{31} - \theta_{31})$$

$$= 1.071^2 \times 0.1591 + 1 \times 1.071 \times 6.098\ \cos\ (\delta_3 - 90.8466°)$$

$$= 0.1823 + 6.5282\ \sin\ (\delta_3 - 0.8466°)$$

Swing Equations—During Fault

$$\frac{d^2\delta_2}{dt^2} = \frac{180f}{H_2}\ (P_{m2} - P_{e2}) = \frac{180f}{H_2}P_{a_2}$$

$$= \frac{180f}{12}\ (3.25 - 0)\ \text{elect deg/s}^2$$

$$\frac{d^2\delta_3}{dt^2} = \frac{180f}{H_3}\ (P_{m3} - P_{e3})$$

$$= \frac{180f}{9}\ [2.1 - \{0.1561 + 5.531\ \sin\ (\delta_3 - 0.755°)\}]$$

$$= \frac{180f}{9}[1.9439 - 5.531\ \sin\ (\delta_3 - 0.755°)]\ \text{elect deg/s}^2$$

Swing Equations—Postfault

$$\frac{d^2\delta_2}{dt^2} = \frac{180f}{11}[3.25 - \{0.6012 + 8.365\ \sin\ (\delta_2 - 1.662°)\}]\ \text{elect deg/s}^2$$

$$\frac{d^2\delta_3}{dt^2} = \frac{180f}{9}[2.10 - \{0.1823 + 6.5282\ \sin(\delta_3 - 0.8466°)\}]\text{elect deg/s}^2$$

It may be noted that in the above swing equations, P_a may be written in general as follows:

$$P_a = P_m - P_c - P_{max} \sin (\delta - \gamma)$$

Digital Computer Solution of Swing Equation

The above swing equations (during fault followed by post fault) can be solved by the point-by-point method presented earlier or by the Euler's method presented in the later part of this section. The plots of δ_2 and δ_3 are given in Fig. 12.42 for a clearing time of 0.275 s and in Fig. 12.43 for a clearing time of 0.08 s. For the case (i), the machine 2 is unstable, while the machine 3 is stable but it oscillates wherein the oscillations are expected to decay if effect of damper winding is considered. For the case (ii), both machines are stable but the machine 2 has large angular swings.

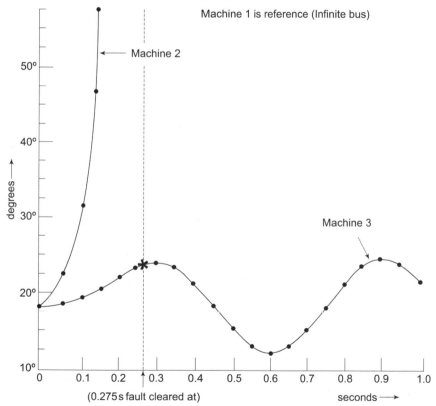

Fig. 12.42 Swing curves for machines 2 and 3 of Example 12.1 for clearing at 0.275 s.

If the fault is a transient one and the line is reclosed, power angle and swing equations are needed for the period after reclosure. These can be computed from the reduced Y_{BUS} matrix after line reclosure.

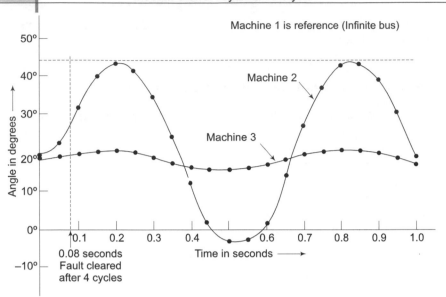

Fig. 12.43 Swing curves for machines 2 and 3 of
Example 12.11 for clearing at 0.08 s

Consideration of Automatic Voltage Regulator (AVR) and Speed Governor Loops

This requires modelling of these two control loops in form of differential equations. At the end of every step in the stability algorithm, the programme computes the modified values of E'_k and P_{mk} and then proceeds to compute the next step. This considerably adds to the dimensionality and complexity of stability calculations. To reduce the computational effort, speed control can continue to be ignored without loss of accuracy of results.

State Variable Formulation of Swing Equations

The swing equation for the kth generator is

$$\frac{d^2\delta_k}{dt^2} = \frac{\pi f}{H_k}(P^0{}_{Gk} - P_{Gk}); \quad k = 1, 2, ..., m \tag{12.73}$$

For the multimachine case, it is more convenient to organise Eq. (12.73) is state variable form. Define

$$x_{1k} = \delta_k = \angle E'_k$$

$$x_{2k} = \dot{\delta}_k$$

Then

$$\dot{x}_{1k} = x_{2k}$$

$$\dot{x}_2 k = \frac{\pi f}{H_k}(P^0{}_{Gk} - P_{Gk}), \ k = 1, 2, ..., m \qquad (12.74)$$

Initial state vector (upon occurrence of fault) is

$$x^0_{1k} = \delta^0_k = \angle E^0_k$$

$$x^0_{2k} = 0 \qquad (12.75)$$

The state form of swing equations (Eq. (12.74)) can be solved by the many available integration algorithms (modified Euler's method is a convenient choice).

Computational Algorithm for Obtaining Swing Curves Using Modified Euler's Method

1. Carry out a load flow study (prior to disturbance) using specified voltages and powers.
2. Compute voltage behind transient reactances of generators (E^0_k) using Eq. (9.31). This fixes generator emf magnitudes and initial rotor angle (reference slack bus voltage V^0_1).
3. Compute, Y_{BUS} (during fault, post fault, line reclosed).
4. Set time count $r = 0$.
5. Compute generator power outputs using appropriate Y_{BUS} with the help of the general form of Eq. (12.27). This gives $P^{(r)}_{Gk}$ for $t = t^{(r)}$.

   Note: After the occurrence of the fault, the period is divided into uniform discrete time intervals (Δt) so that time is counted as $t^{(0)}, t^{(1)}, ...$ A typical value of Δt is 0.05 s.
6. Compute $[(\dot{x}^{(r)}_{1k}, \dot{x}^{(r)}_{2k}), k = 1, 2, ..., m]$ from Eqs. (12.74).
7. Compute the first state estimates for $t = t^{(r+1)}$ as

   $$x^{(r+1)}_{1k} = x^{(r)}_{1k} + \dot{x}^{(r)}_{1k} \Delta t$$
   $$k = 1, 2, ..., m$$
   $$x^{(r+1)}_{2k} = x^{(r)}_{2k} + \dot{x}^{(r)}_{2k} \Delta t$$

8. Compute the first estimates of $E_k^{(r+1)}$

   $$E_k^{(r+1)} = E^0_k (\cos x^{(r+1)}_{1k} + j \sin x^{(r+1)}_{1k})$$

9. Compute $P^{(r+1)}_{Gk}$; (appropriate Y_{BUS} and Eq. (12.72)).
10. Compute $[(\dot{x}^{(r+1)}_{1k}, \dot{x}^{(r+1)}_{2k}), k = 1, 2, ..., m]$ from Eqs. (12.74).
11. Compute the average values of state derivatives

    $$\dot{x}^{(r)}_{1k, \text{ avg}} = \tfrac{1}{2}[\dot{x}_{1k}{}^{(r)} + \dot{x}^{(r+1)}_{1k}]$$
    $$k = 1, 2, ..., m$$
    $$\dot{x}^{(r)}_{2k, \text{avg}} = \tfrac{1}{2}[\dot{x}^{(r)}_{2k} + \dot{x}^{(r+1)}_{2k}]$$

12. Compute the final state estimates for $t = t^{(r+1)}$.

$$x_{1k}^{(r+1)} = x_{1k}^{(r)} + \dot{x}_{1k,\,\text{avg}}^{(r)}\, \varDelta t$$

$$k = 1, 2, \ldots, m$$

$$x_{2k}^{(r+1)} = x_{2k}^{(r)} + \dot{x}_{2k,\,\text{avg}}^{(r)}\, \varDelta t$$

13. Compute the final estimate for E_k at $t = t^{(r+1)}$ using

$$E_k^{(r+1)} = |E_k^0|\cos x_{1k}^{(r+1)} + j \sin x_{1k}^{(r+1)}$$

14. Print $(x_{1k}^{(r+1)}, x_{2k}^{(r+1)})$; $k = 1, 2, \ldots, m$

15. Test for time limit (time for which swing curve is to be plotted), i.e., check if $r > r_{\text{final}}$. If not, $r = r + 1$ and repeat from step 5 above. Otherwise print results and stop.

The swing curves of all the machines are plotted. If the rotor angle of a machine (or a group of machines) with respect to other machines increases without bound, such a machine (or group of machines) is unstable and eventually falls out of step.

The computational algorithm given above can be easily modified to include simulation of voltage regulator, field excitation response, saturation of flux paths and governor action.

Stability Study of Large Systems

To limit the computer memory and the time requirements and for the sake of computational efficiency, a large multi-machine system is divided into a study subsystem and an external system. The study subsystem is modelled in detail whereas approximate modelling is carried out for the external subsystem. The total study is rendered by the modern technique of dynamic equivalencing. In the external subsystem, number of machines is drastically reduced using various methods—coherency based methods being most popular and widely used by various power utilities in the world.

12.11 SOME FACTORS AFFECTING TRANSIENT STABILITY

We have seen in this chapter that the two-machine system can be equivalently reduced to a single machine connected to infinite bus bar. The qualitative conclusions regarding system stability drawn from a two-machine or an equivalent one-machine infinite bus system can be easily extended to a multimachine system. In the last article we have studied the algorithm for determining the stability of a multimachine system.

It has been seen that transient stability is greatly affected by the type and location of a fault, so that a power system analyst must at the very outset of a stability study decide on these two factors. In our examples we have selected a 3–phase fault which is generally more severe from point of view of power transfer. Given the type of fault and its location let us now consider other factors which affect transient stability and therefrom draw the conclusions, regarding methods of improving the transient stability limit of a system and making it as close to the steady state limit as possible.

For the case of one machine connected to infinite bus, it is easily seen from Eq. (12.71) that an increase in the inertia constant M of the machine reduces the angle through which it swings in a given time interval offering thereby a method of improving stability but this cannot be employed in practice because of economic reasons and for the reason of slowing down the response of the speed governor loop (which can even become oscillatory) apart from an excessive rotor weight.

With reference to Fig. 12.30, it is easily seen that for a given clearing angle, the accelerating area decreases but the decelerating area increases as the maximum power limit of the various power angle curves is raised, thereby adding to the transient stability limit of the system. The maximum steady power of a system can be increased by raising the voltage profile of the system and by reducing the transfer reactance. These conclusions along with the various transient stability cases studied, suggest the following method of improving the transient stability limit of a power system.

1. Increase of system voltages, use of AVR.
2. Use of high speed excitation systems.
3. Reduction in system transfer reactance.
4. Use of high speed reclosing breakers (see Fig. 12.32). Modern tendency is to employ single-pole operation of reclosing circuit breakers.

When a fault takes place on a system, the voltages at all buses are reduced. At generator terminals, these are sensed by the automatic voltage regulators which help restore generator terminal voltages by acting within the excitation system. Modern exciter systems having solid state controls quickly respond to bus voltage reduction and can achieve from one-half to one and one-half cycles $(1/2 – 1\frac{1}{2})$ gain in critical clearing times for three-phase faults on the HT bus of the generator transformer.

Reducing transfer reactance is another important practical method of increasing stability limit. Incidentally this also raises system voltage profile. The reactance of a transmission line can be decreased (i) by reducing the conductor spacing, and (ii) by increasing conductor diameter (see Eq. (2.37)). Usually, however, the conductor spacing is controlled by other features such as lightning protection and minimum clearance to prevent the arc from one phase moving to another phase. The conductor diameter can be increased by using material of low conductivity or by hollow cores. However, normally, the conductor configuration is fixed by economic considerations quite apart from stability. The use of bundled conductors is, of course, an effective means of reducing series reactance.

Compensation for line reactance by series capacitors is an effective and economical method of increasing stability limit specially for transmission distances of more than 350 km. The degree of series compensation, however, accentuates the problems of protective relaying, normal voltage profiles, and overvoltages during line-to-ground faults. Series compensation becomes more effective and economical if part of it is switched on so as to increase the degree of compensation upon the occurrence of a disturbance likely to cause instability. Switched series capacitors simultaneously decrease fluctuation of load voltages and raise the transient stability limit to a value almost equal to the steady state limit. Switching shunt capacitors on or switching shunt reactors off also raises stability limits (see Example 12.2) but the MVA rating of shunt capacitors required is three to six times the rating of switched series capacitors for the same increase in stability limit. Thus series capacitors are preferred unless shunt elements are required for other purposes, say, control of voltage profile.

Increasing the number of parallel lines between transmission points is quite often used to reduce transfer reactance. It adds at the same time to reliability of the transmission system. Additional line circuits are not likely to prove economical unit *l* after all feasible improvements have been carried out in the first two circuits.

As the majority of faults are transient in nature, rapid switching and isolation of unhealthy lines followed by reclosing has been shown earlier to be a great help in improving the stability margins. The modern circuit breaker technology has now made it possible for line clearing to be done as fast as in two cycles. Further, a great majority of transient faults are line-to-ground in nature. It is natural that methods have been developed for selective single pole opening and reclosing which further aid the stability limits. With reference to Fig. 12.17, if a transient LG fault is assumed to occur on the generator bus, it is immediately seen that during the fault there will now be a definite amount of power transfer, as different from zero power transfer for the case of a three-phase fault. Also when the circuit breaker pole corresponding to the faulty line is opened, the other two lines (healthy ones) remain intact so that considerable power transfer continues to take place via these lines in comparison to the case of three-pole switching when the power transfer on fault clearing will be reduced to zero. It is, therefore, easy to see why the single pole switching and reclosing aids in stability problem and is widely adopted. These facts are illustrated by means of Example 12.12. Even when the stability margins are sufficient, single pole switching is adopted to prevent large swings and consequent voltage dips. Single pole switching and reclosing is, of course, expensive in terms of relaying and introduces the associated problems of overvoltages caused by single pole opening owing to line capacitances. Methods are available to nullify these capacitive coupling effects.

Recent Trends

Recent trends in design of large alternators tend towards lower short circuit ratio (SCR = $1/X_d$), which is achieved by reducing machine air gap with consequent savings in machine mmf, size, weight and cost. Reduction in the size of rotor reduces inertia constant, lowering thereby the stability margin. The loss in stability margin is made up by such features as lower reactance lines, faster circuit breakers and faster excitation systems as discussed already, and a faster system valving to be discussed later in this article.

A stage has now been reached in technology whereby the methods of improving stability, discussed above, have been pushed to their limits, e.g., clearing times of circuit breakers have been brought down to virtually irreducible values of the order of two cycles. With the trend to reduce machine inertias there is a constant need to determine availability, feasibility and applicability of new methods for maintaining and/or improving system stability. A brief account of some of the recent methods of maintaining stability is given below:

HVDC Links

Increased use of HVDC links employing thyristors would alleviate stability problems. A dc link is asynchronous, i.e., the two ac system at either end do not have to be controlled in phase or even be at exactly the same frequency as they do for an ac link, and the power transmitted can be readily controlled. There is no risk of a fault in one system causing loss of stability in the other system.

Breaking Resistors

For improving stability where clearing is delayed or a large load is suddenly lost, a resistive load called a breaking resistor is connected at or near the generator bus. This load compensates for at least some of the reduction of load on the generators and so reduces the acceleration. During a fault, the resistors are applied to the terminals of the generators through circuit breakers by means of an elaborate control scheme. The control scheme determines the amount of resistance to be applied and its duration. The breaking resistors remain on for a matter of cycles both during fault clearing and after system voltage is restored.

Short Circuit Current Limiters

These are generally used to limit the short circuit duty of distribution lines. These may also be used in long transmission lines to modify favourably the transfer impedance during fault conditions so that the voltage profile of the system is somewhat improved, thereby raising the system load level during the fault.

Turbine Fast Valving or Bypass Valving

The two methods just discussed above are an attempt at replacing the system load so as to increase the electrical output of the generator during fault conditions. Another recent method of improving the stability of a unit is to decrease the mechanical input power to the turbine. This can be accomplished by means of fast valving, where the difference between mechanical input and reduced electrical output of a generator under a fault, as sensed by a control scheme, initiates the closing of a turbine valve to reduce the power input. Briefly, during a fast valving operation, the interceptor valves are rapidly shut (in 0.1 to 0.2 sec) and immediately reopened. This procedure increases the critical switching time long enough so that in most cases, the unit will remain stable for faults with stuck-breaker clearing times. The scheme has been put to use in some stations in the USA.

Full Load Rejection Technique

Fast valving combined with high-speed clearing time will suffice to maintain stability in most of the cases. However, there are still situations where stability is difficult to maintain. In such cases, the normal procedure is to automatically trip the unit off the line. This, however, causes several hours of delay before the unit can be put back into operation. The loss of a major unit for this length of time can seriously jeopardize the remaining system.

To remedy these situations, a full load rejection scheme could be utilized after the unit is separated from the system. To do this, the unit has to be equipped with a large steam bypass system. After the system has recovered from the shock caused by the fault, the unit could be resynchronized and reloaded. The main disadvantage of this method is the extra cost of a large bypass system.

Example 12.12

The system shown in Fig. 12.44 is loaded to 1 pu. Calculate the swing curve and ascertain system stability for:

(i) LG fault three pole switching followed by reclosure, line found healthy.

(ii) LG fault single pole switching followed by reclosure, line found healthy. Switching occurs at 3.75 cycles (0.075 sec) and reclosure occurs at 16.25 cycles (0.325 sec). All values shown in the figure are in pu.

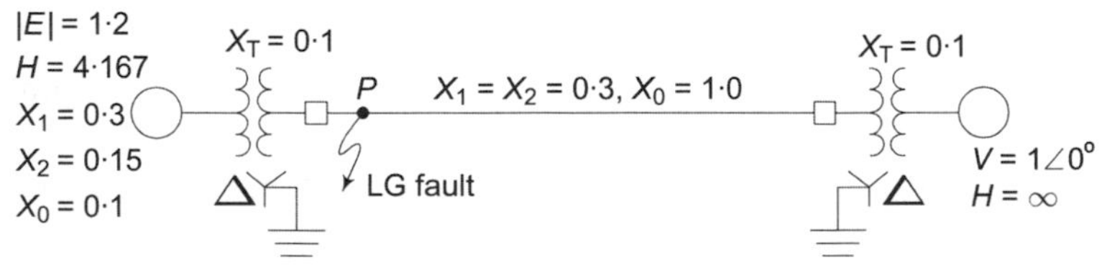

Fig. 12.44

Solution   The sequence networks of the system are drawn and suitably reduced in Figs. 12.45a, b and c.

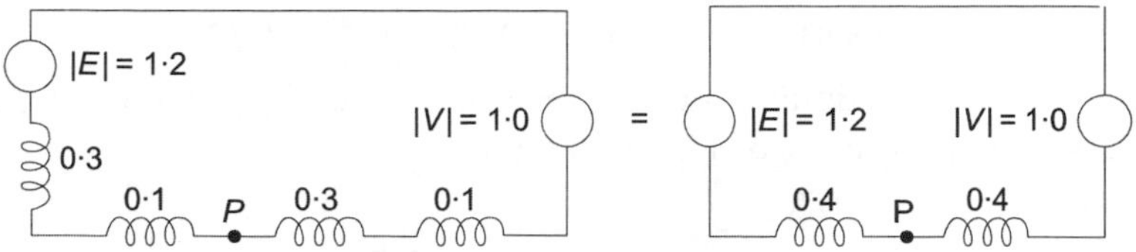

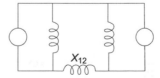

(a) Positive sequence network

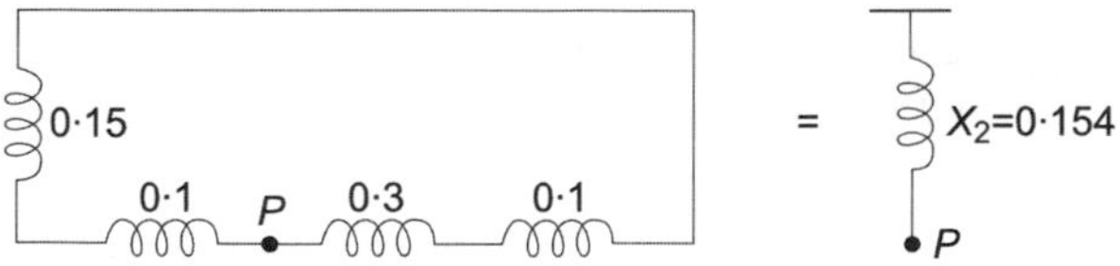

(b) Negative sequence network

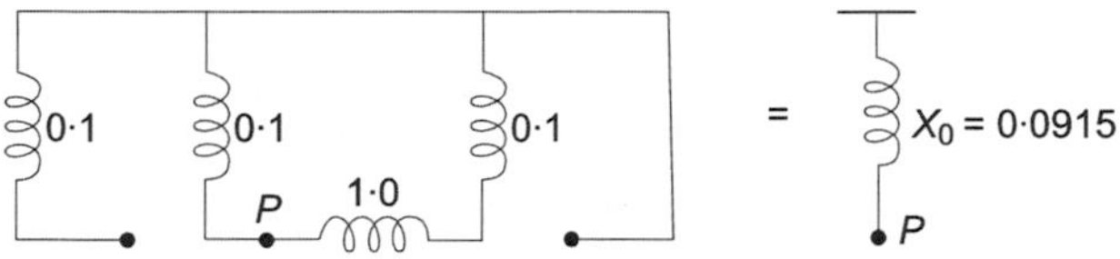

(c) Zero sequence network

Fig. 12.45

For an LG fault at P the sequence networks will be connected in series as shown in Fig. 12.46. A star-delta transformation reduces Fig. 12.38 to that of Fig. 12.47 from which we have the transfer reactance

$$X_{12}(\text{LG fault}) = 0.4 + 0.4 + \frac{0.4 \times 0.4}{0.246} = 1.45$$

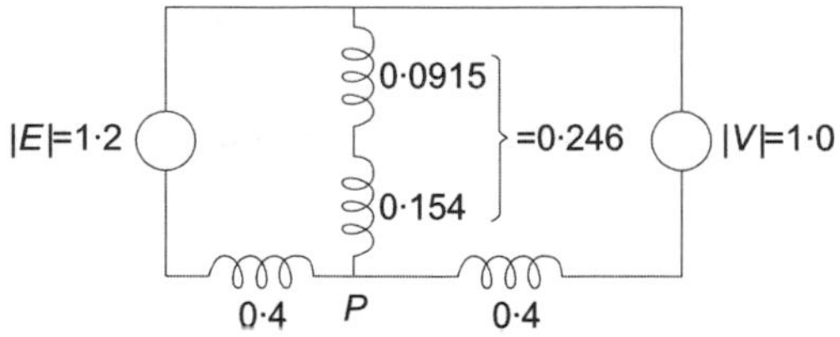

Fig. 12.46 Connection of sequence networks for an LG fault

Fig. 12.47 Transfer impedance for an LG fault

When the circuit breaker poles corresponding to the faulted line are opened (it corresponds to a single-line open fault) the connection of sequence networks is shown in Fig. 12.48. From the reduced network of Fig. 12.49 the transfer reactance with faulted line switched off is

X_{12} (faulted line open) $= 0.4 + 0.42 + 0.4 = 1.22$

Under healthy conditions transfer reactance is easily obtained from the positive sequence network of Fig. 12.45 a as

X_{12}(line healthy) $= 0.8$

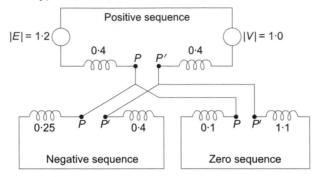

Fig. 12.48 Connection of sequence networks with faulted line switched off

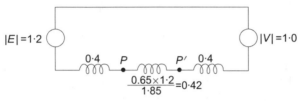

Fig. 12.49 Reduced network of Fig. 12.48 giving transfer reactance

Power angle equations

Prefault

$$P_{eI} = \frac{|E||V|}{X_{12}} \sin \delta = \frac{1.2 \times 1}{0.8} \sin \delta = 1.5 \sin \delta$$

Initial load $= 1.0$ pu

Initial torque angle is given by

$$1 = 1.5 \sin \delta_o$$

or $$\delta_o = 41.8°$$

During fault

$$P_{eII} = \frac{1.2 \times 1}{1.45} \sin \delta = 0.827 \sin \delta$$

During single pole switching

$$P_{eIII} = \frac{1.2 \times 1}{1.22} \sin \delta = 0.985 \sin \delta$$

During three pole switching

$$P_{eIII} = 0$$

Postfault

$$P_{eIV} = P_{eI} = 1.5 \sin \delta$$

Now

$$\Delta\delta_n = \Delta\delta_{n-1} + \frac{(\Delta t)^2}{M} P_{a(n-1)}$$

$$H = 4.167 \text{ MJ/MVA}$$

$$M = \frac{4.167}{180 \times 50} = 4.63 \times 10^{-4} \text{ sec}^2/\text{electrical degree}$$

Taking $\Delta t = 0.05$ sec

$$\frac{(\Delta t)^2}{M} = \frac{(0.05)^2}{4.63 \times 10^{-4}} = 5.4$$

Time when single/three pole switching occurs

$$= 0.075 \text{ sec (during middle of } \Delta t)$$

Time when reclosing occurs = 0.325 (during middle of Δt)

Table 12.9   Swing curve calculation—three pole switching

t sec	P_{max} (pu)	$\sin \delta$	P_e (pu)	P_a (pu)	$5.4P_a$ elec deg	$\Delta\delta$ elec deg	δ elec deg
0_-	1.5	0.667	1.0	0.0			41.8
0_+	0.827	0.667	0.552	0.448			41.8
0_{avg}				0.224	1.2	1.2	41.8
0.05	0.827	0.682	0.564	0.436	2.4	3.6	43.0
$0.075 \rightarrow$							
0.10	0.0	0.726	0.0	1 0	5.4	9.0	46.6
0.15	0.0		0.0	1.0	5.4	14.4	55.6
0.20	0.0		0.0	1.0	5.4	19.8	70.0
0.25	0.0		0.0	1.0	5.4	25.2	89.8
0.30	0.0		0.0	1.0	5.4	30.6	115.0
$0.325 \rightarrow$							
0.35	1.5	0.565	0.85	0.15	0.8	31.4	145.6
0.40	1.5	0.052	0.078	0.922	5.0	36.4	177.0
0.45	1.5	− 0.55	− 0.827	1.827	9.9	46.3	213.4
0.50	1.5	− 0.984	− 1.48	2.48	13.4	59.7	259.7
0.55	1.5	− 0.651	− 0.98	1.98	10.7	70.4	319.4
0.60	1.5	0.497	0.746	0.254	1.4	71.8	389.8
0.65							461.6

The swing curve is plotted in Fig. 12.50 from which it is obvious that the system is unstable.

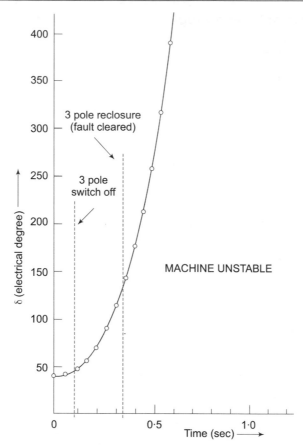

Fig. 12.50 Swing curve for three pole switching with reclosure

Table 12.10 Swing curve calculation—single pole switching

t sec	P_{max} (pu)	sin δ	P_e (pu)	P_a (pu)	$5.4P_a$ elec deg	Δδ elec deg	δ elec deg
0_-	1.5	0.667	1.0	0.0			41.80
0_+	0.827	0.667	0.552	0.448			41.8
0_{avg}				0.224	1.2	1.2	41.8
0.05	0.827	0.682	0.564	0.436	2.4	3.6	43.0
0.075 →							
0.10	0.985	0.726	0.715	0.285	1.5	5.1	46.6
0.15	0.985	0.784	0.77	0.230	1.2	6.3	51.7
0.20	0.985	0.848	0.834	0.166	0.9	7.2	58.0
0.25	0.985	0.908	0.893	0.107	0.6	7.8	65.2
0.30	0.985	0.956	0.940	0.060	0.3	8.1	73.0
0.325 →							
0.35	1.5	0.988	1.485	- 0.485	– 2.6	5.5	81.1

(Contd. ...)

0.40	1.5	0.998	1.5	− 0.5	− 2.7	2.8	86.6
0.45	1.5	1.0	1.5	− 0.5	− 2.7	0.1	89.4
0.50	1.5	1.0	1.5	− 0.5	− 2.7	− 2.6	89.5
0.55	1.5	0.9985	1.5	− 0.5	−2.7	−5.3	86.9
0.60	1.5	0.989	1.485	− 0.485	− 2.6	− 7.9	81.6
0.65	1.5	0.96	1.44	− 0.44	− 2.4	− 10.3	73.7
0.70	1.5	0.894	1.34	− 0.34	− 1.8	− 12.1	63.4
0.75	1.5	0.781	1.17	− 0.17	− 0.9	− 13.0	51.3
0.80	1.5	0.62	0.932	0.068	0.4	− 12.6	38.3
0.85	1.5	0.433	0.65	0.35	1.9	− 10.7	25.7
0.90	1.5	0.259	0.39	0.61	3.3	− 7.4	15.0
0.95	1.5	0.133	0.2	0.8	4.3	− 3.1	7.6
1.00	1.5	0.079	0.119	0.881	4.8	1.7	4.5
1.05	1.5	0.107	0.161	0.839	4.5	6.2	6.2
1.10	1.5	0.214	0.322	0.678	3.7	9.9	12.4
1.15	1.5	0.38	0.57	0.43	2.3	12.2	22.3
1.20	1.5	0.566	0.84	0.16	0.9	13.1	34.5
1.25	1.5	0.738	1.11	− 0.11	− 0.6	12.5	47.6
1.30	1.5	0.867	1.3	− 0.3	− 1.6	10.9	60.1
1.35	1.5	0.946	1.42	− 0.42	− 2.3	8.6	71.0
1.40	1.5	0.983	1.48	− 0.48	− 2.6	6.0	79.6
1.45	1.5	0.997	1.5	− 0.5	− 2.7	3.3	85.6
1.50	1.5						88.9

The swing curve is plotted in Fig. 12.51 from which it follows that the system is stable.

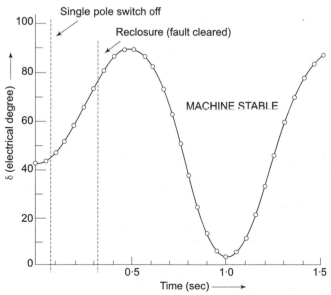

Fig. 12.51 Swing curve for single pole switching with reclosure

Problems

12.1 A two-pole, 50 Hz, 11 kV turboalternator has a rating of 100 MW, power factor 0.85 lagging. The rotor has a moment of inertia of a 10,000 kg-m^2. Calculate H and M.

12.2 Two turboalternators with ratings given below are interconnected via a short transmission line.

Machine 1: 4 pole, 50 Hz, 60 MW, power factor 0.80 lagging, moment of inertia 30,000 kg-m^2

Machine 2: 2 pole, 50 Hz, 80 MW, power factor 0.85 lagging, moment of inertia 10,000 kg-m^2

Calculate the inertia constant of the single equivalent machine on a base of 200 MVA.

12.3 Power station 1 has four identical generator sets each rated 80 MVA and each having an inertia constant 7 MJ/MVA; while power station 2 has three sets each rated 200 MVA, 3 MJ/MVA. The stations are located close together to be regarded as a single equivalent machine for stability studies. Calculate the inertia constant of the equivalent machine on 100 MVA base.

12.4 A 50 Hz transmission line 500 km long with constants given below ties up two large power areas

$R = 0.11$ Ω/km $\qquad\qquad$ $L = 1.45$ mH/km

$C = 0.009$ μF/km $\qquad\qquad$ $G = 0$

Find the steady state stability limit if $|V_S| = |V_R| = 200$ kV (constant). What will the steady state stability limit be if line capacitance is also neglected? What will the steady state stability limit be if line resistance is also neglected? Comment on the results.

12.5 A power deficient area receives 50 MW over a tie line from another area. The maximum steady state capacity of the tie line is 100 MW. Find the allowable sudden load that can be switched on without loss of stability.

12.6 A synchronous motor is drawing 30% of the maximum steady state power from an infinite bus bar. If the load on motor is suddenly increased by 100 per cent, would the synchronism be lost? If not, what is the maximum excursion of torque angle about the new steady state rotor position.

12.7 The transfer reactances between a generator and an infinite bus bar operating at 200 kV under various conditions on the interconnector are:

Prefault $\qquad\qquad$ 150 Ω per phase

During fault $\qquad\qquad$ 400 Ω per phase

Postfault $\qquad\qquad$ 200 Ω per phase

If the fault is cleared when the rotor has advanced 60 degrees electrical from its prefault position, determine the maximum load that could be transferred without loss of stability.

12.8   A synchronous generator is feeding 250 MW to a large 50 Hz network over a double circuit transmission line. The maximum steady state power that can be transmitted over the line with both circuits in operation is 500 MW and is 350 MW with any one of the circuits.

A solid three-phase fault occurring at the network-end of one of the lines causes it to trip. Estimate the critical clearing angle in which the circuit breakers must trip so that synchronism is not lost.

What further information is needed to estimate the critical clearing time?

12.9   A synchronous generator represented by a voltage source of 1.05 pu in series with a transient reactance of $j0.15$ pu and in inertia constant $H = 4.0$ sec, is connected to an infinite inertia system through a transmission line. The line has a series reactance of $j0.30$ pu, while the infinite inertia system is represented by a voltage source of 1.0 pu in series with a transient reactance of $j0.20$ pu.

The generator is transmitting an active power of 1.0 pu when a three-phase fault occurs at its terminals. If the fault is cleared in 100 millisec, determine if the system will remain stable by calculating the swing curve.

12.10   For Problem 12.9 find the critical clearing time from the swing curve for a sustained fault.

12.11   A synchronous generator represented by a voltage of 1.15 pu in series with a transient reactance is connected to a large power system with voltage 1.0 pu through a power network. The equivalent transient transfer reactance X between voltage sources is $j0.50$ pu.

After the occurrence of a three-phase to ground fault on one of the lines of the power network, two of the line circuit breakers A and B operate sequentially as follows with corresponding transient transfer reactance given therein.

   (i)   Short circuit occurs at $\delta = 30°$, A opens instantaneously to make $X = 3.0$ pu.
  (ii)   At $\delta = 60°$, A recloses, $X = 6.0$ pu.
 (iii)   At $\delta = 75°$, A reopens.
  (iv)   At $\delta = 90°$, B also opens to clear the fault making $X = 0.60$ pu

Check if the system will operate stably.

12.12   A 50 Hz synchronous generator with inertia constant $H = 2.5$ sec and a transient reactance of 0.20 pu feeds 0.80 pu active power into an infinite bus (voltage 1 pu) at 0.8 lagging power factor via a network with an equivalent reactance of 0.25 pu.

A three-phase fault is sustained for 150 millisec across generator terminals. Determine through swing curve calculation the torque angle δ, 250 millisec, after fault initiation.

12.13 A 50 Hz, 500 MVA, 400 kV generator (with transformer) is connected to a 400 kV infinite bus bar through an interconnector. The generator has $H = 2.5$ MJ/MVA, voltage behind transient reactance of 450 kV and is loaded 460 MW. The transfer reactances between generator and bus bar under various conditions are:

Prefault	0.5 pu
During fault	1.0 pu
Postfault	0.75 pu

Calculate the swing curve using intervals of 0.05 sec and assuming that the fault is cleared at 0.15 sec.

12.14 Plot swing curves and check system stability for the fault shown on the system of Example 12.10 for fault clearing by simultaneous opening of breakers at the ends of the faulted line at three cycles and eight cycles after the fault occurs. Also plot the swing curve over a period of 0.6 sec if the fault is sustained. For the generator assume $H = 3.5$ pu, $G = 1$ pu and carry out the computations in per unit.

12.15 Solve Example 12.10 for a LLG fault.

References

Books

1. Stevenson, W.D., *Elements of Power System Analysis*, 4th edn., McGraw-Hill, New York, 1982.
2. Elgerd, O.I., *Electric Energy Systems Theory: An Introduction*, 2nd edn., McGraw-Hill, New York, 1982.
3. Anderson, P.M. and A.A. Fund, *Power System Control and Stability*, The Iowa State University Press, Ames, Iowa, 1977.
4. Stagg, G.W. and A.H. O-Abiad, *Computer Methods in Power System Analysis*, Chaps 9 and 10, McGraw-Hill Book Co., New York, 1968.
5. Crary, S.B., *Power System Stability*, Vol. I (Steady State Stability), Vol. II (Transient Stability), Wiley, New York, 1945-1947.
6. Kimbark, E.W., *Power System Stability*, Vols 1, 2 and 3, Wiley, New York, 1948.
7. Venikov, V.A., *Transient Phenomena in Electrical Power System* (translated from the Russian), Mir Publishers, Moscow, 1971.
8. Byerly, R.T. and E.W. Kimbark (Eds.), *Stability of Large Electric Power Systems*, IEEE Press, New York, 1974.
9. Neuenswander, J.R., *Modern Power Systems*, International Text Book Co., 1971.
10. Pai, M.A., *Power System Stability Analysis by the Direct Method of Lyapunov.*, North-Holland, System and Control Services, Vol. 3, 1981.
11. Fouad, A.A and V. Vittal, *Power System Transient Stability Analysis using the Transient Energy Function Method*, Prentice-Hall, New Jersey, 1992.

12. Kundur, P., *Power System Stability and Control*, McGraw-Hill, New York, 1994.

13. Chakrabarti, A., D.P. Kothari and A.K. Mukhopadhyay, *Performance Operation and Control of EHV Power Transmission Systems*, Wheeler Publishing, New Delhi, 1995.

14. Padiyar, K.R., *Power System Dynamics: Stability and Control,* 2nd edn., B.S. Publications, Hyderabad, 2002.

15. Sauer, P.W. and M.A. Pai, *Power System Dynamics and Stability*, Prentice-Hall, New Jersey, 1998.

Papers

16. Cushing, E.W. et al., "Fast Valving as an Aid to Power System Transient Stability and Prompt Resynchronisation and Rapid Reload After Full Load Rejection", *IEEE Trans*, 1972, PAS 91: 1624.

17. Kimbark, E.W., "Improvement of Power System Stability", *IEEE Trans.*, 1969, PAS-88: 773.

18. Dharma Rao, N. "Routh-Hurwitz Condition and Lyapunov Methods for the Transient Stability Problem", *Proc. IEE*, 1969, 116: 533.

19. Shelton, M.L. et al., "BPA 1400 MW Braking Resistor", *IEEE Trans.*, 1975, 94: 602.

20. Nanda, J., D.P. Kothari, P.R. Bijwe and D.L. Shenoy, "A New Approach for Dynamic Equivalents Using Distribution Factors Based on a Moment Concept", *Proc. IEEE Int. Conf. on Computers, Systems and Signal Processing*, Bangalore, Dec. 10–12, 1984.

21. Dillon, T.S., 'Dynamic Modelling and Control of Large Scale System", *Int. Journal of Electric Power and Energy Systems*, Jan. 1982, 4: 29.

22. Patel, R., T.S. Bhatti and D.P. Kothari, "Improvement of Power System Transient Stability using Fast Valving: A Review", *Int. J. of Electric Power Components and Systems*, Vol. 29, Oct 2001, 927–938.

23. Patel, R., T.S. Bhatti and D.P. Kothari, "MATLAB/Simulink Based Transient Stability Analysis of a Multimachine Power System, *IJEEE*, Vol. 39, no. 4, Oct. 2002, pp 339–355.

24. Patel R., T.S. Bhatti and D.P. Kothari, "A Novel Scheme of Fast Valving Control", *IEEE Power Engineering Review*, Oct. 2002, pp. 44–46.

25. Patel, R., T.S. Bhatti and D.P. Kothari, "Improvement of Power System Transient Stability by Coordinated operation of Fast Valving and Braking Resistor", To appear in *IEE proceedings-Gen.*, Trans and Distribution.

13

Power System Security

13.1 INTRODUCTION

In Chapter 7, we have been primarily concerned with the economical operation of a power system. An equally important factor in the operation of a power system is the desire to maintain system security. System security involves practices suitably designed to keep the system operating when components fail. Besides economizing on fuel cost and minimizing emission of gases (CO, CO_2, NOx, SO_2), the power system should be operationally "secure". An operationally "secure" power system is one with low probability of system black out (collapse) or equipment damage. If the process of cascading failures continues, the system as a whole or its major parts may completely collapse. This is normally referred to as *system blackout*. All these aspects require security constrained power system optimization (SCO).

Since security and economy normally have conflicting requirements, it is inappropriate to treat them separately. The final aim of economy is the security function of the utility company. The energy management system (EMS) is to operate the system at minimum cost, with the guaranteed alleviation of emergency conditions. The emergency condition will depend on the severity of violations of operating limits (branch flows and bus voltage limits). The most severe violations result from contingencies. An important part of security study, therefore, moves around the power system's ability to withstand the effects of contingencies. A particular system state is said to be secure only with reference to one or more specific contingency cases, and a given set of quantities monitored for violation. Most power systems are operated in such a way that any single contingency will not leave other components heavily overloaded, so that cascading failures are avoided.

Most of the security related functions deal with static "snapshots" of the power system. They have to be executed at intervals compatible with the rate of change of system state. This quasi-static approach is, to a large extent, the only practical approach at present, since dynamic analysis and optimization are considerably more difficult and computationally more time consuming.

System security can be said to comprise of three major functions that are carried out in an energy control centre: (i) system monitoring, (ii) contingency analysis, and (iii) corrective action analysis.

System monitoring supplies the power system operators or dispatchers with pertinent up-to-date information on the conditions of the power system on real time basis as load and generation change. Telemetry systems measure, monitor and transmit the data, voltages, currents, current flows and the status of circuit breakers and switches in every substation in a transmission network. Further, other critical and important information such as frequency, generator outputs and transformer tap positions can also be telemetered. Digital computers in a control centre then process the telemetered data and place them in a data base form and inform the operators in case of an overload or out of limit voltage. Important data are also displayed on large size monitors. Alarms or warnings may be given if required.

State estimation (Chapter 14) is normally used in such systems to combine telemetered data to give the best estimate (in statistical sense) of the current system condition or "state". Such systems often work with supervisory control systems to help operators control circuit breakers and operate switches and taps remotely. These systems together are called SCADA (supervisory control and data acquisition) systems.

The second major security function is contingency analysis. Modern operation computers have contingency analysis programs stored in them. These foresee possible system troubles (outages) before they occur. They study outage events and alert the operators to any potential overloads or serious voltage violations. For example, the simplest form of contingency analysis can be put together with a standard LF program as studied in Chapter 6, along with procedures to set up the load flow data for each outage to be studied by the LF program. This allows the system operators to locate defensive operating states where no single contingency event will generate overloads and/or voltage violations. This analysis thus evolves operating constraints which may be employed in the ED (economic dispatch) and UC (unit commitment) program. Thus contingency analysis carries out emergency identification and "what if" simulations.

The third major security function, corrective action analysis, permits the operator to change the operation of the power system if a contingency analysis program predicts a serious problem in the event of the occurrence of a certain outage. Thus this provides preventive and post-contingency control. A simple example of corrective action is the shifting of generation from one station to another. This may result in change in power flows and causing a change in loading on overloaded lines.

These three functions together consist of a very complex set of tools that help in the secure operation of a power system.

13.2 SYSTEM STATE CLASSIFICATION

A formal classification of power system security levels was first suggested by DyLiacco [13] and further clarified by Fink and Carlsen [23] in order to define relevant EMS (Energy Management System) functions. Stott et. al [15] have also presented a more practical static security level diagram (see Fig. 13.1) by incorporating correctively secure (Level 2) and correctable emergency (Level 4) security levels.

In the Fig. 13.1, arrowed lines represent involuntary transitions between Levels 1 to 5 due to contingencies. The removal of violations from Level 4 normally requires EMS directed "corrective rescheduling" or "remedial action" bringing the system to Level 3, from where it can return to either Level 1 or 2 by further EMS, directed "preventive rescheduling" depending upon the desired operational security objectives.

Levels 1 and 2 represent normal power system operation. Level 1 has the ideal security but is too conservative and costly. The power system survives any of the credible contingencies without relying on any post-contingency corrective action. Level 2 is more economical, but depends on post-contingency corrective rescheduling to alleviate violations without loss of load, within a specified period of time. Post-contingency operating limits might be different from their pre-contingency values.

13.3 SECURITY ANALYSIS

System security can be broken down into two major functions that are carried out in an operations control centre: (i) security assessment, and (ii) security control. The former gives the security level of the system operating state. The latter determines the appropriate security constrained scheduling required to optimally attain the target security level.

The security functions in an EMS can be executed in 'real time' and 'study' modes. Real time application functions have a particular need for computing speed and reliability.

The static security level of a power system is characterised by the presence or otherwise of emergency operating conditions (limit violations) in its actual (pre-contingency) or potential (post-contingency) operating states. System security assessment is the process by which any such violations are detected.

System assessment involves two functions:

(i) system monitoring and (ii) contingency analysis. System monitoring provides the operator of the power system with pertinent up-to-date information on the current conditions of the power system. In its simplest form, this just detects violations in the actual system operating state. Contingency analysis is much

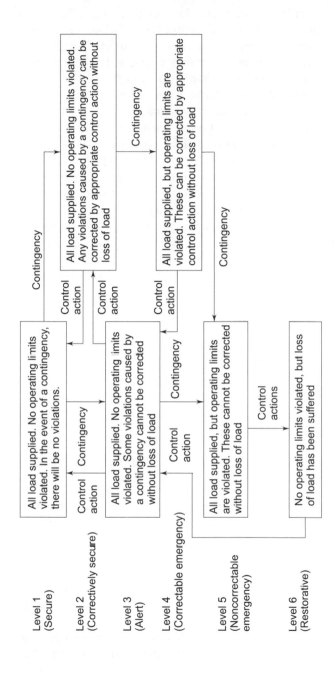

Fig. 13.1 Power system static security levels

more demanding and normally performed in three distinct states, i.e. contingency definition, selection and evaluation. Contingency definition gives the list of contingencies to be processed whose probability of occurrence is high. This list, which is usually large, is in terms of network changes, i.e. branch and/or injection outages. These contingencies are ranked in rough order of severity employing contingency selection algorithms to shorten the list. Limited accuracy results are required, therefore an approximate (linear) system model is utilized for speed. Contingency evaluation is then performed (using AC power flow) on the successive individual cases in decreasing order of severity. The evaluation process is continued up to the point where no post-contingency violations are encountered. Hence, the purpose of contingency analysis is to identify the list of contingencies that, if occur, would create violations in system operating states. They are ranked in order of severity.

The second major security function, security control, allows operating personnel to change the power system operation in the event that a contingency analysis program predicts a serious problem, should a certain outage occur. Normally the security control is achieved through SCO (security constrained optimization) program.

Modelling for Contingency Analysis

The power system limits of most interest in contingency analysis are those on line flows and bus voltages. Since these are soft limits, limited-accuracy models and solutions are justified. The most fundamental approximate load flow model is the NR model discussed in Chapter 6.

$$\begin{bmatrix} \Delta P \\ \Delta Q \end{bmatrix} = [J] \begin{bmatrix} \Delta \delta \\ \Delta |V| \end{bmatrix} \tag{13.1}$$

The DC load flow model in its incremental version is normally preferred.

$$[\Delta P] = [B'][\Delta \delta] \tag{13.2}$$

This model assumes voltages to remain constant after contingencies. However, this is not true for weak systems. The utility has to prespecify whether it wants to monitor post-contingency "steady-state" conditions immediately after the outage (system intertial response) or after the automatic controls*. (governor, AGC, ED) have responded. Depending upon this decision, different participation factors are used to allocate the MW generation among the remaining units. The reactive problem tends to be more non-linear and voltages are also influenced by active power flows.

FDLF (Chapter 6) is normally the best for this purpose since its Jacobian matrix is constant and single-line outages can be modelled using the matrix inversion lemma.

The model often used is

$$[\Delta Q / |V|] = [B''][\Delta|V|] \tag{13.3}$$

* See Chapter 8 for details about these controls.

Contingency Selection

There are two main approaches:

Direct methods: These involve screening and direct ranking of contingency cases. They monitor the appropriate post-contingent quantities (flows, voltages). The severity measure is often a performance index.

Indirect methods: These give the values of the contingency case severity indices for ranking, without calculating the monitored contingent quantities directly.

Simulation of line outage is more complex than a generator outage, since line outage results in a change in system configurations. The Inverse Matrix Modification Lemma (IMML) or 'compensation' method is used throughout the contingency analysis field [12]. The IMML helps in calculating the effects of network changes due to contingencies, without reconstructing and refactorizing or inverting the base case network matrix. It is also possible to achieve computational economy by getting only local solutions by calculating the inverse elements in the vicinity of the contingencies. The question is how far one should go. Some form of sensitivity analysis may be used.

The problem of studying hundreds of possible outages becomes very difficult to solve if it is desired to present the results quickly so that corrective actions can be taken. One of the simplest ways of obtaining a quick calculation of possible overloads is to use network sensitivity factors. These factors show the approximate change in line flows for changes in generation on the network configuration and are derived from the DC load flow [1]. They are of two types:

(i) Generation shift distribution factors

(ii) Line outage distribution factors

These are discussed in detail in the next section.

In a practical situation when a contingency causing emergency occurs, control action to alleviate limit violations is always taken, if such a capability exists and a protective system [Chapter 15 of Ref. 1] permits time to do so.

The security control function (which is normally achieved by SCO) responds to each insecure contingency case (as obtained by contingency analysis), usually in decreasing order of severity by:

(i) Rescheduling the precontingency operating state to alleviate the emergency resulting from the contingency, and/or

(ii) Developing a post-contingency control strategy that will eliminate the emergency, or

(iii) Taking no action, on the basis that post-contingency emergency is small and/or probability of its occurrence is very low.

A specific security control function, then, is designed to

(i) Operate in real time or study mode

(ii) Schedule active or reactive power controls or both

(iii) Achieve a defined security level

(iv) Minimize a defined operational objective

Only a small proportion of work on optimal power flow (OPF) has taken into account the security constraints. The most successful applications have been to the security constrained MW dispatch OPF sub-problem. The contingency-constrained voltage/var rescheduling problem, as of the writing of this text, still remains to be solved to a satisfactory degree.

The total number of contingency constraints imposed on SCO is enormous. The SCO or contingency constrained OPF problem is solved with or without first optimizing with respect to the base case (precontingency) constraints. The general procedure adopted is as follows:

(i) Contingency analysis is carried out and cases with violations or near violations are identified.

(ii) The SCO problem is solved.

(iii) The rescheduling in Step 1 might have created new violations, and therefore Step 1 should be repeated till no violations exist.

Hence, SCO represents a potentially massive additional computing effort. An excellent comprehensive overview of various available methods is presented by Stott et. al [15].

There is still great potential for further improvement in power system security control. Better problem formulations, theory, computer solution methods and implementation techniques are required.

13.4 CONTINGENCY ANALYSIS

In the past many widespread blackouts have occurred in interconnected power systems. Therefore, it is necessary to ensure that power systems should be operated most economically such that power is delivered reliably. Reliable operation implies that there is adequate power generation and the same can be transmitted reliably to the loads. Most power systems are designed with enough redundancy so that they can withstand all major failure events. Here we shall study the possible consequences and remedial actions required by two main failure events: line outages and generating unit failures.

To explain the problem briefly, we consider the five-bus system of Reference [10]. The base case load flow results for the example are given in Fig. 13.2 and show a flow of 24.7 MW and 3.6 MVAR on the line from bus 2 to bus 3. Let us assume that at present, we are only interested in the MW loading of the line. Let us examine what will happen if the line from bus 2 to bus 4 were to open*. The resulting line flows and voltages are shown in Fig. 13.3. It may be noted that the flow on the line 2-3 has increased to 37.5 MW and that most of the other line flows are also changed. It may also be noted that bus voltage magnitudes also get affected, particularly at bus 4, the change is almost 2% less from 1.0236 to 1.0068 pu. Suppose the line from bus 2 to bus 5 were to open. Figure 13.4 shows the resulting flows and voltages. Now the maximum change in voltage is at bus 5 which is almost 10% less.

*Simulation of line outage is more complex than a generator outage, since line outage results in a change in system configurations.

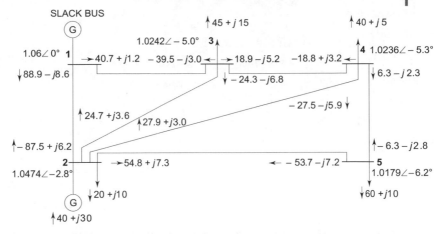

Fig. 13.2 Base Case AC Line flow for sample 5 bus system

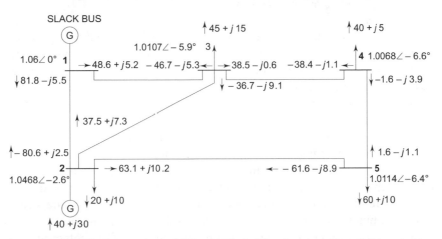

Fig. 13.3 Post outage AC Load Flow (Line between 2 and 4 is open)

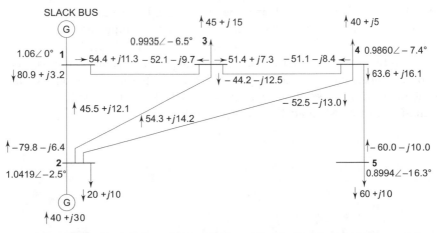

Fig. 13.4 Post outage AC Load Flow (Line between 2 and 5 is open)

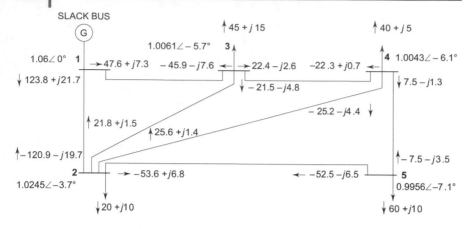

Fig. 13.5 Post outage AC Load Flow (Generator 2 outage, lost generation is picked up by generator 1)

Figure 13.5 is an example of generator outage and is selected to explain the fact that generator outages can also result in changes in line flows and bus voltages. In the example shown in Fig. 13.5 all the generation lost from bus 2 is picked up on the generator at bus 1. Had there been more than 2 generators in the sample system say at bus 3 also, it was possible the loss of generation on bus 2 is made up by an increase in generation at buses 1 and 3. The differences in line flows and bus voltages would show how the lost generation is shared by the remaining units is quite significant.

It is important to know which line or unit outages will render line flows or voltages to cross the limits. To find the effects of outages, contingency analysis techniques are employed. Contingency analysis models single failure events (i.e. one-line outages or one unit outages) or multiple equipment failure events (failure of multiple unit or lines or their combination) one after another until all "credible outages" are considered. For each outage, all lines and voltages in the network are checked against their respective limits. Figure 13.6 depicts a flow chart illustrating a simple method for carrying out a contingency analysis.

One of the important problems is the selection of "all credible outages". Execution time to analyse several thousand outages is typically 1 min based on computer and analytical technology as of 2000. An approximate model such as DC load flow may be used to achieve speedy solution if voltage is also required, then full AC load flow analysis has to be carried out.

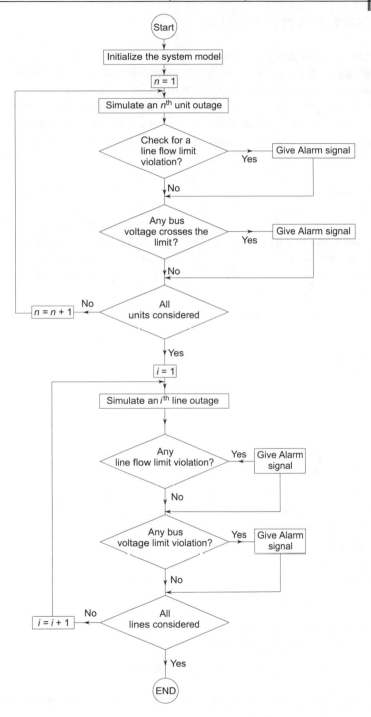

Fig. 13.6 A simple technique for contingency analysis

13.5 SENSITIVITY FACTORS

A security analysis program is run in a load dispatch centre very quickly to help the operators. This can be attempted by carrying out an approximate analysis and using a computer system having multiple processors or vector processors for speedy analysis. The system may be adequately described and an equivalent should be used for neighbours connected through tie-lines. We can eliminate all non-violation cases and run complete exact program for "critical" cases only. This can be achieved by using techniques such as "contingency selection" or "contingency screening", or "contingency ranking". Thus it will be easy to warn the operation staff in advance to enable them to take corrective action if one or more outages will result in serious overloads or any violations. One of the simplest ways to present a quick calculation of possible overloads is to employ network (linear) sensitivity factors. These factors give the approximate change in line flows for changes in generation in the system and can be calculated from the DC load flow. They are mainly of two types:

1. Generation shift factors
2. Line outage distribution factors

Briefly we shall now describe the use of those factors without deriving them. Reference [7] gives their derivation.

The generation shift factors, α_{li} are defined as:

$$\alpha_{li} = \frac{\Delta f_l}{\Delta P_{Gi}} \tag{13.4}$$

where,

Δf_l = Change in MW power flow on line l when a change in generation, ΔP_{Gi} takes place at the ith bus

Here, it is assumed that ΔP_{Gi} is fully compensated by an equal and opposite change in generation at the slack (reference) bus, with all other generators remaining fixed at their original power generations. The factor α_{li} then gives the sensitivity of the lth line flow to a change in generation at ith bus. Let us now study the outage of a large generating unit and assume that all the lost generation (P_{Gi}^0) would be supplied by the slack bus generation. Then

$$\Delta P_{Gi} = -P_{Gi}^o \tag{13.5}$$

and the new power flow on each line could be calculated using a precalculated set of "α" factors as given below.

$$\hat{f}_l = f^o_l + \alpha_{li}\,\Delta P_{Gi} \qquad \text{for all lines } \forall\ l \tag{13.6}$$

where, $\hat{f}_l$ = power flow on lth line after the failure of ith generator

f^o_l = power flow on lth line before the failure or precontingency power flow

The values of line flows obtained from Eq. (13.6) can be compared to their limits and those violating their limit can be informed to the operator for necessary control action.

The generation shift sensitivity factors are linear estimates of the change in line flow with a change in power at a bus. Thus, the effects of simultaneous changes on a given number of generating buses can be computed using the principle of superposition.

Let us assume that the loss of the ith generator is to be made up by governor action on all generators of the interconnected system and pick up in proportion to their maximum MW ratings. Thus, the proportion of generation pick up from unit k $(k \neq i)$ would be

$$\beta_{ki} = \frac{P_{Gk_{max}}}{\underset{\substack{m \\ m \neq i}}{\Sigma} P_{Gm_{max}}} \tag{13.7}$$

where

$\qquad P_{Gm_{max}}$ = maximum MW rating for mth generator

$\qquad \beta_{ki}$ = proportionality factor for pick up on kth unit when ith unit fails.

Now, for checking the lth line flow, we may write

$$\hat{f}_l = f_l^o + \alpha_{li} \, \Delta P_{Gi} - \underset{k \neq i}{\Sigma} \, [\alpha_{lk} \, \beta_{ki} \, \Delta P_{Gi}] \tag{13.8}$$

In Eq. (13.8) it is assumed that no unit will violate its maximum limit. For unit limit violation, algorithm can easily be modified.

Similarly the line outage distribution factors can be used for checking if the line overloads when some of the lines are lost.

The line outage distribution factor is defined as:

$$d_{l,i} = \frac{\Delta f_l}{f_i^o} \tag{13.9}$$

where

$\qquad d_{l,i}$ – line outage distribution factor when monitoring lth line after an outage of ith line.

$\qquad \Delta f_l$ = change in MW flow on lth line.

$\qquad f_i^o$ = precontingency line flow on ith line

If precontingency line flows on lines l and i, the power flow on line l with line i out can be found out employing "d" factors.

$$\hat{f}_l = f_l^o + d_{l,i} f_i^o \tag{13.10}$$

Here,

f_l^o and f_i^o = precontingency or preoutage flows on lines l and i respectively

$\qquad \hat{f}_l$ = power flow on lth line with ith line out.

Thus one can check quickly by precalculating 'd' factors all the lines for overloading for the outage of a particular line. This can be repeated for the outage of each line one by one and overloads can be found out for corrective action.

It may be noted that a line flow can be positive or negative. Hence we must check f against $-f_{l\,max}$ as well as $f_{l\,max}$. Line flows can be found out using telemetry systems or with state estimation techniques. If the network undergoes any significant structural change, the sensitivity factors must be updated.

Example 13.1

Find the generation shift factors and the line outage distribution factors for the five-bus sample network discussed earlier.

Solution Table 13.1 gives the [x] matrix for the five bus sample system, together with the generation shift distribution factors and the line outage distribution factors are given in Tables 13.2 and 13.3 respectively.

Table 13.1 X Matrix for Five-bus Sample System (Bus 1 as a reference)

$$
\begin{bmatrix}
0 & 0 & 0 & 0 & 0 \\
0 & 0.05057 & 0.03772 & 0.04029 & 0.4714 \\
0 & 0.03772 & 0.08914 & 0.07886 & 0.05143 \\
0 & 0.04029 & 0.07886 & 0.09514 & 0.05857 \\
0. & 0.04714 & 0.05143 & 0.05857 & 0.13095
\end{bmatrix}
$$

Table 13.2 Generation Shift Distribution Factor for Five-bus System

	Bus 1	Bus 2
$l = 1$ (line 1-2)	0	-0.8428
$l = 2$ (line 1-3)	0	-0.1572
$l = 3$ (line 2-3)	0	0.0714
$l = 4$ (line 2-4)	0	0.0571
$l = 5$ (line 2-5)	0	0.0286
$l = 6$ (line 3-4)	0	-0.0857
$l = 7$ (line 4-5)	0	-0.0285

Table 13.3 Line Outage Distribution Factors for Five-bus Sample System

	$j=1$ (line 1-2)	$j=2$ (line 1-3)	$j=3$ (line 2-3)	$j=4$ (line 2-4)	$j=5$ (line 2-5)	$j=6$ (line 3-4)	$j=7$ (line 4-5)
$l=1$ (line 1-2)	0.0	1.0001	−0.3331	−0.2685	−0.2094	0.3735	0.2091
$l=2$ (line 1-3)	1.0	0.0	0.3332	0.2686	0.2092	−0.3735	−0.2093
$l=3$ (line 2-3)	−0.4542	0.4545	0	0.4476	0.3488	−0.6226	−0.3488
$l=4$ (line 2-4)	−0.3634	0.3636	0.4443	0.0	0.4418	0.6642	−0.4418
$l=5$ (line 2-5)	−0.1819	0.1818	0.2222	0.2835	0.0	0.3321	1.0
$l=6$ (line 3-4)	0.5451	−0.5451	−0.6662	0.7161	0.5580	0.0	−0.5580
$l=7$ (line 4-5)	0.1816	−0.1818	−0.2222	−0.2835	1.0002	−0.3321	0.0

It has been found that if we calculate the line flows by the sensitivity methods, they come out to be reasonably close to the values calculated by the full AC load flows. However, the calculations carried out by sensitivity methods are faster than those made by full AC load flow methods and therefore are used for real time monitoring and control of power systems. However, where reactive power flows are mainly required, a full AC load flow method (NR/FDLF) is preferred for contingency analysis.

The simplest AC security analysis procedure merely needs to run an AC load flow analysis for each possible unit, line and transformer outage. One normally does ranking or shortlisting of most likely bad cases which are likely to result in an overload or voltage limit violation and other cases need not be analysed. Any good *PI* (performance index can be selected) is used for ranking. One such *PI* is

$$PI = \sum_{\forall l}\left(\frac{P_{\text{flow},l}}{P_{l,\max}}\right)^{2n} \tag{13.11}$$

For large n, *PI* will be a small number if all line flows are within limit, and will be large if one or more lines are overloaded.

For $n=1$ exact calculations can be done for *PI*. *PI* table can be ordered from largest value to least. Suitable number of candidates then can be chosen for further analysis [7].

If voltages are to be included, then the following *PI* can be employed.

$$PI = \sum_{\forall l}\left(\frac{P_{\text{flow},l}}{P_{l,\max}}\right)^{2n} + \sum_{\forall i}\left(\frac{\Delta|V_i|}{\Delta|V|_{\max}}\right)^{2m} \tag{13.12}$$

Here, $\Delta|V_i|$ is the difference between the voltage magnitude as obtained at the end of the 1P1Q FDLF algorithm $\Delta|V|_{\max}$ is the value fixed by the utility.

Largest value of *PI* is placed at the top. The security analysis may now be started for the desired number of cases down the ranking list.

Summary and Further Reading:
Reference [25] has discussed the concept for screening contingencies. Such contingency selection/screening techniques form the foundation for many real-time computer security analysis algorithms.

Reference [15] gives a broad overview of security assessment and contain an excellent bibliography covering the literature on security assessment up to 1987.

Reference [11] gives an excellent bibliography on voltage stability. This topic is discussed briefly in the next section.

13.6 POWER SYSTEM VOLTAGE STABILITY

Power transmission capability has traditionally been limited by either rotor angle (synchronous) stability or by thermal loading capabilities. The blackout problem has been linked with transient stability. Luckily this problem is now not that serious because of fast short circuit clearing, powerful excitation systems, and other special stability controls. Electric companies are now required to squeeze the maximum possible power through existing networks owing to various constraints in the construction of generation and transmission facilities.

Voltage (load) stability, however, is now a main issue in planning and operating electric power systems and is a factor leading to limit power transfers. Voltage stability is concerned with the ability of a power system to maintain acceptable voltages at all buses in the system under normal conditions and after being subjected to a disturbance. A power system is said to have entered a state of voltage instability when a disturbance results in a progressive and uncontrollable decline in voltage.

Inadequate reactive power support from generators and transmission lines leads to voltage instability or voltage collapse, which have resulted in several major system failures in the world. They are:

  (i) South Florida, USA, system disturbance of 17 May 1985, (transient, 4 sec)
 (ii) French system disturbances of December 19, 1978 and January 12, 1987, (longer term).
(iii) Swedish system disturbance of December 27, 1983 (longer term, 55 sec)
 (iv) Japanese (Tokyo) system disturbance of July 23, 1987 (longer term, 20 min)
  (v) NREB grid disturbance in India in 1984 and 1987.
 (vi) Belgium, Aug 4, 1982. (longer term, 4.5 min)
(vii) Baltimore, Washington DC, USA, 5th July 1990 (longer term, insecure for hours)

Hence, a full understanding of voltage stability phenomena and designing mitigation schemes to prevent voltage instability is of great value to utilities. Consequently over the last ten years, utility engineers, consultants and researchers have thoroughly studies voltage stability.

Voltage stability covers a wide range of phenomena. Because of this, voltage stability means different things to different engineers. Voltage instability and voltage collapse are used somewhat interchangeably by many researchers. Voltage instability or collapse is a faster dynamic process. As opposed to angle stability, the dynamics mainly involves the loads and the means for voltage control. Ref [11] provides a comprehensive list of books, reports, workshops and technical papers related to voltage stability and security.

Definitions: [2]

A power system at a given operating state is *small-disturbance voltage stable* if, following any small disturbance, voltages near loads are identical or close to the pre-disturbance values. The concept of small-disturbance voltage stability is related to steady-state stability (Chapter 12) and can be analysed using small-signal (linearised) model of the system.

A power system at a given operating state and subject to a given disturbance is *voltage stable* if voltages near loads approach post-disturbance equilibrium values. The concept of voltage stability is related to the transient stability of a power system. The analysis of voltage stability normally requires simulation of the system modelled by non-linear differential-algebraic equations.

A power system at a given operating state and subject to a given disturbance undergoes *voltage collapse* if post-disturbance equilibrium voltages are below acceptable limits. Voltage collapse may be total (blackout) or partial. The voltage instability and collapse may occur in a time frame of a second. In this case the term *transient voltage* stability is used. Sometimes it may take up to tens of minutes in which case the term *long-term voltage stability* is used.

The term *voltage security* means the ability of a system, not only to operate stably, but also to remain stable following any reasonably credible contingency or adverse system change such as load increases [2].

Voltage stability involves dynamics, but load flow based static analysis methods are generally used for quick and approximate analysis.

Figure 13.7 depicts how voltage stability can be classified into transient and long-term time frame [2].

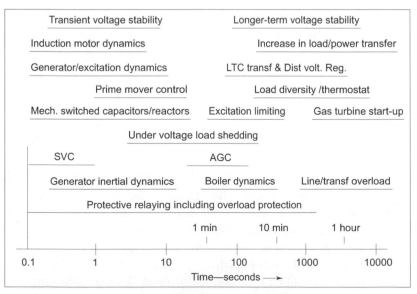

Fig. 13.7 Voltage stability phenomena and time responses

Voltage stability problems normally occur in heavily stressed systems. Voltage stability and rotor angle (or synchronous) stability are more or less interlinked. Rotor angle stability, as well as voltage stability is affected by reactive power control. Voltage stability is concerned with load areas and load characteristics. For rotor angle stability, the main concern is the integration of remote power plants to a large system over long transmission lines. Voltage stability is basically *load stability* and rotor angle stability is basically *generator stability*. In a large inter-connected system, voltage collapse of a load area is possible without loss of synchronism of any generators.

The slower forms of voltage instability are often analysed as steady-state problems. 'Snapshots' in time following an outage or during load buildup are simulated. In addition to post-disturbance load flows, two other load flow based methods are widely used: P-V curves and Q-V curves. PV curves are used for conceptual analysis of voltage stability especially for radial systems.

For carrying out static analysis, PV curves (Fig. 13.8), QV curves (Fig. 13.9), computation of nose point (see Fig. 13.8) and methods to quantify nose point i.e. proximity indicators are computed. Power flow analysis determines how power system equilibrium point values such as voltages and line flows vary as various system parameters and controls are changed. Two values of load voltage exist for each value of load. The upper one indicates stable voltage whereas lower one is the unacceptable value (multiple load flow). At limiting stage of voltage stability i.e. at nose point single load flow solution exists. Nearer the nose point, lesser is the stability margin.

Effective Counter Measures to Prevent or Contain Voltage Instability

(i) Generator terminal voltage should be raised.

(ii) Generator transformer tap value may be increased.

(iii) Q-injection should be carried out at an appropriate location.

(iv) Load-end OLTC (on-load tap changer) should be suitably used.

(v) For under voltage conditions, strategic load shedding should be resorted to.

System reinforcement may be carried out by installing new transmission lines between generation and load centres. Series and shunt compensation may be carried out and SVCs (static var compensation) may be installed. Generation rescheduling and starting-up of gas turbines may be carried out.

Practical aspects of Q-flow problems leading to voltage collapse in EHV lines:

(i) For long lines with uncontrolled buses, receiving-end or load voltages increase for light load conditions and decrease for heavy load conditions.

(ii) For radial transmission lines, if any loss of a line takes place, reactance goes up, I^2X loss increases resulting in increase in voltage drop. This should be suitably compensated by local Q injection. Of course this involves cost. If there is a shortage of local Q source, then import of Q through long line may have to be resorted to. However, this is not desirable.

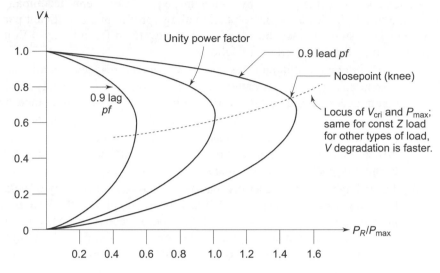

Fig. 13.8 PV curves with different load power factors

Only the operating points above the critical points represent satisfactory operating conditions. At the 'knee' of the V-P curve, the voltage drops rapidly with an increase in load demand. Power-flow solution fails to converge beyond this limit indicating instability. Operation at or near the stability limit is impractical and a satisfactory operating condition is ensured by permitting sufficient "power margin".

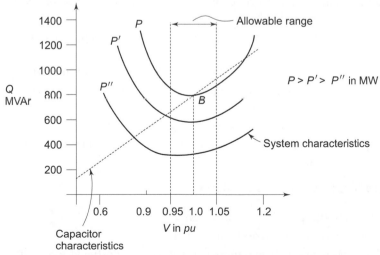

Fig. 13.9 System and shunt capacitor steady-state Q-V characteristics, capacitor MVAr shown at rated voltage

Voltage Collapse

Voltage collapse is the process by which the sequence of events accompanying voltage instability leads to unacceptable voltage profile in a significant part of the power system. It may be manifested in several different ways. Voltage collapse may be characterised as follows:

(i) The initiating event may be due to variety of reasons: Small gradual system changes such as natural increase in system load, or large sudden disturbances such as loss of a generating unit or a heavily loaded line.

(ii) The crux of the problem is the inability of the system to meet its reactive demands. When transport of reactive power from neighbouring areas is difficult, any change that requires additional reactive power support may eventually lead to voltage collapse.

(iii) The voltage collapse generally manifests itself as a slow decay of voltage. It is the result of an accumulative process involving the actions and interactions of many devices, controls, and protective systems. The time frame of collapse in such cases would be of the order of several minutes. Voltage collapse is strongly influenced by system conditions and characteristics.

(iv) Reactive compensation can be made most effective by the judicious choice of a mixture of shunt capacitors, static var system and possibly synchronous condensers.

Methods of Improving Voltage Stability

Voltage stability can be improved by adopting the following means:

(i) Enhancing the localised reactive power support (SVC) is more effective and C-banks are more economical. FACTS devices or synchronous condenser may also be used.

(ii) Compensating the line length reduces net reactance and power flow increases.

(iii) Additional transmission line may be erected. It also improves reliability.

(iv) Enhancing excitation of generator, system voltage improves and Q is supplied to the system.

(v) HVDC tie may be used between regional grids.

(vi) By resorting to strategic load shedding, voltage goes up as the reactive burden is reduced.

Future Trends and Challenges

(i) Optimal siting of FACTs devices.

(ii) Better and probabilistic load modelling.

(iii) Develop techniques and models for study of non-linear dynamics of large size systems. For example, new methods to obtain network equivalents suitable for voltage stability analysis.

(iv) Better and fuller representation of AC system in AC-DC system.

(v) Post-disturbance MW/MVAR margins should be translated to pre-disturbance operating limits that operators can monitor.

(vi) Training in voltage stability basis (a training simulator) for control centre and power plant operators should be imparted.

SUMMARY

Power system security (including voltage stability) is likey to challenge planners, analysts, researchers and operators for the foreseeable future. As load grows, and as new transmission lines and new generations would be increasingly difficult to build or add, more and more utilities will face the security challenge.

Deregulation and socio-economic trends compounded by technological developments have increased the likelihood of voltage instability.

Luckily many creative persons are working tirelessly to find new methods and innovative solutions to meet this challenge.

REFERENCES

Books

1. I.J. Nargath and D.P. Kothari, *Power System Engineering*, Tata McGraw-Hill, New Delhi, 1994.

2. C.W. Taylor, *Power System Voltage Stability*, McGraw-Hill, New York, 1994.

3. P. Kundur, *Power System Stability and Control*, Sections 2.12, 11.2 and Chapter 14, McGraw-Hill, New York, 1994.

4. T.J.E. Miller, Editor, *Reactive Power Control in Electric Systems*, John Wiley and Sons, New York, 1982.

5. A. Chakrabarti, D.P. Kothari and A.K. Mukhopadhyay, *Performance, Operation and Control of EHV Power Transmission Systems*, Wheeler Publishing, New Delhi, 1995.

6. T.V. Cutsem and C. Vournas, *Voltage Stability of Electric Power Systems*, Kluwer Academic Publishers, London, 1998.

7. A.J. Wood and W.F. Wollenberg, *Power Generation, Operation, and Control*, 2nd Edn, John Wiley, New York, 1996.

8. John J. Grainger and W.D. Stevenson, *Power System Analysis*, McGraw-Hill, New York, 1994.

9. G.L. Kusic, *Computer-Aided Power Systems Analysis*, Prentice-Hall, New Jersey, 1986.

10. G.W. Stagg and A.H. El-Abiad, *Computer Methods in Power System Analysis*, McGraw-Hill, New York, 1968.

Papers

11. V. Ajjarapu and B. Lee, "Bibliography on Voltage Stability", *IEEE Trans. on Power Systems*, Vol. 13, No. 1, February 1998, pp 115–125.

12. L.D. Arya, "Security Constrained Power System Optimization", PhD thesis, IIT Delhi, 1990.

13. T.E. DyLiacco, "The Adaptive Reliability Control System", *IEEE Trans. on PAS*, Vol. PAS-86, May 1967, pp 517–531

    (This is a key paper on system security and energy control system)

14. A.A. Fouad, "Dynamic Security Assessment Practices in North America", *IEEE Trans. on Power Systems*, Vol. 3, No. 3, 1988, pp 1310–1321.

15. B. Stott, O. Alsac and A.J. Monticelli, "Security Analysis and Optimization", *Proc IEEE*, Vol. 75, No. 12, Dec. 1987, pp 1623–1644.

16. Special issue of Proc. IEEE, February 2000.

17. P.R. Bijwe, D.P. Kothari and L.D. Arya, "Alleviation of Line Overloads and Voltage Violations by Corrective Rescheduling", *IEE Proc.* C, Vol. 140, No. 4, July 1993, pp 249–255.

18. P.R. Bijwe, D.P. Kothari and L.D. Arya, "Overload Ranking of Line Outages with postoutage generation rescheduling", *Int. J. of Electric Machines and Power Systems*, Vol. 22, No. 5, 1994, pp 557–568.

19. L.D. Arya, D.P. Kothari et al, "Post Contingency Line Switching for Overload Alleviation or Rotation", *Int J. of EMPS*, Vol 23, No. 3, 1995, pp 345–352.

20. P.R. Bijwe, S.M. Kelapure, D.P. Kothari and K.K. Saxena, "Oscillatory Stability Limit Enhancement by Adaptive Control Rescheduling, *Int. J. of Electric Power and Energy Systems*, Vol. 21, No. 7, 1999, pp 507–514.

21. L.D. Arya, S.C. Chaube and D.P. Kothari, "Line switching for Alleviating Overloads under Line Outage Condition Taking Bus Voltage Limits into Account", *Int. J. of EPES*, Vol. 22, No. 3, 2000, pp 213–221.

22. P.R. Bijwe, D.P. Kothari and S. Kelapure, "An Effective Approach to Voltage Security and Enhancement", *Int. J. of EPES*, Vol. 22, No 7, 2000, pp 483–486.

23. L. Fink and K. Carlsen, "Operating under Stress and Strain", *IEEE Spectrum*, March 1978, pp. 48–50.

24. S.M. Kelapure, "Voltage Security Analysis and Enhancement", *Ph.D. thesis, IIT Delhi*, 2000.

25. G.C. Ejebe, et. al, "Fast Contingency Screening and Evaluation for Voltage Security Analysis", *IEEE Trans. on Power Systems*, Vol. 3, No. 4, Nov. 1988, pp 1582–1590.

26. T. Van Cutsen, Voltage Instability: "Phenomena, Counter measures, and Analysis Methods", *Proc. IEEE,* Vol. 88, No. 2, Feb. 2000, pp 208–227.

14

An Introduction to State Estimation of Power Systems

14.1 INTRODUCTION

State estimation plays a very important role in the monitoring and control of modern power systems. As in case of load flow analysis, the aim of state estimation is to obtain the best possible values of the bus voltage magnitudes and angles by processing the available network data. Two modifications are, however, introduced now in order to achieve a higher degree of accuracy of the solution at the cost of some additional computations. First, it is recognised that the numerical values of the data to be processed for the state estimation are generally noisy due to the errors present. Second, it is noted that there are a larger number of variables in the system (e.g. P, Q line flows) which can be measured but are not utilised in the load flow analysis. Thus, the process involves imperfect measurements that are redundant and the process of estimating the system states is based on a statistical criterion that estimates the true value of the state variables to minimize or maximize the selected criterion. A well known and commonly used criterion is that of minimizing the sum of the squares of the differences between the estimated and "true" (i.e. measured) values of a function.

Most state estimation programs in practical use are formulated as overdetermined systems of non-linear equations and solved as weighted least-squares (WLS) problems.

State estimators may be both static and dynamic. Both have been developed for power systems. This chapter will introduce the basic principles of a static-state estimator.

In a power system, the state variables are the voltage magnitudes and phase angles at the buses. The inputs to an estimator are imperfect (noisy) power system measurements. The estimator is designed to give the "best estimate" of the system voltage and phase angles keeping in mind that there are errors in the measured quantities and that there may be redundant measurements. The output data are then used at the energy control centres for carrying out several real-time or on-line system studies such as economic dispatch (Chapter 7), security analysis (Chapter 13).

14.2 LEAST SQUARES ESTIMATION: THE BASIC SOLUTION [7] – [9]

As will be seen later in Section 14.3, the problem of power system state estimation is a special case of the more general problem of estimation of a random vector $\mathbf{x}$ from the numerical values of another related random vector $\mathbf{y}$ with relatively little statistical information being available for both $\mathbf{x}$ and $\mathbf{y}$. In such cases, the method of least-squared-error estimation may be utilised with good results and has accordingly been widely employed.

Assume that $\mathbf{x}$ is a vector of n random variables $x_1, x_2, ..., x_n$, that $\mathbf{y}$ is another vector of m ($> n$) random variables $y_1, y_2, ..., y_m$ and both are related as

$$y = Hx + r \tag{14.1}$$

where H is a known matrix of dimension $m \times n$ and r is a zero mean random variable of the same dimension as $\mathbf{y}$. The vector $\mathbf{x}$ represents the variables to be estimated, while the vector $\mathbf{y}$ represents the variables whose numerical values are available. Equation (14.1) suggests that the measurement vector $\mathbf{y}$ is linearly related to the unknown vector $\mathbf{x}$ and in addition is corrupted by the vector $\mathbf{r}$ (error vector).

The problem is basically to obtain the best possible value of the vector $\mathbf{x}$ from the given values of the vector $\mathbf{y}$. Since the variable r is assumed to be zero mean, one may take the expectation of Eq. (14.1) and get the relation

$$\bar{y} = H\bar{x} \tag{14.2}$$

where $\bar{x}, \bar{y}$ = expected value of x and y, respectively.

This shows that the load flow methods of Chapter 6 could be used to estimate the mean values of the bus voltages. However, one would like to estimate the actual values of bus voltages rather than their averages.

One possible way of obtaining the best possible estimate of the vector $\mathbf{x}$ from $\mathbf{y}$ lies in the use of the method of least square estimation (LSE). To develop this method, assume that $\hat{\mathbf{x}}$ represents the desired estimate of x so that $\hat{\mathbf{y}}$ given by the equation

$$\hat{\mathbf{y}} = H\hat{\mathbf{x}} \tag{14.3}$$

represents the estimate of the vector $\mathbf{y}$. The error $\tilde{y}$ of the estimation of $\mathbf{y}$ is then given by

$$\tilde{y} = y - \hat{y} \tag{14.4}$$

The estimate $\hat{\mathbf{x}}$ is defined to be the LSE if it is computed by minimizing the estimation index J given by

$$J = \tilde{y}' \tilde{y} \tag{14.5}$$

From Eqs. (14.31) and (14.4), one gets the following expression for the index:

$$J = y'y - y'H\hat{x} - \hat{x}'H'y + \hat{x}'H'H\hat{x} \tag{14.6}$$

For minimizing $J = f_{(\hat{x})}$, we must satisfy the following condition.

$$\text{grad}_{\hat{x}} J = 0 \tag{14.7}$$

It is easy to check (see, e.g. [1]) that Eq. (14.7) leads to the following result.

$$H'H\hat{x} - H'y = 0 \tag{14.8}$$

This equation is called the 'normal equation' and may be solved explicitly for the LSE of the vector $\hat{x}$ as

$$\hat{x} = (H'H)^{-1} H'y \tag{14.9}$$

Example 14.1

In order to illustrate the method of LSE, let us consider the simple problem of estimating two random variables x_1 and x_2 by using the data for a three dimensional vector $\mathbf{y}$.

Assume
$$H = \begin{bmatrix} 1 & 0 \\ 0 & 1 \\ 1 & 1 \end{bmatrix}$$

The matrix $H'H$ is then given by

$$H'H = \begin{bmatrix} 2 & 1 \\ 1 & 2 \end{bmatrix} \text{ and its inverse is}$$

$$(H'H)^{-1} = \begin{bmatrix} 2/3 & -1/3 \\ -1/3 & 2/3 \end{bmatrix}$$

It is easy to form the vector $H'y$ and combining this with the inverse of $(H'H)$, the following estimate of x is obtained.

$$\hat{x} = \begin{bmatrix} (2/3)\, y_1 - (1/3)(y_2 - y_3) \\ -(1/3)y_1 + (2/3)y_2 + (1/3)y_3 \end{bmatrix}$$

Weighted LSE

The estimate given in Eq. (14.9) is often referred to as the 'ordinary' least squares estimate and is obtained by minimising the index function that puts equal weightage to the errors of estimation of all components of the vector y. It is often desirable to put different weightages on the different components of y since some of the measurements may be more reliable and accurate than the others and these should be given more importance. To achieve this we define the estimation index as

$$J = \tilde{y}' W \tilde{y} \tag{14.10}$$

where W is a real symmetric weighting matrix of dimension $m \times m$. This is often chosen as a diagonal matrix for simplicity.

It is relatively straightforward to extend the method of LSE to the weighted form of J and to derive the following form of the normal equation.

$$H'WH\hat{x} - H'Wy = 0 \tag{14.11a}$$

This leads to the desired weighted least squares estimate (WLSE)

$$\hat{x} = (H'WH)^{-1} H'Wy \tag{14.11b}$$

This pertains to minimization as the hessian $2H'WH$ is a non-negative definite.

Some Properties:

Rewriting Eq. (14.11b) as

$$\hat{x} = k\, y \tag{14.12a}$$

where $\qquad \mathbf{k} = (H'WH)^{-1} H'W. \tag{14.12b}$

Here the matrix $\mathbf{k}$ depends on the value of H and the choice of W.

Using Eqs. (14.1) and (14.12b) it is easy to get the relation as follows.

$$\hat{x} = KHx + \mathbf{k}r$$
$$= (H'WH)^{-1} (H'WH)\, x + \mathbf{k}r$$

or $\qquad\qquad \hat{x} = x + kr \tag{14.13}$

and $\qquad\qquad E\{\hat{x}\} = E\{x\} \tag{14.14}$

In Eq. (14.14) it is assumed that the error r is statistically independent of columns of H and the vector $\mathbf{r}$ has a zero mean. An estimate that satisfied Eq. (14.14) is called an unbiased estimate. This implies that the estimation error is zero on an average.

$$\tilde{x} = \mathbf{k}r \tag{14.15a}$$

The covariance of the error of estimation is therefore given by

$$P_x = KRK' \tag{14.15b}$$

where R is the covariance of the error vector $\mathbf{r}$. Note that the covariance P_x is a measure of the accuracy of the estimation and a smaller trace of this matrix indicates a better estimate. Eq. (14.15b) suggests that the best possible choice of the weighting matrix is to set $W = R^{-1}$. The optimum value of the error covariance matrix is then given by

$$P_x = (H'R^{-1}H)^{-1} \tag{14.15c}$$

Example 14.2

Assume that in the Example 14.1, we want to obtain the WLSE of the variable x by choosing the following weighting matrix

$$W = \begin{bmatrix} 0.1 & & \\ & 1 & \\ & & 0.1 \end{bmatrix}$$

The matrix $H'WH$ is

$$H'WH = \begin{bmatrix} 0.2 & 0.1 \\ 0.1 & 1.1 \end{bmatrix}$$

and the matrix $H'W$ is obtained as

$$H'W = \begin{bmatrix} 0.1 & 0 & 0.1 \\ 0 & 1 & 0.1 \end{bmatrix}$$

The weighted least squares estimate of the vector **x** is then obtained as (from Eq. (14.11b))

$$\hat{\mathbf{x}} = \begin{bmatrix} (11/21)\,y_1 & -(10/21)\,y_2 & (10/21)\,y_3 \\ -(1/21)\,y_1 & (20/21)\,y_2 & (1/21)\,y_3 \end{bmatrix}$$

If this result is compared with the result in Example 14.1, the effect of introducing the weighting on the estimate is apparent. Note that the choice of W in this case suggests the data for y_2 is considered more valuable and this results in the components of **x** being more heavily dependent on y_2.

The matrix k is in this case found to be (Eq. (14.12b))

$$K = \begin{bmatrix} 11/21 & -10/21 & 10/21 \\ -1/21 & 20/21 & 1/21 \end{bmatrix}$$

If the covariance of the measurement error is assumed to be $R = I$, the covariance of the estimation error is obtained as (Ref. Eq. (14.15c))

$$P_x = (1/147) \begin{bmatrix} 107 & -67 \\ -67 & 134 \end{bmatrix}$$

The choice of W above yields unacceptably large estimation error variances.

Let us now choose the weighting matrix $W = I$. The matrix K is then obtained as

$$K = \begin{bmatrix} 2/3 & -1/3 & 1/3 \\ -1/3 & 2/3 & 1/3 \end{bmatrix}$$

The error covariance matrix is then given by

$$P_x = (1/9) \begin{bmatrix} 6 & -3 \\ -3 & 6 \end{bmatrix}$$

The error variances are now seen to be much smaller as is to be expected.

Non-linear Measurements

The case of special interest to the power system state estimation problem corresponds to the non-linear measurement model.

$$y = h(x) + r \qquad (14.16)$$

where $h(x)$ represents an m dimensional vector of nonlinear functions of the variable x. It is assumed that the components of the vector $h(x)$ are continuous in their arguments and therefore may be differentiated with respect to the components of x. The problem is to extend the method of least squares in order to estimate the vector $\mathbf{x}$ from the data for the vector $\mathbf{y}$ with these two variables being related through Eq. (14.16).

To mimic our treatment of the linear measurement case, assume that $\hat{\mathbf{x}}$ represents the desired estimate so that the estimate of the measurement y could be obtained using the relation

$$\hat{\mathbf{y}} = h(\hat{\mathbf{x}}) \qquad (14.17a)$$

This yields the error of estimation of the vector y

$$\tilde{y} = \mathbf{y} - h(\hat{\mathbf{x}}) \qquad (14.17b)$$

In order to obtain the WLSE of x, we must choose the index of estimation J as follows.

$$J = [y - h(\hat{\mathbf{x}})]' \, W \, [y - h(\hat{\mathbf{x}})] \qquad (14.18)$$

The necessary condition for the index J to have a minimum at x, is given by Eq. (14.19).

$$[y - h(\hat{\mathbf{x}})] \, H \, (\hat{\mathbf{x}}) = 0 \qquad (14.19)$$

where $H (\hat{\mathbf{x}})$ is the Jacobian of $h(x)$ evaluated at $\hat{\mathbf{x}}$. In general this non-linear equation can not be solved for the desired estimate $\hat{\mathbf{x}}$. A way out of this difficulty is to make use of the linearisation technique. Let us assume that an *a priori* estimate x_0 of the vector $\mathbf{x}$ is available (say from the load flow solution).

Using Taylor series approximation, we get

$$y = h (x_0) + H_0 (x - x_0) + r \qquad (14.20)$$

where H_0 stands for the Jacobian evaluated at $x = x_0$ and the noise term r is now assumed to include the effects of the higher order terms in the Taylor series. Equation (14.20) can be rewritten as:

$$\Delta y = y - h (x_0) = H_0 \, \Delta x + r \qquad (14.21)$$

where Δy is the perturbed measurement and Δx is the perturbed value of the vector $\mathbf{x}$. An WLSE of x is then easily obtained as discussed earlier and this leads to the desired expression for the linearized solution of the non-linear estimation problem.

$$\hat{x} = x_0 + [H_0' \, WH_0]^{-1} \, H_0' \, W \, \{y - h (x_0)\} \qquad (14.22)$$

It is not likely that the estimate $\hat{x}$ obtained from Eq. (14.22) is going to be of much use since, in general, the *a priori* estimate x_0 may not be close to the optimal value of the vector $\mathbf{x}$. However, Eq. (14.22) provides us with a very useful result in the sense that it shows a mechanism for improving on the initial estimate by making use of the available measurements. Having obtained new estimate $\hat{x}$, the process of linearisation is repeated as many times as desired and this leads to the following iterative form of the solution of the non-linear estimation problem.

$$\hat{x}\ (l + 1) = \hat{x}\ (l) + K(l)\ \{y - h\ [\hat{x}\ (l)]\} \tag{14.23}$$

where the matrix $K(l)$ is defined as

$$K(l) = [H_l'\ WH_l]^{-1}\ H_l'\ W \tag{14.24}$$

The index l represents the iteration number and H_l represents the value of the Jacobian evaluated at $x = \hat{x}\ (l)$. Usually the iterative process is terminated whenever the norm of the difference of two successive values of the estimate $\hat{x}\ (l + 1) - \hat{x}\ (l)$ reaches a pre-selected threshold level.

A flowchart for implementing the iterative algorithm is shown in Fig. 14.1. A major source of computation in the algorithm lies in the need to update the Jacobian at every stage of iteration. As discussed earlier in Chapter 6 (see Eqs. (6.86) and (6.87)) it is often possible to reduce the computations by holding the value of H a constant, possibly after l exceeds 2 or 3. This is in general, permissible in view of the fact that the change in estimate tends to be rather small after a couple of iterations.

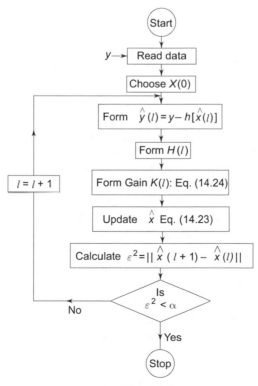

Fig. 14.1

Example 14.3

Consider the simple case of a scalar variable x and assume that the relationship between the measurement y and the variable x is given by

$$y = x^3 + r$$

The Jacobian H_j is easily obtained in this case and the iterative algorithm takes the explicit form

$$\hat{x}\,(l + 1) = \hat{x}\,(l) + [3\,\hat{x}\,(l)]^{-2}\,\{y - [\hat{x}\,(l)]^3\}$$

where we have used $W = 1$.

Let the correct value of x' be equal to 2 and assume that due to the effect of r the measured value of y is found to be 8.5. Also, assume that the initial estimate $x\,(0)$ is taken to be equal to 1. The table below gives the results of the first few iterations.

l	$\hat{x}\,(l)$
0	1.0
1	3.5
2	2.56
3	2.16

It is apparent that the algorithm would yield the correct solution after several iterations.

14.3 STATIC STATE ESTIMATION OF POWER SYSTEMS [10]–[12]

As noted earlier, for a system with N buses, the state vector **x** may be defined as the $2N - 1$ vector of the $N - 1$ voltage angles δ_2, ..., δ_N and the N voltage magnitudes V_1, V_2, ..., V_N. The load flow data, depending on type of bus, are generally corrupted by noise and the problem is that of processing an adequate set of available data in order to estimate the state vector. The readily available data may not provide enough redundancy (the large geographical area over which the system is spread often prohibits the telemetering of all the available measurements to the central computing station). The redundancy factor, defined as the ratio m/n should have a value in the range 1.5 to 2.8 in order that the computed value of the state may have the desired accuracy. It may be necessary to include the data for the power flows in both the directions of some of the tie lines in order to increase the redundancy factor. In fact, some 'pseudo measurements' which represent the computed values of such quantities as the active and reactive injections at some remote buses may also be included in the vector **y** (k).

It is thus apparent that the problem of estimation of the power system state is a non-linear problem and may be solved using either the batch processing or sequential processing formula [see Section 3.3 of Reference 1]. Also, if the system is assumed to have reached a steady-state condition, the voltage angles and magnitudes would remain more or less constant. The state estimation problem is then a static problem and the methods of Sec. 14.2 may be used, if so desired. To develop explicit solutions, it is necessary to start by noting the exact forms of the model equations for the components of the vector $\mathbf{y}(k)$.

Let P_i and Q_i denote the active and reactive power injections of ith bus. These are related to the components of the state vector through the following equations.

$$P_i = \sum_{j=1}^{N} |V_i| |V_j| |Y_{ij}| \cos(-\delta_i + \delta_j + \theta_{ij}) \tag{14.25}$$

$$Q_i = -\sum_{j=1}^{N} |V_i| |V_j| |Y_{ij}| \sin(-\delta_i + \delta_j + \theta_{ij}) \tag{14.26}$$

where $|Y_{ij}|$ represents the magnitude and θ_{ij} represents the angle of the admittance of the line connecting the ith and jth buses. The active and reactive components of the power flow from the ith to the jth bus, on the other hand are given by the following relations.

$$P_{ij} = |V_i| |V_j| |Y_{ij}| \cos(\delta_i - \delta_j + \theta_{ij}) - |V_i|^2 |Y_{ij}| \cos\theta_{ij} \tag{14.27}$$

$$Q_{ij} = |V_i| |V_j| |Y_{ij}| \sin(\delta_i - \delta_j + \theta_{ij}) - |V_i|^2 |Y_{ij}| \sin\theta_{ij} \tag{14.28}$$

Let us assume that the vector $\mathbf{y}$ has the general form

$$y = [P_1 \ \dots \ P_N \ Q_1 \ \dots \ Q_N \ P_{12} \ \dots \ P_{N-1,N} \ Q_{12} \dots \ Q_{N-1,N},$$
$$\delta_2 \ \dots \ \delta_N \ |V_1|, \ \dots, \ |V_N|]' \tag{14.29}$$

The Jacobian H will then have the form

$$H = \begin{bmatrix} H_1 & H_2 \\ H_3 & H_4 \\ H_5 & H_6 \\ H_7 & H_8 \\ \dots & \dots \\ I_{N-1} & 0 \\ 0 & I_N \end{bmatrix} \tag{14.30}$$

where I_N is the identity matrix of dimension N, H_1 is the $N \times (N-1)$ submatrix of the partial derivatives of the active power injections wrt δ's, H_2 is the

$N \times N$ sub-matrix of the partial-derivatives of the active power injections wrt $|V|^s$ and so on. Jacobian H will also be a sparse matrix since Y is a sparse matrix.

Two special cases of interest are those corresponding to the use of only the active and reactive injections and the use of only the active and reactive line flows in the vector $\mathbf{y}$. In the first case, there are a total of $2N$ components of $\mathbf{y}$ compared to the $2N - 1$ components of the state $\mathbf{x}$. There is thus almost no redundancy of measurements. However, this case is very close to the case of load flow analysis and therefore provides a good measure of the relative strengths of the methods of load flow and state estimation. In the second case, it is possible to ensure a good enough redundancy if there are enough tie lines in the system. One can obtain two measurements using two separate meters at the two ends of a single tie-line. Since these two data should have equal magnitudes but opposite signs, this arrangement also provides with a ready check of meter malfunctioning. There are other advantages of this arrangement as will be discussed later.

The Injections Only Algorithm

In this case, the model equation has the form
$$y = h\,[x] + r \tag{14.31}$$
with the components of the non-linear function given by

$$h_i\,[x] = \sum_{j=1}^{N} |V_i|\,|V_j|\,|Y_{ij}| \cos(\delta_i - \delta_j + \theta_{ij}),\ i = 1, \ldots, N \tag{14.32a}$$

$$= \sum_{j=1}^{N} |V_{N-i}|\,|V_j|\,|Y_{N-i,j}| \sin(\delta_i - \delta_j + \theta_{ij})$$
$$i = N + 1,\ N + 2 \ldots,\ 2N \tag{14.32b}$$

The elements of the sub-matrices H_1, H_2, H_3 and H_4 are then determined easily as follows.

$$H_1\,(i, j) = |V_i|\,|V_j|\,|Y_{ij}| \sin(\delta_i - \delta_j + \theta_{ij})$$
$$i = 1, 2, \ldots, N,$$
$$j = 1, 2, \ldots, N - 1$$
$$H_2\,(i, j) = |V_i|\,|Y_{ij}| \cos(\delta_i - \delta_j + \theta_{ij})\ i = 1, 2, \ldots, N,$$
$$j = 1, 2, \ldots, N.$$
$$H_3\,(i, j) = -|V_i|\,|V_j|\,|Y_{ij}| \cos(\delta_i - \delta_j + \theta_{ij})\ i = 1, 2, \ldots, N,$$
$$j = 1, 2, \ldots, N - 1$$
$$H_4\,(i, j) = |V_i|\,|Y_{ij}| \sin(\delta_i - \delta_j + \theta_{ij})\ i = 1, 2, \ldots, N,$$
$$j = 1, 2, \ldots, N. \tag{14.33}$$

Equation (14.33) may be used to determine the Jacobian at any specified value of the system state vector. The injections only state estimation algorithm is then obtained directly from the results of sec 14.2. Since the problem is non-linear, it is convenient to employ the iterative algorithm given in Eq. (14.22).

Applying the principle of decoupling, the submatrices H_2 and H_3 become null with the result that the linearised model equation may be approximated as:

$$\Delta y = \begin{bmatrix} H_1 & 0 \\ 0 & H_4 \end{bmatrix} \Delta x + r \tag{14.34}$$

If we partition the vectors **y**, **x** and **r** as

$$\Delta y = \begin{bmatrix} \Delta y_p \\ \Delta y_q \end{bmatrix}; \quad x = \begin{bmatrix} \Delta x_\delta \\ \Delta x_v \end{bmatrix}; \quad r = \begin{bmatrix} r_p \\ r_q \end{bmatrix}$$

then, Eq. (14.34) may be rewritten in the decoupled form as the following two separate equations for the two partitioned components of the state vector

$$\Delta y_p = H_1 \, \Delta x_\delta + r_p \tag{14.35}$$
$$\Delta y_q = H_4 \, \Delta x_v + r_q \tag{14.36}$$

Based on these two equations, we obtain the following nearly decoupled state estimation algorithms.

$$\hat{x}_\delta (j + 1) = x_\delta (j) + [H_1' (j) \, W_p \, H_1 (j)]^{-1} \, H_1 (j) \, \{v_p - h_p \, [\hat{x} (j)]\}$$
$$j = 0, 1, 2, \ldots \tag{14.37}$$

$$\hat{x}_v (j + 1) = \hat{x}_v (j) + [H_4' (j) \, W_q \, H_4 (j)]^{-1} \, H_4' (j) \, W_q \, \{y_q - h_q \, [\hat{x} (j)]\}$$
$$j = 0, 1, 2, \ldots \tag{14.38}$$

where the subscripts p and q are used to indicate the partitions of the weighting matrix W and the non-linear function h (.) which correspond to the vectors y_p and y_q respectively. As mentioned earlier, if the covariances R_p and R_q of the errors r_p and r_q are assumed known, one should select $W_p = R_p^{-1}$ and $W_q = R_q^{-1}$.

Note that Eqs. (14.37) and (14.38) are not truly decoupled because the partitions of the non-linear function depend on the estimate of the entire state vector. It may be possible to assume that $v_i (j) = 1$ for all i and j while, Eq. (14.37) is being used in order to estimate the angle part of the state vector. Similarly one may assume $\delta_i (j) = 0°$ for all i and j while using Eq. (14.38) in order to estimate the voltage part of the state vector. Such approximations allow the two equations to be completely decoupled but may not yield very good solutions. A better way to decouple the two equations would be to use the load flow solutions for x_v and x_δ as their supposedly constant values in Eq. (14.37) and Eq. (14.38) respectively. There are several forms of fast decoupled estimation algorithms based on such considerations (see e.g. [13], [14]). A flow chart for one scheme of fast decoupled state estimation in shown in Fig. 14.2.

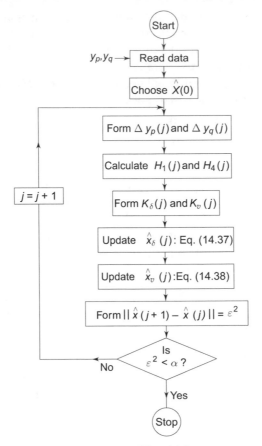

Fig. 14.2

Example 14.4

In order to illustrate an application of the injections only algorithm, consider the simple 2-bus system shown in Fig. 14.3.

Assuming lossless line, $\theta_{ij} = 90°$. Also let $Y_{11} = \gamma_{22} = 2$ and $Y_{12} = Y_{21} = 1$. The power relations in this case would be

$$P_1 = -|V_1||V_2||Y_{12}|\sin \delta_2$$
$$P_2 = |V_1||V_2||Y_{12}|\sin \delta_2$$
$$Q_1 = |Y_{11}||V_1|^2 - |Y_{12}||V_1||V_2|\cos \delta_2$$
$$Q_2 = |Y_{22}||V_2|^2 - |Y_{12}||V_1||V_2|\cos \delta_2$$

If we choose the initial values $|V_1^o| = |V_2^o| = 1$, $\delta_2^o = 0°$, the corresponding power values are $P_1^o = P_2^o = 0$, $Q_1^o = Q_2^o = 1$. The value of the Jacobian matrix evaluated at the above nominal values of the variables turns out to be

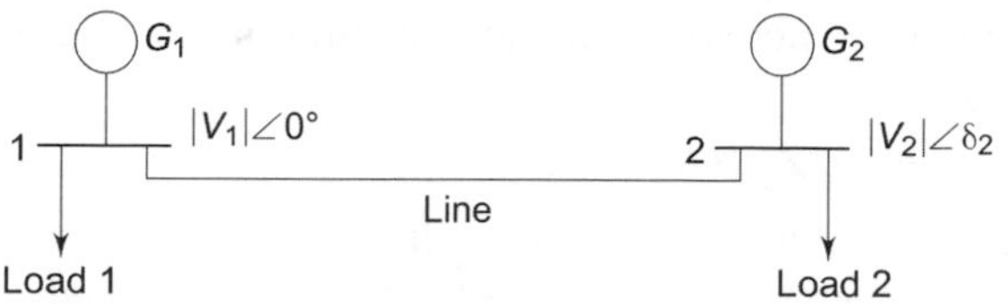

Fig. 14.3

$$H_0 = \begin{bmatrix} -1 & 0 & 0 \\ 1 & 0 & 0 \\ 0 & 3 & -1 \\ 0 & -1 & 3 \end{bmatrix}$$

Application of the LSE then yields the following expressions for the estimates of the perturbations in the three state variables around their chosen initial values:

$$\Delta \hat{\delta}_2 = \Delta P_2 - \Delta P_1$$

$$\Delta \hat{V}_1 = 0.78 \ \Delta Q_2 - 0.26 \ \Delta Q_1$$

$$\Delta \hat{V}_2 = 0.38 \ \Delta Q_2 + 0.14 \ \Delta Q_1$$

These equations should be used in order to translate the measured values of the perturbations in the active and reactive power injections into the estimates of the perturbations of the state variables.

It is interesting to note that for the simple example, partitions H_2 and H_3 are null matrices so that the decoupled state estimators are the same as those given above.

The Line Only Algorithm

This algorithm has been developed in order to avoid the need for solving a non-linear estimation problem, which as seen earlier, requires some approximation or other. In the line flow only algorithm, the data for the active and reactive tie line flows are processed in order to generate the vector of the voltage difference across the tie-lines. Let z denote this vector. A model equation for this vector may be expressed as

$$z = Bx + r_z \tag{14.39}$$

where B is the node-element incidence matrix and r_z is the vector of the errors in the voltage data. Since this is a linear equation, one may use the WLSE technique to generate the estimate as

$$\hat{x} = [B' \ WB]^{-1} \ B' \ Wz \tag{14.40}$$

where the weighting matrix may be set equal to the inverse of the covariance of r_z if this is known. The main problem with Eq. (14.40) is that the vector z is not directly measurable but needs to be generated from the tie line flow data.

V_{ij} denotes the voltage across the line connecting the ith and the jth buses, the following relation holds.

$$V_{ij} = Z_{ij} \, [(P_{ij} - j \, Q_{ij})/V_j^* - V_j \, Y_{ij}] \tag{14.41}$$

Here Z_{ij} stands for the impedance of the line.

This shows that the vector **z** is related to the vectors **x** and **y** in a non-linear fashion and one may use the notation

$$\mathbf{z} = g \, (\mathbf{x}, \mathbf{y}) \tag{14.42}$$

In view of this non-linear relation, Eq. (14.40) may be expressed in the form

$$\hat{x} = [B' \, WB]^{-1} \, B'W \, g \, (\hat{x}, y) \tag{14.43}$$

This, being a non-linear relationship, can not be solved except through a numerical approach (iterative solution). The iterative form of Eq. (14.43) is

$$\hat{x} \, (j + 1) = [B' \, WB]^{-1} \, B' \, W \, g \, [\hat{x} \, (j), y], \tag{14.44}$$
$$j = 0, 1, 2, \ldots$$

Note that the original problem of estimation of x from the data for z is a linear problem so that the solution given by Eq. (14.40) is the optimal solution. However, the data for z need to be generated using the non-linear transformation in Eq. (14.42), which in turn has necessitated the use of iterative Eq. (14.44). Compared to the injections only iterative algorithm, the present algorithm has the advantage of a constant gain matrix $[B' \, W \, B]^{-1} \, B' \, W$. This result in a considerable computational simplification. The concept of decoupled estimation is easily extended to the case of the line flows [15].

14.4 TRACKING STATE ESTIMATION OF POWER SYSTEMS [16]

Tracking the state estimation of a given power system is important for real time monitoring of the system. As is well known, the voltages of all real system vary randomly with time and should therefore be considered to be stochastic processes. It is thus necessary to make use of the sequential estimation techniques of Ref. [1] in order to obtain the state estimate at any given time point. The power relations in Eqs. (14.25) and (14.26) are still valid but must be rewritten after indicating that the voltage magnitudes and angles are new functions of the discrete time index k.

14.5 SOME COMPUTATIONAL CONSIDERATIONS

Both the static and the tracking estimation algorithms presented in the preceding sections are computationally intensive, particularly for large power networks which may have more than 200 important buses. It is, therefore, very important to pay attention to such computational issues as illconditioning, computer storage and time requirements. However, we need to first consider the question of existence of a solution of the state estimation problem.

Network Observability [17]

Consider the static WLSE formula [Eq. (14.11b)] which serves as the starting point for all the algorithms. Inverse of information matrix $M_{n \times n} = H'\,WH$ should exist otherwise there is no state estimate. This will happen if rank of H is equal to n (no. of state variables). Since one can always choose a non-singular W, so if H has a rank n, the power network is said to be *observable*.

Problem of Ill-conditioning

Even if the given power system is an observable system in terms of the measurements selected for the state estimation purposes, there is no guarantee that the required inversion of the information matrix will exist. During multiplications of the matrices, there is some small but definite error introduced due to the finite word length and quantisation. Whether or not these errors create ill-conditioning of the information matrix may be determined from a knowledge of the condition number of the matrix. This number is defined as the ratio of the largest and the smallest eigen values of the information matrix. The matrix M becomes more and more ill-conditioned as its condition number increases in magnitude. Some detailed results on power system state estimation using Cholesky factorization techniques may be found in [18]. Factorization helps to reduce ill-conditioning but may not reduce the computational burden. A technique to reduce computational burden is described in Ref. [19].

14.6 EXTERNAL SYSTEM EQUIVALENCING [20]

One of the widely practiced methods used for computational simplification is to divide the given system into three subsystems as shown in Fig. 14.4. One of these is referred to as the 'internal' subsystem and consists of those buses in which we are really interested. The second subsystem consists of those buses which are not of direct interest to us and is referred to as the 'external' subsystem. Finally, the buses which provide links between these internal and external subsystems constitute the third subsystem referred to as the 'boundary' subsystem. For any given power network, the identification of the three subsystems may be done either in a natural or in an artificial way.

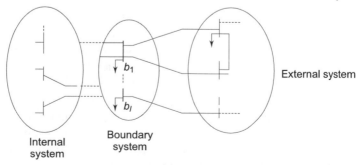

Internal system   Boundary system   External system

Fig. 14.4

To illustrate the simplification of the state estimation algorithm, consider the linearised measurement equation for the injections only case. Since the system is partitioned into three subsystems, this equation may be written as

$$\begin{bmatrix} \Delta y_i \\ \Delta y_b \\ \Delta y_e \end{bmatrix} = \begin{bmatrix} H_{ii} & H_{ib} & 0 \\ H_{bi} & H_{bb} & H_{be} \\ 0 & H_{eb} & H_{ee} \end{bmatrix} \begin{bmatrix} \Delta x_i \\ \Delta x_b \\ \Delta x_e \end{bmatrix} + \begin{bmatrix} r_i \\ r_b \\ r_e \end{bmatrix} \tag{14.45}$$

It may be noted that the internal measurement vector Δy_i is not completely independent of the external subsystem state Δx_e since Δy_i depends on the boundary subsystem state Δx_b and Δx_b depends on Δx_e.

$$\Delta y_b = \Delta y_{bi} + \Delta y_{bb} + \Delta y_{be} + r_b \tag{14.46}$$

where Δy_{be} represents the injections into the boundary buses from the external buses, Δy_{bb} is the injection from the boundary buses and Δy_{bi} is the injection from the internal buses. It is assumed that the term Δy_{be} $(= H_{be} \Delta x_e)$, may be approximated as $\hat{H} \Delta x_b$ where $\hat{H}$ is estimated from the relation

$$\hat{H} = \Delta y_{be} / \Delta x_b \tag{14.47}$$

The component Δy_{be} may be estimated if the terms Δy_{bi} and Δy_{bb} are computed as $H_{bi} \Delta x_i$ and $H_{bb} \Delta x_b$ and then subtracted from the measured value of Δy_b. This would result in part of Eq. (14.45) to be rewritten as

$$\begin{bmatrix} \Delta y_i \\ \Delta y_b \end{bmatrix} = \begin{bmatrix} H_{ii} & H_{ib} \\ H_{bi} & H_{bb}^* \end{bmatrix} \begin{bmatrix} \Delta x_i \\ \Delta x_b \end{bmatrix} + \begin{bmatrix} r_i \\ r_b \end{bmatrix} \tag{14.48}$$

where $H_{bb}^* = H_{bb} + \hat{H}$ represents the effective Jacobian if the boundary subsystem that accounts for the effects of the external subsystem on the boundary subsystem. Equation (14.48) has a lower dimension than the original measurement equation and would therefore involve less computations. The concept of external system equivalencing may be employed with the line or mixed data situations also.

14.7 TREATMENT OF BAD DATA [21, 22]

The ability to detect and identify bad measurements is extremely valuable to a load dispatch centre. One or more of the data may be affected by malfunctioning of either the measuring instruments or the data transmission system or both. Transducers may have been wired incorrectly or the transducer itself may be malfunctioning so that it simply no longer gives accurate readings.

If such faulty data are included in the vector Δy, the estimation algorithm will yield unreliable estimates of the state. It is therefore important to develop techniques for detecting the presence of faulty data in the measurement vector at any given point of time, to identify the faulty data and eliminate these from the vector **y** before it is processed for state estimation. It is also important to modify the estimation algorithms in a way that will permit more reliable state estimation in the presence of bad data.

Bad Data Detection [23]

A convenient tool for detecting the presence of one or more bad data in the vector $\mathbf{y}$ at any given point of time is based on the 'Chi Square Test'. To appreciate this, first note that the method of least square ensures that the performance index $J(x) = [y - h(x)]' W [y - h(x)] = r'Wr$ has its minimum value when $x = \hat{x}$. Since the variable r is random, the minimum value $J_{\min}$ is also a random quantity. Quite often, r may be assumed to be a Gaussian variable and then $J_{\min}$ would follow a chi square distribution with $L = m - n$ degrees of freedom. It turns out that the mean of $J_{\min}$ is equal to L and its variance is equal to $2L$. This implies that if all the data processed for state estimation are reliable, then the computed value of $J_{\min}$ should be close to the average value $(=L)$. On the other hand, if one or more of the data for y are unreliable, then the assumptions of the least squares estimation are violated and the computed value of $J_{\min}$ will deviate significantly from L.

It is thus possible to develop a reliable scheme for the detection of bad data in y by computing the value of $[y - h(\hat{x})]' W [y - h(\hat{x})]$, $\hat{x}$ being the estimate obtained on the basis of the concerned y. If the scalar so obtained exceeds some threshold $T_j = cL$, c being a suitable number, we conclude that the vector $\mathbf{y}$ includes some bad data. (Note that the data for the component y_i, $i = 1, 2, ...,$ m will be considered bad if it deviates from the mean of y_i by more than $\pm 3\sigma_i$, where σ_i is the standard deviation of r_i). Care must be exercised while choosing the value of threshold parameter c. If it is close to 1, the test may produce many 'false alarms' and if it is too large, the test would fail to detect many bad data.

To select an appropriate value of c, we may start by choosing the significance level d of the test by the relation.

$$P\{J(x) > cL / J(x) \text{ follows chi square distribution}\} = d$$

We may select, for example, $d = 0.05$ which corresponds to a 5% false alarm situation. It is then possible to find the value of c by making use of the table $\chi(L)$. Once the value of c is determined, it is simple to carry out the test whether or not $J(x)$ exceeds cL.

Identification of Bad Data [23]

Once the presence of bad data is detected, it is imperative that these be identified so that they could be removed from the vector of measurements before it is processed. One way of doing this is to evaluate the components of the measurement residual $\tilde{y}_i = y_i - h_i(x)$, $i = 1, 2, ..., m$. If we assume that the residuals have the Gaussian distribution with zero mean and the variance σ_i^2, then the magnitude of the residual y_i should lie in the range $-3\sigma_i < y_i < 3\sigma_i$ with 95% confidence level. Thus, if any one of the computed residual turns out to be significantly larger in magnitude than three times its standard deviation, then corresponding data is taken to be a bad data. If this happens for more than one component of y, then the component having the largest residual is assumed to be the bad data and is removed from y. The estimation algorithm is re-run with the remaining data and the bad data detection and identification tests are performed again to find out if there are additional bad data in the measurement set. As we will see later bad measurement data are detected, eliminated and replaced by pseudo or calculated values.

Suppression of Bad Data [24]

The procedures described so far in this section are quite tedious and time consuming and may not be utilized to remove all the bad data which may be present in the vector **y** at a given point of time. It may often be desirable on the other hand to modify the estimation algorithms in a way that will minimise the influence of the bad data on the estimates of the state vector. This would be possible if the estimation index $J(x)$ is chosen to be a non-quadratic function. The reason that the LSE algorithm does not perform very well in the presence of bad data is the fact that because of the quadratic nature of $J(x)$, the index assumes a large value for a data that is too far removed from its expected value. To avoid this overemphasis on the erroneous data and at the same time to retain the analytical tractability of the quadratic performance index, let us choose

$$J(\hat{x}) = g'\,(\tilde{y})\,W\,g(\tilde{y}) \tag{14.49a}$$

where $g(\tilde{y})$ is a non-linear function of the residual $\tilde{y}$. There may be several possible choices for this function. A convenient form is the so-called 'quadratic flat' form. In this case, the components of the function $g(y)$ are defined by the following relation.

$$g_i\,(\tilde{y}) = \tilde{y}_i, \quad \text{for } \tilde{y}_i/\sigma_i \le a_i$$
$$= a_i, \quad \text{for } \tilde{y}_i/\sigma_i > a_i \tag{14.49b}$$

where a_i is a pre-selected constant threshold level. Obviously, the performance index $J(x)$ may be expressed as

$$J(\hat{x}) = \sum_{i=1}^{m} J_i\,(\hat{x}) \tag{14.50}$$

and each component has a quadratic nature for small values of the residual but has a constant magnitude for residual magnitudes in excess of the threshold. Figure 14.5 shows a typical variation of $J_i\,(x)$ for the quadratic and the non-quadratic choices.

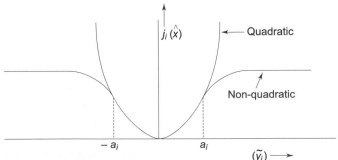

Fig. 14.5

The main advantage of the choice of the form (14.49) for the estimation index is that it is still a quadratic in the function $g(\tilde{y})$ and so the LSE theory may be mimicked in order to derive the following iterative formula for the state estimate.

$$\hat{x}(l+1) = \hat{x}(l) + [H'(l) \; C' WCH(l)] \; H'(l) \; WCg \; [\hat{y}(l)]$$

$$(14.51a)$$

where the matrix C is diagonal and its elements are computed as

$$C_i = 1, \text{ for } \tilde{y}_i/\sigma_i \leq a_i \qquad (14.51b)$$
$$= 0, \text{ for } \tilde{y}_i/\sigma_i > a_i$$

Comparing this solution with that given in Eq. (14.22), it is seen that the main effect of the particular choice of the estimation index in Eq. (14.49) is to ensure that the data producing residuals in excess of the threshold level will not change the estimate. This is achieved by the production of a null value for the matrix C for large values of the residual.

14.8   NETWORK OBSERVABILITY AND PSEUDO-MEASUREMENTS

A minimum amount of data is necessary for State Estimation (SE) to be effective. A more analytical way of determining whether a given data is enough for SE is called observability analysis. It forms an integral part of any real time state estimator.The ability to perform state estimation depends on whether sufficient measurements are well distributed throughout the system. When sufficient measurements are available SE can obtain the state vector of the whole system. In this case the network is observable. As explained earlier in Sec. 14.5 this is true when the rank of measurement Jacobian matrix is equal to the number of unknown state variables. The rank of the measurement Jacobian matrix is dependent on the locations and types of available measurements as well as on the network topology.

An auxiliary problem in state estimation is where to add additional data or pseudo measurements to a power system in order to improve the accuracy of the calculated state i.e. to improve observability. The additional measurements represent a cost for the physical transducers, remote terminal or telemetry system, and software data processing in the central computer. Selection of pseudo measurements, filling of missing data, providing appropriate weightage are the functions of the observability analysis algorithm.

Observability can be checked during factorization. If any pivot becomes very small or zero during factorization, the gain matrix may be singular, and the system may not be observable.

To find the value of an injection without measuring it, we must know the power system beyond the measurements currently being made. For example, we normally know the generated MWs and MVARs at generators through telemetry channels (i.e. these measurements would generally be known to the state estimator). If these channels are out, we can perhaps communicate with the operators in the plant control room by telephone and ask for these values and enter them manually. Similarly, if we require a load-bus MW and MVAR for a pseudo measurement, we could use past records that show the relationship between an individual load and the total system load. We can estimate the total system load quite accurately by finding the total power being generated and estimating the line losses. Further, if we have just had a telemetry failure, we could use the most recently estimated values from the estimator (assuming that it is run periodically) as pseudo measurements. Thus, if required, we can give the state estimator with a reasonable value to use as a pseudo measurement at any bus in the system.

Pseudo measurements increase the data redundancy of SE. If this approach is adapted, care must be taken in assigning weights to various types of measurements. Techniques that can be used to determine the meter or pseudo measurement locations for obtaining a complete observability of the system are available in Ref. [25]. A review of the principal observability analysis and meter placement algorithms is available in Ref. [26].

14.9 APPLICATION OF POWER SYSTEM STATE ESTIMATION

In real-time environment the state estimator consists of different modules such as network topology processor, observability analysis, state estimation and bad data processing. The network topology processor is required for all power system analysis. A conventional network topology program uses circuit breaker status information and network connectivity data to determine the connectivity of the network.

Figure 14.6 is a schematic diagram showing the information flow between the various functions to be performed in an operations control centre computer system. The system gets information from remote terminal unit (RTU) that encode measurement trunsducer outputs and opened/closed status information into digital signals which are sent to the operation centre over communications circuits. Control centre can also transmit commands such as raise/lower to generators and open/close to circuit breakers and switches. The analog measurements of generator output would be directly used by the AGC program (Chapter 8). However, rest of the data will be processed by the state estimator before being used for other functions such as OLF (Optimal Load Flow) etc.

Before running the SE, we must know how the transmission lines are connected to the load and generator buses i.e. network topology. This keeps on changing and hence the current telemetered breaker/switch status must be used to restructure the electrical system model. This is called the *network topology program* or *system status processor* or *network configurator*.

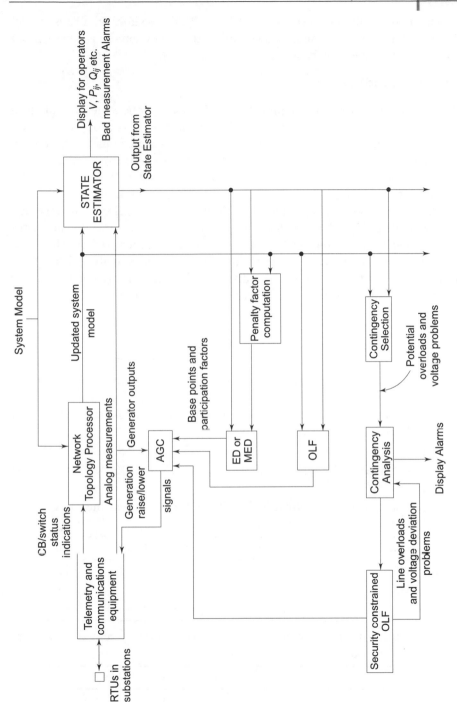

Fig. 14.6 A power system control centre

The output of the state estimator i.e. $|V|$, δ, P_{ij}, Q_{ij} together with latest model form the basis for the economic dispatch (ED) or minimum emission dispatch (MED), contingency analysis program etc.

Further Reading

The weighted least-squares approach to problems of static state estimation in power systems was introduced by Schweppe [1969–74]. It was earlier originated in the aerospace industry. Since 1970s, state estimators have been installed on a regular basis in new energy (power system or load dispatch) control centres and have proved quite helpful. Reviews of the state of the art in state estimation algorithms based on this modelling approach were published by Bose and Clements [27] and Wu [28]. Reviews of external system modelling are available in [29]. A generalised state estimator with integrated state, status and parameter estimation capabilities has recently been proposed by Alsac et al [30]. The new role of state estimation and other advanced analytical functions in competitive energy markets was discussed in Ref. [31]. A comprehensive bibliography on SE from 1968–89 is available in Ref. [32].

Problems

14.1 For Ex. 6.6 if the power injected at buses are given as $S_1 = 1.031 - j0.791$, $S_2 = 0.5 + j1.0$ and $S_3 = -1.5 - j0.15$ pu. Consider $W_1 = W_2 = W_3 = 1$. Bus 1 is a reference bus. Using flat start, find the estimates of $|V_i|$ and δ_i. Tolerance = 0.0001.

[*Ans:* $V_1^1 = 1\angle0°$, $V_2^1 = 1.04223 \angle0.4297°$, $V_3^1 = 0.99824 \angle-2.1864°$; Final values: $V_1 = 1.04\angle0°$, $V_2 = 1.080215 \angle-1.356°$, $V_3 = 1.03831 \angle- 3.736°$].

14.2 For sample system shown in Fig. P. 14.2, assume that the three meters have the following characteristics.

Meter	Full scale (MW)	Accuracy (MW)	σ (pu)
M_{12}	100	± 8	0.02
M_{13}	100	± 4	0.01
M_{32}	100	± 0.8	0.002

Calculate the best estimate for the phase angles δ_1 and δ_2 given the following measurements

Meter	Measured value (MW)
M_{12}	70.0
M_{13}	4.0
M_{32}	30.5

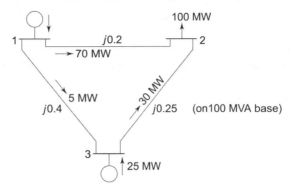

Fig. P. 14.2

14.3 Given a single line as shown in Fig. P 14.3, two measurements are available. Using DC load flow, calculate the best estimate of the power flowing through the line.

$$\delta_1 = 0 \text{ rad.}$$

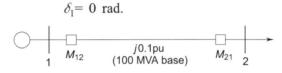

Fig. P. 14.3

Meter	Full scale (MW)	Meter Standard Deviation (σ) in full scale	Meter Reading (MW)
M_{12}	100	1	32
M_{21}	100	4	– 26

References

Books

1. Mahalanabis, A.K., D.P. Kothari and S.I. Ahson, *Computer Aided Power System Analysis and Control*, Tata McGraw-Hill, New Delhi, 1988.
2. Nagrath, I.J. and D.P. Kothari, *Power System Engineering*, Tata McGraw-Hill, New Delhi, 1994.
3. Monticelli, A., *State Estimation in Electric Power Systems A Generalised Approach*, Kluwer Academic Publishers, Boston, 1999.
4. Kusic, G.L., *Computer-Aided Power Systems Analysis*, Prentice-Hall, N.J. 1986.
5. Wood, A.J. and B.F. Wollenberg, *Power Generation, Operation and Control*, 2nd Ed., John Wiley, NY, 1996.

6. Grainger, J.J. and W.D. Stevenson, *Power System Analysis*, McGraw-Hill, NY., 1994.
7. Deautsch, R., *Estimation Theory*, Prentice-Hall Inc. NJ, 1965
8. Lawson, C.L. and R.J. Hanson, *Solving Least Squares Problems*, Prentice-Hall. Inc., NJ., 1974.
9. Sorenson, H.W., *Parameter Estimation*, Mercel Dekker, NY, 1980.

Papers

10. Schweppe, F.C., J. Wildes, D. Rom, "Power System Static State Estimation, Parts I, II and III", *IEEE Trans.*, Vol. PAS-89, 1970, pp 120–135.
11. Larson, R.E., *et. al.*, "State Estimation in Power Systems", *Parts I and II*, ibid., pp 345–359.
12. Schweppe, F.C. and E.J. Handschin, "Static State Estimation in Electric Power System", *Proc. of the IEEE*, 62, 1975, pp 972–982.
13. Horisberger, H.P., J.C. Richard and C. Rossier, "A Fast Decoupled Static State Estimator for Electric Power Systems", *IEEE Trans.* Vol. PAS-95, Jan/Feb 1976, pp 208–215.
14. Monticelli, A. and A. Garcia, "Fast Decoupled Estimators", *IEEE Trans. Power Syst*, 5, May 1990, pp 556–564.
15. Dopazo, J.F. *et. al.*, "State Calculation of Power Systems from Line Flow Measurements, Parts I and II", *IEEE Trans.*, 89, pp. 1698–1708, 91, 1972, pp 145–151.
16. Debs, A.S. and R.E. Larson, "A Dynamic Estimator for Tracking the State of a Power System", *IEEE Trans.* 89, 1970, pp 1670–1678.
17. Krumpholz, G.R. *et. al*, "Power System Observability: A Practical Algorithm Using Network Topology", *IEEE Trans.* 99, 1980, pp 1534–1542.
18. Simoes-Costa, A and V.H. Quintana, "A Robust Numerical Technique for Power System State Estimation", *IEEE Trans.* 100, 1981, pp 691–698.
19. Simoes-Costa, A and V.H. Quintana, "An Orthogonal Row Processing Algorithm for Power System Sequential State Estimation", *IEEE Trans.*, 100, 1981, pp 3791–3799.
20. Debs, A.S., "Estimation of External Network Equivalents from Internal System Data", *IEEE Trans.*, 94, 1974, pp 1260–1268.
21. Garcia, A., A. Monticelli and P. Abreu, "Fast Decoupled State Estimation and Bad Data Processing", *IEEE Trans.* PAS-98, Sept/Oct 1979, pp 1645–1652.
22. Handschin, E. *et. al.*, "Bad Data Analysis for Power System State Estimation", *IEEE Trans.*, PAS–94, 1975, pp 329–337.
23. Koglin, H.J. *et. al.*, "Bad Data Detection and Identification", *Int. J. Elec. Power*, Vol. 12, No. 2, April 1990, pp 94–103.
24. Merril, H.M. and F.C. Schweppe, "Bad Data Suppression in Power System State Estimation", *IEEE Trans.* PAS–90, 1971, pp 2718–2725.
25. Mafaakher, F., *et. al*, "Optimum Metering Design Using Fast Decoupted Estimator", *IEEE Trans.* PAS–98, 1979, pp 62–68.

26. Clements, K.A., "Observability Methods and optimal Meter Placement", *Int. J. Elec. Power*, Vol. 12, no. 2, April 1990, pp 89–93.

27. Bose, A. and Clements, K.A., "Real-time Modelling of Power Networks", *IEEE Proc., Special Issue on Computers in Power System Operations*, Vol. 75, No. 12, Dec 1987, pp 1607–1622.

28. Wu, F.F., "Power System State Estimation: A Survey", *Int. J. Elec. Power and Energy Syst.*, Vol. 12, Jan 1990, pp 80–87.

29. Wu, F.F. and A. Monticelli, "A Critical Review on External Network Medelling for On-line Security Analysis", *Int. J. Elec. Power and Energy Syst.*, Vol. 5, Oct 1983, pp 222–235.

30. Alsac, O., *et. al.*, "Generalized State Estimation", *IEEE Trans. on Power Systems*, Vol. 13, No. 3, Aug. 1998, pp 1069–1075.

31. Shirmohammadi, D. *et. al.*, "Transmission Dispatch and Congestion Management in the Emerging Energy Market Structures", *IEEE Trans. Power System.*, Vol 13, No. 4, Nov 1998, pp 1466–1474.

32. Coutto, M.B. *et. al*, "Bibliography on Power System State Estimation (1968–1989)", *IEEE Trans. Power Syst.*, Vol. 7, No. 3, Aug. 1990, pp 950–961.

15

Compensation in Power Systems

15.1 INTRODUCTION

For reduction of cost and improved reliability, most of the world's electric power systems continue to be interconnected. Interconnections take advantage of diversity of loads, availability of sources and fuel price for supplying power to loads at minimum cost and pollution with a required reliability. In a deregulated electric service environment, an effective electric grid is essential to the competitive environment of reliable electric service.

Now-a-days, greater demands have been placed on the transmission network, and these demands will continue to rise because of the increasing number of nonutility generators and greater competition among utilities themselves. It is not easy to acquire new rights of way. Increased demands on transmission, absence of long-term planning, and the need to provide open access to generating companies and customers have resulted in less security and reduced quality of supply.

Compensation in power systems is, therefore, essential to alleviate some of these problems. Series/shunt compensation has been in use for past many years to achieve this objective.

In a power system, given the insignificant electrical storage, the power generation and load must balance at all times. To some extent, the electrical system is self-regulating. If generation is less than load, voltage and frequency drop, and thereby reducing the load. However, there is only a few percent margin for such self-regulation. If voltage is propped up with reactive power support, then load increase with consequent drop in frequency may result in system collapse. Alternatively, if there is inadequate reactive power, the system may have voltage collapse.

This chapter is devoted to the study of various methods of compensating power systems and various types of compensating devices, called compensators, to alleviate the problems of power system outlined above. These compensators can be connected in the system in two ways, in series and in shunt at the line ends (or even in the midpoint).

Apart from the well-known technologies of compensation, the latest technology of Flexible AC Transmission System (FACTS) will be introduced towards the end of the chapter.

15.2 LOADING CAPABILITY

There are three kinds of limitations for loading capability of transmission system:
(i) Thermal (ii) Dielectric (iii) Stability
Thermal capability of an overhead line is a function of the ambient temperature, wind conditions, conditions of the conductor, and ground clearance.

There is a possibility of converting a single-circuit to a double-circuit line to increase the loading capability.
Dieletric Limitations From insulation point of view, many lines are designed very conservatively. For a given nominal voltage rating it is often possible to increase normal operating voltages by 10% (i.e. 400 kV – 440 kV). One should, however, ensure that dynamic and transient overvoltages are within limits. [See Chapter 13 of Ref. 7].
Stability Issues. There are certain stability issues that limit the transmission capability. These include steady-state stability, transient stability, dynamic stability, frequency collapse, voltage collapse and subsynchronous resonance.

Several good books [1, 2, 6, 7, 8] are available on these topics. The FACTS technology can certainly be used to overcome any of the stability limits, in which case the final limits would be thermal and dielectric.

15.3 LOAD COMPENSATION

Load compensation is the management of reactive power to improve power quality i.e. V profile and pf. Here the reactive power flow is controlled by installing shunt compensating devices (capacitors/reactors) at the load end bringing about proper balance between generated and consumed reactive power. This is most effective in improving the power transfer capability of the system and its voltage stability. It is desirable both economically and technically to operate the system near unity power factor. This is why some utilities impose a penalty on low pf loads. Yet another way of improving the system performance is to operate it under near balanced conditions so as to reduce the flow of negative sequence currents thereby increasing the system's load capability and reducing power loss.

A transmission line has three critical loadings (i) natural loading (ii) steady-state stability limit and (iii) thermal limit loading. For a compensated line the natural loading is the lowest and before the thermal loading limit is reached, steady-state stability limit is arrived.

15.4 LINE COMPENSATION

Ideal voltage profile for a transmission line is flat, which can only be achieved by loading the line with its surge impedance loading while this may not be achievable, the characteristics of the line can be modified by line compensators so that

(i) Ferranti effect is minimized.

(ii) Underexcited operation of synchronous generators is not required.

(iii) The power transfer capability of the line is enhanced. Modifying the characteristics of a line(s) is known as *line compensation*.

Various compensating devices are:

Capacitors

Capacitors and inductors

Active voltage source (synchronous generator)

When a number of capacitors are connected in parallel to get the desired capacitance, it is known as a bank of capacitors, similarly a bank of inductors. A bank of capacitors and/or inductors can be adjusted in steps by switching (mechanical).

Capacitors and inductors as such are passive line compensators, while synchronous generator is an active compensator. When solid-state devices are used for switching off capacitors and inductors, this is regarded as active compensation.

Before proceeding to give a detailed account of line compensator, we shall briefly discuss both shunt and series compensation.

Shunt compensation is more or less like load compensation with all the advantages associated with it and discussed in Section 15.3. It needs to be pointed out here that shunt capacitors/inductors can not be distributed uniformally along the line. These are normally connected at the end of the line and/or at midpoint of the line.

Shunt capacitors raise the load pf which greatly increases the power transmitted over the line as it is not required to carry the reactive power. There is a limit to which transmitted power can be increased by shunt compensation as it would require very large size capacitor bank, which would be impractical. For increasing power transmitted over the line other and better means can be adopted. For example, series compensation, higher transmission voltage, HVDC etc.

When switched capacitors are employed for compensation, these should be disconnected immediately under light load conditions to avoid excessive voltage rise and ferroresonance in presence of transformers.

The purpose of series compensation is to cancel part of the series inductive reactance of the line using series capacitors. This helps in (i) increase of maximum power transfer (ii) reduction in power angle for a given amount of power transfer (iii) increased loading. From practical point of view, it is desirable not to exceed series compensation beyond 80%. If the line is 100% compensated, it will behave as a purely resistive element and would cause series resonance even at fundamental frequency. The location of series capacitors is decided by economical factors and severity of fault currents. Series capacitor reduces line reactance thereby level of fault currents.

A detailed discussion on various issues involved in series and shunt compensators now follows.

15.5 SERIES COMPENSATION

A capacitor in series with a line gives control over the effective reactance between line ends. This effective reactance is given by

$$X_l' = X - X_c$$

where

X_l = line reactance

X_C = capacitor reactance

It is easy to see that capacitor reduces the effective line reactance*.
This results in improvement in performance of the system as below.

(i) Voltage drop in the line reduces (gets compensated) i.e. minimization of end-voltage variations.

(ii) Prevents voltage collapse.

(iii) Steady-state power transfer increases; it is inversely proportional to X_l'.

(iv) As a result of (ii) transient stability limit increases.

The benefits of the series capacitor compensator are associated with a problem. The capacitive reactance X_C forms a series resonant circuit with the total series reactance

$$X = X_l + X_{gen} + X_{trans}$$

The natural frequency of oscillation of this circuit is given by.

$$f_C = \frac{1}{2\pi\sqrt{LC}}$$

$$= \frac{1}{2\pi\sqrt{\dfrac{X}{2\pi f}\dfrac{2\pi f C}{2\pi f}}} = f\sqrt{\frac{X_C}{X}}$$

where f = system frequency

---

*Reactive voltage drops of a series reactance added in a line is I^2X
It is positive if X is inductive and negative if X is capacitive. So a series capacitive reactance reduces the reactance voltage drop of the line, which is an alternative way of saying that

$$X_l' = X_l - X_c.$$

$$\frac{X_C}{X} = \text{degree of compensation}$$

$$= 25 \text{ to } 75\% \text{ (recommended)}$$

For this degree of compensation

$$f_C < f$$

which is subharmonic oscillation.

Even though series compensation has often been found to be cost-effective compared to shunt compensation, but sustained oscillations below the fundamental system frequency can cause the phenomenon, referred to as subsynchronous resonance (SSR) first observed in 1937, but got world-wide attention only in the 1970s, after two turbine-generator shaft failures occurred at the Majave Generating station in Southern Nevada. Theoretical studies pointed out that interaction between a series capacitor-compensated line, oscillating at subharmonic frequency, and torsional mechanical oscillation of turbine-generator set can result in negative damping with consequent mutual reinforcement of the two oscillations. Subsynchronous resonance is often not a major problem, and low cost countermeasures and protective measures can be applied. Some of the corrective measures are:

(i) Detecting the low levels of subharmonic currents on the line by use of sensitive relays, which at a certain level of currents triggers the action to bypass the series capacitors.

(ii) Modulation of generator field current to provide increased positive damping at subharmonic frequency.

Series inductors are needed for line compensation under light load conditions to counter the excessive voltage rise (Ferranti effect).

As the line load and, in particular the reactive power flow over the line varies, there is need to vary the compensation for an acceptable voltage profile. The mechanical switching arrangement for adjusting the capacitance of the capacitor bank in series with the line is shown in Fig. 15.1. Capacitance is varied by opening the switches of individual capacitances with the capacitance C_1, being started by a bypass switch. This is a step-wise arrangement. The whole bank can also be bypassed by the starting switch under any emergent conditions on the line. As the switches in series with capacitor are current carrying suitable circuit breaking arrangement are necessary. However, breaker switched capacitors in series are generally avoided these days the capacitor is either fixed or thyristor switched.

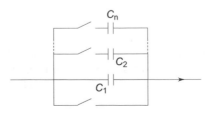

Fig. 15.1

With fast advancement in thyristor devices and associated switching control technology, the capacitance of the series capacitance bank can be controlled much more effectively; both stepwise and smooth control. This is demonstrated by the schematic diagram of Fig. 15.2 wherein the capacitor is shunted by two thyristors in antiparallel. Upon firing the thyristors alternately one carries current in positive half cycle and the other in negative half cycle.

In each half cycle when the thyristor is fired (at an adjustable angle), it conducts current for the rest of the half cycle till natural current zero. During the off-time of the thyristor current is conducted by the capacitor and capacitor voltage is v_c. During on-time of the thyristor capacitor is short circuited i.e. $v_c = 0$ and current is conducted by the thyristor. The same process is repeated in the other half cycle. This means that v_c can be controlled for any given i, which is equivalent of reducing the capacitance as $C = v_c/i$. By this scheme capacitance can be controlled smoothly by adjusting the firing angle.

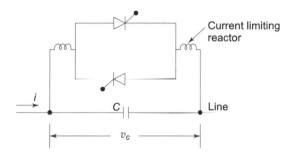

Fig. 15.2

Thyristors are now available to carry large current and to withstand (during off-time) large voltage encountered in power systems. The latest device called a Gate Turn Off (GTO) thyristor has the capability that by suitable firing circuit, angle (time) at which it goes on and off can both be controlled.

This means wider range and finer control over capacitance. Similarly control is possible over series reactor in the line.

All controlled and uncontrolled (fixed) series compensators require a protective arrangement. Protection can be provided externally either by voltage arrester or other voltage limiting device or an approximate bypass switch arrangement. In no case the VI rating of the thyristors should be exceeded.

Depending on (i) kind of solid-state device to be used (ii) capacitor and/or reactor compensation and (iii) switched (step-wise) or smooth (stepless) control, several compensation schemes have been devised and are in use. Some of the more common compensation schemes are as under.

  (i) Thyristor Controlled Series Capacitor (TCSC)

 (ii) Thryristor Switched Series Capacitor (TSSC)

(iii) Thyristor Controlled Reactor with Fixed Capacitor (TCR + FC)

(iv) GTO thyristor Controlled Series Capacitor (GCSC)

(v) Thyristor Controlled reactor (TCR)

Capacitor and/or reactor series compensator act to modify line impedance. An alternative approach is to introduce a controllable voltage source in series with the line. This scheme is known as static synchronous series compensator (SSSC). SSSC has the capability to induce both capacitive and inductive voltage in series with line, thereby widening the operating region of the scheme. It can be used for power flow control both increasing or decreasing reactive flow on the line. Further this scheme gives better stability and is more effective in damping out electromechanical oscillations.

Though various types of compensators can provide highly effective power flow control, their operating characteristics and compensating features are different. These differences are related to their inherent attributes of their control circuits; also they exhibit different loss characteristics.

From the point of view of almost maintenance free operation impedance modifying (capacitors and/or reactors) schemes are superior. The specific kind of compensator to be employed is very much dependent on a particular application.

15.6 SHUNT COMPENSATORS

As already explained in Sec. 15.4 and in Ch. 5 (Sec. 5.10) shunt compensators are connected in shunt at various system nodes (major substations) and sometimes at mid-point of lines. These serve the purposes of voltage control and load stabilization. As a result of installation of shunt compensators in the system, the nearby generators operate at near unity pf and voltage emergencies mostly do not arise. The two kinds of compensators in use are:

(i) *Static var compensators (SVC):*   These are banks of capacitors (sometimes inductors also for use under light load conditions)

(ii) STATCOM: static synchronous compensator

(iii) *Synchronous condenser:*   It is a synchronous motor running at no-load and having excitation adjustable over a wide range. It feeds positive VARs into the line under overexcited conditions and negative VARS when underexcited. (For details see Sec. 5.10.)

It is to be pointed out here that SVC and STATCOM are static var generators which are thyristor controlled. In this section SVC will be detailed while STATCOM forms a part of FACTS whose operation is explained in Sec. 15.10.

Static VAR Compensator (SVC)

These comprise capacitor bank fixed or switched (controlled) or fixed capacitor bank and switched reactor bank in parallel. These compensators draw reactive (leading or lagging) power* from the line thereby regulating voltage, improve

---

*A reactance connected in shunt to line at voltage V draws reactive power V^2/X. It is negative (leading) if reactance is capacitive and positive (lagging) if reactance is inductive.

stability (steady-state and dynamic), control overvoltage and reduce voltage flicker. These also reduce voltage and current unbalances. In HVDC application these compensators provide the required reactive power and damp out subharmonic oscillations.

Since static var compensators use switching for var control. These are also called static var switches or systems. It means that terminology wise

$$SVC = SVS$$

and we will use these interchangeably.

Basic SVC Configurations (or Designs)

Thyristors in antiparallel can be used to switch on a capacitor/reactor unit in stepwise control. When the circuitry is designed to adjust the firing angle, capacitor/reactor unit acts as continuously variable in the power circuit.

Capacitor or capacitor and inductor bank can be varied stepwise or continuously by thyristor control. Several important SVS configurations have been devised and are applied in shunt line compensation. Some of the static compensators schemes are discussed in what follows.

(i) *Saturated reactor*

This is a multi-core reactor with the phase windings so arranged as to cancel the principal harmonics. It is considered as a constant voltage reactive source. It is almost maintenance free but not very flexible with respect to operating characteristics.

(ii) *Thyristor-controlled reactor (TCR)*

A thyristor-controlled-reactor (Fig. 15.3) compensator consists of a combination of six pulse or twelve pluse thyristor-controlled reactors with a fixed shunt capacitor bank. The reactive power is changed by adjusting the thyristor firing angle. TCRs are characterised by continuous control, no transients and generation of harmonics*. The control system consists of voltage (and current)

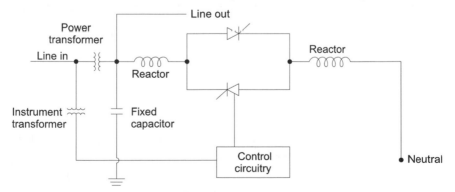

Fig. 15.3 Thyristor controlled reactor (TCR) with fixed capacitor

*Though λ –connected TCR's are used here, it is better to use Δ-connected TCR's since it is better configuration.

measuring devices, a controller for error-signal conditioning, a linearizing circuit and one or more synchronising circuits.

(iii)   *Thyristor switched capacitor (TSC)*

It consists of only a thyristor-switched capacitor bank which is split into a number of units of equal ratings to achieve a stepwise control (Fig. 15.4).

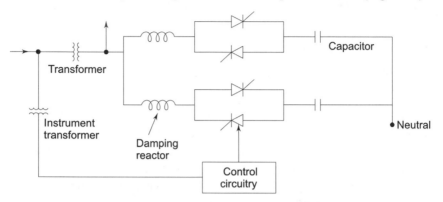

Fig. 15.4   Thyristor switched capacitor (TSC)

As such they are applied as a discretly variable reactive power source, where this type of voltage support is deemed adequate. All switching takes place when the voltage across the thyristor valve is zero, thus providing almost transient free switching. Disconnection is effected by suppressing the firing plus to the thyristors, which will block when the current reaches zero. TSCs are characterised by step wise control, no transients, very low hormonics, low losses, redundancy and flexibility.

(iv)   *Combined TCR and TSC Compensator*

A combined TSC and TCR (Fig. 15.5) is the optimum solution in majority of cases. With this, continuous variable reactive power is obtained throughout the complete control range. Furthermore full control of both inductive and capacitive parts of the compensator is obtained. This is a very advantageous

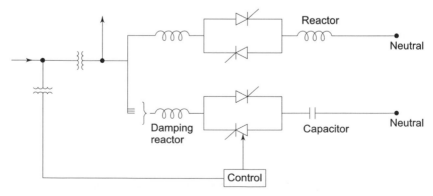

Fig. 15.5   A combined TCR/TSC compensator

feature permitting optimum performance during large disturbances in the power system (e.g. line faults, load rejection etc) TSC/TCR combinations are characterised by continuous control, no transients, low generations of harmonics, low losses, redundancy, flexible control and operation.

The basic characteristics of the main static var generator schemes are given in Table 15.1.

Table 15.1 Comparison of Static Var Generators

Type of Var Generator	TCR-FC (1)	TSC-(TSR) (2)	TCR-TSC (3)
VI and VQ characteristics	Max comp. current is proportional to system voltage. Max cap. var output decreases with the square of the voltage decrease.	Max. Comp. current is proportional to system voltage. Max. cap. var output decreases with the square of the voltage decrease.	Same as in (1) or (2)
Loss Vs var output.	High losses at zero output. Losses decrease smoothly with cap. output, increase with inductive output	Low losses at zero ouput. Losses increase step-like with cap. output	Low losses at zero output. Losses increase step-like with cap. output, smoothly with ind. output
Harmonic generation	Internally high (large pu TCR) Requires significant filtering	Internally very low Resonance may necessitate tuning reactors	Internally low (small pu TCR) Filtering required
Max. theoret. delay	1/2 cycle	1 cycle	1 cycle
Transient behaviour under system voltage disturbances	Poor (FC causes transient over-voltages in response to step disturbances)	Can be neutral. (Capacitors can be switched out to minimise transient over-voltages)	Same as in (2)

15.7 COMPARISON BETWEEN STATCOM AND SVC

It may be noted that in the normal linear operating range of the V-I characteristic and functional compensation capability of the STATCOM and the SVC are similar [2]. However, the basic operating principles of the STATCOM, which, with a converter based var generator, functions as a shunt-connected synchronous voltage source, are basically different from those of the SVC, since SVC functions as a shunt-connected, controlled reactive admittance. This basic operational difference renders the STATCOM to have overall superior functional characteristics, better performance, and greater application flexibility as compared to SVC. The ability of the STATCOM to maintain full capacitive output current at low system voltage also makes it more effective than the SVC in improving the transient (first swing) stability.

Comparison between series and shunt compensation:

Advantages of series compensation:

(i) Series capacitors are inherently self regulating and a control system is not required.

(ii) For the same performance, series capacitors are often less costly than SVCs and losses are very low.

(iii) For voltage stability, series capacitors lower the critical or collapse voltage.

(iv) Series capacitors possess adequate time-overload capability.

(v) Series capacitors and switched series capacitors can be used to control loading of paralled lines to minimise active and reactive losses.

Disadvantages of series compensation:

(i) Series capacitors are line connected and compensation is removed for outages and capacitors in parallel lines may be overloaded.

(ii) During heavy loading, the voltage on one side of the series capacitor may be out-of range.

(iii) Shunt reactors may be needed for light load compensation.

(iv) Subsynchronous resonance may call for expensive countermeasures.

Advantages of SVC

(i) SVCs control voltage directly.

(ii) SVCs control temporary overvoltages rapidly.

Disadvantages of SVC

(i) SVCs have limited overload capability.

(ii) SVCs are expensive.

The best design perhaps is a combination of series and shunt compensation. Because of higher initial and operating costs, synchronous condensers are normally not competitive with SVCs. Technically, synchronous condensers are better than SVCs in voltage-weak networks. Following a drop in network voltage, the increase in condenser reactive power output is instantaneous. Most synchronous condenser applications are now associated with HVDC installations.

15.8 FLEXIBLE AC TRANSMISSION SYSTEMS (FACTS)

The rapid development of power electronics technology provides exciting opportunities to develop new power system equipment for better utilization of existing systems. Since 1990, a number of control devices under the term FACTS technology have been proposed and implemented. FACTS devices can be effectively used for power flow control, load sharing among parallel corridors, voltage regulation, enhancement of transient stability and mitigation of system oscillations. By giving additional flexibility, FACTS controllers can enable a line to carry power closer to its thermal rating. Mechanical switching has to be supplemented by rapid response power electronics. It may be noted that FACTS is an enabling technology, and not a one-on-one substitute for mechanical switches.

FACTS employ high speed thyristors for switching in or out transmission line components such as capacitors, reactors or phase shifting transformer for some desirable performance of the systems. The FACTS technology is not a single high-power controller, but rather a collection of controllers, which can be applied individually or in coordination with others to control one or more of the system parameters.

Before proceeding to give an account of some of the important FACTS controllers the principle of operation of a switching converter will be explained, which forms the heart of these controllers.

15.9 PRINCIPLE AND OPERATION OF CONVERTERS

Controllable reactive power can be generated by dc to ac switching converters which are switched in synchronism with the line voltage with which the reactive power is exchanged. A switching power converter consists of an array of solid-state switches which connect the input terminals to the output terminals. It has no internal storage and so the instantaneous input and output power are equal. Further the input and output terminations are complementary, that is, if the input is terminated by a voltage source (charged capacitor or battery), output is a current source (which means a voltage source having an inductive impedance) and vice versa. Thus, the converter can be voltage sourced (shunted by a capacitor or battery) or current sourced (shunted by an inductor).

Single line diagram of the basic voltage sourced converter scheme for reactive power generation is drawn in Fig. 15.6. For reactive power flow bus voltage V and converter terminal voltage V_0 are in phase.

Then on per phase basis

$$I = \frac{V - V_0}{X}$$

The reactive power exchange is

$$Q = VI = \frac{V(V - V_0)}{X}$$

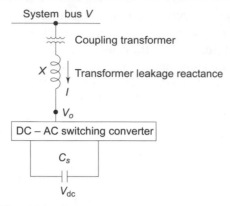

Fig. 15.6   Static reactive power generator

The switching circuit is capable of adjusting V_0, the output voltage of the converter. For $V_0 < V$, I lags V and Q drawn from the bus is inductive, while for $V_0 > V$, I leads V and Q drawn from the bus is leading. Reactive power drawn can be easily and smoothly varied by adjusting V_0 by changing the on-time of the solid-state switches. It is to be noted that transformer leakage reactance is quite small (0.1–0.15 pu), which means that a amall difference of voltage $(V - V_0)$ causes the required I and Q flow. Thus the converter acts like a static synchronous condenser (or var generator).

A typical converter circuit is shown in Fig. 15.7. It is a 3-phase two-level, six-pulse H-bridge with a diode in antiparallel to each of the six thyristors (Normally, GTO's are used). Timings of the triggering pulses are in synchronism with the bus voltage waves.

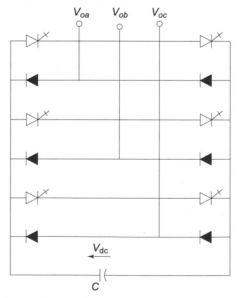

Fig. 15.7   Three-phase, two-level six-pulse bridge

As the converter draws only reactive power, the real power drawn from the capacitor is zero. Also at dc (zero frequency) the capacitor does not supply any reactive power. Therefore, the capacitor voltage does not change and the capacitor establishes only a voltage level for the converter. The switching causes the converter to interconnect the 3-phase lines so that reactive current can flow between them.

The converter draws a small amount of real power to provide for the internal loss (in switching). If it is required to feed real power to the bus, the capacitor is replaced by a storage battery. For this the circuit switching has to be modified to create a phase difference δ between V_0 and V with V_0 leading V.

The above explained converter is connected in shunt with the line. On similar lines a converter can be constructed with its terminals in series with the line. It has to carry the line current and provide a suitable magnitude (may also be phase) voltage in series with the line. In such a connection it would act as an impedance modifier of the line.

15.10 FACTS CONTROLLERS

The development of FACTS controllers has followed two different approaches. The first approach employs reactive impedances or a tap changing transformer with thyristor switches as controlled elements, the second approach employs self-commutated static converters as controlled voltage sources.

In general, FACTS controllers can be divided into four categories.
(i) series (ii) shunt (iii) combined series-series (iv) combined series-shunt controllers.

The general symbol for a FACTS controller is given in Fig. 15.8(a). which shows a thyristor arrow inside a box. The *series controller* of Fig. 15.8b could be a variable impedance, such as capacitor, reactor, etc. or a power electronics based variable source. All series controllers inject voltage in series with the line. If the voltage is in phase quadrature with the line, the series controller only supplies or consumes variable reactive power. Any other phase relationship will involve real power also.

The *shunt controllers* of Fig. 15.8c may be variable impedance, variable source or a combination of these. All shunt controllers inject current into the system at the point of connection. Combined series-series controllers of Fig. 15.8d could be a combination of separate series controllers which are controlled in a coordinated manner or it could be a unified controller.

Combined series-shunt controllers are either controlled in a coordinated manner as in Fig. 15.8e or a unified Power Flow Controller with series and shunt elements as in Fig. 15.8f. For unified controller, there can be a real power exchange between the series and shunt controllers via the dc power link.

Storage source such as a capacitor, battery, superconducting magnet, or any other source of energy can be added in parallel through an electronic interface to replenish the converter's dc storage as shown dotted in Fig. 15.8 (b). A controller with storage is much more effective for controlling the system dynamics than the corresponding controller without storage.

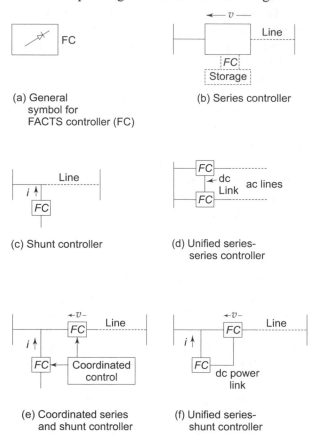

(a) General symbol for FACTS controller (FC)

(b) Series controller

(c) Shunt controller

(d) Unified series-series controller

(e) Coordinated series and shunt controller

(f) Unified series-shunt controller

Fig. 15.8 Different FACTS controllers

The group of FACTS controllers employing switching converter-based synchronous voltage sources include the STATic synchronous COMpensator (STATCOM), the static synchronous series compensator (SSCC), the unified power flow controller (UPFC) and the latest, the Interline Power Flow Controller (IPFC).

STATCOM

STATCOM is a static synchronous generator operated as a shunt-connected static var compensator whose capacitive or inductive output current can be controlled independent of the ac system voltage. The STATCOM, like its conventional counterpart, the SVC, controls transmission voltage by reactive shunt compensation. It can be based on a voltage-sourced or current-sourced converter. Figure 15.9 shows a one-line diagram of STATCOM based on a voltage-sourced converter and a current sourced converter. Normally a voltage-source converter is preferred for most converter-based FACTS controllers. STATCOM can be designed to be an active filter to absorb system harmonics.

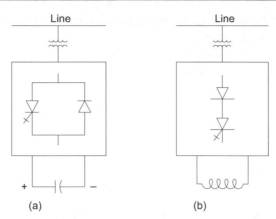

Fig. 15.9 (a) STATCOM based on voltage-sourced and (b) current-sourced converters.

A combination of STATCOM and any energy source to supply or absorb power is called static synchronous generator (SSG). Energy source may be a battery, flywheel, superconducting magnet, large dc storage capacitor, another rectifier/inverter etc.

Static Synchronous Series Compensator (SSCC)

It is a series connected controller. Though it is like STATCOM, but its output voltage is in series with the line. It thus controls the voltage across the line and hence its impedance.

Interline Power Flow Controller (IPFC)

This is a recently introduced controller [2, 3]. It is a combination of two or more static synchronous series compensators which are coupled via a common dc link to facilitate bi-directional flow of real power between the ac terminals of the SSSCs, and are controlled to provide independent reactive series compensation for the control of real power flow in each line and maintain the desired distribution of reactive power flow among the lines. Thus it manages a comprehensive overall real and reactive power management for a multi-line transmission system.

Unified Power Flow Controller (UPFC)

This controller is connected as shown in Fig. 15.10. It is a combination of STATCOM and SSSC which are coupled via a common dc link to allow bi-directional flow of real power between the series output terminals of the SSSC and the shunt output terminals of the STATCOM. These are controlled to provide concurrent real and reactive series line compensation without an external energy source. The UPFC, by means of angularly unconstrained series voltage injection, is able to control, concurrently/simultaneously or selectively, the transmission line voltage, impedance, and angle or, alternatively, the real and reactive line flows. The UPFC may also provide independently controllable shunt reactive compensation.

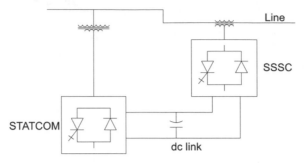

Fig. 15.10 Unified Power Flow Controller UPFC

Thyristor-Controlled Phase-Shifting Transformer (TCPST)

This controller is also called Thyristor-controlled Phase Angle Regulator (TCPAR). A phase shifting transformer controlled by thyristor switches to give a rapidly variable phase angle.

Thyristor-Controlled Voltage Regulator (TCVR)

A thyristor controlled transformer which can provide variable in-phase voltage with continuous control.

Interphase Power Controller (IPC)

A series-connected controller of active and reactive power consisting, in each phase, of inductive and capacitive branches subjected to separately phase-shifted voltages. The active and reactive power can be set inpendently by adjusting the phase shifts and/or the branch impedances, using mechanical or electronic switches.

Thyristor Controlled Braking Resistor (TCBR)

It is a shunt-connected thyristor-switched resistor, which is controlled to aid stabilization of a power system or to minimise power acceleration of a generating unit during a disturbance.

Thyristor-controlled Voltage Limiter (TCVL)

A thyristor-switched metal-oxide varistor (MOV) used to limit the voltage across its terminals during transient conditions.

HVDC

It may be noted that normally HVDC and FACTS are complementary technologies. The role of HVDC, for economic reasons, is to interconnect ac systems where a reliable ac interconnection would be too expensive. HVDC transmission as well as back-to-back HVDC system can improve transient stability and control line flows. Voltage source converter based (self-commutated) HVDC system may have the same features as those of STATCOM or UPFC. This system also regulates voltage and provides system damping.

A comparative performance of major FACTS controllers in ac system is given in Table 15.2 [14].

Table 15.2 A comparative performance of major FACTS controller

Type of FACTS Controller	Load flow control	V control	Transient stability	Oscillation Damping
SVC/STATCOM	×	×××	×××	××
TCSC	××	×	×××	××
SSSC	×××	×	×××	××
TCPAR	×××	××	×	××
UPFC	×××	×××	×××	×××

* ×××—strong influence; ××—average influence; ×—small influence

SUMMARY

Since the 1970s, energy cost, environmental restrictions, right-of-way difficulties, along with other legislative social and cost problems have postponed the construction of both new generation and transmission systems in India as well as most of other countries. Recently, because of adoption of power reforms or restructuring or deregulation, competitive electric energy markets are being developed by mandating open access transmission services.

In the late 1980s, the vision of FACTS was formulated. In this various power electronics based controllers (compensators) regulate power flow and transmission voltage and through fast control action, mitigate dynamic disturbances. Due to FACTS, transmission line capacity was enhanced. Two types of FACTS controllers were developed. One employed conventional thyristor-switched capacitors and reactors, and quadrature tap-changing transformers such as SVC and TCSC. The second category was of self-commutated switching converters as synchronous voltage sources, e.g. STATCOM, SSSC, UPFC and IPFC. The two groups of FACTS controllers have quite different operating and performance characteristics. The second group uses self-commutated dc to ac converter. The converter, supported by a dc power supply or energy storage device can also exchange real power with the ac system besides controlling reactive power independently.

The increasing use of FACTS controllers in future is guaranteed. What benefits are required for a given system would be a principal justification for the choice of a FACTS controller. Its final form and operation will, ofcourse, depend not only on the successful development of the necessary control and communication technologies and protocols, but also on the final structure of the evolving newly restructured power systems.

REFERENCES

Books

1. Chakrabarti, A., D.P. Kothari and A.K. Mukhopadhyay, *Performance Operation and Control of EHV Power Transmission Systems*, Wheeler, New Delhi, 1995.
2. Hingorani, N.G. and Laszlo Gyugyi, *Understanding FACTS*, IEEE Press, New York, 2000.
3. Song, Y.H. and A.T. Johns, *Flexible AC Transmission Systems*, IEE, London, 1999.
4. Miller, T.J.E., *Reactive Power Control in Electric Systems*, John Wiley and Sons, NY, 1982.
5. Nagrath. I.J. and D.P. Kothari, *Electric Machines*, 2nd edn, Tata McGraw-Hill, New Delhi, 1997.
6. Taylor, C.W., *Power System Voltage Stability*, McGraw-Hill, Singapore, 1994.
7. Nagrath, I.J and D.P. Kothari, *Power System Engineering*, Tata McGraw-Hill, New Delhi, 1994.
8. Indulkar, C.S. and D.P. Kothari, *Power System Transients A Statistical Approach*, Prentice-Hall of India, New Delhi, 1996.
9. Mathur, R.M. and R.K. Verma, *Thyristor-Based FACTS Controllers for Electrical Transmission Systems*, John Wiley, New York, 2002.

Papers

10. Edris, A., "FACTS Technology Development: An Update", *IEEE Power Engineering Review*, Vol. 20, March 2000, pp 4–9.
11. Iliceto F. and E. Cinieri, "Comparative Analysis of Series and Shunt Compensation Schemes for AC Transmission Systems", *IEEE Trans. PAS 96 (6)*, 1977, pp 1819–1830.
12. Kimbark, E.W., "How to Improve System Stability without Risking Subsynchronous Resonance", *IEEE Trans. PAS, 96 (5)*, Sept/Oct 1977, pp 1608–19.
13. CIGRE' 'WG 38-01, *Static Var Compensators*, CIGRE', Paris, 1986
14. Povh, D., "Use of HDVC and FACTS", *IEEE Proceedings*, Vol. 88, 2, Feb. 2000, pp 235–245.
15. Kimbark, E.W, "A New Look At Shunt Compensation", *IEEE Trans*, Vol PAS-102, No. 1, Jan 1983, pp 212–218.

16

Load Forecasting Technique

16.1 INTRODUCTION

Load forecasting plays an important role in power system planning, operation and control. Forecasting means estimating active load at various load buses ahead of actual load occurrence. Planning and operational applications of load forecasting requires a certain 'lead time' also called forecasting intervals. Nature of forecasts, lead times and applications are summarised in Table 16.1.

Table 16.1

Nature of forecast	Lead time	Application
Very short term	A few seconds to several minutes	Generation, distribution schedules, contingency analysis for system security
Short term	Half an hour to a few hours	Allocation of spinning reserve; operational planning and unit commitment; maintenance scheduling
Medium term	A few days to a few weeks	Planning for seasonal peak-winter, summer
Long term	A few months to a few years	Planning generation growth

A good forecast reflecting current and future trends, tempered with good judgement, is the key to all planning, indeed to financial success. The accuracy of a forecast is crucial to any electric utility, since it determines the timing and characteristics of major system additions. A forecast that is too low can result in low revenue from sales to neighbouring utilities or even in load curtailment. Forecasts that are too high can result in severe financial problems due to excessive investment in a plant that is not fully utilized or operated at low capacity factors. No forecast obtained from analytical procedures can be strictly relied upon the judgement of the forecaster, which plays a crucial role in arriving at an acceptable forecast.

Choosing a forecasting technique for use in establishing future load requirements is a nontrivial task in itself. Depending on nature of load variations, one particular method may be superior to another.

The two approaches to load forecasting namely total load approach and component approach have their own merits and demerits. Total load approach has the merit that it is much smoother and indicative of overall growth trends and easy to apply. On the other hand, the merit of the component approach is that abnormal conditions in growth trends of a certain component can be detected, thus preventing misleading forecast conclusions. There is a continuing need, however, to improve the methodology for forecasting power demand more accurately.

The aim of the present chapter is to give brief expositions of some of the techniques that have been developed in order to deal with the various load forecasting problems. All of these are based on the assumption that the actual load supplied by a given system matches the demands at all points of time (i.e., there has not been any outages or any deliberate shedding of load). It is then possible to make a statistical analysis of previous load data in order to set up a suitable model of the demand pattern. Once this has been done, it is generally possible to utilize the identified load model for making a prediction of the estimated demand for the selected lead time. A major part of the forecasting task is thus concerned with that of identifying the best possible model for the past load behaviour. This is best achieved by decomposing the load demand at any given point of time into a number of distinct components. The load is dependent on the industrial, commercial and agricultural activities as well as the weather condition of the system/area. The weather sensitive component depends on temperature, cloudiness, wind velocity, visibility and precipitation. Recall the brief discussions in Ch. 1 regarding the nature of the daily load curve which has been shown to have a constant part corresponding to the base load and other variable parts. For the sake of load forecasting, a simple decomposition may serve as a convenient starting point. Let $y(k)$ represent the total load demand (either for the whole or a part of the system) at the discrete time $k = 1, 2, 3,$ .... It is generally possible to decompose $y(k)$ into two parts of the form

$$y(k) = y_d(k) + y_s(k) \tag{16.1}$$

where the subscript d indicates the deterministic part and the subscript s indicates the stochastic part of the demand. If k is considered to be the present time, then $y(k + j), j > 0$ would represent a future load demand with the index j being the lead time. For a chosen value of the index j, the forecasting problem is then the same as the problem of estimating the value of $y(k + j)$ by processing adequate data for the past load demand.

16.2 FORECASTING METHODOLOGY

Forecasting techniques may be divided into three broad classes. Techniques may be based on extrapolation or on correlation or on a combination of both. Techniques may be further classified as either deterministic, probabilistic or stochastic.

Extrapolation

Extrapolation techniques involve fitting trend curves to basic historical data adjusted to reflect the growth trend itself. With a trend curve the forecast is obtained by evaluating the trend curve function at the desired future point. Although a very simple procedure, it produces reasonable results in some instances. Such a technique is called a deterministic extrapolation since random errors in the data or in analytical model are not accounted for.

Standard analytical functions used in trend curve fitting are [3].

(i) Straight line $\quad y = a + bx$
(ii) Parabola $\quad y = a + bx + cx^2$
(iii) S-curve $\quad y = a + bx + cx^2 + dx^3$
(iv) Exponential $\quad y = ce^{dx}$
(v) Gempertz $\quad y = ln^{-1}(a + ce^{dx})$

The most common curve-fitting technique for fitting coefficients and exponents $(a$–$d)$ of a function in a given forecast is the method of least squares. If the uncertainty of extrapolated results is to be quantified using statistical entities such as mean and variance, the basic technique becomes probabilistic extrapolation. With regression analysis the best estimate of the model describing the trend can be obtained and used to forecast the trend.

Correlation

Correlation techniques of forecasting relate system loads to various demographic and economic factors. This approach is advantageous in forcing the forecaster to understand clearly the interrelationship between load growth patterns and other measurable factors. The disadvantage is the need to forecast demographic and economic factors, which can be more difficult than forecasting system load. Typically, such factors as population, employment, building permits, business, weather data and the like are used in correlation techniques.

No one forecasting method is effective in all situations. Forecasting techniques must be used as tools to aid the planner; good judgement and experience can never be completely replaced.

16.3 ESTIMATION OF AVERAGE AND TREND TERMS

The simplest possible form of the deterministic part of $y(k)$ is given by

$$y_d(k) = \bar{y}_d + bk + e(k) \tag{16.2}$$

where $\bar{y}_d$ represents the average or the mean value of $y_d(k)$, bk represents the 'trend' term that grows linearly with k and $e(k)$ represents the error of modelling the complete load using the average and the trend terms only. The question is one of estimating the values of the two unknown model parameters $\bar{y}_d$ and b to ensure a good model. As seen in Ch. 14, when little or no statistical information is available regarding the error term, the method of LSE is helpful. If this method is to be used for estimating $\bar{y}_d$ and b, the estimation index J is defined using the relation

$$J = E\{e^2(k)\} \tag{16.3}$$

where $E(\cdot)$ represents the expectation operation. Substituting for $e(k)$ from Eq. (16.2) and making use of the first order necessary conditions for the index J to have its minimum value with respect to y_d and b, it is found that the following conditions must be satisfied [2].

$$E\{\bar{y}_d - y_d(k) + bk\} = 0 \tag{16.4a}$$

$$E\{bk^2 - y_d(k)k + \bar{y}_d k\} = 0 \tag{16.4b}$$

Since the expectation operation does not affect the constant quantities, it is easy to solve these two equations in order to get the desired relations.

$$\bar{y}_d = E\{y_d(k)\} - b\{E(k)\} \tag{16.5a}$$

$$b = [E\{y_d(k)k\} - \bar{y}_d E\{k\}]/E\{k^2\} \tag{16.5b}$$

If $y(k)$ is assumed to be stationary (statistics are not time dependent) one may involve the ergodic hypothesis and replace the expectation operation by the time averaging formula. Thus, if a total of N data are assumed to be available for determining the time averages, the two relations may be equivalently expressed as follows.

$$\bar{y}_d = \left(\frac{1}{N}\right)\left[\sum_{k=1}^{N} y_d(k) - b\sum_{k=1}^{N} k\right] \tag{16.6a}$$

$$b = \frac{N\left[\sum_{k=1}^{N} y_d(k)k\right] - \left[\sum_{k=1}^{N} k\right]\left[\sum_{k=1}^{N} y_d(k)\right]}{N\sum_{k=1}^{N} k^2 - \left[\sum_{k=1}^{N} k\right]^2} \tag{16.6b}$$

These two relations may be fruitfully employed in order to estimate the average and the trend coefficient for any given load data.

Note that Eqs. (16.6a) and (16.6b) are not very accurate in case the load data behaves as a non-stationary process since the ergodic hypothesis does not hold for such cases. It may still be possible to assume that the data over a finite window is stationary and the entire set of data may then be considered as the juxtaposition of a number of stationary blocks, each having slightly different statistics. Equations (16.6a) and (16.6b) may then be repeated over the different blocks in order to compute the average and the trend coefficient for each window of data.

Example 16.1

In order to illustrate the nature of results obtainable from Eqs. (16.6a) and (16.6b), consider the data shown in the graphs of Fig. 16.1 which give the population in millions. The cash values of the agricultural and the industrial outputs in millions of rupees and the amount of electrical energy consumption (load demand) in MWs in Punjab over a period of seven years starting from 1968. A total of 85 data have been generated from the graphs by sampling the graphs at intervals of 30 days. These have been substituted in Eqs. (16.6a) and (16.6b) in order to compute the average and the trend coefficients of the four variables. The results are given in Table 16.2.

Table 16.2

Variable	Average	Trend Coefficient
Population	13 million	0.2
Industrial output	Rs 397 million	0.54
Agricultural output	Rs 420.9 million	0.78
Load demand	855.8 MW	1.34

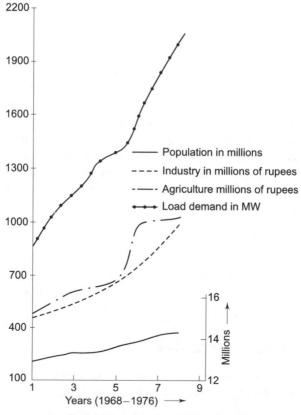

Fig. 16.1   Graphs of the data

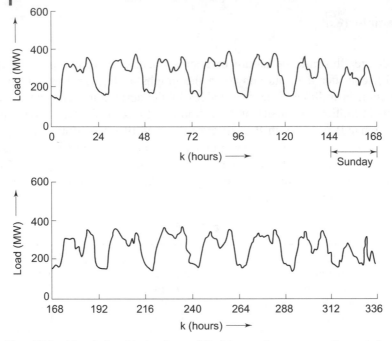

Fig. 16.2 Hourly load behaviour of Delhi over two consecutive weeks

Caution

The 85 data, used in Example 16.1, are generally not adequate for making statistical calculations so that the values given above may not be entirely adequate. In addition, the statistical characteristics of the set of variables concerned may have changed (i.e., the data may in fact be non-stationary) and this also may introduce some error in the results. Finally, the graphs in Fig. 16.1 are actually based on half yearly data obtained from the planning commission document and an interpolation process has been employed in order to generate the monthly data. This may add some unspecified errors to the data which will also affect the accuracy of the estimates.

Prediction of $y_d(k+j)$

Once the model for the deterministic component of the load has been determined, it is simple to make the prediction of its future value. For the simple model in Eq. (16.2), the desired prediction is computed using the relation

$$y_d(k + j) = \bar{y}_d + b(k + j) \qquad (16.7)$$

More General Forms of Models

Before leaving this section, it may be pointed out that the load model may be generalized by including second and higher order terms on the right hand side of Eq. (16.2) in order to represent more complex load behaviour. For example, the load model may be assumed to be

$$y(k) = \sum_{i=1}^{L} b_i k_i + e(k) \qquad (16.8a)$$

where the coefficient b_i need to be estimated from the past load data. The load model above is obviously a non-linear function of the time index k and would need L coefficients to be estimated. A much simpler approach to non-linear modelling of the load is to introduce an exponential form

$$y(k) = c \ \exp \ [bk] + e(k) \qquad (16.8b)$$

which involves only two unknown coefficients. Besides reducing the number of unknowns, the exponential model has the additional advantage of being readily transformed into a linear form. All that is required is to take the natural log of the given data. In either case, the method of LSE is easily extended to estimate the model parameters from the given historical data.

16.4 ESTIMATION OF PERIODIC COMPONENTS

The deterministic part of the load may contain some periodic components in addition to the average and the polynomial terms. Consider for example the curve shown in Fig. 16.2 which gives the variation of the active power supplied by a power utility over a period of two weeks. It is observed that the daily load variations are repetitive from day to day except for some random fluctuations. It is also seen that the curve for Sundays differ significantly from those of the week days in view that Sundays are holidays. It turns out that the curve for the entire weekly period starting from, say, the mid night of one Sunday till the mid night of the next Sunday behaves as a distinctly periodic waveform with superposed random variations.

If it is assumed that the load data are being sampled at an hourly interval, then there are a total of 168 load data in one period so that the load pattern may be expressed in terms of a Fourier series with the fundamental frequency ω being equal to $2\pi/168$ rads. A suitable model for the load $y(k)$ is then given by

$$y(k) = y + \sum_{i=1}^{L} [a_i \sin i\omega k + b_i \cos i\omega k] + e(k) \qquad (16.9)$$

where L represents the total number of harmonics present and a_i and b_i are the amplitudes of respectively the sinusoidal and the cosinusoidal components. Only dominant harmonics need to be included in the model.

Once the harmonic load model is identified, it is simple to make a prediction of the future load $y_d(k + j)$ using the relation

$$\hat{y}_d (k + j) = h' (k + j)\hat{x} (k) \qquad (16.10)$$

16.5 ESTIMATION OF $y_S(k)$: TIME SERIES APPROACH

If $y_d(k)$ is subtracted from the sequence $y(k)$, the result would be a sequence of data for the stochastic part of the load. We have to identify the model for $y_s(k)$ and then use it to make the prediction $y_s(k + j)$. A convenient method for this is based on the use of the stochastic time series models. The simplest form of such a model is the so called auto-regressive model which has been widely used to represent the behaviour of a zero mean stationary stochastic sequence. The method proposed for the generation of data for $y_s(k)$ ensures that the sequence $y_s(k)$ will have a zero mean. If it is also assumed to be stationary, it may be possible to identify a suitable autoregressive model for this sequence.

Auto-regressive Models

The sequence $y_s(k)$ is said to satisfy an AR model of order n i.e. it is $[AR(n)]$, if it can be expressed as

$$y_s(k) = \sum_{i=1}^{n} a_i y_s(k - i) + w(k) \tag{16.11}$$

where a_i are the model parameters and $w(k)$ is a zero mean white sequence. In order that the solution of this equation may represent a stationary process, it is required that the coefficients a_i make the roots of the characteristics equation

$$1 - a_1 z^{-1} - a_2 z^{-2} - \ldots - a_n z^{-n} = 0$$

lie inside the unit circle in the z-plane.

The problem in estimating the value of n is referred to as the *problem of structural identification*, while the problem of estimation of the parameters a_i is referred to as the *problem of parameter estimation*. An advantage of the *AR* model is that both these problems are solved relatively easily if the autocorrelation functions are first computed using the given data. Once the model order n and the parameter vector **a** have been estimated, the next problem is that of estimating the statistics of the noise process $w(k)$. The best that can be done, is based on the assumption that an estimate of $w(k)$ is provided by the residual $e(k) = y_s(k) - \hat{y}_s(k)$, the estimate $\hat{y}_s(k)$ having been determined from

$$\hat{y}_s(k) = -\sum_{i=1}^{n} a_i y_s(k - i) \tag{16.12}$$

The variance σ^2 of $w(k)$ is then estimated using the relation

$$\sigma^2 = (1/N) \sum_{k=1}^{N} e^2(k) \tag{16.13}$$

Auto-Regressive Moving-Average Models

In some cases, the *AR* model may not be adequate to represent the observed load behaviour unless the order n of the model is made very high. In such a case ARMA (n, m) model is used.

$$y_s(k) = -\sum_{i=1}^{n} a_i \, y_s \, (k - i) + \sum_{j=1}^{m} b_j \, w(k - j) + w(k) \quad (16.14)$$

Estimation of two structural parameters n and m as well as model parameters a_i, b_j and the variance σ^2 of the noise term $w(k)$ is required. More complex behaviour can be represented. The identification problem is solved off-line. The acceptable load model is then utilized on-line for obtaining on-line load forecasts. ARMA model can easily be modified to incorporate the temperature, rainfall, wind velocity and humidity data [2]. In some cases, it is desirable to show the dependence of the load demand on the weather variables in an explicit manner. The time series models are easily generalized in order to reflect the dependence of the load demand on one or more of the weather variables.

16.6 ESTIMATION OF STOCHASTIC COMPONENT: KALMAN FILTERING APPROACH

The time series approach has been widely employed in dealing with the load forecasting problem in view of the relative simplicity of the model forms. However, this method tends to ignore the statistical information about the load data which may often be available and may lead to improved load forecasts if utilized properly. In ARMA model, the model identification problem is not that simple. These difficulties may be avoided in some situations if the Kalman filtering techniques are utilized.

Application to Short-term Forecasting

An application of the Kalman filtering algorithm to the load forecasting problem has been first suggested by Toyada et. al. [11] for the very short-term and short-term situations. For the latter case, for example, it is possible to make use of intuitive reasonings to suggest that an acceptable model for load demand would have the form

$$y_s(k) = y_t(k) + v(k) \quad (16.15)$$

where $y_s(k)$ is the observed value of the stochastic load at time k, $y_t(k)$ is the true value of this load and $v(k)$ is the error in the observed load. In addition, the dynamics of the true load may be expressed as

$$y_t(k + 1) = y_t(k) + z(k) + u_1(k) \quad (16.16)$$

where $z(k)$ represents the increment of the load demand at time k and $u_1(k)$ represents a disturbance term which accounts for the stochastic perturbations in $y_t(k)$. The incremental load itself is assumed to remain constant on an average at every time point and is modelled by the equation

$$z(k + 1) = z(k) + u_2(k) \quad (16.17)$$

where the term $u_2(k)$ represents a stochastic disturbance term.

In order to make use of the Kalman filtering techniques, the noise terms $u_1(k)$, $u_2(k)$ and $v(k)$ are assumed to be zero mean independent white Gaussian sequences. Also, the model equations are rewritten in the form

$$\mathbf{x}\ (k+1) = F\mathbf{x}(k) + G\mathbf{u}(k) \tag{16.18a}$$

$$y_s(k) = h'\ \mathbf{x}(k) + \mathbf{v}(k) \tag{16.18b}$$

where the vectors $\mathbf{x}(k)$ and $\mathbf{u}(k)$ are defined as

$$x(k) = [y_t(k)\ \neq\ (k)]^T \text{ and } u(k) = [u_1(k)\ u_2(k)]^T$$

The matrices F, G and h' are then obtained from Eqs. (16.15)–(16.17) easily and have the following values.

$$F = \begin{bmatrix} 1 & 1 \\ 0 & 1 \end{bmatrix},\ G = \begin{bmatrix} 1 & 0 \\ 0 & 1 \end{bmatrix},\ h = \begin{bmatrix} 1 \\ 0 \end{bmatrix}$$

Based on model (16.18), it is possible to make use of the Kalman filtering algorithm to obtain the minimum variance estimate of the vector $\mathbf{x}(k)$ based on the data $y_s(k)$: $\{y_s(1), y_s(2) \ldots y_s(k)\}$. This algorithm consists of the following equations.

$$\hat{x}\ (k/k) = \hat{x}\ (k/k - 1) + K_x\ (k)\ [y_s(k) - h'\hat{x}\ (k/k-1)] \tag{16.19a}$$

$$\hat{x}\ (k/k - 1) = F\ \hat{x}\ ((k-1)/(k-1)) \tag{16.19b}$$

$$K_x(k) = P_x(k/k-1)\ h[h'\ P_x(k/k-1)h + R(k)]^{-1} \tag{16.19c}$$

$$P_x(k/k) = [I - K_x(k)\ h']\ P_x(k/k-1) \tag{16.19d}$$

$$P_x(k/k - 1) = FP_x(k-1/k-1)F' + GQ(k-1)G' \tag{16.19e}$$

where,

$Q(k) =$ covariance of $u(k)$

$R(k) =$ covariance of $v(k)$

$\hat{x}\ (k/k) =$ filtered estimate of $x(k)$

$\hat{x}\ (k/k-1) =$ single step prediction of $x(k)$

$K_x(k) =$ filter gain vector of same dimension as $x(k)$

$P_x(k/k) =$ filtering error covariance

$P_x(k/k-1) =$ prediction error covariance

From Eq. (16.18b) obtain the prediction $\hat{x}\ ((k+1)/k)$
From this the one step ahead load forecast is obtained as

$$\hat{y}_s\ (k + 1) = h'\hat{x}\ ((k+1)/k) \tag{16.19f}$$

It may be noted that filtering implies removal of disturbance or stochastic term with zero mean.

It is also possible to obtain a multi-step ahead prediction of the load from the multi-step ahead prediction of the vector $\mathbf{x}(k)$. For example, if the prediction

$x(k + d)$ by processing the data set $Y_s(k)$ is required for any $d > 1$, it is possible to use the solution of Eq. (16.18a) for the vector x $(k + d)$ to get the result

$$\hat{x}(k + d) = F^d \hat{x}(k/k) \tag{16.19g}$$

In order to be able to make use of this algorithm for generating the forecast of the load $y_s(k + d)$, it is necessary that the noise statistics and some other information be available. The value of $R(k)$ may often be estimated from a knowledge of the accuracy of the meters employed. However, it is very unlikely that the value of the covariance $Q(k)$ will be known to start with and will therefore have to be obtained by some means. An adaptive version of the Kalman filtering algorithm may be utilized in order to estimate the noise statistics alongwith the state vector $x(k)$ [2]. Now let it be assumed that both $R(k)$ and $Q(k)$ are known quantities. Let it also be assumed that the initial estimate $\hat{x}(0/0)$ and the covariance $P_x(0/0)$ are known. Based on these *a priori* information, it is possible to utilize Eq. (16.19a)–(16.19e) recursively to process the data for $y_s(1), y_s(2), ..., y_s(k)$ to generate the filtered estimate $\hat{x}(k/k)$. Once this is available, Eq. (16.19g) may be utilized to generate the desired load forecast.

Example 16.2

To illustrate the nature of the results obtainable through the algorithm just discussed, the data for the short term load behaviour for Delhi have been processed. A total of 1030 data collected at the interval of 15 minutes have been processed. It has been assumed that, in view of the short time interval over which the total data set lies, the deterministic part of the load may be assumed to be a constant mean term. Using the sample average Formula (16.7) (with $b = 0$), we get $\bar{y} = 220$ MW. The data for $y_s(k)$ have then been generated by subtracting the mean value from the measured load data.

To process these stochastic data, the following *a priori* information have been used:

$$R(k) = 3.74, \qquad Q(k) = \begin{bmatrix} 20 & 0 \\ 0 & 0.386 \end{bmatrix}$$

$$\hat{x}(0/0) = \begin{bmatrix} 0 \\ 0 \end{bmatrix} \qquad P_x(0/0) = \begin{bmatrix} 0.1 & 0 \\ 0 & 0.01 \end{bmatrix}$$

The results of application of the prediction Algorithm (16.19) are shown in Fig. 16.3. It is noted that the error of 15 minutes ahead load prediction is around 8 MW which is about 3% of the average load and less than 2% of the daily maximum load.

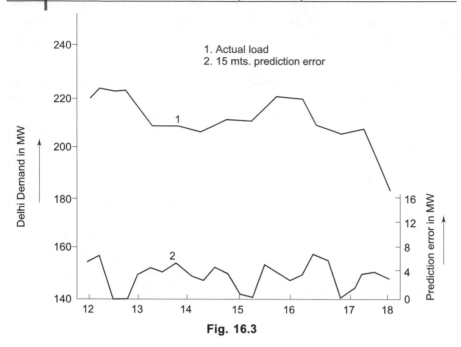

Fig. 16.3

Comment

Application of Kalman filtering and prediction techniques is often hampered by the non-availability of the required state variable model of the concerned load data. For the few cases discussed in this section, a part of the model has been obtainable from physical considerations. The part that has not been available include the state and output noise variances and the data for the initial state estimate and the corresponding covariance. In a general load forecasting situation, none of the model parameters may be available to start with and it would be necessary to make use of system identification techniques in order to obtain the required state model. It has been shown that the Gauss-Markov model described by Eq. (16.18) in over-parameterised from the model identification point of view in the sense that the data for $y_s(k)$ do not permit the estimation of all the parameters of this model. It has been shown in Ref. [12] that a suitable model that is identifiable and is equivalent to the Gauss-Markov model for state estimation purposes is the innovation model used for estimation of the stochastic component.

On-line Techniques for Non-stationary Load Prediction

Most practical load data behave as non-stationary and it is therefore important to consider the question of adapting the techniques discussed so far to the non-stationary situation. Ref. [2] has discussed the three models for this purpose viz. (i) ARIMA Models, (ii) Time varying model and (iii) Non-dynamic models.

16.7   LONG-TERM LOAD PREDICTIONS USING ECONOMETRIC MODELS

If the load forecasts are required for planning purposes, it is necessary to select the lead time to lie in the range of a few months to a few years. In such cases, the load demand should be decomposed in a manner that reflects the dependence of the load on the various segments of the economy of the concerned region. For example, the total load demand $y(k)$ may be decomposed as

$$y(k) = \sum_{i=1}^{M} a_i \, y_i(k) + e(k) \tag{16.20a}$$

where a_i are the regression coefficients, $y_i(k)$ are the chosen economic variables and $e(k)$ represents the error of modelling. A relatively simple procedure is to rewrite the model equation in the familiar vector notation

$$y(k) = h'(k)x + e(k) \tag{16.20b}$$

where $h'(k) = [y_1(k) \; y_2(k) \; ... \; y_M(k)]$ and $x = [a_1 \; a_2 \; ... \; a_M]$.

The regression coefficients may then be estimated using the convenient least squares algorithm. The load forecasts are then possible through the simple relation

$$\hat{y}(k + 1) = \hat{x}'(k) \; \hat{h}(k + 1/k) \tag{16.21}$$

where $\hat{x}(k)$ is the estimate of the coefficient vector based on the data available till the kth sampling point and $\hat{h}(k + 1/k)$ is the one-step-ahead prediction of the regression vector $h(k)$.

16.8   REACTIVE LOAD FORECAST

Reactive loads are not easy to forecast as compared to active loads, since reactive loads are made up of not only reactive components of loads, but also of transmission and distribution networks and compensating VAR devices such as SVC, FACTs etc. Therefore, past data may not yield the correct forecast as reactive load varies with variations in network configuration during varying operating conditions. Use of active load forecast with power factor prediction may result in somewhat satisfactory results. Of course, here also, only very recent past data (few minutes/hours) may be used, thus assuming steady-state network configuration. Forecasted reactive loads are adapted with current reactive requirements of network including var compensation devices. Such forecasts are needed for security analysis, voltage/reactive power scheduling etc. If control action is insufficient, structural modifications have to be carried out, i.e., new generating units, new lines or new var compensating devices normally have to be installed.

SUMMARY

Load forecasting is the basic step in power system planning. A reasonably self-contained account of the various techniques for the load prediction of a modern power system has been given in this chapter with an emphasis on the short-term prediction problems. Applications of time series, Gauss-Markov and innovation models in setting up a suitable dynamic model for the stochastic part of the load data have been discussed. The time series model identification problem has also been dealt with through the least squares estimation techniques developed in Chapter 14.

In an interconnected power system, load forecasts are usually needed at all the important load buses. A great deal of attention has in recent years been given to the question of setting up the demand models for the individual appliances and their impact on the aggregated demand. It may often be necessary to make use of non-linear forms of load models and the question of identification of the non-linear models of different forms is an important issue.

Finally, a point may be made that no particular method or approach will work for all utilities. All methods are strung on a common thread, and that is the judgement of the forecaster. In no way the material presented here is exhaustive. The intent has been to introduce some ideas currently used in forecasting system load requirements.

Future Trends

Forecasting electricity loads had reached a comfortable state of performance in the years preceding the recent waves of industry restructuring. As discussed in this chapter adaptive time-series techniques based on ARIMA, Kalman Filtering, or spectral methods are sufficiently accurate in the short term for operational purposes, achieving errors of 1-2%. However, the arrival of competitive markets has been associated with the expectation of greater consumer participation. Overall we can identify the following trends.

(i) Forecast errors have significant implications for profits, market shares, and ultimately shareholder value.

(ii) Day ahead, weather-based, forecasting is becoming the most crucial activity in a deregulated market.

(iii) Information is becoming commercially sensitive and increasingly trade secret.

(iv) Distributed, embedded and dispersed generation may increase.

A recent paper [7] takes a selective look at some of the forecasting issues which are now associated with decision-making in a competitive market. Forecasting loads and prices in the wholesale markets are mutually intertwined activities. Models based on simulated artificial agents may eventually become as important on supply side as artificial neural networks have already become important for demand prediction. These would, however, need further research and integration with conventional time-series methods in order to provide a more precise forecasting.

REVIEW QUESTIONS

16.1 Which method of load forecasting would you suggest for long term and why?

16.2 Which method of load forecasting would you suggest for very short term and why?

16.3 What purpose does medium term forecasting serve?

16.4 How is the forecaster's knowledge and intuition considered superior to any load forecasting method? Should a forecaster intervene to modify a forecast, when, why and how?

16.5 Why and what are the non-stationary components of load changes during very short, short, medium and long terms?

REFERENCES

Books

1. Nagrath I.J. and D.P. Kothari, *Power System Engineering*, Tata McGraw-Hill, New Delhi, 1994.
2. Mahalanabis, A.K., D.P. Kothari and S.I. Ahson, *Computer Aided Power System Analysis and Control*, Tata McGraw-Hill, New Delhi, 1988.
3. Sullivan, R.L., *Power System Planning*, McGraw-Hill Book Co., New York, 1977.
4. Pabla, A.S., *Electrical Power Systems Planning*, Macmillan India Ltd., New Delhi, 1998.
5. Pabla, A.S., *Electric Power Distribution*, 4th Edn, Tata McGraw-Hill, New Delhi, 1997.
6. Wang, X and J.R. McDonald (Eds), *Modern Power System Planning*, McGraw-Hill, Singapore, 1994.

Papers

7. Bunn, D.W., "Forecasting Loads and Prices in Competitive Power Markets", *Proc. of the IEEE*, Vol. 88, No. 2, Feb 2000, pp 163–169.
8. Dash, P.K. *et. al.*, "Fuzzy Neural Network and Fuzzy Expert System for Load Forecasting", *Proc. IEE*, Vol. 143, No. 1, 1996, 106–114.
9. Ramanathan, R. *et. al.*, "Short-run Forecasts of Electricity Loads and Peaks", *Int. J. Forecasting*, Vol. 13, 1997, pp 161–174.

10. Mohammed, A. *et. al.*, "Short-term Load Demand Modelling and Forecasting A Review," *IEEE Trans. SMC*, Vol. SMC-12, No. 3, 1982, pp 370–382.

11. Tyoda, J. et. al., "An Application of State Estimation to Short-term Load Forecasting", *IEEE Trans.*, Vol. PAS-89, 1970, pp 1678-1688.

12. Mehra, R.K., "On-line Identification of Linear Dynamic Systems with Application to Kalmann Filtering", *IEEE Trans.* Vol. AC-16, 1971, pp 12–21.

17

Voltage Stability

17.1 INTRODUCTION

Voltage control and stability problems are very much familiar to the electric utility industry but are now receiving special attention by every power system analyst and researcher. With growing size alongwith economic and environmental pressures, the possible threat of voltage instability is becoming increasingly pronounced in power system networks. In recent years, voltage instability has been responsible for several major network collapses in New York, France, Florida, Belgium, Sweden and Japan [4, 5]. Research workers, R and D organizations and utilities throughout the world, are busy in understanding, analyzing and developing newer and newer strategies to cope up with the menace of voltage instability/collapse.

Voltage stability* covers a wide range of phenomena. Because of this, voltage stability means different things to different engineers. Voltage stability is sometimes also called load stability. The terms voltage instability and voltage collapse are often used interchangeably. The voltage instability is a dynamic process wherein contrast to rotor angle (synchronous) stability, voltage dynamics mainly involves loads and the means for voltage control. Voltage collapse is also defined as a process by which voltage instability leads to very low voltage profile in a significant part of the system. Voltage instability limit is not directly correlated to the network maximum power transfer limit.

A CIGRE Task Force [25] has proposed the following definitions for voltage stability.

Small-disturbance voltage stability

A power system at a given operating state is *small-disturbance voltage stable* if, following any small disturbance, voltages near loads do not change or remain

---

*The problem of voltage stability has already been briefly tackled in Ch. 13. Here it is again discussed in greater details by devoting a full chapter.

close to the pre-disturbance values. The concept of small-disturbance voltage stability is related to steady-state stability and can be analysed using small-signal (linearised) model of the system.

Voltage Stability

A power system at a given operating state is voltage stable if on being subjected to a certain disturbance, the voltages near loads approach the post-disturbance equilibrium values.

The concept of voltage stability is related to transient stability of a power system. The analysis of voltage stability normally requires simulation of the system modelled by non-linear differential-algebraic equations.

Voltage Collapse

Following voltage instability, a power system undergoes voltage collapse if the post-disturbance equilibrium voltages near loads are below acceptable limits. Voltage collapse may be total (blackout) or partial.

Voltage security is the ability of a system, not only to operate stably, but also to remain stable following credible contingencies or load increases.

Although voltage stability involves dynamics, power flow based static analysis methods often serve the purpose of quick and approximate analysis.

17.2   COMPARISON OF ANGLE AND VOLTAGE STABILITY

The problem of rotor angle (synchronous) stability (covered in Ch. 12) is well understood and documented [3]. However, with power system becoming overstressed on account of economic and resource constraint on addition of generation, transformers, transmission lines and allied equipment, the voltage instability has become a serious problem. Therefore, voltage stability studies have attracted the attention of researchers and planners worldwide and is an active area of research.

Real power is related to rotor angle instability. Similarly reactive power is central to voltage instability analyses. Deficit or excess reactive power leads to voltage instability either locally or globally and any increase in loadings may lead to voltage collapse.

Voltage Stability Studies

The voltage stability can be studied either on static (slow time frame) or dynamic (over long time) considerations. Depending on the nature of distur-bance and system/subsystem dynamics voltage stability may be regarded a slow or fast phenomenon.

Static Voltage Analysis

Load flow analysis reveals as to how system equilibrium values (such as voltage and power flow) vary as various system parameters and controls are changed. Power flow is a static analysis tool wherein dynamics is not explicitly considered. Many of the indices used to assess voltage stability are related to NR load flow study. Details of static and dynamic voltage stability will be considered further in Section 17.5.

Some Counter Measures

Certain counter measures to avoid voltage instability are:

(i) generator terminal voltage increase (only limited control possible)

(ii) increase of generator transformer tap

(iii) reactive power injection at appropriate locations

(iv) load-end OLTC blocking

(v) strategic load shedding (on occurrence of undervoltage)

Counter measures to prevent voltage collapse will be taken up in Section 17.6.

17.3 REACTIVE POWER FLOW AND VOLTAGE COLLAPSE

Certain situations in power system cause problems in reactive power flow which lead to system voltage collapse. Some of the situations that can occur are listed and explained below.

(i) *Long Transmission Lines:* In power systems, long lines with voltage uncontrolled buses at the receiving ends create major voltage problems during light load or heavy load conditions.

(ii) *Radial Transmission Lines:* In a power system, most of the parallel EHV networks are composed of radial transmission lines. Any loss of an EHV line in the network causes an enhancement in system reactance. Under certain conditions the increase in reactive power delivered by the line(s) to the load for a given drop in voltage, is less than the increase in reactive power required by the load for the same voltage drop. In such a case a small increase in load causes the system to reach a voltage unstable state.

(iii) *Shortage of Local Reactive Power:* There may occur a disorganised combination of outage and maintenance schedule that may cause localised reactive power shortage leading to voltage control problems. Any attempt to import reactive power through long EHV lines will not be successful. Under this condition, the bulk system can suffer a considerable voltage drop.

17.4 MATHEMATICAL FORMULATION OF VOLTAGE STABILITY PROBLEM

The slower forms of voltage instability are normally analysed as steady state problems using power flow simulation as the primary study method. "Snapshots" in time following an outage or during load build up are simulated. Besides these post-disturbance power flows, two other power flow based methods are often used; *PV* curves and *VQ* curves. (*See* also Sec. 13.6) These two methods give steady-state loadability limits which are related to voltage stability. Conventional load flow programs can be used for approximate analysis.

P–V. curves are useful for conceptual analysis of voltage stability and for study of radial systems.

The model that will be employed here to judge voltage stability is based on a single line performance. The voltage performance of this simple system is qualitatively similar to that of a practical system with many voltage sources, loads and the network of transmission lines.

Consider the radial two bus system of Fig. 17.1. This is the same diagram as that of Fig. 5.26 except that symbols are simplified. Here E is V_S and V is V_R and E and V are magnitudes with E leading V by δ. Line angle $\phi = \tan^{-1} X/R$ and $|z| \approx X$.

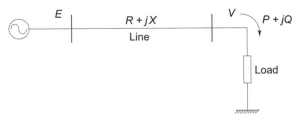

Fig. 17.1

In terms of P and Q, the system load end voltage can be expressed as [1].

$$V = \left[\frac{-2QX + E^2}{2} \pm \frac{1}{2} \sqrt{(2QX - E^2)^2 - 4X^2(P^2 + Q^2)} \right]^{1/2} \quad (17.1)$$

It is seen from Eq. (17.1) that V is a double-valued function (i.e. it has two solutions) of P for a particular pf which determines Q in terms of P. The PV curves for various values of pf are plotted in Fig. 17.2. For each value of pf, the higher voltage solution indicates stable voltage case, while the lower voltage lies in the unstable voltage operation zone. The changeover occurs at V_{cri} (critical) and P_{max}. The locus of V_{cri}–P_{max} points for various pfs is drawn in dotted line in the figure. Any attempt to increase the load above P_{max} causes a reversal of voltage and load. Reducing voltage causes an increasing current to be drawn by the load. In turn the larger reactive line drop causes the voltage to dip further. This being unstable causes the system to suffer voltage collapse. This is also brought out by the fact that in upper part of the curve $\frac{dP}{dV} < 0$ and in the lower part (unstable part) $\frac{dP}{dV} > 0$ (reducing load means reducing voltage and vice-versa). It may be noted here that the type of load assumed in Fig. 17.2 is constant impedance. In practical systems the type of loads are mixed or predominantly constant power type such that system voltage degradation is more and voltage instability occurs much prior to the theoretical power limit.

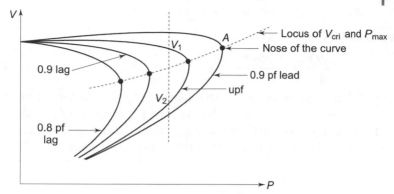

Fig. 17.2 PV curves for various power factors

As in the case of single line system, in a general power system, voltage instability occurs above certain bus loading and certain Q injections. This condition is indicated by the singularity of the Jacobian of Load Flow equations and level of voltage instability is assessed by the minimum singular value.

Certain results that are of significance for voltage stability are as under.

• Voltage stability limit is reached when

$$\left| \frac{S}{Y_{LL}^* V^2} \right| = 1 \tag{17.2}$$

where

$S =$ complex power at load bus

$Y_{LL} =$ load bus admittance

$V =$ load bus voltage

Nearer the magnitude in Eq. (17.2) to unity, lesser the stability margin.

• The loading limit of a transmission line can be determined from

$$|S| = V_{\text{cri}}^2 / X_{\text{cri}} \tag{17.3}$$

X_{cri} is the critical system reactance beyond which voltage stability is lost. It can be expressed as

$$X_{\text{cri}} = \frac{E^2}{2P} (-\tan \phi + \sec \phi) \tag{17.4}$$

We have so far considered how the PV characteristics with constant load power factor affect the voltage stability of a system. A more meaningful characteristic for certain aspects of voltage stability is the QV characteristic, which brings out the sensitivity and variation of bus voltage with respect to reactive power injections (+ve or −ve).

Consider once again the simple radial system of Fig. 17.1. For Q flow it is sufficiently accurate to assume $X \gg R$ i.e. $\phi \approx 90°$. It then follows that

$$Q = \frac{EV}{X} \cos \delta - \frac{V^2}{X} \tag{17.5}$$

or
$$V^2 - EV \cos \delta + QX = 0 \qquad (17.6)$$

Taking derivative wrt V gives

$$\frac{dQ}{dV} = \frac{E \cos \delta - 2V}{X} \qquad (17.7)$$

The QV characteristic on normalized basis (Q/P_{max}, V/E) for various values of P/P_{max} are plotted in Fig. 17.3. The system is voltage stable in the region where dQ/dV is positive, while the voltage stability limit is reached at $dQ/dV = 0$ which may also be termed as the critical operating point.

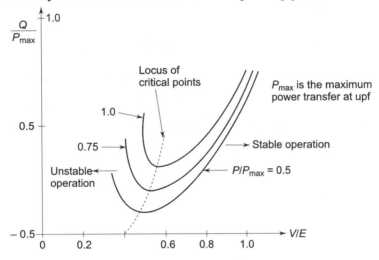

Fig. 17.3 *QV characteristics for the system of Fig. 17.1 for various values of P/P_{max}.*

The limiting value of the reactive power transfer at the limiting stage of voltage stability is given by

$$Q_{\lim} = \frac{V^2}{X} \cos 2\delta \qquad (17.8)$$

The inferences drawn from the simple radial system qualitatively apply to a practical size system. Other factors that contribute to system voltage collapse are: strength of transmission system, power transfer levels, load characteristics, generator reactive power limits and characteristics of reactive power compensating devices.

Other Criteria of Voltage Stability

(i) $\dfrac{dE}{dV}$ criterion: (E=generator voltage; V=load voltage). Using this criterion, the voltage stability limit is reached when

$$\cos \delta \left\{ \frac{dQ}{dV} + \frac{2V}{X} \right\} + \sin \delta \frac{dP}{dV} - \frac{E}{X} = 0 \qquad (17.9)$$

Using the decoupling principle i.e. $\dfrac{\mathrm{d}P}{\mathrm{d}V} = 0$, we get

$$\frac{E}{X} = \cos \delta \left[\frac{\mathrm{d}Q}{\mathrm{d}V} + \frac{2V}{X} \right]$$

or

$$I_{SC} = \cos \delta \left[\frac{\mathrm{d}Q}{\mathrm{d}V} + \frac{2V}{X} \right]$$

or

$$EI_{SC} = E \cos \delta \left[\frac{\mathrm{d}Q}{\mathrm{d}V} + \frac{2V}{X} \right]$$

Voltage stability is achieved when

$$E \cos \delta \left(\frac{\mathrm{d}Q}{\mathrm{d}V} + \frac{2V}{X} \right) > EI_{SC} \quad \text{(short circuit MVA of power source)}$$

(17.10)

(ii) $\dfrac{\mathrm{d}Z}{\mathrm{d}V}$ criterion

Voltage instability occurs when the system Z is such that

$$\frac{\mathrm{d}V}{\mathrm{d}Z} = \infty \quad \text{or} \quad \frac{\mathrm{d}Z}{\mathrm{d}V} = 0 \tag{17.11}$$

Application of this criterion gives value of Z_{cri}.

(iii) Ratio of source to load reactance is very important and for voltage stability

$$\frac{x_{\text{source}}}{x_{\text{load}}} < a^2 \tag{17.12}$$

a indicates the off-nominal tap ratio of the OLTC transformer at the load end.

17.5 VOLTAGE STABILITY ANALYSIS

The voltage stability analysis for a given system state involves examining following two aspects.

(i) *Proximity to voltage instability*: Distance to instability may be measured in terms of physical quantities, such as load level, real power flow through a critical interface, and reactive power reserve. Possible contingencies such as a line outage, loss of a generating unit or a reactive power source must be given due consideration.

(ii) *Mechanism of voltage instability*: How and why does voltage instability take place? What are the main factors leading to instability? What are the voltage-weak areas? What are the most effective ways to improve voltage stability?

The static analysis techniques permit examination of a wide range of system conditions and can describe the nature of the problem and give the main contributing factors. Dynamic analysis is useful for detailed study of specific voltage collapse situations, coordination of protection and controls, and testing of remedial measures. Dynamic simulations further tell us whether and how the steady-state equilibrium point will be reached.

Modelling Requirements of Various Power System Components

Loads

Load modelling is very critical in voltage stability analysis. Detailed subtransmission system representation in a voltage-weak area may be required. This may include transformer ULTC action, reactive power compensation, and voltage regulators.

It is essential to consider the voltage and frequency dependence of loads. Induction motors should also be modelled.

Generators and their excitation controls

It is necessary to consider the droop characteristics of the AVR, load compensation, SVSs (static var system), AGC, protection and controls should also be modelled appropriately [4].

Dynamic Analysis

The general structure of the system model for voltage stability analysis is similar to that for transient stability analysis. Overall system equations may be expressed as

$$\dot{X} = f(X, V) \tag{17.13}$$

and a set of algebraic equations

$$I(X, V) = Y_N V \tag{17.14}$$

with a set of known initial conditions (X_0, V_0).

where $X =$ system state vector

   $V =$ bus voltage vector

   $I =$ current injection vector

   $Y_N =$ network node admittance matrix.

Equations (17.13) and (17.14) can be solved in time domain by employing any of the numerical integration methods described in Ch. 12 and power flow analysis methods described in Ch. 6. The study period is of the order of several minutes. As the special models representing the "slow system dynamics" leading to voltage collapse have been included, the stiffness of the system differential equations is considerably higher than that of transient stability models. Stiffness is also called synchronizing coefficient as discussed in Ch. 12.

Static Analysis

The static approach captures *snapshots* of system conditions at various time frames along the time-domain trajectory. At each of these time frames, $\dot{X}$ in Eq. (17.13) is assumed to be zero, and the state variables take on values appropriate to the specific time frame. Thus, the overall system equations reduce to purely algebraic equations allowing the use of static analysis techniques.

In static analysis voltage stability is determined by computing VP and VQ curves at selected load buses. Special techniques using static analysis have been reported in literature. Methods based on VQ sensitivity such as eigenvalue (or modal) analysis have been devised. These methods give stability-related information from a system-wide perspective and also identify areas of potential problems [13–15].

Proximity to Instability

Proximity to small-disturbance voltage instability is determined by increasing load-generation in steps until the system becomes unstable or the load flow fails to converge. Refs. [16–18] discuss special techniques for determining the point of voltage collapse and proximity to voltage instability.

The Continuation Power-flow Analysis

The Jacobian matrix becomes singular at the voltage stability limit. As a result, conventional load-flow algorithms may have convergence problems at operating conditions near the stability limit. The continuation power-flow analysis overcomes this problem by reformulating the load-flow equations so that they remain well-conditioned at all possible loading conditions. This allows the solution of load-flow problem for both upper and lower portions of the P-V curve [17].

The continuation method of power-flow analysis is robust and flexible and suited for solving load flow problems with convergence difficulties. However, the method is very slow and time-consuming. Hence the better approach is to use combination of conventional load flow method (NR/FDLF) and continuation method. Starting from the base case, LF is solved using a conventional method to compute power flow solutions for successively increasing load levels until a solution cannot be obtained. Hereafter, the continuation method is resorted to obtain the load-flow solutions. Normally, the continuation method is required only if solutions are required exactly at and past the critical point.

Voltage Stability with HVDC Links

High voltage direct current* (HVDC) links are used for extremely long distance transmission and for asynchronous interconnections. An HVDC link can be

---

*For detailed account of HVDC, the reader may refer to [3].

either a back-to-back rectifier/inverter link or can include long distance dc transmission. Multi-terminal HVDC links are also feasible.

The technology has come to such a level that HVDC terminals can be connected even at voltage-weak points in power systems. HVDC links may present unfavourable "load" characteristics to the power system as HVDC converter consumes reactive power equal to 50–60% of the dc power.

HVDC-related voltage control (voltage stability and fundamental frequency temporary over voltages) may be studied using a transient stability program. Transient stability is often interrelated with voltage stability. Ref. [2] considers this problem in greater detail.

17.6 PREVENTION OF VOLTAGE COLLAPSE

(i) *Application of reactive power-compensating devices.*

Adequate stability margins should be ensured by proper selection of compensation schemes in terms of their size, ratings and locations.

(ii) *Control of network voltage and generator reactive output*

Several utilities in the world such as EDF (France), ENEL (Italy) are developing special schemes for control of network voltages and reactive power.

(iii) *Coordination of protections/controls*

Adequate coordination should be ensured between equipment protections/ controls based on dynamic simulation studies. Tripping of equipment to avoid an overloaded condition should be the last alternative. Controlled system separation and adaptive or intelligent control could also be used.

(iv) *Control of transformer tap changers*

Tap changers can be controlled, either locally or centrally, so as to reduce the risk of voltage collapse. Microprocessor-based OLTC controls offer almost unlimited flexibility for implementing ULTC control strategies so as to take advantage of the load characteristics.

(v) *Under voltage load shedding*

For unplanned or extreme situations, it may be necessary to use undervoltage load-shedding schemes. This is similar to under frequency load shedding, which is a common practice to deal with extreme situations resulting from generation deficiency.

Strategic load shedding provides cheapest way of preventing widespread voltage collapse. Load shedding schemes should be designed so as to differentiate between faults, transient voltage dips and low voltage conditions leading to voltage collapse.

(vi) *Operators' role*

Operators must be able to recognise voltage stability-related symptoms and take required remedial actions to prevent voltage collapse. On-line monitoring and analysis to identify potential voltage stability problems and appropriate remedial measures are extremely helpful.

17.7 STATE-OF-THE-ART, FUTURE TRENDS AND CHALLENGES

The present day transmission networks are getting more and more stressed due to economic and environmental constraints. The trend is to operate the existing networks optimally close to their loadability limit. This consequently means that the system operation is also near voltage stability limit (nose point) and there is increased possibility of voltage instability and even collapse.

Off-line and on-line techniques of determining state of voltage stability and when it enters the unstable state, provide the tools for system planning and real time control. Energy management system (EMS) provide a variety of measured and computer processed data. This is helpful to system operators in taking critical decisions *inter alia* reactive power management and control. In this regard automation and specialized software relieve the operator of good part of the burden of system management but it does add to the complexity of the system operation.

Voltage stability analysis and techniques have been pushed forward by several researchers and several of these are in commercial use as outlined in this chapter. As it is still hot topic, considerable research effort is being devoted to it.

Pai *et al.* [8] considered an exponential type voltage dependent load model and a new index called condition number for static voltage stability prediction. Eigenvalue analyses has been used to find critical group of buses responsible for voltage collapse. Some researchers [26] have also investigated aspects of bifurcations (local, Hopf, global) and chaos and their implications on power system voltage stability. FACTS devices can be effectively used for controlling the occurrence of dynamic bifurcations and chaos by proper choice of error signal and controller gains.

Tokyo Electric Power Co. has developed a μP-based controller for coordinated control of capacitor bank switching and network transformer tap changing. HVDC power control is used to improve stability.

More systematic approach is still required for optimal siting and sizing of FACTS devices. The availability of FACTS controllers allow operation close to the thermal limit of the lines without jeopardizing security. The reactive power compensation close to the load centres and at the critical buses is essential for overcoming voltage instability. Better and probabilistic load modelling [11] should be tried. It will be worthwhile developing techniques and models for study of non-linear dynamics of large size systems. This may require exploring new methods to obtain network equivalents suitable for the voltage stability analysis. AI is another approach to centralized reactive power and voltage control. An expert system [9] could assist operators in applying C-banks so that generators operate near upf. The design of suitable protective measures in the event of voltage instability is necessary.

So far, computed PV curves are the most widely used method of estimating voltage security, providing MW margin type indices. Post-disturbance MW or MVAr margins should be translated to predisturbance operating limits that operators can monitor. Both control centre and power plant operators should be trained in the basics of voltage stability. For operator training simulator [10] a real-time dynamic model of the power system that interfaces with EMS controls such as AGC is of great help.

Voltage stability is likely to challenge utility planners and operators for the foreseable future. As load grows and as new transmission and load area generation become increasingly difficult to build, more and more utilities will face the voltage stability challenge. Fortunately, many creative researchers and planners are working on new analysis methods and an innovative solutions to the voltage stability challenge.

Example 17.1

A load bus is composed of induction motor where the nominal reactive power is 1 pu. The shunt compensation is K_{sh}. Find the reactive power sensitivity at the bus wrt change in voltage.

Solution

$$Q_{load} = Q_{nom} \, V^2 \text{ [given]}$$

$$Q_{comp} = - K_{sh} \, V^2 \qquad\qquad \text{[–ve sign denotes inductive reactive power injection]}$$

$$Q_{net} = Q_{load} + Q_{comp}$$

$\therefore$ Here, $\qquad Q_{net} = V^2 - K_{sh} \, V^2 \; [Q_{nom} = 1.0 \text{ given}]$

$\therefore \qquad\qquad \dfrac{dQ_{net}}{dV} = 2V\text{–}2V \, K_{sh}$

Sensitivity increases or decreases with K_{sh} as well as the magnitude of the voltage. Say at $V = 1.0$ pu, $K_{sh} = 0.8$

$\therefore \qquad\qquad \dfrac{dQ_{net}}{dV} = 2 - 1.6 = 0.4 \text{ pu.}$

Example 17.2

Find the capacity of a static VAR compensator to be installed at a bus with ± 5% voltage fluctuation. The short circuit capacity is 5000 MVA.

Solution For the switching of static shunt compensator,

Let $\quad \Delta V =$ voltage fluctuation

$\quad\quad \Delta Q =$ reactive power variation
(i.e. the size of the compensator)

$\quad\quad S_{s/c} =$ system short circuit capacity

Then $\quad \Delta V = \dfrac{\Delta Q}{S_{s/c}}$

$$\Delta Q = \Delta V S_{s/c}$$

$$= \pm (0.05 \times 5000)$$

$$= \pm 250 \text{ MVAR}$$

The capacity of the static VAR compensator is +250 MVAR.

References

Books

1. Chakrabarti, A., D.P. Kothari and A.K. Mukhopadhyay, *Performance, Operation and Control of EHV Power Transmission Systems*, Wheeler Publishing, New Delhi, 1995.
2. Taylor, C.W., *Power System Voltage Stability*, McGraw-Hill, New York, 1994.
3. Nagrath, I.J. and D.P. Kothari, *Power System Engineering*, Tata McGraw-Hill, New Delhi, 1994.
4. Kundur, P., *Power System Stability and Control*, McGraw-Hill, New York, 1994.
5. Padiyar, K.R., *Power System Dynamics: Stability and Control*, John Wiley, Singapore, 1996.
6. Cutsem, T Van and Costas Vournas, *Voltage Stability of Electric Power Systems*, Kluwer Int. Series, 1998.

Papers

7. Concordia, C (Ed.), "Special Issue on Voltage Stability and Collapse", *Int. J. of Electrical Power and Energy Systems*, Vol. 15, no. 4, August 1993.
8. Pai, M.A. and M.G.O. Grady,"Voltage Collpase Analysis with Reactive Generation and Voltage Dependent Constraints", *J. of Elect Machines and Power Systems*, Vol. 17, No. 6, 1989, pp 379–390.
9. CIGRE' Task Force 38–06–01,"Expert Systems Applied to voltage and Var Control ″, 1991.
10. "Operator Training Simulator", *EPRI Final Report* EL–7244, May 1991, prepared by EMPROS Systems International.
11. Xu, W and Y Mansour, "Voltage Stability using Generic Dynamic Load Models", *IEEE Trans. on Power Systems*, Vol 9, No 1, Feb 1994, pp 479–493.

12. Verma, H.K., L.D. Arya and D.P. Kothari, "Voltage Stability Enhancement by Reactive Power Loss Minimization", *JIE (I)*, Vol. 76, May 1995, pp. 44–49.

13. IEEE, special Publication 90 TH 0358–2 PWR, "Voltage Stability of Power Systems: *Concepts, Analytical Tools, and Industry Experience*" 1990.

14. Flatabo, N., R. Ognedal and T. Carlsen, "Voltage Stability Condition in a Power Transmission System calculated by Sensitivity Methods", *IEEE Trans.* Vol. PWRS-5, No. 5, Nov 1990, pp 1286–93.

15. Gao, B., G.K. Morison and P. Kundur, "Voltage Stability Evaluation Using Modal Analysis", *IEEE Trans.* Vol. PWRS-7, No. 4, Nov. 1992, pp 1529–1542.

16. Cutsem, T. Van, "A Method to Compute Reactive Power Margins wrt Voltage Collapse", *IEEE Trans.*, Vol. PWRS-6, No. 2, Feb 1991, pp 145–156.

17. Ajjarapu, V. and C. Christy, "The continuation Power Flow: A Tool for Steady State Voltage Stability Analysis", *IEEE PICA Conf. Proc.*, May 1991, pp 304–311.

18. Löf, A-P, T. Sined, G. Anderson and D.J. Hill, "Fast Calculation of a Voltage Stability Index", *IEEE Trans.*, Vol. PWRS-7, No. 1, Feb 1992, pp 54–64.

19. Arya, L.D., S.C. Chaube and D.P. Kothari, "Line Outage Ranking based on Estimated Lower Bound on Minimum Eigen Value of Load Flow Jacobian", *JIE (I)*, Vol. 79, Dec 1998, pp 126–129.

20. Bijwe, P.R., S.M. Kelapure, D.P. Kothari and K.K. Saxena, "Oscillatory Stability Limit Enhancement by Adaptive Control Rescheduling", *Int J. of Electrical Power and Energy Systems*, Vol. 21, No. 7, 1999, pp 507–514.

21. Arya, L.D., S.C. Chaube and D.P. Kothari, "Line Switching for Alleviating Overloads under Line Outage Condition taking Bus Voltage Limits into Account", *Int J. of Electric Power and Energy System*, Vol. 22, No. 3, 2000, pp 213–221.

22. Bijwe, P.R., D.P. Kothari and S. Kelapure, "An Efficient Approach to Voltage Security Analysis and Enhancement", *Int J. of EP and ES.*, Vol. 22, No. 7, Oct. 2000, pp 483–486.

23. Arya, L.D., S.C., Chaube and D.P. Kothari, "Reactive Power Optimization using Static Stability Index (VSI)", *Int J. of Electric Power Components and Systems*, Vol. 29, No. 7, July 2001, pp 615–628.

24. Arya, L.D., S.C. Chaube and D.P. Kothari, "Line Outage Ranking for Voltage Limit Violations with Corrective Rescheduling Avoiding Masking", *Int. J. of EP and ES*, Vol. 23, No. 8, Nov. 2001, pp 837–846.

25. CIGRE Task Force 38–02–10, "Cigre Technical Brochure: Modelling of Voltage Collapse Including Dynamic Phenomena", *Electra*, No. 147, April 1993, pp. 71–77.

26. Mark J. Laufenberg and M.A. Pai, "Hopf bifurcation control in power system with static var compensators", *Electrical Power and Energy Systems*, Vol. 19, No. 5, 1997, pp 339–347.

Introduction to Vector and Matrix Algebra

In this appendix, our aim is to present definitions and elementary operations of vectors and matrices necessary for power system analysis.

VECTORS

A vector **x** is defined as an ordered set of numbers (real or complex), i.e.

$$\mathbf{x} \triangleq \begin{bmatrix} x_1 \\ x_2 \\ \vdots \\ x_n \end{bmatrix} \tag{A-1}$$

$x_1, ..., x_n$ are known as the components of the vector x. Thus the vector **x** is a n-dimensional column vector. Sometimes transposed form is found to be more convenient and is written as the row vector.

$$x^{\mathrm{T}} \triangleq [x_1, x_2, ..., x_n] \tag{A-2}$$

Some Special Vectors

The null vector **0** is one whose each component is zero, i.e.

$$\mathbf{0} \triangleq \begin{bmatrix} 0 \\ 0 \\ \vdots \\ 0 \end{bmatrix}$$

The sum vector **1** has each of its components equal to unity, i.e.

$$1 \triangleq \begin{bmatrix} 1 \\ 1 \\ \vdots \\ 1 \end{bmatrix}$$

The unit vector e_k is defined as the vector whose kth component is unity and the rest of the components are zero, i.e.

$$e_k \triangleq \begin{bmatrix} 0 \\ \vdots \\ 0 \\ 1 \\ 0 \\ \vdots \\ 0 \end{bmatrix} \quad k\text{th component}$$

Some Fundamental Vector Operations

Two vectors **x** and **y** are known as equal if, and only if, $x_k = y_k$ for $k = 1, 2,$ ..., n. Then we say

$$\mathbf{x} = \mathbf{y}$$

The product of a vector by a scalar is carried out by multiplying each component of the vector by that scalar, i.e.

$$\alpha \mathbf{x} = \mathbf{x}\alpha \triangleq \alpha \begin{bmatrix} x_1 \\ x_2 \\ \vdots \\ x_n \end{bmatrix}$$

If a vector **y** is to be added to or subtracted from another vector **x** of the same dimension, then each component of the resulting vector will consist of addition or subtraction of the corresponding components of the vectors **x** and **y**, i.e.

$$x \pm y \triangleq \begin{bmatrix} x_1 \pm y_1 \\ x_2 \pm y_2 \\ \vdots \\ x_n \pm y_n \end{bmatrix}$$

The following properties are applicable to the vector algebra:

$$x + y = y + x$$
$$x + (y + z) = (x + y) + z$$
$$\alpha_1 (\alpha_2 x) = (\alpha_1 \alpha_2)x$$
$$(\alpha_1 + \alpha_2)x = \alpha_1 x + \alpha_2 x$$

$$\alpha(x + y + z) = \alpha x + \alpha y + \alpha z$$

$$0x = 0$$

The multiplication of two vectors $\mathbf{x}$ and $\mathbf{y}$ of same dimensions results in a very important product known as *inner* or *scalar* product[*], i.e.

$$\mathbf{x}^T\mathbf{y} \triangleq \sum_{i=1}^{n} x_i y_i \triangleq \mathbf{y}^T\mathbf{x} \tag{A-3}$$

Also, it is interesting to note that

$$\mathbf{x}^T\mathbf{x} = |\mathbf{x}|^2 \tag{A-4}$$

$$\cos \phi \triangleq \frac{\mathbf{x}^T\mathbf{y}}{|\mathbf{x}||\mathbf{y}|} \tag{A-5}$$

where ϕ is angle between vectors, $|\mathbf{x}|$ and $|\mathbf{y}|$ are the geometric lengths of vectors $\mathbf{x}$ and $\mathbf{y}$, respectively. Two non-zero vectors are said to be *orthogonal*, if

$$\mathbf{x}^T\mathbf{y} = 0 \tag{A-6}$$

MATRICES

Definitions

Matrix

An $m \times n$ (or m, n) matrix is an ordered rectangular array of elements which may be real numbers, complex numbers, functions or operators. The matrix

$$A \triangleq \begin{bmatrix} a_{11} & a_{12} & \cdots & a_{1n} \\ a_{21} & a_{22} & \cdots & a_{2n} \\ \vdots & \vdots & & \vdots \\ a_{m1} & a_{m2} & \cdots & a_{mn} \end{bmatrix} = [a_{ij}] \tag{A-7}$$

is a rectangular array of mn elements.

a_{ij} denotes the (i, j)th element, i.e. the element located in the ith row and the jth column. The matrix A has m rows and n columns and is said to be of order $m \times n$.

When $m = n$, i.e. the number of rows is equal to that of columns, the matrix is said to be a *square matrix* of order n.

An $m \times 1$ matrix, i.e. a matrix having only one column is called a *column vector*. An $1 \times n$ matrix, i.e. a matrix having only one row is called a *row vector*.

---

[*]Sometimes inner product is also represented by the following alternative forms $x \cdot y$, $(x, y), \langle x, y \rangle$.

Diagonal matrix

A diagonal matrix is a square matrix whose elements off the main diagonal are all zeros ($a_{ij} = 0$ for $i \neq j$).

Example

$$D \triangleq \begin{bmatrix} 4 & 0 & 0 \\ 0 & 2 & 0 \\ 0 & 0 & 9 \end{bmatrix}$$

Null matrix

If all the elements of the square matrix are zero, the matrix is a *null* or *zero* matrix.

Example

$$\mathbf{0} \triangleq \begin{vmatrix} 0 & \cdots & 0 \\ \cdots & \cdots & \cdots \\ 0 & \cdots & 0 \end{vmatrix} \tag{A-8}$$

Unit (identity) matrix

A unit matrix I is a diagonal matrix with all diagonal elements equal to unity. If a unit matrix is multiplied by a constant (λ), the resulting matrix is a diagonal matrix with all diagonal elements equal to λ. This matrix is known as a *scalar* matrix.

Examples

$$I = \begin{bmatrix} 1 & 0 & 0 & 0 \\ 0 & 1 & 0 & 0 \\ 0 & 0 & 1 & 0 \\ 0 & 0 & 0 & 1 \end{bmatrix}; 4\begin{bmatrix} 1 & 0 & 0 \\ 0 & 1 & 0 \\ 0 & 0 & 1 \end{bmatrix} = \begin{bmatrix} 4 & 0 & 0 \\ 0 & 4 & 0 \\ 0 & 0 & 4 \end{bmatrix}$$

$\qquad\qquad$ = 4 × 4 unit matrix $\qquad\qquad\qquad$ = 3 × 3 scalar matrix

Determinant of a matrix

For each square matrix, there exists a determinant which is formed by taking the determinant of the elements of the matrix.

For example, if

$$A = \begin{bmatrix} 2 & -1 & 1 \\ -1 & 3 & 2 \\ 1 & 2 & 4 \end{bmatrix} \tag{A-9}$$

then,

$$\det (A) = |A| = 2\begin{vmatrix} 3 & 2 \\ 2 & 4 \end{vmatrix} - (-1)\begin{vmatrix} -1 & 2 \\ 1 & 4 \end{vmatrix} + \begin{vmatrix} -1 & 3 \\ 1 & 2 \end{vmatrix}$$

$$= 2(8) + (-6) + (-5) = 5 \tag{A-10}$$

Transpose of a matrix

The transpose of matrix A denoted by A^T is the matrix formed by interchanging the rows and columns of A.

Note that $(A^T)^T = A$

Symmetric matrix

A square matrix is symmetric, if it is equal to its transpose, i.e.

$$A^T = A$$

Notice that the matrix A of Eq. (A-9) is a symmetric matrix.

Minor

The minor M_{ij} of an $n \times n$ matrix is the determinant of $(n-1) \times (n-1)$ matrix formed by deleting the ith row and the jth column of the $n \times n$ matrix.

Cofactor

The cofactor A_{ij} of element a_{ij} of the matrix A is defined as

$$A_{ij} = (-1)^{i+j} M_{ij}$$

Adjoint matrix

The adjoint matrix of a square matrix A is found by replacing each element a_{ij} of matrix A by its cofactor A_{ij} and then transposing.

For example, if A is given by Eq. (A-9), then

$$\text{adj } A = \begin{bmatrix} \begin{vmatrix} 3 & 2 \\ 2 & 4 \end{vmatrix} & -\begin{vmatrix} -1 & 2 \\ 1 & 4 \end{vmatrix} & \begin{vmatrix} -1 & 3 \\ 1 & 2 \end{vmatrix} \\ -\begin{vmatrix} -1 & 1 \\ 2 & 4 \end{vmatrix} & \begin{vmatrix} 2 & 1 \\ 1 & 4 \end{vmatrix} & -\begin{vmatrix} 2 & -1 \\ 1 & 2 \end{vmatrix} \\ \begin{vmatrix} -1 & 1 \\ 3 & 2 \end{vmatrix} & -\begin{vmatrix} 2 & 1 \\ -1 & 2 \end{vmatrix} & \begin{vmatrix} 2 & -1 \\ -1 & 3 \end{vmatrix} \end{bmatrix}^T$$

$$= \begin{bmatrix} 8 & 6 & -5 \\ 6 & 7 & -5 \\ -5 & -5 & 5 \end{bmatrix}^{\mathrm{T}} = \begin{bmatrix} 8 & 6 & -5 \\ 6 & 7 & -5 \\ -5 & -5 & 5 \end{bmatrix} \qquad (A\text{-}11)$$

Singular and non-singular matrices

A square matrix is called singular, if its associated determinant is zero, and non-singular, if its associated determinant is non-zero.

ELEMENTARY MATRIX OPERATIONS

Equality of matrices

Two matrices $A(m \times n)$ and $B(m \times n)$ are said to be equal, if the only if
$$a_{ij} = b_{ij} \quad \text{for } i = 1, 2, \ldots, m, \quad j = 1, 2, \ldots, n$$
Then we write
$$A = B$$

Multiplication of a matrix by a scalar

A matrix is multiplied by a scalar α if all the mn elements are multiplied by α, i.e.

$$\alpha A = A\alpha = \begin{bmatrix} \alpha a_{11} & \cdots & \alpha a_{1n} \\ \cdots & \cdots & \cdots \\ \alpha a_{m1} & \cdots & \alpha a_{mn} \end{bmatrix} \qquad (A\text{-}12)$$

Addition (or subtraction) of matrices

To add (or subtract) two matrices of the same order (same number of rows, and same number of columns), simply add (or subtract) the corresponding elements of the two matrices, i.e. when two matrices A and B of the same order are added, a new matrix C results such that
$$C = A + B;$$
whose ijth element equals
$$c_{ij} = a_{ij} + b_{ij}$$

Example

Let $\qquad A = \begin{bmatrix} 3 & 0 \\ 2 & -1 \end{bmatrix} \quad B = \begin{bmatrix} 2 & -1 \\ 0 & 3 \end{bmatrix}$

then,

$$C = A + B = \begin{bmatrix} 5 & -1 \\ 2 & 2 \end{bmatrix}$$

Addition and subtraction are defined only for matrices of the same order. The following laws hold for addition:

(i) The *commutative law*: $A + B = B + A$

(ii) The *associative law*: $A + (B + C) = (A + B) + C$

Further

$$(A \pm B)^{\mathrm{T}} = A^{\mathrm{T}} \pm B^{\mathrm{T}}$$

Matrix Multiplication

The product of two matrices $A \times B$ is defined if A has the same number of columns as the number of rows in B. The matrices are then said to be *conformable*. If a matrix A is of order $m \times n$ and B is an $n \times q$ matrix, the product $C = AB$ will be an $m \times q$ matrix. The element c_{ij} of the product is given by

$$c_{ij} = \sum_{k=1}^{n} a_{ik} b_{kj} \tag{A-13}$$

Thus the elements c_{ij} are obtained by multiplying the elements of the ith row of A with the corresponding elements of the jth column of B and then summing these element products.

For example

$$\begin{bmatrix} a_{11} & a_{12} \\ a_{21} & a_{22} \end{bmatrix} \begin{bmatrix} b_{11} & b_{12} \\ b_{21} & b_{22} \end{bmatrix} = \begin{bmatrix} c_{11} & c_{12} \\ c_{21} & c_{22} \end{bmatrix}$$

where

$$c_{11} = a_{11}b_{11} + a_{12}b_{21}$$

$$c_{12} = a_{11}b_{12} + a_{12}b_{22}$$

$$c_{21} = a_{21}b_{11} + a_{22}b_{21}$$

$$c_{22} = a_{21}b_{12} + a_{22}b_{22}$$

If the product AB is defined, the product BA may or may not be defined. Even if BA is defined, the resulting products of AB and BA are not, in general, equal. Thus, it is important to note that in general matrix multiplication is not commutative, i.e.

$$AB \neq BA$$

The associative and distributive laws hold for matrix multiplication (when the appropriate operations are defined), i.e.

Associative law: $(AB)C = A(BC) = ABC$

Distributive law: $A(B + C) = AB + AC$

Given the two matrices

$$A = \begin{bmatrix} 1 & 0 \\ 2 & 3 \\ 0 & 1 \end{bmatrix}; \ B = \begin{bmatrix} 1 & -1 & 3 \\ 0 & 2 & 1 \end{bmatrix}$$

Find AB and BA.

A and B are conformable (A has two columns and B has two rows), thus we have

$$AB = \begin{bmatrix} 1 & -1 & 3 \\ 2 & 4 & 9 \\ 0 & 2 & 1 \end{bmatrix}; \ \ BA = \begin{bmatrix} -1 & 0 \\ 4 & 7 \end{bmatrix}$$

A matrix remains unaffected, if a null matrix, defined by Eq. (A-8) is added to it, i.e.

$$A + 0 = A$$

If a null matrix is multiplied to another matrix A, the result is a null matrix

$$A0 = 0A = 0$$

Also

$$A - A = 0$$

Note that equation $AB = 0$ does not mean that either A or B necessarily has to be a null matrix, e.g.

$$\begin{bmatrix} 1 & 3 \\ 0 & 0 \end{bmatrix}\begin{bmatrix} 3 & 0 \\ -1 & 0 \end{bmatrix} = \begin{bmatrix} 0 & 0 \\ 0 & 0 \end{bmatrix}$$

Multiplication of any matrix by a unit matrix results in the original matrix, i.e.

$$AI = IA = A$$

The transpose of the product of two matrices is the product of their transposes in reverse order, i.e.

$$(AB)^{\mathrm{T}} = B^{\mathrm{T}}A^{\mathrm{T}}$$

The concept of matrix multiplication assists in the solution of simultaneous linear algebraic equations. Consider such a set of equations

$$a_{11}x_1 + a_{12}x_2 + \ldots + a_{1n}x_n = c_1$$

$$a_{21}x_1 + a_{22}x_2 + \ldots + a_{2n}x_n = c_2$$

$$\vdots$$

$$a_{m1}x_1 + a_{m2}x_2 + \ldots + a_{mn}x_n = c_m \qquad \text{(A-14)}$$

or

$$\sum_{i=1}^{n} a_{ij}x_j = c_i; \ i = 1, 2, \ldots, m$$

Using the rules of matrix multiplication defined above, Eqs (A-14) can be written in the compact notation as

$$Ax = c \qquad\qquad\qquad\qquad\qquad\qquad\text{(A-15)}$$

where

$$A = \begin{bmatrix} a_{11} & a_{12} & \cdots & a_{1n} \\ a_{21} & a_{22} & \cdots & a_{2n} \\ \vdots & \vdots & & \vdots \\ a_{m1} & a_{m2} & \cdots & a_{mn} \end{bmatrix}$$

$$x = \begin{bmatrix} x_1 \\ x_2 \\ \vdots \\ x_n \end{bmatrix}; \quad c = \begin{bmatrix} c_1 \\ c_2 \\ \vdots \\ c_m \end{bmatrix}$$

It is clear that the *vector-matrix* Eq. (A-15) is a useful shorthand representation of the set of linear algebraic equations (A-14).

Matrix Inversion

Division does not exist as such in matrix algebra. However, if A is a square non-singular matrix, its inverse (A^{-1}) is defined by the relation

$$A^{-1}A = AA^{-1} = I \qquad\qquad\qquad\qquad\text{(A-16)}$$

The conventional method for obtaining an inverse is to use the following relation

$$A^{-1} = \frac{\text{adj } A}{\det A} \qquad\qquad\qquad\qquad\text{(A-17)}$$

It is easy to prove that the inverse is unique

The following are the important properties characterizing the inverse:

$$(AB)^{-1} = B^{-1}A^{-1}$$

$$(A^{-1})^{\text{T}} = (A^{\text{T}})^{-1} \qquad\qquad\qquad\qquad\text{(A-18)}$$

$$(A^{-1})^{-1} = A$$

Example

If A is given by Eq. (A-9), then from Eqs. (A-10), (A-11), (A-17), we get

$$A^{-1} = \frac{\text{adj } A}{\det A} = \frac{1}{5}\begin{bmatrix} 8 & 6 & -5 \\ 6 & 7 & -5 \\ -5 & -5 & 5 \end{bmatrix} = \begin{bmatrix} 8/5 & 6/5 & -1 \\ 6/5 & 7/5 & -1 \\ -1 & -1 & 1 \end{bmatrix}$$

SCALAR AND VECTOR FUNCTIONS

A scalar function of n scalar variables is defined as

$$y \triangleq f(x_1, x_2, \ldots, x_n) \tag{A-19}$$

It can be written as a scalar function of a vector variable **x**, i.e.

$$y = f(\mathbf{x}) \tag{A-20}$$

where **x** is an n-dimension vector,

$$x = \begin{bmatrix} x_1 \\ x_2 \\ \vdots \\ x_n \end{bmatrix}$$

In general, a scalar function could be a function of several vector variables, e.g.

$$y = f(\mathbf{x}, \mathbf{u}, \mathbf{p}) \tag{A-21}$$

where $\mathbf{x}$, $\mathbf{u}$ and $\mathbf{p}$ are vectors of various dimensions.

A vector function is defined as

$$\mathbf{y} = \mathbf{f}(x) \triangleq \begin{bmatrix} f_1(\mathbf{x}) \\ f_2(\mathbf{x}) \\ \vdots \\ f_m(\mathbf{x}) \end{bmatrix} \tag{A-22}$$

In general, a vector function is a function of several vector variables, e.g.

$$\mathbf{y} = \mathbf{f}(\mathbf{x}, \mathbf{u}, \mathbf{p}) \tag{A-23}$$

DERIVATIVES OF SCALAR AND VECTOR FUNCTIONS

A derivative of a scalar function (A-20) with respect to a vector variable $\mathbf{x}$ is defined as

$$\frac{\partial f}{\partial \mathbf{x}} \triangleq \begin{bmatrix} \dfrac{\partial f}{\partial x_1} \\ \dfrac{\partial f}{\partial x_2} \\ \vdots \\ \dfrac{\partial f}{\partial x_n} \end{bmatrix} \tag{A-24}$$

It may be noted that the derivative of a scalar function with respect to a vector of dimension n is a vector of the same dimension.

The derivative of a vector function (A-22) with respect to a vector variable x is defined as

$$\frac{\partial f}{\partial x} \triangleq \begin{bmatrix} \dfrac{\partial f_1}{\partial x_1} & \dfrac{\partial f_1}{\partial x_2} & \cdots & \dfrac{\partial f_1}{\partial x_n} \\ \dfrac{\partial f_2}{\partial x_1} & \dfrac{\partial f_2}{\partial x_2} & \cdots & \dfrac{\partial f_2}{\partial x_n} \\ \dfrac{\partial f_m}{\partial x_1} & \dfrac{\partial f_m}{\partial x_2} & \cdots & \dfrac{\partial f_m}{\partial x_n} \end{bmatrix} \tag{A-25}$$

$$= \begin{bmatrix} \left[\dfrac{\partial f_1}{\partial x}\right]^{\mathrm{T}} \\ \left[\dfrac{\partial f_2}{\partial x}\right]^{\mathrm{T}} \\ \vdots \\ \left[\dfrac{\partial f_m}{\partial x}\right]^{\mathrm{T}} \end{bmatrix} \tag{A-26}$$

Consider now a scalar function defined as

$$s = \lambda^{\mathrm{T}} f(x, u, p) \tag{A-27}$$

$$= \lambda_1 f_1(x, u, p) + \lambda_2 f_2(x, u, p) + \ldots + \lambda_m f_m(x, u, p) \tag{A-28}$$

Let us find $\dfrac{\partial s}{\partial \lambda}$. According to Eq. (A-24), we can write

$$\frac{\partial s}{\partial \lambda} = \begin{bmatrix} f_1(x, u, p) \\ f_2(x, u, p) \\ \vdots \\ f_m(x, u, p) \end{bmatrix} = f(x, u, p) \tag{A-29}$$

Let us now find $\dfrac{\partial s}{\partial x}$. According to Eq. (A-24), we can write

$$\frac{\partial s}{\partial x} = \begin{bmatrix} \dfrac{\partial s}{\partial x_1} \\ \dfrac{\partial s}{\partial x_2} \\ \vdots \\ \dfrac{\partial s}{\partial x_n} \end{bmatrix} = \begin{bmatrix} \lambda_1 \dfrac{\partial f_1}{\partial x_1} + \lambda_2 \dfrac{\partial f_2}{\partial x_1} + \ldots + \lambda_m \dfrac{\partial f_m}{\partial x_1} \\ \lambda_1 \dfrac{\partial f_1}{\partial x_2} + \lambda_2 \dfrac{\partial f_2}{\partial x_2} + \ldots + \lambda_m \dfrac{\partial f_m}{\partial x_2} \\ \lambda_1 \dfrac{\partial f_1}{\partial x_n} + \lambda_2 \dfrac{\partial f_2}{\partial x_n} + \ldots + \lambda_m \dfrac{\partial f_m}{\partial x_n} \end{bmatrix}$$

$$\frac{\partial s}{\partial \mathbf{x}} = \begin{bmatrix} \dfrac{\partial f_1}{\partial x_1} & \dfrac{\partial f_2}{\partial x_1} & \cdots & \dfrac{\partial f_m}{\partial x_1} \\[2mm] \dfrac{\partial f_1}{\partial x_2} & \dfrac{\partial f_2}{\partial x_2} & \cdots & \dfrac{\partial f_m}{\partial x_2} \\[2mm] \cdots & \cdots & \cdots & \cdots \\[2mm] \dfrac{\partial f_1}{\partial x_n} & \dfrac{\partial f_2}{\partial x_n} & \cdots & \dfrac{\partial f_m}{\partial x_n} \end{bmatrix} \begin{bmatrix} \lambda_1 \\[2mm] \lambda_2 \\[2mm] \vdots \\[2mm] \lambda_n \end{bmatrix}$$

$$= \left[\frac{\partial \mathbf{f}}{\partial \mathbf{x}}\right]^T \boldsymbol{\lambda} \tag{A-30}$$

REFERENCES

1. Shipley, R.B., *Introduction to Matrices and Power Systems*, Wiley, New York, 1976.
2. Hadley, G., *Linear Algebra*, Addison-Wesley Pub. Co. Inc., Reading, Mass., 1961.
3. Bellman, R., *Introduction to Matrix Analysis*, McGraw-Hill Book Co., New York, 1960.

Generalized Circuit Constants

We can represent, as we saw in Chapter 5, a three-phase transmission line* by a circuit with two input terminals (sending-end, where power enters) and two output terminals (receiving-end, where power exits). This two-terminal pair circuit is *passive* (since it does not contain any electric energy sources), *linear* (impedances of its elements are independent of the amount of current flowing through them), and *bilateral* (impedances being independent of direction of current flowing). It can be shown that such a two-terminal pair network can be represented by an equivalent T- or π-network.

Consider the unsymmetrical T-network of Fig. B-1, which is equivalent to the general two-terminal pair network.

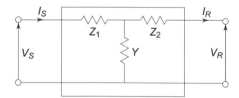

Fig. B-1 Unsymmetrical T-circuit equivalent to a general two terminal pair network

For Fig. B-1, the following circuit equations can be written

$$I_S = I_R + Y(V_R + I_R Z_2)$$

or   (B-1)

$$I_S = YV_R + (1 + YZ_2)I_R$$

$$V_S = V_R + I_R Z_2 + I_S Z_1$$

$$= V_R + I_R Z_2 + Z_1 YV_R + I_R Z_1 + I_R YZ_1 Z_2$$

---

*A transformer is similarly represented by a circuit with two input and two output terminals.

or

$$V_S = (1 + YZ_1) V_R + (Z_1 + Z_2 + YZ_1Z_2)I_R \qquad \text{(B-2)}$$

Equations (B-1) and (B-2) can be simplified by letting

$$A = 1 + YZ_1 \qquad B = Z_1 + Z_2 + YZ_1Z_2 \qquad \text{(B-3)}$$

$$C = Y \qquad D = 1 + YZ_2$$

Using these, Eqs. (B-1) and (B-2) can be written in matrix form as

$$\begin{bmatrix} V_S \\ I_S \end{bmatrix} = \begin{bmatrix} A & B \\ C & D \end{bmatrix} \begin{bmatrix} V_R \\ I_R \end{bmatrix} \qquad \text{(B-4)}$$

This equation is the same as Eq. (5.1) and is valid for any linear, passive and bilateral two-terminal pair network. The constants A, B, C and D are called the *generalized circuit constants* or the *ABCD* constants of the network, and they can be calculated for any such two-terminal pair network.

It may be noted that *ABCD* constants of a two-terminal pair network are complex numbers in general, and always satisfy the following relationship

$$AD - BC = 1 \qquad \text{(B-5)}$$

Also, for any symmetrical network the constants A and D are equal. From Eq. (B-4) it is clear that A and D are dimensionless, while B has the dimensions of impedance (ohms) and C has the dimensions of admittance (mhos).

The *ABCD* constants are extensively used in power system analysis. A general two-terminal pair network is often represented as in Fig. B-2.

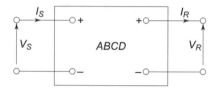

Fig. B-2 Schematic representation of a two-terminal pair network using *ABCD* constants

ABCD CONSTANTS FOR VARIOUS SIMPLE NETWORKS

We have already obtained the *ABCD* constants of an unsymmetrical T-network. The *ABCD* constants of unsymmetrical π-network shown in Fig. B-3 may be obtained in a similar manner and are given below:

$$A = 1 + Y_2Z$$

$$B = Z$$

$$C = Y_1 + Y_2 + ZY_1Y_2 \qquad \text{(B-6)}$$

$$D = 1 + Y_1Z$$

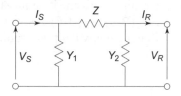

Fig. B-3 Unsymmetrical π-circuit

A series impedance often represents short transmission lines and transformers. The *ABCD* constants for such a circuit (as shown in Fig. B-4) can immediately be determined by inspection of Eqs. (B-1) and (B-2), as follows:

$$A = 1$$
$$B = Z$$
$$C = 0 \hspace{3cm} \text{(B-7)}$$
$$D = 1$$

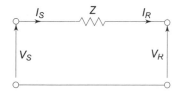

Fig. B-4 Series impedance

Another simple circuit of Fig. B-5 consisting of simple shunt admittance can be shown to possess the following *ABCD* constants

$$A = 1$$
$$B = 0$$
$$C = Y \hspace{3cm} \text{(B-8)}$$
$$D = 1$$

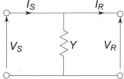

Fig. B-5 Shunt admittance

It may be noted that whenever *ABCD* constants are computed, it should be checked that the relation *AD–BC* = 1 is satisfied. For example, using Eq. (B-8) we get

$$AD\text{–}BC = 1 \times 1 - 0 \times Y = 1$$

If $ABCD$ constants of a circuit are given, its equivalent T- or π-circuit can be determined by solving Eq. (B-3) or (B-6) respectively, for the values of series and shunt branches. For the equivalent π-circuit of Fig. B-3, we have

$$Z = B$$

$$Y_1 = \frac{D-1}{B} \qquad \text{(B-9)}$$

$$Y_2 = \frac{A-1}{B}$$

ABCD CONSTANTS OF NETWORKS IN SERIES AND PARALLEL

Whenever a power system consists of series and parallel combinations of networks, whose $ABCD$ constants are known, the overall $ABCD$ constants for the system may be determined to analyze the overall operation of the system.

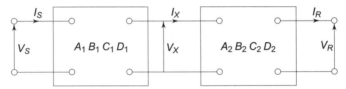

Fig. B-6 Networks in series

Consider the two networks in series, as shown in Fig. B-6. This combination can be reduced to a single equivalent network as follows:
For the first network, we have

$$\begin{bmatrix} V_S \\ I_S \end{bmatrix} = \begin{bmatrix} A_1 & B_1 \\ C_1 & D_1 \end{bmatrix} \begin{bmatrix} V_X \\ I_X \end{bmatrix} \qquad \text{(B-10)}$$

For the second network, we can write

$$\begin{bmatrix} V_X \\ I_X \end{bmatrix} = \begin{bmatrix} A_2 & B_2 \\ C_2 & D_2 \end{bmatrix} \begin{bmatrix} V_R \\ I_R \end{bmatrix} \qquad \text{(B-11)}$$

From Eqs. (B-10) and (B-11), we can write

$$\begin{bmatrix} V_S \\ I_S \end{bmatrix} = \begin{bmatrix} A_1 & B_1 \\ C_1 & D_1 \end{bmatrix} \begin{bmatrix} A_2 & B_2 \\ C_2 & D_2 \end{bmatrix} \begin{bmatrix} V_R \\ I_R \end{bmatrix}$$

$$= \begin{bmatrix} A_1 A_2 + B_1 C_2 & A_1 B_2 + B_1 D_2 \\ C_1 A_2 + D_1 C_2 & C_1 B_2 + D_1 D_2 \end{bmatrix} \begin{bmatrix} V_R \\ I_R \end{bmatrix} \qquad \text{(B-12)}$$

If two networks are connected in parallel as shown in Fig. B-7, the $ABCD$ constants of the combined network can be found out similarly with some simple manipulations of matrix algebra. The results are presented below:

$$A = (A_1 B_2 + A_2 B_1)/(B_1 + B_2)$$

$$B = B_1 B_2/(B_1 + B_2)$$ 　　　　　　　　　　　　　　　(B-13)

$$C = (C_1 + C_2) + (A_1 - A_2)(D_2 - D_1)/(B_1 + B_2)$$

$$D = (B_2 D_1 + B_1 D_2)/(B_1 + B_2)$$

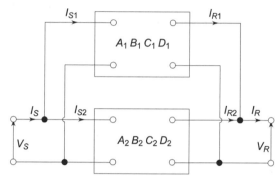

Fig. B-7 Networks in parallel

Measurement of ABCD Constants

The generalized circuit constants may be computed for a transmission line which is being designed from a knowledge of the system impedance/admittance parameters using expressions such as those developed above. If the line is already built, the generalized circuit constants can be measured by making a few ordinary tests on the line. Using Eq. (B-4), these constants can easily be shown to be ratios of either voltage or current at the sending-end to voltage or current at the receiving-end of the network with the receiving-end open or short-circuited. When the network is a transformer, generator, or circuit having lumped parameters, voltage and current measurements at both ends of the line can be made, and the phase angles between the sending and receiving-end quantities can be found out. Thus the *ABCD* constants can be determined.

It is possible, also, to measure the magnitudes of the required voltages and currents simultaneously at both ends of a transmission line, but there is no simple method to find the difference in phase angle between the quantities at the two ends of the line. Phase difference is necessary because the *ABCD* constants are complex. By measuring two impedances at each end of a transmission line, however, the generalized circuit constants can be computed.

The following impedances are to be measured:

Z_{SO} = sending-end impedance with receiving-end open-circuited

Z_{SS} = sending-end impedance with receiving-end short circuited

Z_{RO} = receiving-end impedance with sending-end open-circuited

Z_{RS} = receiving-end impedance with sending-end short-circuited

The impedances measured from the sending-end can be determined in terms of the *ABCD* constants as follows:

From Eq. (B-4), with $I_R = 0$,

$$Z_{SO} = V_S/I_S = A/C \tag{B-14}$$

and with $V_R = 0$,

$$Z_{SS} = V_S/I_S = B/D \tag{B-15}$$

When the impedances are measured from the receiving-end, the direction of current flow is reversed and hence the signs of all current terms in Eq. (5.25). We can therefore rewrite this equation as

$$V_R = DV_S + BI_S \tag{B-16}$$

$$I_R = CV_S + AI_S$$

From Eq. (B-16), with $I_S = 0$,

$$Z_{RO} = V_R/I_R = D/C \tag{B-17}$$

and when $V_S = 0$,

$$Z_{RS} = V_R/I_R = B/A \tag{B-18}$$

Solving Eqs. (B-14), (B-15), (B-17) and (B-18) we can obtain the values of the $ABCD$ constants in terms of the measured impedances as follows.

$$Z_{RO} - Z_{RS} = \frac{AD - BC}{AC} = \frac{1}{AC} \text{ [using Eq. (B-5)]}$$

$$\frac{Z_{RO} - Z_{RS}}{Z_{SO}} = \frac{1}{AC} \cdot \frac{C}{A} = \frac{1}{A^2}$$

$$\therefore \qquad A = \left(\frac{Z_{SO}}{Z_{RO} - Z_{RS}} \right)^{1/2} \tag{B-19}$$

By substituting this value of A in Eqs. (B-14), (B-18) and substituting the value of C so obtained in Eq. (B-17), we get

$$B = Z_{RS} \left(\frac{Z_{SO}}{Z_{RO} - Z_{RS}} \right)^{1/2} \tag{B-20}$$

$$C = \frac{1}{(Z_{SO}(Z_{RO} - Z_{RS}))^{1/2}} \tag{B-21}$$

$$D = \frac{Z_{RO}}{(Z_{SO}(Z_{RO} - Z_{RS}))^{1/2}} \tag{B-22}$$

REFERENCE

1. Cotton, H. and H. Barber, *The Transmission and Distribution of Electrical Energy*, 3rd edn., B.I. Publishers, New Delhi, 1970.

Triangular Factorization and Optimal Ordering

We know that the nodal matrix Y_{BUS} and its associated Jacobian are very sparse, whereas their inverse matrices are full. For large power systems the sparsity of these matrices may be as high as 98% and must be exploited. Apart from reducing storage and time of computation, sparsity utilization limits the round-off computational errors. In fact, straight-forward application of the iterative procedure for system studies like load flow is not possible for large systems unless the sparsity of the Jacobian is dealt with effectively.

GAUSS ELIMINATION

One of the recent techniques of solving a set of linear algebraic equations, called *triangular factorization*, replaces the use of matrix inverse which is highly inefficient for large sparse systems. In triangularization the elements of each row below the main diagonal are made zero and the diagonal element of each row is normalized as soon as the processing of that row is completed. It is possible to proceed columnwise but it is computationally inefficient and is therefore not used. After triangularization the solution is easily obtained by *back substitution*. The technique is illustrated in the example below.

Example C-1

Consider the linear vector-matrix equation

$$\begin{bmatrix} 2 & 1 \\ 3 & 5 \end{bmatrix} \begin{bmatrix} x_1 \\ x_2 \end{bmatrix} = \begin{bmatrix} 1 \\ 0 \end{bmatrix}$$

Procedure

1. Divide row 1 by the self-element of the row, in this case 2.
2. Eliminate the element (2, 1) by multiplying the modified row 1, by element (2, 1) and subtract it from row 2.
3. Divide the modified row 2 by its self-element $(\frac{7}{2})$, and stop.

Following this procedure, we get the upper triangular equation as

$$\begin{bmatrix} 1 & \frac{1}{2} \\ 0 & \frac{\frac{7}{2}}{\frac{7}{2}} = 1 \end{bmatrix} \begin{bmatrix} x_1 \\ x_2 \end{bmatrix} = \begin{bmatrix} \frac{1}{2} \\ \frac{0 - \frac{3}{2}}{\frac{7}{2}} \end{bmatrix}$$

Upon back substituting, that is first solving for x_2 and then for x_1, we get

$$x_2 = \frac{0 - \frac{3}{2}}{\frac{7}{2}} = -\frac{3}{7}$$

$$x_1 = \frac{1}{2} - \frac{1}{2}x_2 = \frac{1}{2} - \frac{1}{2}(-\frac{3}{7}) = \frac{5}{7}$$

Check

$$2x_1 + x_2 = 2(\frac{5}{7}) - \frac{3}{7} = 1$$

$$3x_1 + 5x_2 = 3(\frac{5}{7}) - 5(\frac{3}{7}) = 0$$

Thus, we have demonstrated the use of the basic Gauss elimination and back substitution procedure for a simple system, but the same procedure applies to any general system of linear algebraic equations, i.e.

$$Ax = b \tag{C-1}$$

An added advantage of row processing (elimination of row elements below the main diagonal and normalization of the self-element) is that it is easily amenable to the use of low storage compact storage schemes—avoiding storage of zero elements.

GAUSS ELIMINATION USING TABLE OF FACTORS

Where repeated solution of vector-matrix Eq. (C-1) with constant A but varying values of vector b is required, it is computationally advantageous to split the matrix A into triangular factor (termed as 'Table of factors' or 'LU decomposition') using the Gauss elimination technique. If the matrix A is sparse, so is the table of factors which can be compactly stored thereby not only reducing core storage requirements, but also the computational effort. Gauss elimination using the table of factors is illustrated in the following example.

Example C-2

Consider the following system of linear equations:

$$(2)x_1 + (1)x_2 + (3)x_3 = 6$$
$$(2)x_1 + (3)x_2 + (4)x_3 = 9 \qquad \text{(C-2)}$$
$$(3)x_1 + (4)x_2 + (7)x_3 = 14$$

For computer solution, maximum efficiency is attained when elimination is carried out by rows rather than the more familiar column order. The successive reduced sets of equations are as follows:

$$(1)x_1 + (\tfrac{1}{2})x_2 + (\tfrac{3}{2})x_3 = \tfrac{6}{2}$$
$$(2)x_1 + (3)x_2 + (4)x_3 = 9 \qquad \text{(C-3)}$$
$$(3)x_1 + (4)x_2 + (7)x_3 = 14$$

$$(1)x_1 + (\tfrac{1}{2})x_2 + (\tfrac{3}{2})x_3 = 3$$
$$(2)x_2 + (1)x_3 = 3 \qquad \text{(C-4)}$$
$$(3)x_1 + (4)x_2 + (7)x_3 = 14$$

$$(1)x_1 + (\tfrac{1}{2})x_2 + (\tfrac{3}{2})x_3 = 3$$
$$(1)x_2 + (\tfrac{1}{2})x_3 = \tfrac{3}{2} \qquad \text{(C-5)}$$
$$(3)x_1 + (4)x_2 + (7)x_3 = 14$$

$$(1)x_1 + (\tfrac{1}{2})x_2 + (\tfrac{2}{3})x_3 = 3$$
$$(1)x_2 + (\tfrac{1}{2})x_3 = \tfrac{3}{2} \qquad \text{(C-6)}$$
$$(\tfrac{5}{2})x_2 + (\tfrac{5}{2})x_3 = 5$$

$$(1)x_1 + (\tfrac{1}{2})x_2 + (\tfrac{3}{2})x_3 = 3$$
$$(1)x_2 + (\tfrac{1}{2})x_3 = \tfrac{3}{2} \qquad \text{(C-7)}$$
$$(\tfrac{5}{4})x_3 = \tfrac{5}{4}$$

$$(1)x_1 + (\tfrac{1}{2})x_2 + (\tfrac{3}{2})x_3 = 3$$
$$(1)x_2 + (\tfrac{1}{2})x_3 = \tfrac{3}{2} \qquad \text{(C-8)}$$
$$x_3 = 1$$

These steps are referred to as 'elimination' operations. The solution x may be immediately determined by 'back substitution' operation using Eq. (C-8).

The solution for a new set of values for **b** can be easily obtained by using a table of factors prepared by a careful examination of Eqs. (C-3) to (C-8). We can write the table of factors **F** as below for the example in hand.

$$
\mathbf{F} = \begin{bmatrix} f_{11} & f_{12} & f_{13} \\ f_{21} & f_{22} & f_{23} \\ f_{31} & f_{32} & f_{33} \end{bmatrix} = \begin{bmatrix} \frac{1}{2} & \frac{1}{2} & \frac{3}{2} \\ 2 & \frac{1}{2} & \frac{1}{2} \\ 3 & \frac{5}{2} & \frac{4}{5} \end{bmatrix} \tag{C-9}
$$

The row elements of **F** below the diagonal are the multipliers of the normalized rows required for the elimination of the row element, e.g. $f_{32} = \frac{5}{2}$, the multiplier of normalized row 2 [Eq. (C-6)] to eliminate the element (3, 2), i.e. $(\frac{5}{2})x_2$. The diagonal elements of **F** are the multipliers needed to normalize the rows after the row elimination has been completed, e.g. $f_{22} = \frac{1}{2}$, the factor by which row 2 of Eq. (C-4) must be multiplied to normalize the row. The elements of **F** above the diagonal can be immediately written down by inspection of Eq. (C-8). These are needed for the back substitution process.

In rapidly solving Eq. (C-1) by use of the table of factors **F**, successive steps appear as columns (left to right) in Table C.1 below:

Table C.1

b	1, 1	2, 1	2, 2	3, 1	3, 2	3, 3	2, 3	1, 3	1, 2	x
6	3	3	3	3	3	3	3	5/2	1	1
9	9	3	3/2	3/2	3/2	3/2	3	1	1	1
14	14	14	14	5	5/4	1	1	1	1	1

The heading row (i, j) of Table C-1 represents the successive elimination and back substitution steps. Thus,

1, 1   represents normalization of row 1
2, 1   represents elimination of element (2, 1)
2, 2   represents normalization of row 2
3, 1; 3, 2   represent elimination of elements (3, 1) and (3, 2) respectively
3, 3   represents normalization of row 3
2, 3   represents elimination of element (2, 3) by back substitution
1, 3; 1, 2   represent elimination of elements (1, 3) and (1, 2) respectively by back substitution.

The solution vector at any stage of development is denoted by

$$
[y_1 \ y_2 \ y_3]^T = \mathbf{y}
$$

The modification of solution vector from column to column (left to right) is carried out for the heading (i, j) as per the operations defined below:

$$
y_i = f_{ii} y_i \qquad \text{if } j = i \tag{C-10}
$$

$$
y_i = y_i - f_{ij} y_j \qquad \text{if } j \neq i \tag{C-11}
$$

Thus for heading 3, 2

$$y_3 = y_3 - f_{32}y_2$$

$$= 5 - (\tfrac{5}{2})(\tfrac{3}{2}) = \tfrac{5}{4}$$

and for heading 3, 3

$$y_3 = f_{33}y_3$$

$$= (\tfrac{4}{5})(\tfrac{5}{4}) = 1$$

In fact, operation (C-10) represents row normalization and (C-11) represents elimination and back substitution procedures.

Optimal Ordering

In power system studies, the matrix **A** is quite sparse so that the number of non-zero operations and non-zero storage required in Gauss elimination is very sensitive to the sequence in which the rows are processed. The row sequence that leads to the least number of non-zero operations is not, in general, the same as the one which yields least storage requirement. It is believed that the absolute optimum sequence of ordering the rows of a large network matrix (this is equivalent to renumbering of buses) is too complicated and time consuming to be of any practical value. Therefore, some simple yet effective schemes have been evolved to achieve near optimal ordering with respect to both the criteria. Some of the schemes of near optimal ordering the sparse matrices, which are fully symmetrical or at least symmetric in the pattern of non-zero off-diagonal terms, are described below [4].

Scheme 1

Number the matrix rows in the order of the fewest non-zero terms in each row. If more than one unnumbered row has the same number of non-zero terms, number these in any order.

Scheme 2

Number the rows in the order of the fewest non-zero terms in a row at each step of elimination. This scheme requires updating the count of non-zero terms after each step.

Scheme 3

Number the rows in order of the fewest non-zero off-diagonal terms generated in the remaining rows at each step of elimination. This scheme also involves an updating procedure.

The choice of scheme is a trade-off between speed of execution and the number of times the result is to be used. For Newton's method of load flow solution, scheme 2 seems to be the best. The efficiency of scheme 3 is not sufficiently established to offset the increased time required for its execution.

Scheme 1 is useful for problems requiring only a single solution with no iteration.

Compact Storage Schemes

The usefulness of the Newton's method depends largely upon conserving computer storage and reducing the number of non-zero computations. To effect these ideas on the computer, elimination of lower triangle elements is carried out a row at a time using the concept of compact working row. The non-zero modified upper triangle elements and mismatches are stored in a compact and convenient way. Back substitution progresses backwards through the compact upper triangle table. A properly programmed compact storage scheme results in considerable saving of computer time during matrix operations.

Naturally, there are as many compact working rows and upper triangle storage schemes as there are programmers. One possible scheme for a general matrix stores the non-zero elements of successive rows in a linear array. The column location of these non-zero elements and the location where the next row starts (row index) is stored separately. The details of this and various other schemes are given in [2].

REFERENCES

1. Singh, L.P., *Advanced Power System Analysis and Dynamics*, 2nd edn., Wiley Eastern, New Delhi, 1986.
2. Agarwal, S.K., 'Optimal Power Flow Studies', *Ph.D. Thesis*, B.I.T.S., Pilani, 1970.
3. Tinney, W.F. and J.W. Walker, "Direct Solutions of Sparse Network Equations by Optimally Ordered Triangular Factorizations", *Proc. IEEE*, Nov. 1967, 55: 1801.
4. Tinney, W.F. and C.E. Hart, "Power Flow Solution by Newton's Method", *IEEE Trans.*, Nov. 1967, No. 11, PAS-86: 1449.

Elements of Power System Jacobian Matrix

Expressions to be used in evaluating the elements of the Jacobian matrix of a power system are derived below:

From Eq. (6.25b)

$$P_i - jQ_i = V_i^* \sum_{k=1}^{n} Y_{ik} V_k$$

$$= |V_i| \exp(-j\delta_i) \sum_{k=1}^{n} |Y_{ik}| \exp(j\theta_{ik}) |V_k| \exp(j\delta_k) \qquad (D\text{-}1)$$

Differentiating partially with respect to δ_m ($m \neq i$)

$$\frac{\partial P_i}{\partial \delta_m} - j\frac{\partial Q_i}{\partial \delta_m} = j|V_i| \exp(-j\delta_i)\, (|Y_{im}| \exp(j\theta_{im}) |V_m| \exp(j\delta_m))$$

$$= j(e_i - jf_i)\,(a_m + jb_m) \qquad (D\text{-}2)$$

where

$$Y_{im} = G_{im} + jB_{im}$$

$$V_i = e_i + jf_i$$

$$(a_m + jb_m) = (G_{im} + jB_{im})\,(e_m + jf_m)$$

Although the polar form of the NR method is being used, rectangular complex arithmetic is employed for numerical evaluation as it is faster.

From Eq. (D-2), we can write

$$\frac{\partial P_i}{\partial \delta_m} = (a_m f_i - b_m e_i) = H_{im}$$

$$\frac{\partial Q_i}{\partial \delta_m} = -(a_m e_i + b_m f_i) = J_{im}$$

For the case of $m = i$, we have

$$\frac{\partial P_i}{\partial \delta_i} - j\frac{\partial Q_i}{\partial \delta_i} = -j|V_i| \exp(-j\delta_i) \sum_{k=1}^{n} |Y_{ik}| \exp(j\theta_{ik}) |V_k| \exp(j\delta_k)$$

$$+ j|V_i| \exp(-j\delta_i) (|Y_{ii}| \exp(j\theta_{ii})|V_i| \exp(j\delta_i))$$

$$= -j(P_i - jQ_i) + j|V_i|^2(G_{ii} + jB_{ii}) \qquad \text{(D-3)}$$

From Eq. (D-3), we can write

$$\frac{\partial P_i}{\partial \delta_i} = -Q_i - B_{ii} |V_i|^2 = H_{ii}$$

$$\frac{\partial Q_i}{\partial \delta_i} = P_i - G_{ii} |V_i|^2 = J_{ii}$$

Now differentiate Eq. (D-1) partially with respect to $|V_m|$ $(m \neq i)$. We have

$$\frac{\partial P_i}{\partial |V_m|} - j\frac{\partial Q_i}{\partial |V_m|} = |V_i| \exp(-j\delta_i) (|Y_{im}| \exp(j\theta_{im}) \exp(j\delta_m))$$

Multiplying by $|V_m|$ on both sides,

$$\frac{\partial P_i}{\partial |V_m|}|V_m| - j\frac{\partial Q_i}{\partial |V_m|} |V_m|$$

$$= |V_i| \exp(-j\delta_i) |Y_{im}| \exp(j\theta_{im}) |V_m| \exp(j\delta_m)$$

$$= (e_i - jf_i) (a_m + jb_m) \qquad \text{(D-4)}$$

It follows from Eq. (D-4) that

$$\frac{\partial P_i}{\partial |V_m|}|V_m| = a_m e_i + b_m f_i = N_{im}$$

$$\frac{\partial Q_i}{\partial |V_m|}|V_m| = a_m f_i - b_m e_i = L_{im}$$

Now for the case of $m = i$, we have

$$\frac{\partial P_i}{\partial |V_i|} - j\frac{\partial Q_i}{\partial |V_i|} = \exp(-j\delta_i) \sum_{k=1}^{n} |Y_{ik}| \exp(j\theta_{ik}) |V_k| \exp(j\delta_k)$$

$$+ |V_i| \exp(-j\delta_i) |Y_{ii}| \exp(j\theta_{ii}) \exp(j\delta_i)$$

Multiplying by $|V_i|$ on both sides

$$\frac{\partial P_i}{\partial |V_i|}|V_i| - j\frac{\partial Q_i}{\partial |V_i|} |V_i|$$

$$= |V_i| \exp(-j\delta_i) \sum_{k=1}^{n} |Y_{ik}| \exp(j\theta_{ik}) |V_k| \exp(j\delta_k)$$

$$+ |V_i|^2 |Y_{ii}| \exp(j\theta_{ii})$$

$$= (P_i - jQ_i) + |V_i|^2 (G_{ii} + jB_{ii}) \tag{D-5}$$

It follows from Eq. (D-5) that

$$\frac{\partial P_i}{\partial |V_i|} |V_i| = P_i + G_{ii}|V_i|^2 = N_{ii}$$

$$\frac{\partial Q_i}{\partial |V_i|} |V_i| = Q_i - B_{ii}|V_i|^2 = L_{ii}$$

The above results are summarized below:

Case 1 $\qquad m \neq i$

$$H_{im} = L_{im} = a_m f_i - b_m e_i$$

$$N_{im} = -J_{im} = a_m e_i + b_m f_i \tag{D-6}$$

where

$$Y_{im} = G_{im} + jB_{im}$$

$$V_i = e_i + jf_i \tag{D-7}$$

$$(a_m + jb_m) = (G_{im} + jB_{im})(e_m + jf_m)$$

Case 2 $\qquad m = i$

$$H_{ii} = -Q_i - B_{ii}|V_i|^2$$

$$N_{ii} = P_i + G_{ii}|V_i|^2 \tag{D-8}$$

$$J_{ii} = P_i \cdot G_{ii}|V_i|^2$$

$$L_{ii} = Q_i - B_{ii}|V_i|^2$$

REFERENCES

1. Tinney, W.F. and C.E. Hart "Power Flow Solution by Newton's Method", *IEEE Trans.*, Nov 1967, No. 11, PAS-86: 1449.
2. Van Ness, J.E., "Iteration Methods for Digital Load Flow Studies", *Trans. AIEE*, Aug 1959, 78A: 583.

Kuhn-Tucker Theorem

The Kuhn-Tucker theorem makes it possible to solve the general non-linear programming problem with several variables wherein the variables are also constrained to satisfy certain equality and inequality constraints.

We can state the minimization problem with inequality constraints for the control variables as

$$\min_{\mathbf{u}} f(\mathbf{x}, \mathbf{u}) \tag{E-1}$$

subject to equality constraints

$$g(x, u, p) = 0 \tag{E-2}$$

and to the inequality constraints

$$u - u_{\max} \leq 0 \tag{E-3}$$

$$u_{\min} - u \leq 0 \tag{E-4}$$

The Kuhn-Tucker theorem [1] gives the necessary conditions for the minimum, assuming convexity for the functions (E-1)-(E-4), as

$$\Delta L = 0 \text{ (gradient with respect to } \mathbf{u}, \mathbf{x}, \boldsymbol{\lambda}) \tag{E-5}$$

where L is the Lagrangian formed as

$$L = f(\mathbf{x}, \mathbf{u}) + \boldsymbol{\lambda}^{\mathrm{T}} g(x, u, p) + \alpha^{\mathrm{T}}_{\max}(u - u_{\max}) + \alpha^{\mathrm{T}}_{\min}(u_{\min} - u)$$

and

$$\tag{E-6}$$

$$\left. \begin{array}{c} \alpha_{\max}^{\mathrm{T}}(\mathbf{u} - \mathbf{u}_{\max}) = 0 \\ \alpha_{\min}^{\mathrm{T}}(\mathbf{u}_{\min} - \mathbf{u}) = 0 \\ \alpha_{\max} \geq \mathbf{0}, \alpha_{\min} \geq \mathbf{0} \end{array} \right\} \tag{E-7}$$

Equations (E-7) are known as *exclusion equations*.

The multipliers $\alpha_{\max}$ and $\alpha_{\min}$ are the *dual variables* associated with the upper and lower limits on control variables. They are auxiliary variables similar to the Lagrangian multipliers λ for the equality constraints case.

If u_i violates a limit, it can either be upper or lower limit and not both simultaneously. Thus, either inequality constraint (E-3) or (E-4) is active at a time, that is, either $\alpha_{i,\max}$ or $\alpha_{i,\min}$ exists, but never both. Equation (E-5) can be written as

$$\frac{\partial L}{\partial \mathbf{x}} = \frac{\partial f}{\partial \mathbf{x}} + \left(\frac{\partial g}{\partial \mathbf{x}}\right)^T \boldsymbol{\lambda} = \mathbf{0} \tag{E-8}$$

$$\frac{\partial L}{\partial \mathbf{u}} = \frac{\partial f}{\partial \mathbf{u}} + \left(\frac{\partial g}{\partial \mathbf{u}}\right)^T \boldsymbol{\lambda} + \boldsymbol{\alpha} = \mathbf{0} \tag{E-9}$$

In Eq. (E-9),

$$\alpha_i = \alpha_{i,\max} \quad \text{if } u_i - u_{i,\max} > 0$$

$$\alpha_i = -\alpha_{i,\min} \quad \text{if } u_{i,\min} - u_i > 0$$

$$\frac{\partial L}{\partial \lambda} = g\,(\mathbf{x},\,\mathbf{u},\,\mathbf{p}) = 0 \tag{E-10}$$

It is evident that $\boldsymbol{\alpha}$ computed from Eq. (E-9) at any feasible solution, with λ from Eq. (E-8) is identical with negative gradient, i.e.

$$\boldsymbol{\alpha} = -\frac{\partial f}{\partial \mathbf{u}} = \text{negative of gradient with respect to } \mathbf{u} \tag{E-11}$$

At the optimum, α must also satisfy the exclusion equations (E-7), which state that

$$\alpha_i = 0 \qquad\qquad \text{If } u_{i,\min} < u_i < u_{i,\max}$$

$$\alpha_i = \alpha_{i,\max} \geq 0 \qquad \text{If } u_i = u_{i,\max}$$

$$\alpha_i = -\alpha_{i,\min} \leq 0 \qquad \text{If } u_i = u_{i,\min}$$

which can be rewritten in terms of the gradient using Eq. (E-11) as follows:

$$\frac{\partial L}{\partial u_i} = 0 \qquad \text{if } u_{i,\min} < u_i < u_{i,\max}$$

$$\frac{\partial L}{\partial u_i} \leq 0 \qquad \text{if } u_i = u_{i,\max} \tag{E-12}$$

$$\frac{\partial L}{\partial u_i} \geq 0 \qquad \text{if } u_i = u_{i,\min}$$

REFERENCE

1. Kuhn, H.W. and A.W. Tucker, "Nonlinear Programming", *Proceedings of the Second Berkeley Symposium on Mathematical Statistics and Probability*, University of California Press, Berkeley, 1951.

Real-time Computer Control of Power Systems

In developed countries the focus is shifting in the power sector from the creation of additional capacity to better capacity utilization through more effective management and efficient technology. This applies equally to developing countries where this focus will result in reduction in need for capacity addition.

Immediate and near future priorities now are better plant management, higher availability, improved load management, reduced transmission losses, revamps of distribution system, improved billing and collection, energy efficiency, energy audit and energy management. All this would enable an electric power system to generate, transmit and distribute electric energy at the lowest possible economic and ecological cost.

These objectives can only be met by use of information technology (IT) enabled services in power systems management and control. Emphasis is therefore, being laid on computer control and information transmission and exchange.

The operations involved in power systems require geographically dispersed and functionally complex monitoring and control system. The monitory and supervisory control that is constantly developing and undergoing improvement in its control capability is schematically presented in Fig. F.1 which is easily seen to be distributed in nature.

Starting from the top, control system functions are:

EMS   Energy Management System — It exercises overall control over the total system.

SCADA Supervisory Control and Data Acquisition System — It covers generation and transmission system.

DAC   Distribution Automation and Control System — It oversees the distribution system including connected loads.

         Automation, monitoring and real-time control have always been a part of SCADA system. With enhanced emphasis on IT in power systems, SCADA has been receiving a lot of attention lately.

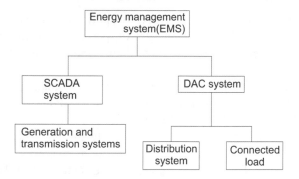

Fig. F.1 Real time monitoring and controlling of an electric power system.

SCADA refers to a system that enables an electricity utility to remotely monitor, coordinate, control and operate transmission and distribution components, equipment and devices in a real-time mode from a remote location with acquisition of data for analysis and planning from one control location. Thus, the purpose of SCADA is to allow operators to observe and control the power system. The specific tasks of SCADA are:

α Data acquisition, which provides measurements and status information to operators.

α Trending plots and measurements on selected time scales.

α Supervisory control, which enables operators to remotely control devices such as circuit breakers and relays.

Capability of SCADA system is to allow operators to control circuit breakers and disconnect switches and change transformer taps and phase-shifter position remotely. It also allows operators to monitor the generation and high-voltage transmission systems and to take action to correct overloads or out-of-limit voltages. It monitors all status points such as switchgear position (open or closed), substation loads and voltages, capacitor banks, tie-line flows and interchange schedules. It detects through telemetry the failures and errors in bilateral communication links between the digital computer and the remote equipment. The most critical functions, mentioned above, are scanned every few seconds. Other noncritical operations, such as the recording of the load, forecasting of load, unit start-ups and shut-downs are carried out on an hourly basis.

Most low-priority programs (those run less frequently) may be executed on demand by the operator for study purposes or to initialize the power system. An operator may also change the digital computer code in the execution if a parameter changes in the system. For example, the MW/min capability of a generating unit may change if one of its throttle values is temporarily removed for maintenance, so the unit's share of regulating power must accordingly be decreased by the code. The computer software compilers and data handlers are designed to be versatile and readily accept operator inputs.

DAC is a lower level version of SCADA applicable in distribution system (including loads), which of course draws power from the transmission/subtransmission levels. Obviously then there is no clear cut demarcation between DAC and SCADA.

In a distribution network, computerisation can help manage load, maintain quality, detect theft and tampering and thus reduce system losses. Computerisation also helps in centralisation of data collection. At a central load dispatch centre, data such as current, voltage, power factor and breaker status are telemetered and displayed. This gives the operator an overall view of the entire distribution network. This enables him to have effective control on the entire network and issue instructions for optimising flow in the event of feeder overload or voltage deviation. This is carried out through switching in/out of shunt capacitors, synchronous condensers and load management. This would help in achieving better voltage profile, loss reduction, improved reliability, quick detection of fault and restoration of service.

At a systems level, SCADA can provide status and measurements for distribution feeders at the substation. Distribution automation equipment can monitor selectionalising devices like switches, interrupters and fuses. It can also operate switches for circuit reconfuration, control voltage, read customers' meters, implement time-of-day pricing and switch customer equipment to manage load. This equipment significantly improves the functionality of distribution control centres.

SCADA can be used extensively for compilation of extensive data and management of distribution systems. Pilferage points too can be zeroed in on, as the flow of power can be closely scrutinised. Here again, trippings due to human errors can be avoided. Modern metering systems using electronic meters, automatic meter readers (AMRs), remote metering and spot billing can go a long way in helping electric utility. These systems can bring in additional revenues and also reduce the time lag between billing and collection.

Distribution automation through SCADA systems directly leads to increased reliability of power for consumers and lower operating costs for the utility. It results in forecasting accurate demand and supply management, faster restoration of power in case of a failure and alternative routing of power in an emergency.

A key feature of these systems is the remote control facility that allows faster execution of decisions. Manual errors and oversights are eliminated. Besides on line and real-time information, the system provides periodic reports that help in the analysis of performance of the power system. Distribution automation combines distribution network monitoring functions with geographical mapping, fault location, and so on, to improve availability. It also integrates load management, load despatch and intelligent metering.

Data Acquisition Systems and Man-Machine Interface

The use of computers nowadays encompasses all phases of power system operation: planning, forecasting, scheduling, security assessment, and control.

An energy control centre manages these tasks and provides optimal operation of the system. A typical control centre can perform the following functions:

    (i) Short, medium and long-term load forecasting (LF)
    (ii) System planning (SP)
    (iii) Unit commitment (UC) and maintenance scheduling (MS)
    (iv) Security monitoring (SM)
    (v) State estimation (SE)
    (vi) Economic dispatch (ED)
    (vii) Load frequency control (LFC)

The above monitoring and control functions are performed in the hierarchical order classified according to time scales. The functions performed in the control centre are based on the availability of a large information base and require extensive software for data acquisition and processing.

At the generation level, the philosophy of 'distributed control' has dramatically reduced the cabling cost within a plant and has the potential of replacing traditional control rooms with distributed CRT/keyboard stations.

Data acquisition systems provide a supporting role to the application software in a control centre. The data acquisition system (DAS) collects raw data from selected points in the power system and converts these data into engineering units. The data are checked for limit violations and status changes and are sent to the data base for processing by the application software. The real-time data base provides structured information so that application programs needing the information have direct and efficient access to it.

The Man-Machine interface provides a link between the operator and the software/hardware used to control/monitor the power system. The interface generally is a colour graphic display system. The control processors interface with the control interface of the display system. The DAS and Man-Machine interface support the following functions:

    (i) Load/Generation Dispatching
    (ii) Display and CRT control
    (iii) Data Base Maintenance
    (iv) Alarm Handling
    (v) Supervisory control
    (vi) Programming functions
    (vii) Data logging
    (viii) Event logging
    (ix) Real-time Network Analysis

With the introduction of higher size generating units, the monitoring requirements have gone up in power plants also. In order to improve the plant performance, now all the utilities have installed DAS in their generating units of sizes 200 MW and above. The DAS in a thermal power station collects the following inputs from various locations in the plant and converts them into engineering units.

Analog Inputs

    (i) Pressures, flows, electrical parameters, etc.

    (ii) Analog input of 0–10 V DC

    (iii) Thermocouple inputs

    (iv) RTD input

Digital Inputs:

    (i) Contract outputs

    (ii) Valve position, pressure and limit switches

All these process inputs are brought from the field through cables to the terminals. The computer processes the information and supplies to the Man-Machine interface to perform the following functions.

    (i) Display on CRT screen

    (ii) Graphic display of plant sub-systems

    (iii) Data logging

    (iv) Alarm generation

    (v) Event logging

    (vi) Trending of analogue variables

    (vii) Performance calculation

    (viii) Generation of control signals

Some of the above functions are briefly discussed as follows.

There are generally 2-3 CRTs in an operator's console in the control room which facilitate the operator to have display of alarms, plant variables and other desired information simultaneously. To give a full view of the plant alongwith various real-time variables and status information, displays of mimic diagrams on a graphic CRT screen are very useful. The data can be logged on printers at a fixed interval automatically as well as on-demand. The limits of all the variables are checked by the software and if any of the variables is outside its pre-defined range, an alarm is raised to draw the operator's attention. The selected important status changes are logged and a record of time of occurrences is maintained.

The DAS software contains programs to calculate periodically the efficiency of various equipment like boiler, turbine, generator, condenser, fans, heaters, etc.

The DAS can also carry out complete on-line control of power station equipment. The DAS for modern power plants employ distributed system configuration. The system is made up of a number of microcomputers each performing an assigned task and exchanging information among others through interconnection via a coaxial cable — the so-called Highway.

Distributed processing is a term that refers to any computing environment, in which multiple loosely coupled computer systems implement a given application. Automatic control by distributed intelligence has, among other advantages, great design flexibility together with improved reliability and performance. The information technology has now captured the imagination of power system engineers who are adopting the low-cost and relatively powerful computing devices in implementing their distributed DAS and control systems.

Computer control brings in powerful algorithms with the following advantages: (i) increase in capacity utilization in generation, (ii) savings in energy and so in raw materials due to increased operational efficiency, (iii) flexibility and modifiability, (iv) reduction in human drudgery, (v) improved operator effectiveness.

Intelligent database processors will become more common in power systems since the search, retrieval and updating activity can be speeded up. New functional concepts from the field of Artificial Intelligence (AI) will be integrated with power system monitoring, automatic restoration of power networks, and real-time control.

Personal computers (PCs) are being used in a wide range of power system operations including power station control, load management, SCADA systems, protection, operator training, maintenance functions, administrative data processing, generator excitation control and control of distribution networks. IT enabled systems thus not only monitor and control the grid, but also improve operational efficiencies and play a key part in maintaining the security of the power system.

REFERENCES

1. *Power Line Magazine*, Vol. 7, No. 1, October 2002, pp 65–71.
2. A.K. Mahalanabis, D.P. Kothari and S.I. Ahson, *Computer Aided Power System Analysis and Control*, TMH, New Delhi, 1988.
3. IEEE Tutorial Course, *Fundamentals of Supervisory Control System*, 1981.
4. IEEE Tutorial Course, *Energy Control Centre Design*, 1983.

APPENDIX G

Introduction to MATLAB and SIMULINK

MATLAB has been developed by MathWorks Inc. It is a powerful software package used for high performance scientific numerical computation, data analysis and visualization. MATLAB stands for MATrix LABoratory. The combination of analysis capabilities, flexibility, reliability and powerful graphics makes MATLAB the main software package for power system engineers. This is because unlike other programming languages where you have to declare matrices and operate on them with their indices, MATLAB provides matrix as one of the basic elements. It provides basic operations, as we will see later, like addition, subtraction, multiplication by use of simple mathematical operators. Also, we need not declare the type and size of any variable in advance. It is dynamically decided depending on what value we assign to it. But MATLAB is case sensitive and so we have to be careful about the case of variables while using them in our programs.

MATLAB gives an interactive environment with hundreds of reliable and accurate built-in functions. These functions help in providing the solutions to a variety of mathematical problems including matrix algebra, linear systems, differential equations, optimization, non-linear systems and many other types of scientific and technical computations. The most important feature of MATLAB is its programming capability, which supports both types of programming—object oriented and structured programming and is very easy to learn and use and allows user developed functions. It facilitates access to FORTRAN and C codes by means of external interfaces. There are several optional toolboxes for simulating specialized problems of different areas and extensions to link up MATLAB and other programs. SIMULINK is a program build on top of MATLAB environment, which along with its specialized products, enhances the power of MATLAB for scientific simulations and visualizations.

For a detailed description of commands, capabilities, MATLAB functions and many other useful features, the reader is referred to MATLAB User's Guide/Manual.

G.1   HOW TO START MATLAB?

You can start MATLAB by double clicking on MATLAB icon on your Desktop of your computer or by clicking on Start Menu followed by 'Programs' and then clicking appropriate program group such as 'MATLAB Release 12'. You will visualize a screen shown in Fig. G.1.

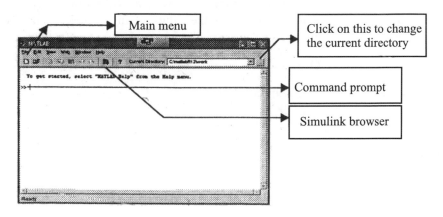

Fig. G.1

The command prompt (characterised by the symbol >>) is the one after which you type the commands. The main menu contains the submenus such as File, Edit, Help, etc. If you want to start a new program file in MATLAB (denoted by .m extension) one can click on **File** followed by **new** and select the desired **M-file**. This will open up a MATLAB File Editor/Debugger window where you can enter your program and save it for later use. You can run this program by typing its name in front of command prompt. Let us now learn some basic commands.

Matrix Initialization

A matrix can be initialized by typing its name followed by = sign and an opening square bracket after which the user supplies the values and closes the square brackets. Each element is separated from the other by one or more spaces or tabs. Each row of matrix is separated from the other by pressing Enter key at the end of each row or by giving semicolon at its end. Following examples illustrate this.

```
>>  A  =  [1   2   5]
>>  B  =  [ 1   5   7
            7   2  -9
           -3   2  -5 ]
```

The above operation can also be achieved by typing

```
>>  B  =  [1   5   7;   7   2  -9;  -3   2  -5]
```

If we do not give a semicolon at the end of closing square brackets, MATLAB displays the value of matrices in the command window. **If you do not want MATLAB to display the results, just type semicolon(;) at the end of the statement**. That is why you will find that in our programmes, whenever we want to display value of variables say voltages, we have just typed the name of the variable without semicolon at the end, so that user will see the values during the program. After the program is run with no values of the variables being displayed, if the user wants to see the values of any of these variables in the program he can simply type the variable name and see the values.

Common Matrix Operations

First let us declare some matrices
```
>>  A  =  [1  5;  6  9]
>>  B  =  [2  7;  9  -3]
```

Addition

```
>>  C  =  A  +  B
```
Adds matrices A and B and stores them in matrix C

Subtraction

```
>>  D  =  A  -  B
```
Subtracts matrix B from matrix A and stores the result in D

Multiplication

```
>>  E  =  A  *  B
```
Multiplies two conformable matrices A and B and stores the result in E

Inverse

```
>>  F  =  inv(A)
```
This calculates the inverse of matrix A by calculating the co-factors and stores the result in F.

Transpose

Single quote ($'$) operator is used to obtain transpose of matrix. In case the matrix element is complex, it stores the conjugate of the element while taking the transpose, e.g.
```
>>  G  =  A'
```
or
```
>> G1 = [1 3; 4 5; 6 3]'
```
also stores the transpose of matrix given in square brackets in matrix G1

In case one does not want to take conjugate of elements while taking transpose of complex matrix, one should use . 'operator instead of' operator.

Determinant

```
>>  H  =  det  (A)
```
This stores determinant of matrix A in H.

Eigen Values

```
>>  K  =  eig(A)
```
obtains eigen values of matrix A and stores them in K.

G.2 SPECIAL MATRICES AND PRE-DEFINED VARIABLES AND SOME USEFUL OPERATORS

MATLAB has some preinitialised variables. These variables can directly be used in programs.

pi

This gives the value of π
```
>>  r  =  d*pi/180
```
converts degrees d into radians and stores it in variable r.

inf

You can specify a variable to have value as ∞.
```
>>  gain  =  inf
```

i and j

These are predefined variables whose value is equal to sqrt (-1). This is used to define complex numbers. One can specify complex matrices as well.
```
>>  P  =  20
>>  Q  =  - 10
>>  S  =  P  +  j*Q
```
The above statement defines complex power S.

A word of caution here is that if we use i and j variables as loop counters then we cannot use them for defining complex numbers. Hence you may find that in some of the programs we have used $i1$ and $j1$ as loop variables instead of i and j.

eps

This variable is preinitilized to 2^{-52}.

Identity Matrix

To generate an identity matrix and store it in variable K, give the following command
```
>>  K  =  eye  (3,  3)
```

So K after this becomes

K = [1   0   0
      0   1   0
      0   0   1]

Zeros Matrix

```
>> L = zeros (3, 2)
```
generates a 3 × 2 matrix whose all elements are zero and stores them in L.

Ones Matrix

```
>> M = ones (3, 3)
```
generates a 3 × 2 matrix whose all elements are one and stores them in M.

: (Colon) operator

This is an important operator which can extract a submatrix from a given matrix. Matrix A is given as below

A= [1   5   7   8
      2   6   9   10
      5   4   3   1
      9   3   1   2]

```
>> B = A(2,:)
```
    This command extracts all columns of matrix A corresponding to row 2 and stores them in B.

So B becomes [2 6 9 10]

Now try this command.
```
>> C = A (:, 3)
```
The above command extracts all the rows corresponding to column 3 of matrix A and stores them in matrix C.

So C becomes

[7

9

3

1]

Now try this command
```
>> D = A(2:4, 1:3)
```
This command extracts a submatrix such that it contains all the elements corresponding to row number 2 to 4 and column number 1 to 3.

So D = [2   6   9
        5   4   3
        9   3   1]

G.3 ELEMENT BY ELEMENT MULTIPLICATION (.* OPERATOR)

This operation unlike complete matrix multiplication, multiplies element of one matrix with corresponding element of other matrix having same index. However in latter case both the matrices must have equal dimensions.

We have used this operator in calculating complex powers at the buses.

Say V = [0.845 + *j**0.307 0.927 + *j**0.248 0.966 + *j**0.410]'

And I = [0.0654 – *j**0.432 0.876 – *j**0.289 0.543 + *j**0.210]'

Then the complex power S is calculated as

>> S = V.* conj (I)

Here, **conj** is a built-in function which gives complex conjugate of its argument. So S is obtained as [– 0.0774 + 0.3851*i* 0.7404 + 0.4852*i* 0.6106 + 0.0198*i*] Note here, that if the result is complex MATLAB automatically assigns *i* in the result without any multiplication sign(*). But while giving the input as complex number we have to use multiplication (*) along with *i* or *j*.

G.4 COMMON BUILT-IN FUNCTIONS

sum

sum(A) gives the sum or total of all the elements of a given matrix A.

min

This function can be used in two forms

(a) For comparing two scalar quantities

>> a = 4

>> b = 7

>> min (*a*, *b*) results in 4

if either *a* or *b* is complex number, then its absolute value is taken for comparison

(b) For finding minimum amongst a matrix or array

e.g. if A = [6 –3; 2 –5]

>> min (A) results in – 5

abs

If applied to a scalar, it gives the absolute positive value of that element (magnitude). For example,

>> x = 3 + *j**4

abs(x) gives 5

If applied to a matrix, it results in a matrix storing absolute values of all the elements of a given matrix.

G.5 CONTROL STRUCTURES

IF Statement

The general from of the IF statement is

```
IF    expression
  statements
ELSE    expression
  statements
ELSEIF
  statements
END
```

Expression is a logical expression resulting in an answer 'true' (1) or 'false'(0). The logical expression can consist of

(i) an expression containing relational operators tabulated along with their meanings in Table G.1.

Table G.1

Relational Operator	Meaning
>	Greater than
> =	Greater than or Equal to
<	Less than
< =	Less than or Equal to
= =	Equal to
~=	Not equal to

(ii) or many logical expressions combined with Logical Operators. Various logical operators and their meanings are given in Table G.2.

Table G.2

Logical Operator	Meaning
&	AND
!	OR
!	NOT

FOR Loops

This repeats a block of statements predetermined number of times.
The most common form of **FOR** loop used is

```
for  k = a : b : c,
  statements
end
```

where k is the loop variable which is initialised to value of initial variable a. If the final value (i.e. c) is not reached, the statements in the body for the loop are executed. The value of k is then incremented by step variable b. The process comes to an end when k reaches or exceeds the final value c. For example,

```
for   i   =   1:1:10,
   a(i) =   1
end
```

This initializes every element of a to 1. If increment is of 1, as in this case, then the increment part may as well be omitted and the above loop could be written as

```
for  i   =  1  :   10,
 a(i)   =   1
end
```

While Loop

This loop repeats a block of statements till the condition given in the loop is true

```
while      expression
   statements
end
```

For example,

```
i   =  1
while   i   <=   10
   a(i)   =   1
   i   =  i  +  1;
end
```

This loop makes first ten elements of array a equal to 1.

break statement

This statement allows one to exit prematurely from a **for** or **while** loop.

G.6   HOW TO RUN THE PROGRAMS GIVEN IN THIS APPENDIX?

1. Copy these programs into the work subfolder under MATLAB folder.
2. Just type the name of the program without '.m' extension and the program will run.
3. If you want to copy them in some other folder say c:\power, then after copying those files in c:\power, change the work folder to c:\power. You can do this by clicking on toolbar containing three periods ⋯ which is on the right side to the Current Directory on the top right corner.
4. You can see or edit these programs by going through File — Open menu and opening the appropriate file. However do not save those programs, unless you are sure that you want the changes you have made to these original files.

5. You can see which are the variables already defined by typing **whos** in front of command prompt. That is why you will normally find a **clear** command at the beginning of our programs. This clears all the variables defined so far from the memory, so that those variables do not interfere or maloperate our programs.

G.7 SIMULINK BASICS

SIMULINK is a software package developed by MathWorks Inc. which is one of the most widely used software in academia and industry for modeling and simulating dynamical systems. It can be used for modeling linear and nonlinear systems, either in continuous time frame or sampled time frame or even a hybrid of the two. It provides a very easy drag-drop type Graphical user interface to build the models in block diagram form. It has many built-in block-library components that you can use to model complex systems. If these built-in models are not enough for you, SIMULINK allows you to have user defined blocks as well. However, in this short appendix, we will try to cover some of the very common blocks that one comes across while simulating a system. You can try to construct the models given in the examples.

How to Start?

You can start SIMULINK by simply clicking the simulink icon in the tools bar or by typing *Simulink* in front of the MATLAB command prompt >>. This opens up SIMULINK library browser, which should look similar to the one shown in Fig. G2. There may be other tool boxes depending upon the license you have. The plus sign that you see in the right half of the window indicates that there are more blocks available under the icon clicking on the (+) sign will expand the library. Now for building up a new model click on File and select New Model. A blank model window is opened. Now all you have to do is to select the block in the SIMULINK library browser and drop it on your model window. Then connect them together and run the simulation. That is all.

An Example

Let us try to simulate a simple model where we take a sinusoidal input, integrate it and observe the output. The steps are outlined as below.

1. Click on the **Sources** in the SIMULINK library browser window.
2. You are able to see various sources that SIMULINK provides. Scroll down and you will see a Sine Wave sources icon.
3. Click on this sources icon and without releasing the mouse button drag and drop it in your model window which is currently named as 'untitled'.
4. If you double-click on this source, you will be able to see Block parameters for sine wave which includes amplitude, frequency, phase, etc. Let not change these parameters right now. So click on **cancel** to go back.

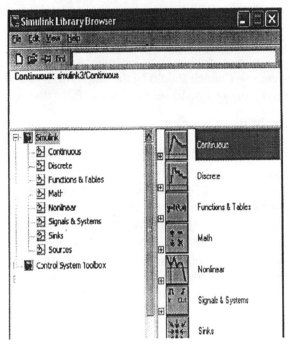

Fig. G.2

5. Similarly click on **continuous** library icon. You can now see various built-in blocks such as derivative, integrator, transfer-function, state-space etc. Select integrator block and drag-drop it in your model window.

6. Now click on sinks and drag-drop scope block into your model. This is one of the most common blocks used for displaying the values of the blocks.

7. Now join output of sine-wave source to input of Integrator block. This can be done in two ways. Either you click the left button and drag mouse from output of sine-wave source to input of Integrator block and leave left button or otherwise click on right button and drag the mouse to form connection from input of integrator block to output of sine-wave source.

8. Now in the main menu, click on **Simulation** and click **Start**. The simulation runs and stops after the time specified by giving ready prompt at bottom left corner.

9. Now double-click on scope to see the output. Is something wrong? The result is a sine wave of magnitude 2. Is there something wrong with SIMULINK software?

   No, we have in fact forgotten to specify the integration constant! Integration of $\sin \theta$ is $- \cos \theta + C$. At $\theta = 0$, $C = -1$. If we do not specify any initial condition for output of the integrator, simulink assumes it to be 0 and calculates the constant. So it calculates

$-\cos\theta + C = 0$ at $t = 0$ giving $C = 1$. So the equation for output becomes $-\cos\theta + 1$. Thus naturally, it starts from 0 at $t = 0$ and reaches its peak value of 2 at $\theta = \pi$, i.e. 3.14.

10. To rectify this error, double click on Integrator block and in the initial conditions enter -1 which should be the output of the block at $t = 0$. Now run the simulation again and see for yourself that the result is correct.

Some Commonly used Blocks

1. *Integrator*   We have already described its use in the above example.

2. *Transfer function*   Using this block you can simulate a transfer function of the form $T(s) = N(s)/D(s)$, where $N(s)$ and $D(s)$ are polynomials in s. You can double click the block and enter the coefficients of s in numerator and denominator of the expression in ascending order of s which are to be enclosed inside square brackets and separated by a space.

3. *Sum*   You can find this block under **Math** in **Simulink** block. By default, it has two inputs with both plus signs. You can modify it to have required number of inputs to be summed up by specifying a string of + or − depending upon the inputs. So if there are 3 inputs you can give the list of signs as + − − . This will denote one positive input and two inputs with − signs which are often used to simulate negative feedback.

4. *Gain*   This block is also found under **Math** in **Simulink** block. It is used to simulate static gain. It can even have fractional values to act as attenuator.

5. *Switch*   This block is available under **Non-linear** block in **Simulink**. It has 3 inputs with the top input being numbered 1. When the input number 2 equals or exceeds the threshold value specified in the properties of this block, it allows input number 1 to pass through, else it allows input number 3 to pass through.

6. *Mux and Demux*   These blocks are available under **Signals and systems** block in **Simulink.** The Mux block combines its inputs into a single output and is mostly used to form a vector out of input scalar quantities. Demux block does the reverse thing. It splits the vector quantity into multiple scalar outputs.

7. *Scope*   We have already described its use. However, if you are plotting a large number of points, click on properties toolbar and select **Data History** tab. Then uncheck the Limit data points box so that all points are plotted. Also when the waveform does not appear smooth, in the **general** tab of properties toolbar, select sample time instead of decimation in the sample time box and enter a suitable value like le-3 (10^{-3}). The scope then uses the value at sample-time interval to plot.

8. *Clock*   This block is used to supply time as a source input and is available in **Sources block** under **Simulink.**

9. *Constant* This block is used to provide constant input and is available in **Sources block** under **Simulink**. We have made use of this block in stability studies to provide constant mechanical input.

10. *Sinewave* This block we have already discussed. It can be used to generate a sine wave of any amplitude, frequency and phase.

The reader is encouraged to work out, the examples given in this Appendix to gain greater insight into the software.

G.8   SCRIPT AND FUNCTION FILES

Types of m-files

There are two types of m-files used in Matlab programming:

(i) Script m-file — This file needs no input parameters and does not return any values as output parameters. It is just a set of Matlab statements which is stored in a file so that one can execute this set by just typing the file name in front of the command prompt, eg. programmes in G2 to G18 are script files.

(ii) Function m-files — This file accepts input agruments and return values as output parameters. It can work with variables which belong to the workspace as well as with the variables which are local to the functions. These are useful for making your own function for a particular application, eg. PolarTorect.m in G1 is a function m-file.

The basic structure of function m-file is given below:

(a) Function definition line — This is the first line of a function. It specifies function name, number and order of input variables.

Its syntax is- function [output variables] = function-name (list of input variables)

(b) First line of help — Whenever help is requested on this funciton or look for is executed MATLAB displays this line.

Its syntax is - % function-name help

(c) Help text — Whenever help is requested on the function-name help text is displayed by Matlab in addition to first help line.

Its syntax is - % function-name (input variables)

(d) Body of the function — This consists of codes acting on the set of input variables to produce the output variables.

User can thus develop his/her own programs and functions and add them to the existing library of functions and blocks.

G.9   SOME SAMPLE EXAMPLES SOLVED BY MATLAB

In this section 18 solved examples of this book are solved again using MATLAB/SIMULINK to encourage the reader to solve more power system problems using MATLAB.

Ex G.1

```
% This function converts polar to rectangular coordinates
% The argument to this function is 1) Magnitude & 2)Angle is in degrees
% This function has been used in many problems which are solved in this
  appendix
function rect=polarTorect(a,b)
rect=a*cos(b*pi/180)+j*a*sin(b*pi/180);
```

Note: This function is used in solution of many of the examples solved later.

Ex G.2 (Example 5.6)

```
% This illustrates the Ferranti effect
% It simulates the effect by varying the length of line from
% zero(receiving end) to 5000km in steps of 10 km
% and plots the sending end voltage phasor
% This corresponds to Fig. 5.13 and data from Example 5.6
clear
VR=220e3/sqrt(3);
alpha=0.163e-3;
beta=1.068e-3;
l=5000;
k=1;
for i=0:10:l,
   VS=(VR/2)*exp(alpha*i)*exp(j*beta*i)+(VR/2)*exp(-alpha*i)*exp(j*beta*i);
   X(k)=real(VS);
   Y(k)=imag(VS);
   k=k+1;
end
plot(X,Y)
```

Ex. G.3 (Example 5.7)

```
% This Program illustrates the use of different line models as
% in Example 5.7
clear
f=50
ℓ=300
Z=40+j*125
Y=1e-3
PR=50e6/3
VR=220e3/(sqrt(3))
pfload=0.8
IR=PR/(VR*pfload)
z=Z/l
y=Y/l
```

```
% Now we calculate the sending-end voltage, sending-end current
% and sending-end pf by following methods for various lengths of line
```

```
% varying line lengths from 10 to 300 km in steps of 10 km
% and compare them as a function of line lengths.
% The methods are
% 1) Short Line approximatiom
% 2) Nominal-PI method
% 3) Exact transmission line equations
% 4) Approximation of exact equations
i=1
for l=10:10:300,
    % Short line approximation
    Vs_shortline(i)=VR+(z*l)*IR
    Is_shortline(i)=IR
     Spf_shortline(i)=cos(angle(Vs_shortline(i)-angle(Is_shortline(i))))
     Spower_shortline(i)=real(Vs_shortline(i)*conj(Is_shortline(i)))

    %Nominal PI method
    A=1+(y*l)*(z*l)/2
    D=A
    B=z*l
    C=y*l*(1+(y*l)*(z*l)/4)
    Vs_nominalpi(i)=A*VR+B*IR
    Is_nominalpi(i)=C*VR+D*IR
     Spf_nominalpi(i)=cos(angle(Vs_nominalpi(i)-angle(Is_nominalpi(i))))
     Spower_nominalpi(i)=real(Vs_nominalpi(i)*conj(Is_nominalpi(i)))

    % Exact transmission Line Equations
    Zc=sqrt(z/y)
    gamma=sqrt(y*z)
    Vs_exact(i)=cosh(gamma*l)*VR + Zc*sinh(gamma*l)*IR
    Is_exact(i)=(1/Zc)*sinh(gamma*l) + cosh(gamma*l)*IR
     Spf_exact(i)=cos(angle(Vs_exact(i)-angle(Is_exact(i))))
     Spower_exact(i)=real(Vs_exact(i)*conj(Is_exact(i)))

    % Approximation of above exact equations
    A=1+(y*l)*(z*l)/2
    D=A
    B=(z*l)*(1+(y*l)*(z*l)/6)
    C=(y*l)*(1+(y*l)*(z*l)/6)
    Vs_approx(i)=A*VR+B*IR
    Is_approx(i)=C*VR+D*IR
     Spf_approx(i)=cos(angle(Vs_approx(i)-angle(Is_approx(i))))
     Spower_approx(i)=real(Vs_approx(i)*conj(Is_approx(i)))
    point(i)=i
    i=i+1

end
% The reader can uncomment any of the four plot statements given below
% by removing the percentage sign against that statement
% for ex. in the plot statement uncommented below
% it plots the sending end voltages for short line model in red
```

```
% by nominal pi-model in green , by exact parameters model in black
% and by approx.  Pi model in blue
plot(point,abs(Vs_shortline),'r',point,abs(Vs_nominalpi),'g',point,abs…
(Vs_exact), 'b', point,abs(Vs_approx),'k')
%plot(point,abs(Is_shortline),'r',point,abs(Is_nominalpi),'g',point,abs…
(Is_exact),'b',point,abs(Is_approx),'k')
%plot(point,abs(Spf_shortline),'r',point,abs(Spf_nominalpi),'g',point,abs…
(Spf_exact),'b',point,abs(Spf_approx),'k')
plot(point,abs(Spower_shortline),'r',point,abs(Spower_nominalpi),'g',point,abs…
(Spower_exact),'b',point,abs(Spower_approx),'k')
```

Ex G.4 (Example 6.2)

```
% This program forms YBUS by Singular Transformation
% The data for this program is a primitive admittance matrix y
% which is to be given in the following format and stored in ydata.
% Ground is given as bus no 0.
% If the element is not mutually coupled with any other element,
% then the entry corresponding to 4th and 5th column of ydata
% has to be zero
% ---------------------------------------------------------------
% element no | connected  |   y (self)    |Mutually  |  y(mutual)  |
%            | From | To   |               |coupled to|             |
%            |Busno |Busno |
% --------------------------------- --------------------------------
ydata=[ 1      1    2    1/(0.05+j*0.15)    0           0
        2      1    3    1/(0.1 +j*0.3 )    0           0
        3      2    3    1/(0.15+j*0.45)    0           0
        4      2    4    1/(0.10+j*0.30)    0           0
        5      3    4    1/(0.05+j*0.15)    0           0];

%  form primitive y matrix from this data and initialize it to zero.
% to start with
elements=max(ydata(:,1))
yprimitive=zeros(elements,elements)
% Process ydata matrix rowwise to form yprimitive
for i=1:elements,
    yprimitive(i,i)=ydata(i,4)
% Also if the element is mutually coupled with any other element
% (whose element no is indicated in 5th column of ydata above)
% the corresponding column no in the ith row is made equal to y(mutual)
    if (ydata(i,5) ~= 0 )
%         j is the element no with which ith element is mutually coupled
        j=ydata(i,5)
        ymutual=ydata(i,6)
        yprimitive(i,j) = ymutual
    end
end

% Form Bus incidence matrix A from ydata
```

```
% this gives a matrix of size element x buses
buses=max(max(ydata(2,:)),max(ydata(3,:)))
A=zeros(elements,buses)

for i=1:elements,
% ydata(i,2) gives the 'from' bus no.
% The entry corrosponding to column corrosponding to this bus
% in A matrix is made 1 if this is not ground bus
    if ydata(i,2) ~= 0
    A(i,ydata(i,2))=1;
    end
% ydata(i,3) gives the 'to' bus no.
% The entry corrosponding to column corresponding to this bus
% in A matrix is made 1 if this is not ground bus

    if ydata(i,3)~= 0
    A(i,ydata(i,3))=-1
    end
end


YBUS=A'*yprimitive*A
```

Ex G.5 (data same as Example 6.2)

```
% This program forms YBUS by adding one element at a time
% The data for this program is a primitive admittance matrix y
% which is to be given in the following format and stored in ydata
% Ground is given as bus no. 0
% If the element is not mutually coupled with any other element,
% then the entry corresponding to 4th and 5th column of ydata
% has to be zero
% The data must be arranged in ascending order of element no.
% ----------------------------------------------------------------
```

% element no	connected		y (self)	Mutually	y(mutual)	
%	From	To		coupled to		
%	Busno	Busno				

```
% -------------------------------   -------------------------------
ydata=[ 1      1     2    1/(0.05+j*0.15)    0         0
        2      1     3    1/(0.1 +j*0.3 )    0         0
        3      2     3    1/(0.15+j*0.45)    0         0
        4      2     4    1/(0.10+j*0.30)    0         0
        5      3     4    1/(0.05+j*0.15)    0         0];

% Following statement gives maximum no. of elements by calculating the
% maximum out of all elements (denoted by :) of first column of ydata
elements=max(ydata(:,1))
% this gives no. of buses which is nothing but the maximum entry
% out of 2nd and 3rd column of ydata which is 'from' and 'to' columns
buses=max(max(ydata(:,2)),max(ydata(:,3)));
```

```
YBUS=zeros(buses,buses);

for row=1:elements,
%  if ydata(row,5) is zero that means the corresponding element
% is not mutually coupled with any other element

    if ydata(row,5) == 0
        i1 =ydata(row,2);
        j1 =ydata(row,3);
        if i1 ~= 0 & j1 ~= 0
            YBUS(i1,i1) = YBUS(i1,i1) + ydata(row,4);
            YBUS(i1,j1) = YBUS(i1,j1) - ydata(row,4);
            YBUS(j1,i1) = YBUS(i1,j1);
            YBUS(j1,j1) = YBUS(j1,j1) + ydata(row,4);
        end
        if i1 == 0 & j1 ~=0
            YBUS(i1,i1) = YBUS(i1,i1) + ydata(row,4);
        end
        if i1 ~= 0 & j1 ==0
            YBUS(j1,j1) = YBUS(j1,j1) + ydata(row,4);
        end
    end
%  if ydata(row,5) is NOT zero that means the corresponding element
%  is   mutually coupled with element given in ydata(row,5)

    if ydata(row,5) ~= 0
        i1 =ydata(row,2);
        j1 =ydata(row,3);
        % mutualwith gives the element no with which the current element
        % is mutually coupled with k and l give the bus nos between
        % which the mutually coupled element is connected
        mutualwith=ydata(i1,5);
        k=ydata(mutualwith,2);
        l=ydata(mutualwith,3);

        zs1=1/ydata(row,4);
        zs2=1/ydata(mutualwith,4);
        zm =1/ydata(row,6);
        zsm=[zs1 zm
             zm  zs2];
        ysm=inv(zsm);
        % Following if block gives the necessary modifications in YBUS
        % when none of the buses is reference (ground) bus.
        if i1 ~= 0 & j1 ~= 0 & k ~=0 & l~=0
            YBUS(i1,i1) = YBUS(i1,i1) + ysm(1,1);
            YBUS(j1,j1) = YBUS(j1,j1) + ysm(1,1);
            YBUS(k,k) = YBUS(k,k) + ysm(2,2);
            YBUS(l,l) = YBUS(l,l) + ysm(2,2);
            YBUS(i1,j1) = YBUS(i1,j1) - ysm(1,1);
            YBUS(j1,i1) = YBUS(i1,j1);
```

```
            YBUS(k,1) = YBUS(k,1) - ysm(2,2);
            YBUS(1,k) = YBUS(k,1);
            YBUS(i1,k) = YBUS(i1,k) + ysm(1,2);
            YBUS(k,i1) = YBUS(i1,k);
            YBUS(j1,1) = YBUS(j1,1) + ysm(1,2);
            YBUS(1,j1) = YBUS(j1,1);
            YBUS(i1,1) = YBUS(i1,1) - ysm(1,2);
            YBUS(1,i1) = YBUS(i1,1);
            YBUS(j1,k) = YBUS(j1,k) - ysm(1,2);
            YBUS(k,j1) = YBUS(j1,k);
        end
        if i1 == 0 & j1 ~=0 & k~=0 & 1~=0
            YBUS(j1,j1) = YBUS(j1,j1) + ysm(1,1);
            YBUS(k,k) = YBUS(k,k) + ysm(2,2);
            YBUS(1,1) = YBUS(1,1) + ysm(2,2);
            YBUS(k,1) = YBUS(k,1) - ysm(2,2);
            YBUS(1,k) = YBUS(k,1);
            YBUS(j1,1) = YBUS(j1,1) + ysm(1,2);
            YBUS(1,j1) = YBUS(j1,1);
            YBUS(j1,k) = YBUS(j1,k) - ysm(1,2);
            YBUS(k,j1) = YBUS(j1,k);
        end
        if i1~=0 & j1==0 & k~=0 & 1~=0
            YBUS(i1,i1) = YBUS(i1,i1) + ysm(1,1);
            YBUS(k,k) = YBUS(k,k) + ysm(2,2);
            YBUS(1,1) = YBUS(1,1) + ysm(2,2);
            YBUS(k,1) = YBUS(k,1) - ysm(2,2);
            YBUS(1,k) = YBUS(k,1);
            YBUS(i1,k) = YBUS(i1,k) + ysm(1,2);
            YBUS(k,i1) = YBUS(i1,k);
            YBUS(i1,1) = YBUS(i1,1) - ysm(1,2);
            YBUS(1,i1) = YBUS(i1,1);
        end
        if i1~=0 & j1~=0 & k==0 & 1~=0
            YBUS(i1,i1) = YBUS(i1,i1) + ysm(1,1);
            YBUS(j1,j1) = YBUS(j1,j1) + ysm(1,1);
            YBUS(1,1) = YBUS(1,1) + ysm(2,2);
            YBUS(i1,j1) = YBUS(i1,j1) - ysm(1,1);
            YBUS(j1,i1) = YBUS(i1,j1);
            YBUS(j1,1) = YBUS(j1,1) + ysm(1,2);
            YBUS(1,j1) = YBUS(j1,1);
            YBUS(i1,1) = YBUS(i1,1) - ysm(1,2);
            YBUS(1,i1) = YBUS(i1,1);
        end

        if i1~=0 & j1==0 & k~=0 & 1==0
            YBUS(i1,i1) = YBUS(i1,i1) + ysm(1,1);
            YBUS(j1,j1) = YBUS(j1,j1) + ysm(1,1);
            YBUS(k,k) = YBUS(k,k) + ysm(2,2);
            YBUS(i1,j1) = YBUS(i1,j1) - ysm(1,1);
```

```
            YBUS(j1,i1) = YBUS(i1,j1);
            YBUS(i1,k) = YBUS(i1,k) + ysm(1,2);
            YBUS(k,i1) = YBUS(i1,k);
            YBUS(j1,k) = YBUS(j1,k) - ysm(1,2);
            YBUS(k,j1) = YBUS(j1,k);

        end
     end
end


    YBUS
```

Ex G.6

```
% This is program for gauss siedel Load flow. The data is from Example 6.5
clear
n=4
V=[1.04 1.04 1 1]
```

$$Y = \begin{bmatrix} 3 - j*9 & -2 + j*6 & -1 + j*3 & 0 \\ -2 + j*6 & 3666 - j*11 & -0.666 + j*2 & -1 + j*3 \\ -1 + j*3 & -0.666 + j*2 & 3.666 - j*11 & -2 + j*6 \\ 0 & -1 + j*3 & -2 + j*6 & 3 - j*9 \end{bmatrix}$$

```
type=ones(n,1)
typechanged=zeros(n,1)Qlimitmax=zeros(n,1)
Qlimitmin=zeros(n,1)
Vmagfixed=zeros(n,1)
type(2)=2
Qlimitmax(2)=1.0
Qlimitmin(2)=0.2
Vmagfixed(2)=1.04
diff=10;noofiter=1
Vprev=V;
while (diff>0.00001 | noofiter==1),
abs(V)
abs(Vprev)
%pause
Vprev=V;
P=[inf 0.5 -1  0.3];
Q=[inf 0  0.5 -0.1] ;
S=[inf+j*inf (0.5-j*0.2) (-1.0 + j*0.5) (0.3-j*0.1)];
for i=2:n,
        if type(i)==2 |typechanged(i)==1,
            if (Q(i)>Qlimitmax(i)| Q(i)<Qlimitmin(i)),
                if(Q(i)<Qlimitmin(i)),
                    Q(i)=Qlimitmin(i);
                else
                    Q(i)=Qlimitmax(i);
                end
                type(i)=1;
```

```
                        typechanged(i)=1;
                else
                        type(i)=2;
                        typechanged(i)=0;
                end

        end
    end
        sumyv=0;
        for k=1:n,
            if(i ~= k)
                    sumyv=sumyv+Y(i,k)*V(k);
            end
        end
        V(i)=(1/Y(i,i))*((P(i)-j*Q(i))/conj(V(i))-sumyv);
        if type(i)==2 & typechanged(i)~=1,
            V(i)=PolarTorect(Vmagfixed(i),angle(V(i))*180/pi);
        end
end
diff=max(abs(abs(V(2:n))-abs(Vprev(2:n))));
noofiter=noofiter+1;
end
V
```

Ex. G.7 (Example 6.6)

```
% Program for load flow by Newton-Raphson Method.
clear;
% n stands for number of buses
n=3;
% V ,voltages at those buses are initialised
V=[1.04 1.0 1.04];
% Y is YBus
Y=[  5.88228-j*23.50514 -2.9427+j*11.7676    -2.9427+j*11.7676
     -2.9427+j*11.7676   5.88228-j*23.50514  -2.9427+j*11.7676
     -2.9427+j*11.7676  -2.9427+j*11.7676     5.88228-j*23.50514];
%Bus types are initialised in type array to code 1 which stands for PQ
%bus.
%code 2 stands for PV bus
type=ones(n,1);
%When Q limits are exceeded for a PV bus Bus type is changed to PQ
%temporarily
%an element i of typechanged is set to 1 in case its bus status
%is temporarily changed from PQ to PV. Otherwise it is zero
typechanged=zeros(n,1);
% Since max and min Q limits are checked only for PV buses,
% max & min  Q limits for other types of buses can be set to any values.
% here we have set them to zeros for convenience
Qlimitmax=zeros(n,1);
Qlimitmin=zeros(n,1);
Vmagfixed=zeros(n,1);
```

```
% Here we change type of PV buses to 2 and also set the Q limits for them
type(3)=2;
Qlimitmax(3)=1.5;
Qlimitmin(2)=0;
Vmagfixed(2)=1.04;
diff=10;noofiter=1;
Pspec=[inf 0.5 -1.5];
Qspec=[inf 1    0 ];
S=[inf+j*inf (0.5-j*0.2) (-1.0 + j*0.5) (0.3-j*0.1)];
% Here for all the buses depending on bustype, associated variables array
%element for each equation (deltaP or deltaQ) is initialised
%Also associated variables with each col( ddelta or dv) is formed
while (diff>0.00001 | noofiter==1),
eqcount=1;
for i=2:n,
        Scal(i)=0;
        sumyv=0;
        for k=1:n,
            sumyv=sumyv+Y(i,k)*V(k);
        end
        Scal(i)=V(i)*conj(sumyv);
        P(i)=real(Scal(i));
        Q(i)=imag(Scal(i));
% If the bus is a PV bus and the calculated Q is exceeding the
% limits, the bus type is temporarily changed to PQ and
% type changed is made 1 for that bus
% Otherwise its switched back to PV bus and
% type changed is resetted to zero for that bus
        if type(i)==2 |typechanged(i)==1,
            if (Q(i)>Qlimitmax(i)| Q(i)<Qlimitmin(i)),
                if(Q(i)<Qlimitmin(i)),
                    Q(i)=Qlimitmin(i);
                else
                    Q(i)=Qlimitmax(i);
                end
                type(i)=1;
                typechanged(i)=1;
            else
                type(i)=2;
                typechanged(i)=0;
            end

        end

% The mismatch equations are arranged and solved in matrix form
% as indicated below
% |dPi|              |dDelta|
% |dQi|       = [J]  |dV|
% |dP(i + 1)|        |dDelta|
% |dQ(i + 1)|        |dV|
```

```
% assoeqvar (i) [associated equation variable] indicates if the equation deal
% with   the mismatch for `P' real
% power or `Q' reactive power
% This decides the numerator quantity of the jacobian element
% i.e. the quantity we want to differentiate will be dP/d* or dQ/d*
% where * may stand for either Delta or V
% mismatch array stores the mismatch quantities for P or Q
% assoeqbus (i) indicates which is the bus associated with equation i
% assocolvar (i) decides the denominator quantity of Jacobian element
% along each column of the Jacobian for a given equation
% i.e. for a jacobian element in the ith column of a given equation
% with respect to what we are differentiating
% So we decide here if the element is d*/dDelta or d*/dV
% where * may be either P or Q
% assocolbus decides the bus associated
% with each element in the column corresponding to given equation
        if type(i)==1,
        assoeqvar(eqcount)='P';
        assoeqbus(eqcount)=i;
        mismatch(eqcount)=Pspec(i)-P(i);
        assoeqvar(eqcount+1)='Q';
        assoeqbus(eqcount+1)=i;
        mismatch(eqcount+1)=Qspec(i)-Q(i);

        assocolvar(eqcount)='d';
        assocolbus(eqcount)=i;
        assocolvar(eqcount+1)='V';
        assocolbus(eqcount+1)=i;
        eqcount=eqcount+2;

    else
    assoeqvar(eqcount)='P';
    assoeqbus(eqcount)=i;
    assocolvar(eqcount)='d';
    assocolbus(eqcount)=i;
    mismatch(eqcount)=Pspec(i)-P(i);
    eqcount=eqcount+1;
    end
end
mismatch
eqcount=eqcount-1;
noofeq=eqcount;
Update=zeros(eqcount,1);
Vprev=V

abs(V);
abs(Vprev)
pause
Vprev=V;
% ceq stands for current equation being processed, which is varied from
```

```
% 1 to total no. of equations (eqcount)
for ceq=1:eqcount,
        for ccol=1:eqcount,

            am=real(Y(assoeqbus(ceq),assocolbus(ccol))*V(assocolbus(ccol)));
            bm=imag(Y(assoeqbus(ceq),assocolbus(ccol))*V(assocolbus(ccol)));
                ei=real(V(assoeqbus(ceq)));
                fi=imag(V(assoeqbus(ceq)));

          if assoeqvar(ceq)=='P' & assocolvar(ccol)=='d',

                if assoeqbus(ceq)~=assocolbus(ccol),
                    H=am*fi-bm*ei;
                else
              H=-Q(assoeqbus(ceq))imag(Y(assoeqbus(ceq),assocolbus(ceq)) ...
                        *abs(V(assoeqbus(ceq)))^2);
                end
                Jacob(ceq,ccol)=H
            end
            if assoeqvar(ceq)=='P' & assocolvar(ccol)=='V',
                if assoeqbus(ceq)~=assocolbus(ccol),
                    N=am*ei+bm*fi;
                else
                    N=P(assoeqbus(ceq))+real(Y(assoeqbus(ceq),assocolbus(ceq))...
                                            *abs(V(assoeqbus(ceq)))^2);
                end
                 Jacob(ceq,ccol)=N
            end
          if assoeqvar(ceq)=='Q' & assocolvar(ccol)=='d',
                if assoeqbus(ceq)~=assocolbus(ccol),
                     J=am*ei+bm*fi;
                else
                     J=P(assoeqbus(ceq))real(Y(assoeqbus(ceq),assocolbus(ceq))
                     *abs(V(assoeqbus(ceq)))^2);
                end
                Jacob(ceq,ccol)=J
            end
            if assoeqvar(ceq)=='Q' & assocolvar(ccol)=='V',
                if assoeqbus(ceq)~=assocolbus(ccol),
                    L=am*fi-bm*ei;
                else
                    L=Q(assoeqbus(ceq))-...
imag(Y(assoeqbus(ceq),assocolbus(ceq))*abs(V(assoeqbus(ceq)))^2);
                end
                Jacob(ceq,ccol)=L
            end
        end
    end

    %New Update vector is calculated from Inverse of the Jacobian
```

```
      Jacob
      pause
      update=inv(Jacob)*mismatch';
      noofeq=1;
      for i=2:n,
          if type(i)==1
              newchinangV=update(noofeq);
              newangV=angle(V(i))+newchinangV;
              newchinmagV=update(noofeq+1)*abs(V(i));
              newmagV=abs(V(i))+newchinmagV;
              V(i)=polarTorect(newmagV,newangV*180/pi);
              noofeq=noofeq+2;
          else
              newchinangV=update(noofeq);
              newangV=angle(V(i))+newchinangV;
              V(i)=polarTorect(abs(V(i)),newangV*180/pi);
              noofeq=noofeq+1;
          end
      end
% All the following variables/arrays are cleared from
% memory. This is because their dimensions may change due to
% bus switched and once updates are calculated, the variables
% are of no use as they are being reformulated at the
% end of each iteration
clear mismatch Jacob update assoeqvar assoeqbus assocolvar assocolbus;
diff=min(abs(abs(V(2:n))-abs(Vprev(2:n))));
noofiter=noofiter+1;
end
```

Ex. G.8 (Table 7.1)

```
% MATLAB Program for optimum loading of generators
% The data are from Example 7.1
% It finds lamda by the algorithm given on the same page, once the demand
% is specified
% We have taken the demand as 231.25MW corrosponding to the last but one
% row of Table 7.1 and calculated lamda and the load sharing
%  n is no of generators
n=2
% Pd stands for load demand.
% alpha and beta arrays denote alpha beta coefficients
% for given generators.

Pd=231.25
alpha=[0.20
       0.25]
beta=[40
      30]
% initial guess for lamda
lamda=20
```

```
lamdaprev=lamda
% tolerance is eps and increment in lamda is deltalamda
eps=1
deltalamda=0.25
% the minimum and maximum limits of each generating unit
%are stored in arrays Pgmin and Pgmax.
% In real life large scale problems, we can first initialse the Pgmax
% array to inf using for loop and
% Pgmin array to zero using Pgmin=zeros(n,1) command
% Later we can change the limits individually
Pgmax=[125 125]
Pgmin=[20 20]
Pg=100*ones(n,1)
while abs(sum(Pg)-Pd)>eps
for i=1:n,
    Pg(i)=(lamda-beta(i))/alpha(i);
    if Pg(i)>Pgmax(i)
        Pg(i)=Pgmax(i);
    end
    if Pg(i)<Pgmin(i)
        Pg(i)=Pgmin(i);
    end

end
if (sum(Pg)-Pd )<0
    lamdaprev=lamda;
    lamda=lamda+deltalamda;
else
    lamdaprev=lamda;
    lamda=lamda-deltalamda;
end
end
disp(' The final value of Lamda is')
lamdaprev
disp(' The distribution of load shared by two units is')
Pg
```

Ex. G.9 (Table 7.2)

```
% MATLAB Program for optimum unit committment by Brute Force method
% The data for this program corresponds to Table 7.2

clear;
% alpha and beta arrays denote alpha beta coefficients for given generators
alpha=[0.77 1.60 2.00 2.50]';
beta=[23.5 26.5 30.0 32.0]';
Pgmin=[1 1 1 1]';
Pgmax=[12 12 12 12]';
  n=9
% n denotes total MW to be committed
```

```
min = inf;
cost=0;
for i=0:n,
    for j=0:n,
        for k=0:n,
            for l=0:n,
                unit = [0 0 0 0];
                % Here we elimnate straightaway those combinations which
                % dont make up the
                % n MW demand or such combinations where maximum generation
                % on individual
                % generation is exceeding the maximum capacity of any of the
                % generators
                if(i+j+k+l)==n & i<Pgmax(1) & j<Pgmax(2) & k<Pgmax(3) &l < ...
                Pgmax(4)
                    if i~=0
                        unit(1,1)=i;
                        % Find out the cost of generating these units and
                        % add it up to total cost
                        cost=cost+0.5*alpha(1)*i*i+beta(1)*i;
                    end
                    if j~=0
                        unit(1,2)=j;
                        cost=cost+0.5*alpha(2)*j*j+beta(2)*j    ;
                     end
                    if k~=0
                        unit(1,3)=k;
                        cost=cost+0.5*alpha(3)*k*k+beta(3)*k ;
                    end
                    if l~=0
                        unit(1,4)=l;
                        cost=cost+0.5*alpha(4)*l*l+beta(4)*l;
                    end
                    % If the total cost is coming out to be less than
                    % minimum of the cost in
                    % previous combinations then make min equal to cost and
                    % cunit (stand for committed units) equal to units
                    % committed in this iteration
                    % (denoted by variable units)
                    if cost < min

                        cunit = unit;
                        min=cost;
                    else
                    cost=0;

                    end
                end
            end
        end
end
```

```
      end
end
disp ('cunit display the no of committed units on each of the four genera-
tors')
disp(' If cunit for a particular generator is 0 it means the unit is not
committed')
disp (' The total no of units to be committed are')
cunit
```

Ex G.10 (Ex. 7.4)

```
clear
% MATLAB Program for optimum loading of generators
% This program finds the optimal loading of generators including
% penalty factors
% It implements the algorithm given just before Example 7.4.
% The data for this program are taken from Example 7.4
% Here we give demand Pd and alpha, beta and B-coefficients
% We calculate load shared by each generator
%   n is no of generators
n=2
% Pd stands for load demand
% alpha and beta arrays denote alpha beta coefficients for given
% generators

Pd=237.04;
alpha=[0.020
      0.04];
beta=[16
      20];
% initial guess for lamda
lamda=20;
lamdaprev=lamda;
% tolerance is eps and increment in lamda is deltalamda
eps=1;
deltalamda=0.25;
% the minimum and maximum limits of each generating unit
%are stored in arrays Pgmin and Pgmax.
% In actual large scale problems, we can first initialise the Pgmax array
% to inf using for loop
% and Pgmin array to zero using Pgmin=zeros(n,1) command
% Later we should can change the limits individually
Pgmax=[200 200];
Pgmin=[0 0];
B=[0.001 0
    0    0];
Pg=zeros(n,1);
noofiter=0;
PL=0;
Pg=zeros(n,1);
```

```
while abs(sum(Pg)-Pd-PL)>eps
for i=1:n,
    sigma=B(i,:)*Pg-B(i,i)*Pg(i);
    Pg(i)=(1-(beta(i)/lamda)-(2*sigma))/(alpha(i)/lamda+2*B(i,i));
    PL=Pg'*B*Pg;
    if Pg(i)>Pgmax(i)
        Pg(i)=Pgmax(i);
    end
    if Pg(i)<Pgmin(i)
        Pg(i)=Pgmin(i);
    end

end
    PL=Pg'*B*Pg;


if (sum(Pg)-Pd-PL )<0
    lamdaprev=lamda;
    lamda=lamda+deltalamda;
else
    lamdaprev=lamda;
    lamda=lamda-deltalamda;
end
noofiter=noofiter+1;
Pg;
end
disp('The no of iterations required are')
noofiter
disp('The final value of lamda is')
lamdaprev
disp('The optimal loading of generators including penalty factors is')
Pg
disp('The losses are')
PL
```

Ex. G.11

```
% In this example the parameters for all the blocks for the system in
% Fig. 8.8 are initialized . This program has to be run prior to the
simulation both for Figs. G.3 and G.4.
Tsg=.4
Tt=0.5
Tps=20
Kps=100
R=3
Ksg=10
Kt=0.1
Ki=0.09
```

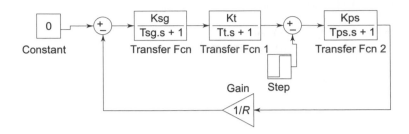

Fig. G.3 First order approximation for load frequency control

Ex. G.12

The system of Fig. 8.10 is simulated using Simulink as was done in Example G.11.

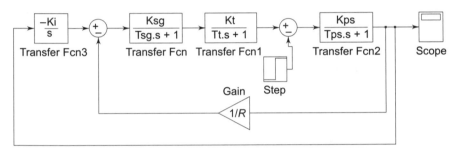

Fig. G.4 Proportional plus Integral load frequency control

Ex. G.13 (Example 9.8)

```
% Program for building of Zbus by addition of branch or link
% Zprimary=[elementno from   to    value
%             --         --    --     --
%             --         --    --     --
%             --         --    --     --
%             --         --    --    --]
% Here care should be taken that to begin with an element is added to
% reference and both from and to nodes should not be new nodes
clear
zprimary=[
    1 1 0 0.25
    2 2 1 0.1
    3 3 1 0.1
    4 2 0 0.25
    5 2 3 0.1]

[elements columns]=size(zprimary)
% To begin with zbus matrix is a null matrix
```

```
zbus=[]
% currentbusno indicates maximum no. of buses added until now
currentbusno=0


% Process each row of zprimary
for count=1:elements,
  [rows cols]=size(zbus)
  from=zprimary(count,2)
    to=zprimary(count,3)
    value=zprimary(count,4)
  % newbus variable indicates the maximum of the two buses
  % newbus bus may or may not already be a part of existing zbus

    newbus=max(from,to)
  % ref variable indicates the minimum of the two buses
  % & not necessarily the reference bus
  % ref bus must always exist in the existing zbus
    ref=min(from,to)
%  Modification of type1
% A new element is added from new bus to reference bus
if newbus >currentbusno & ref ==0
    zbus=[zbus zeros(rows,1)
        zeros(1,cols) value]
    currentbusno=newbus
    continue
end
% Modification of type2
% A new element  is added from new bus to old bus other than reference
bus
if newbus >currentbusno & ref ~=0
    zbus=[zbus zbus(:,ref)
        zbus(ref,:) value+zbus(ref,ref)]
      currentbusno=newbus
          continue
end
% Modificaiton of type3
% A new element is added between an old bus and reference bus
if newbus <=currentbusno & ref==0
    zbus=zbus-1/(zbus(newbus,newbus)+value)*zbus(:,newbus)*zbus(newbus,:)
        continue
end
% Modification of type4
% A new element is added betwen two old buses
if newbus <= currentbusno & ref ~=0
    zbus=zbus- 1/(value+zbus(from,from)+zbus(to,to)-
2*zbus(from,to))*((zbus(:,from)-zbus(:,to))*((zbus(from,:)-zbus(to,:))))
        continue
end
end
```

Ex. G.14 (Example 12.10)

```
% Program for transient stability of single machine connected to infinite
% bus this program simulates Example 12.10 using point by point method
clear
t=0
tf=0
tfinal=0.5
tc=0.125
tstep=0.05
M=2.52/(180*50)
i=2
delta=21.64*pi/180
ddelta=0
time(1)=0
ang(1)=21.64
Pm=0.9
Pmaxbf=2.44
Pmaxdf=0.88
Pmaxaf=2.00

while t<tfinal,
    if (t==tf),
        Paminus=0.9-Pmaxbf*sin(delta)
        Paplus=0.9-Pmaxdf*sin(delta)
        Paav=(Paminus+Paplus)/2
        Pa=Paav
    end
    if (t==tc),
        Paminus=0.9-Pmaxdf*sin(delta)
        Paplus=0.9-Pmaxaf*sin(delta)
        Paav=(Paminus+Paplus)/2
        Pa=Paav
    end

    if(t>tf & t<tc),
        Pa=Pm-Pmaxdf*sin(delta)
    end
    if(t>tc),
        Pa=Pm-Pmaxaf*sin(delta)
    end
        t,Pa
        ddelta=ddelta+(tstep*tstep*Pa/M)
        delta=(delta*180/pi+ddelta)*pi/180
        deltadeg=delta*180/pi
        t=t+tstep
        pause
        time(i)=t
        ang(i)=deltadeg
```

```
    i=i+1
end
axis([0 0.6 0 160])
plot(time,ang,'ko-')
```

Ex. G.15

Here the earlier Example G.14 is solved again using SIMULINK.
Before running simulation shown in Fig. G.5 integrater 1 has to be initialized
to prefault value of δ, i.e. δ_0. This can be done by double-clicking on integrater
1 block and changing the initial value from 0 to δ_0 (in radius). Also double click
the switch block and change the threshold value from 0 to the fault clearing time
(in sec.).

Ex. G.16 (Ex. 12.11)

```
% This program simulates transient stability of multimachine systems
% The data is from Example 12.11
clear all
format long
%Step 1 Initialisation with load flow and machine data
f=50;tstep=0.01; H=[12 9]';
Pgnetterm=[ 3.25 2.10]';
Qgnetterm=[ 0.6986 0.3110]';
Xg=[0.067 0.10]';
% Note the use of .' operator here
% This does a transpose without taking the conjugate of each element

V0=[polarTorect(1.03, 8.235) polarTorect(1.02, 7.16)].'
% m is no of generators other than slack bus
m=2;
%Step 2
V0conj=conj(V0);
Ig0=conj((Pgnetterm+j*Qgnetterm)./V0);
Edash0=V0+j*(Xg.*Ig0);
Pg0=real(Edash0.*conj(Ig0));
x1_r=angle(Edash0);
% Initialisation of state vector
Pg_r=Pg0;
% Pg_rplus1=Pg0;
x2_r=[0 0]';
x1dot_r=[0 0]';
x2dot_r=[0 0]';

x1dotrplus1=[0 0]';
x2dotrplus1=[0 0]';

%Step 3
```

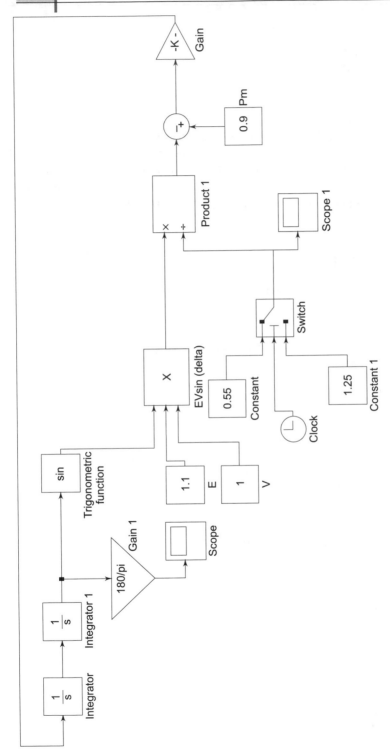

Fig. G.5 Transient stability for single machine connected to infinite Bus

```
% Here in this example we have not really calculated YBus But one
% can write separate program.
YBusdf=[
5.7986-j*35.6301              0                 -0.0681+j*5.1661
        0                -j*11.236              0
 -0.0681+j*5.1661            0                  0.1362-j*6.2737];

YBuspf=[
1.3932-j*13.8731       -0.2214+j*7.6289        -0.0901+j*6.0975
 -0.2214+j*7.6289       0.5+j*7.7898                   0
 -0.0901+j*6.0975            0                  0.1591-j*6.1168];
%Step 4
% Set the values for initial time t(occurance of fault) and
% tc is time at which fault is cleared
t=0;tc=0.08;tfinal=1.0;
r=1;
Edash_r=Edash0;Edash_rplus1=Edash0;
while t < tfinal,
%Step 5 Compute Generator Powers using appropriate YBus
% the YBus chosen in the following step is set according to the current
% time
if t <= tc YBus=YBusdf; else YBus=YBuspf;end
% Note that here we obtain the currents injected at generator bus
% by multiplying the corresponding row of the Ybus with the vector of
% voltages behind the transient reactances. This should also include
% slack bus voltage
% and hence the entry 1 appears in the bus voltage vector in addition to
% generator bus voltages
 I=YBus(2:m+1,:)*[1
                Edash_r];
 Pg_r=real(Edash_r.*conj(I));


%Step 6 compute x1dot_r and x2dot_r
    x1dot_r=x2_r;
    for k=1:m,
        x2dot_r(k,1)=(pi*f/H(k))*(Pg0(k)-Pg_r(k));
    end
%Step 7 Compute first state estimates for t=t(r+1)
    x1_rplus1=x1_r+x1dot_r*tstep;
    x2_rplus1=x2_r+x2dot_r*tstep;
%Step 8 Compute first estimates of E'_r+1
Edash_rplus1=abs(Edash0).*(cos(x1_rplus1)+j*sin(x1_rplus1));
%Step 9 Compute Pg for t=t(r+1)
 I=YBus(2:m+1,:)*[1
                Edash_rplus1];
 Pg_rplus1=real(Edash_rplus1.*conj(I));

%Step 10 Compute State derivatives at t=t(r+1)
x1dot_rplus1=[0 0]';
```

```
x2dot_rplus1=[0 0]';
for k=1:m,
x1dot_rplus1(k,1)=x2_rplus1(k,1);
x2dot_rplus1(k,1)=pi*f/H(k)*(Pg0(k)-Pg_rplus1(k));
end
%step 11 Compute average values of state derivatives
x1dotav_r=(x1dot_r+x1dot_rplus1)/2.0;
x2dotav_r=(x2dot_r+x2dot_rplus1)/2.0;
%step 12 Compute final State estimates for t=t(r+1)
x1_rplus1=x1_r+x1dotav_r*tstep;
x2_rplus1=x2_r+x2dotav_r*tstep;
%step 13 Compute final estimate for Edash at t=t(r+1)
Edash_rplus1=abs(Edash0).*(cos(x1_rplus1)+j*sin(x1_rplus1));
%Step 14 Print State Vector
    x2_r=x2_rplus1;
    x1_r=x1_rplus1;
    Edash_r=Edash_rplus1;
%Step 15
time(r)=t;
for k=1:m,
    ang(r,k)=(x1_r(k)*180)/pi;
end
t=t+tstep;
r=r+1;
end
plot(time,ang)
```

Ex. G. 17 (Example 12.11)

```
%Example 12.11 is solved using SIMULINK
% The code given below should be run prior to simulation shown in Fig. G.6.
clear all
global n r y yr
global Pm f H E  Y ngg
global rtd dtr %conversion factor rad/degree
global Ybf Ydf Yaf

f=50;
ngg=2;
r=5;
nbus=r;

rtd=180/pi;
dtr=pi/180;

%          Gen.  Ra   Xd'       H
gendata=[  1    0    0         inf
           2    0    0.067    12.00
           3    0    0.100     9.00 ];

Ydf=[
```

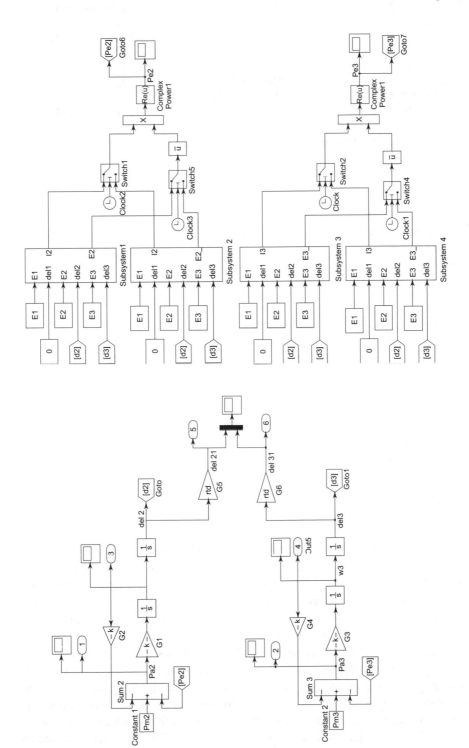

Fig. G.6 Multimachine Transient Stability

```
5.7986-j*35.6301                    0                   -0.0681+j*5.1661
        0                       -j*11.236                      0
-0.0681+j*5.1661                    0                    0.1362-j*6.2737];

Yaf=[
1.3932-j*13.8731            -0.2214+j*7.6289            -0.0901+j*6.0975
-0.2214+j*7.6289             0.5+j*7.7898                      0
-0.0901+j*6.0975                    0                    0.1591-j*6.1168];

fct= input('fault clearing time fct = ');

%Damping factors
damp2=0.0;
damp3=0.0;

%Initial generator Angles
d2=0.3377*rtd;
d3=0.31845*rtd;

%Initial Powers
Pm2=3.25;
Pm3=2.10;

%Generator Internal Voltages
E1=1.0;
E2=1.03;
E3=1.02;

%Machine Inertia Constants;
H2=gendata(2,4);
H3=gendata(3,4);

%Machine Xd';
Xdd2=gendata(2,3);
Xdd3=gendata(3,3);
```

Note : For the simulation for multimachine stability the two summation boxes sum 2 & sum 3 give the net acceleating powers P_{a2} and P_{a3}. The gains of the gain blocks G1, G2 G3 and G4 are set equal to pi*f/H2, damp2, pi*f/H3 and damp3, respectively. The accelerating power P_a is then integrated twice for each mahcine to give the rotor angles δ_2 and δ_3. The initial conditions for integrator blocks integrater 1, integrator 2, integrator 3 and integrator 4 are set to 0, d2/r+d, 0 and d3/rtd, respectively. The gain blocks G5 and G6 convert the angles δ_2 and δ_3 into degrees and hence their gains are set to rtd. The electrical power P_{e2} is calculated by using two subsystems 1 and 2. The detailed diagram for subsystem 1 is shown in Fig. G.7. Subsystem 1 gives two outputs (i) complex voltage $E_2\angle\delta_2$ and (ii) current of generator I_2 which is equal to $E_1\angle\delta1 *Y_{af}(2,1) + E_2\angle\delta_2*Y_{af}(2,2) + E_3\angle\delta_3 Y_{af}(2,3)$. The switches are used to

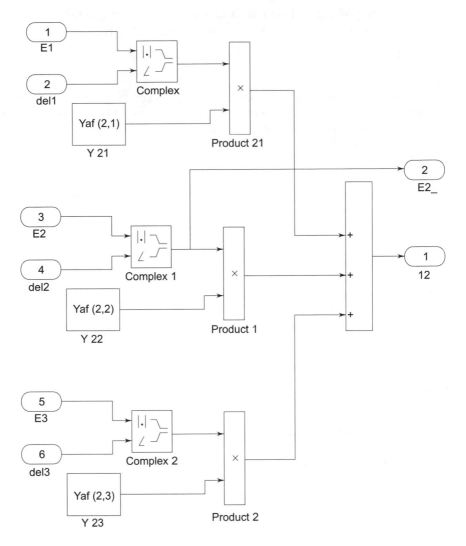

Fig. G.7 Multimachine transient Stability : Subsystem 1

switch between prefault and post fault conditions for each machine and their threshold values are adjusted to fct, i.e. the fault clearing time.

Ex. G.18

```
clear
% This Program finds the reduced matrix for stability studies
% which elimnates the load buses and retains only the generator buses
% |Ig1|      |Y11   Y12 . . . Y1n   Y1n+1   Y1n+2. . . Y1n+m|   |V1   |
% |Ig2|      |Y21   Y21 . . . Y2n   Y2n+1   Y2n+2. . . Y1n+m|   |V2   |
% | . |      |                                            |   | .   |
% | . |      |                                            |   | .   |
% |Ign|   =  |Yn1   Yn2 . . . Ynn   Ynn + 1 Ynn + 2  Ynn + m|  |Vn   |
% |IL1|      |Yn+12 .  . . . .                       Yn+1m|    |Vn+1 |
% |IL2|      |                                       |       |Vn+1 |
% | . |      |                                            |   | .   |
% | . |      |                                            |   | .   |
% |ILm|      |Yn+m1  .  . . .                   Yn+mn+m|       |Vn+m|

Yfull= -j*5      0          7*5
          0     -j*7.5     j*2.5
         j*5     j*2.5     -j*12.5]
[row columns]=size (Yfull)
n=2
YAA=Yfull (1:n,1:n)
YAB=Yfull (1:n, n+1;columns)
YBA=Yfull (n+1:rows,n+1:columns)
YBB=Yfull (n+1:rows,n+1:colums)
% This gives the reduced matrix

Yreduced=YAA-YAB*inv(YBB)*YBA
```

Answers to Problems

CHAPTER 2

2.1 $L_{\text{int}} = \dfrac{1}{2} \times 10^{-7} \times \dfrac{1}{(r_2^2 - r_1^2)^2}\left[(r_2^4 - r_1^4) - 4r_1^2(r_2^2 - r_1^2) + 4r_1^4 \ln\dfrac{r_2}{r_1}\right]$

2.2 0.658 ohm/km

2.3 $L = \dfrac{\mu}{2\pi}\ln\dfrac{R}{r}$ H/m

2.4 260.3 V/km

2.5 $H_p = -\dfrac{I}{3\pi d}$ AT/m^2 (directed upwards)

2.6 $X = \dfrac{(X_1 - X_{12})(X_2 - X_{12})}{X_1 + X_2 - 2X_{12}}$

2.7 0.00067 mH/km, 0.0314 V/km

2.8 0.0044 $\angle 140°$ mH/km, 0.553 $\angle 140°$ V/km

2.9 0.346 ohm/km   2.10 1.48 m

2.11 0.191 ohm/km/phase 2.12 0.455 mH/km/phase

2.13 2.38 m

2.14 (i) 0.557 $d^{1/2}\ A^{1/4}$ (ii) 0.633 $d^{2/3}\ A^{1/6}$ (iii) 0.746 $d^{3/4}\ A^{1/8}$

Chapter 3

3.1 $q_a = \dfrac{2\pi k\,|V|(\ln(r/2D)\angle 30° - \ln(D/2r)\angle -30°)}{2\ln(D/r)\ln(r/2D) - \ln(2D/r)\ln(D/2r)}$ F/m,

$I_a = 2\pi f q_a \angle 90°$ A

3.2 0.0204 μF/km   3.3 0.0096 μF/km

3.4 0.0103 μF/km to neutral   3.5 3.08 × 10^{-5} coulomb/km

3.6 5.53 × 10^{-6} mhos/km   3.7 8.54 × 10^3 ohms/km

3.8 8.72 × 10^{-3} μF/km   3.9 71.24 kV

Chapter 4

4.1 12 kV

Chapter 5

5.1 (a) 992.75 kW   (b) No solution possible

5.2 $A' = 0.9\angle1.5°$, $B' = 239.9\angle66.3°$, $C' = 0.001\angle102.6°$, $D' = 0.85\angle1.96°$

5.3 (a) $0.978\angle0.5°$, $86.4\angle68.6°$, $0.00056\angle90.2°$, $0.978\angle0.5°$

    (b) 165.44 kV, $0.244\angle-28.3°$ kA, 0.808 lagging, 56.49 MW

    (c) 70.8%   (d) 28.15%

5.4 (a) 273.5 MVA   (b) 1,174 A   (c) 467.7 MVA

5.5 133.92 kV, 23.12 MW   5.6 202.2 kV

5.7 At $x = 0$, $i_{x1} = 0.314 \cos(\omega t - 21.7°)$, $i_{x2} = 0.117 \cos(\omega t + 109°)$,

    At $x = 200$ km, $i_{x1} = 0.327 \cos(\omega t - 9.3°)$, $i_{x2} = 0.112 \cos(\omega t + 96.6°)$

5.8 $135.8\angle7.8°$ kV, $0.138\angle15.6°$ kA, 0.99 leading, 55.66 MW, 89.8%, $373.1\angle-1.5°$, 3,338 km, 1,66,900 km/sec

5.9 $Z' = 128.3\angle72.6°$, $\dfrac{Y'}{2} = 0.00051\angle89.5°$.

5.10 7.12°, $pf_1 = 0.7$ lagging, $pf_2 = 0.74$ lagging

5.11 47.56 MVAR lagging

5.12 10.97 kV, 0.98 leading, -0.27%, 86.2%

5.13 51.16 kV, 38.87 MVAR leading, 40 MW

5.14 238.5 kV, $P_s + jQ_s = 53 - j10$, $pf = 0.983$ leading

5.15 17.39 MVAR leading, 3.54 MW

Chapter 6

6.1 For this network tree is shown in Fig. 6.3a; A is given by Eq. (6.17). The matrix is not unique. It depends upon the orientation of the elements.

6.2 $V_2^1 = 0.972\angle-8.15°$ 6.3 $V_2^1 = 1.26\angle-74.66°$

6.4 (a)

$$A =$$

bus \ e	1	2	3	4
1	1	0	0	0
2	0	1	0	0
3	0	0	1	0
4	0	0	0	1
5	0	0	1	-1
6	-1	0	0	1
7	1	-1	0	0
8	0	-1	0	1
9	-1	0	1	0

Note: Elements joining each bus to the ground node form the tree.

(b)

$$Y = \begin{bmatrix} j0.3049 \\ & j0.1694 \\ & & j0.1948 \\ & & & j0.3134 \\ & & & & 0.807 - j5.65 \\ & & & & & 0.645 - j4.517 \\ & & & & & & 0.968 - j6.776 \\ & & & & & & & 0.968 - j6.776 \\ & & & & & & & & 0.880 - j6.160 \end{bmatrix}$$

(c)

$$Y_{BUS} = \begin{matrix} & 1 & 2 & 3 & 4 \\ 1 & 2.493 - j17.148 & -0.968 + j6.776 & -0.880 + j6.160 & -0.645 + j4.517 \\ 2 & -0.968 + j6.776 & 1.936 - j13.383 & 0 & -0.968 + j6.776 \\ 3 & -0.880 + j6.160 & 0 & 1.687 - j11.615 & -0.807 + j5.650 \\ 4 & -0.645 + j4.517 & -0.968 + j6.776 & -0.807 + j5.650 & -2.420 - j16.630 \end{matrix}$$

6.5 $P_{12} = -0.598$ pu, $P_{13} = 0.2$ pu, $P_{23} = 0.796$ pu

$Q_{12} = Q_{21} = 0.036$ pu, $Q_{13} = Q_{31} = 0.004$ pu, $Q_{23} = Q_{32} = 0.064$ pu

6.6 (a) $P_{12} = -0.58$ pu, $P_{13} = 0.214$ pu, $P_{23} = 0.792$ pu

$Q_{12} = -0.165$ pu, $Q_{21} = 0.243$ pu, $Q_{13} = 0.204$ pu

$Q_{31} = -0.188$ pu, $Q_{23} = 0.479$ pu, $Q_{32} = -0.321$ pu

(b) $P_{12} = -0.333$ pu, $P_{23} = 0.664$ pu, $P_{13} = -0.333$ pu

$Q_{12} = Q_{21} = 0.011$ pu, $Q_{13} = Q_{31} = 0.011$ pu, $Q_{23} = Q_{32} = 0.044$ pu

6.7 (a) (i) $\begin{bmatrix} -j10.1015 & j5.0505 & j5 \\ j5.0505 & -j10 & j5 \\ j5 & j5 & -j10 \end{bmatrix}$ (ii) $\begin{bmatrix} -j10 & 5\angle 93° & j5 \\ 5\angle 87° & -j10 & j5 \\ j5 & j5 & -j10 \end{bmatrix}$

(b) (i) $P_{12} = 0.600$ pu, $P_{13} = 0.202$ pu, $P_{23} = 0.794$ pu

$Q_{12} = 0.087$ pu, $Q_{21} = -0.014$ pu

$Q_{13} = Q_{31} = 0.004$ pu, $Q_{23} = Q_{32} = 0.064$ pu

(ii) $P_{12} = -0.685$ pu,

$P_{13} = 0.287$ pu, $P_{23} = 0.711$ pu

$Q_{12} = 0.047$ pu, $Q_{13} = 0.008$ pu, $Q_{23} = 0.051$ pu

6.8 $V_3^1 = 1.025 - j0.095 = 1.029\angle -5.3°$ pu

Chapter 7

7.1 Rs 22.5/hr

7.2 (a) P_{G1} = 140.9 MW, P_{G2} = 159.1 MW

(b) Net saving = Rs 218.16/day

7.3 (i) Gen A will share more load than Gen B

(ii) Gen A and Gen B will share load of P_G each

(iii) Gen B will share more load than Gen A.

7.4 P_{G1} = 148 MW, P_{G2} = 142.9 MW, P_{G3} = 109.1 MW

7.5 (dC/dP_G) = $0.175P_G$ + 23

7.6 (a) P_{G1} = 138.89 MW, P_{G2} = 150 MW, P_D = 269.6 MW

(b) P_{G1} = 310.8 MW, P_{G2} = 55.4 MW

(c) For part (a): C_T = Rs 6,465.14/hr

For part (b): C_T = Rs 7,708.15/hr

7.7 B_{11} = 0.03387 pu or 0.03387 × 10^{-2} MW^{-1}

B_{12} = 9.6073 × 10^{-5} pu or 9.6073 × 10^{-7} MW^{-1}

B_{22} = 0.02370 pu or 0.02370 × 10^{-2} MW^{-1}.

7.8 Economically optimum UC

Time	Load (MW)	Unit Number
		1 2 3 4
0–4	20	1 1 1 1
4–8	14	1 1 1 0
8–12	6	1 1 0 0
12–16	14	1 1 1 0
16–20	4	1 0 0 0
20–24	10	1 1 0 0

Optimal and secure UC

Period	Unit Number
	1 2 3 4
0–4	1 1 1 1
4–8	1 1 1 0
8–12	1 1 0 0
12–16	1 1 1 0
16–20	1 1 0 0
20–24	1 1 0 0

7.9 Total operating cost (both units in service for 24 hrs) = Rs 1,47,960

Total operating cost (unit 1 put off in light load period) = Rs 1,45,840

Chapter 8

8.1 Load on G_1 = 123 MW, Load on G_2 = 277 MW, 50.77 Hz,

$$f_{10} = 51\frac{1}{3} \text{ Hz}, f_{20} = 51\frac{2}{3} \text{ Hz}$$

8.2 $\Delta f(t) = -0.029 - 0.04e^{-0.58t} \cos(1.254t + 137.8°)$

8.3 $1/(50K_i)$ sec

8.4 $\Delta P_{\text{tie, 1}} =$

$$\frac{(1/K_{ps1} + K_{i1}b_1 + 1/R_1) - (1/K_{ps2} + K_{i2}b_2 + 1/R_2)}{a_{12}(K_{i2}+1)(1/K_{ps1} + K_{i1}b_1 + 1/R_1) + (K_{i1}+1)(1/K_{ps2} + K_{i2}b_2 + 1/R_2)}$$

8.5 $\Delta P_{\text{tie, 1}}(s) = - \dfrac{100(0.2s^2 + 0.9s + 1)}{80s^5 + 364s^4 + 458s^3 + 866s^2 + 1050s + 85}$

System is found to be unstable.

Chapter 9

9.1 $i_t = 3.14 \sin(314t - 66°) + 2.87e^{-50t}$, $i_{\text{mm}} = 5A$

9.2 (a) 81° (b) – 9°

9.3 (i) I_A = 2.386 kA, I_B = 1.75 kA (ii) I_A = 4.373 kA, I_B = 1.75 kA

9.4 8.87 kA, 4.93 kA   9.5 26.96 kA

9.6 6.97 kA

9.7 (a) 0.9277 kA (b) 1.312 kA (c) 1.4843 kA (d) 1.0205 kV, 53.03 MVA
(e) 0.1959 kA

9.8 8.319 kA   9.9 2.39 pu

9.10 132.1, 47.9; 136.9, 45.6   9.11 0.6 pu

9.12 $I^f = -j8.006$ pu, $I^f_{13} = -j4.004$ pu

Chapter 10

10.1 (i) $1.732\angle 210°$ (ii) $2\angle 0°$ (iii) $1.732\angle 150°$ (iv) $1\angle 210°$

10.2 $I_A = j1.16$ pu, $V_{AB} = 1.17\angle 109.5°$ pu
$V_{BC} = 0.953\angle - 65.4°$ pu, $V_{CB} = 0.995\angle -113.1°$ pu.

10.3 $V_{a1} = 197.8\angle - 3.3°$ V, $V_{a2} = 20.2\angle 158.1°$ V, $V_{a0} = 21.61\angle 10.63°$ V

10.4 $I_{a1} = 19.23\angle - 30°$ A, $I_{a2} = 19.23\angle 150°$ A, $I_{a0} = 0$ A

10.5 $I_{A1} = 27.87\angle -30°$ A, $I_{A2} = 13\angle - 44.93°$ A, $I_{A0} = 0$ A
$I_{ab1} = 16.1$ A, $I_{ab2} = 7.5\angle - 75°$ A, $I_{ab0} = 7.5\angle 75°$ A

10.6 $I_a = 16.16 + j1.335$ A, $I_b = -9.24 - j10.66$ A
$I_c = -6.93 + j9.32$ A, $|V_{Nn}| = |V_{a0}| = 40.75$ V

10.7 1,500.2 W

Chapter 11

11.1 $- j6.56$ kA, $|V_{bc}| = 12.83$ kV, $|V_{ab}| = 6.61$ kV, $|V_{ca}| = 6.61$ kV

11.2 (a) $V_{ab} = V_{ac} = 1.8$ pu, $I_b = I_c = -2\sqrt{3}$ pu
    (b) $V_{ab} = V_{ac} = 0.816$ pu, $|I_b| = |I_c| = 5.69$ pu

11.3 (i) $- j6.25$ (ii) $- 4.33$ (iii) 6.01 (iv) $- j5$ pu
    In order of decreasing magnitude of line currents the faults can be listed as
    (a) LG (b) LLG (c) 3-phase (d) LL

11.4 0.1936 ohm, 0.581 ohm, $- 4.33$ pu, $j5$ pu

11.5 (a) 3.51 pu (b) $V_b = 1.19\angle -159.5°$ pu, $V_c = 1.68\angle 129.8°$ pu
    (c) 0.726 pu

11.6 $I_b = - I_c = - 2.887$ pu

11.7 (a) $I_Y = - 5.79 + j5.01$ kA, $I_B = 5.79 + j5.01$ kA, $I_G = j10.02$ kA
    (b) $I_B = - I_Y = - 6.111$ kA, $I_G = 0$

11.8 $I_{ag} = 0$ \qquad $I_{am} = - j3.51$ pu
    $I_{bg} = - j2.08$ pu \qquad $I_{bm} = - j1.2$ pu
    $I_{cg} = j2.08$ pu \qquad $I_{cm} = - j1.2$ pu

11.9 $5,266$ A

11.10 $j2.0$ pu

11.11 $I^f = - j6.732$ pu, I_a (A) $= - j4.779$ pu,
    I_b(A) $= - j0.255$ pu, I_c (A) $= - j0.255$ pu

11.12 0.42 pu, $- j9.256$ pu

11.13 $-j11.152$ pu, $-j2.478$ pu, $-j1.239$ pu

11.14 4.737 pu, 1 pu

11.15 $I_2^f = - j12.547$ pu, $I_{12}^f(b) = - j0.0962$ pu

Chapter 12

12.1 4.19 MJ/MVA, 0.0547 MJ-sec/elec deg

12.2 4.315 MJ/MVA \qquad 12.3 40.4 MJ/MVA

12.4 140.1 MW, 130.63 MW, 175.67 MW

12.5 72.54 MW \qquad 12.6 $\delta_3 = 58°$

12.7 127.3 MW

12.8 $53°$. We need to know the inertia constant M to determine t_c.

12.9 The system is stable \qquad 12.10 $70.54°$, 0.1725 sec

12.11 The system is unstable \qquad 12.12 $63.36°$

12.13 The system is stable \qquad 12.14 The system is stable

12.15 The system is unstable for both three pole and single pole switching

Index